UNDERSTANDING SEMICONDUCTOR DEVICES

PRENTICE HALL UNDERSTANDING ELECTRONICS TECHNOLOGY SERIES

Book 1 Understanding Electricity and Electronics: Concepts, Experiments, and Troubleshooting

Book 2 Understanding DC Circuits: Concepts, Experiments, and Troubleshooting

Book 3 Understanding AC Circuits: Concepts, Experiments, and Troubleshooting

Book 4 Understanding Semiconductor Devices: Concepts, Experiments, and Troubleshooting

Book 5 Understanding Circuits and Op-Amps: Concepts, Experiments, and Troubleshooting

Book 6 Understanding Digital Electronics: Concepts, Experiments, and Troubleshooting

UNDERSTANDING SEMICONDUCTOR DEVICES

Concepts, Experiments, and Troubleshooting

Dale R. Patrick
Stephen W. Fardo

Eastern Kentucky University

Prentice Hall
Englewood Cliffs, New Jersey 07632

Library of Congress Cataloging-in-Publication Data

Patrick, Dale R.
Understanding semiconductor devices : concepts, experiments, and troubleshooting / Dale R. Patrick, Stephen W. Fardo.
p. cm.—(Prentice Hall understanding electronics technology series ; bk. 4)
Includes index.
ISBN 0-13-943192-6
1. Semiconductors. I. Fardo, Stephen W. II. Title.
III. Series.
TK7871.85.P384 1989 88-31901
621.3815'2—dc19 CIP

Editorial/production supervision
and interior design: *Denise Gannon*
Cover design: *Lundgren Graphics, Ltd.*
Manufacturing buyer: *Robert Anderson*

Printed in the United States of America

10 9 8 7 6 5 4 3 2 1

ISBN 0-13-943192-6

Prentice-Hall International (UK) Limited, *London*
Prentice-Hall of Australia Pty. Limited, *Sydney*
Prentice-Hall Canada Inc., *Toronto*
Prentice-Hall Hispanoamericana, S.A., *Mexico*
Prentice-Hall of India Private Limited, *New Delhi*
Prentice-Hall of Japan, Inc., *Tokyo*
Simon & Schuster Asia Pte., Ltd., *Singapore*
Editora Prentice-Hall do Brasil, Ltda., *Rio de Janeiro*

Contents

UNIT TWO SEMICONDUCTOR DIODES 33

UNIT THREE ZENER DIODES 81

UNIT FOUR SPECIAL SEMICONDUCTOR DIODES 111

UNIT FIVE BIPOLAR TRANSISTORS 141

UNIT SIX BIPOLAR TRANSISTOR AMPLIFICATION 181

Preface

Understanding Semiconductor Devices: Concepts, Experiments, and Troubleshooting is an introductory text that provides coverage of the various topics in the study of semiconductor devices. The key concepts presented in this book are discussed using a simplified approach that greatly enhances learning. The use of mathematics is kept to the very minimum and is discussed clearly through applications and illustrations.

Every unit is organized in a step-by-step progression of concepts and theory. Each chapter begins with a *unit introduction* and *unit objectives.* A discussion of important *concepts* and *theories* follows. Numerous *self-examinations,* with answers provided, are integrated into each chapter to reinforce learning. In addition, *experimental activities* are included with each unit to help students learn electronics through practical experimental activities. The final learning activity for each unit is a *unit examination,* which includes at least twenty multiple choice objective questions.

Definitions of *important terms* are presented at the beginning of each unit. Several *appendices* are used to aid students in performing experimental activities. It should be noted that the expense of the equipment required for the experiments is kept to a minimum. A comprehensive parts list is provided. The experiments suggested are low-cost activities that can be performed in the home or a school laboratory. They are very simple and easy to understand and stress troubleshooting concepts. Electronics can be learned experimentally at a low cost by completing these labs. Appendices to supplement lab activities are included for easy reference.

This book is the fourth in Prentice Hall's "Understanding Electronics Technology" series. This book covers electronic devices, an important *foundation* in the study of electronics. Prentice Hall's "Understanding Electronics Technology" series includes the following books:

Book 1—*Understanding Electricity and Electronics: Concepts, Experiments, and Troubleshooting*
Book 2—*Understanding DC Circuits: Concepts, Experiments, and Troubleshooting*
Book 3—*Understanding AC Circuits: Concepts, Experiments, and Troubleshooting*
Book 4—*Understanding Semiconductor Devices: Concepts, Experiments, and Troubleshooting*
Book 5—*Understanding Circuits and Op-amps; Concepts, Experiments, and Troubleshooting*
Book 6—*Understanding Digital Electronics: Concepts, Experiments, and Troubleshooting*

Each of these books is organized in the same easy-to-understand format. They can be used to acquire an understanding of electronics in the home, school, or work place. The sequence of the books allows the student to progress at a desired pace in the study of electronics and to perform low-cost experiments with equipment and supplies (depending on availability). As students progress, they may wish to purchase various types of test equipment with varying degrees of expense. The experiments allow the student to further develop an understanding of the topics discussed in each unit. They are intended as an important supplement to student learning.

Several *supplemental materials* are available to provide an aid to effective learning. These include the following:

1. *Instructor's Resource Manual*—provides the instructor with answers to all unit examinations and suggested data for experimental activities, including comprehensive analysis of each experiment.
2. *Instructor's Transparency Masters*—enlarged reproductions of selected illustrations used in the textbook, which are suitable for use for transparency preparation for class presentation.
3. *Instructor's Test Item File*—provides the instructor with many suggested multiple choice objective questions that may be used with each unit of instruction.

These supplements should be extremely valuable for instructors in organizing electronics classes. The complete instructional cycle, from objectives to evaluation, is included in this series of books. The authors hope you will find the Understanding Electronics Technology Series easy to understand and that you are successful in your pursuit of knowledge in an exciting technical area. Electronics is an extremely vast and interesting field of study. This book provides a *foundation* for *understanding electronics technology.*

Dale R. Patrick
Stephen W. Fardo

Richmond, Kentucky

Course Objectives

Upon completion of this course on Understanding Semiconductor Devices, you will be able to:

1. Explain the electrical differences between conductors, semiconductors, and insulators.
2. Use semiconductor components in the construction of an electronic circuit.
3. Describe the characteristics of basic semiconductor devices.
4. Interpret the graphic data of basic semiconductor devices.
5. Identify semiconductor devices by schematic symbols, package styles, and characteristic data.
6. Interpret information from the data sheet of a semiconductor device.
7. Identify the leads of semiconductor devices with a VOM.
8. Determine if a semiconductor device is good or defective with meters or instruments.
9. Analyze the data of a semiconductor device and show how it can be used to accomplish an operation.
10. Explain how amplification is achieved with semiconductor devices.

Parts List for Experiments

The parts included in this list are needed to perform the experimental activities of this course. They can be obtained through local electronic component dealers and mail order suppliers. As a rule, a number of manufacturers have these parts available. Some substitutions can be made without altering the desired results. This includes generalized components such as resistors, capacitors, potentiometers, power sources, and some solid-state devices.

In some activities, a specific component is needed to achieve the desired results. If a substitution is made it may alter the data to some extent. Keep this in mind if different parts are used.

RESISTORS

Film type, 5% tolerance preferred.

- 47-Ω, $\frac{1}{4}$-W, 5% tolerance, (yellow-violet-black-gold)
- 100-Ω, $\frac{1}{4}$-W, 5% tolerance, (brown-black-brown-gold)
- 220-Ω, $\frac{1}{4}$-W, 5% tolerance, (red-red-brown-gold)
- 470-Ω, $\frac{1}{4}$-W, 5% tolerance, (yellow-violet-brown-gold)
- 1-kΩ, $\frac{1}{4}$-W, 5% tolerance, (brown-black-red-gold) (2)
- 4.7-kΩ, $\frac{1}{4}$-W, 5% tolerance, (yellow-violet-red-gold)
- 10-kΩ, $\frac{1}{4}$-W, 5% tolerance, (brown-black-orange-gold)

56-kΩ, $\frac{1}{4}$-W, 5% tolerance, (green-blue-orange-gold)
100-kΩ, $\frac{1}{4}$-W, 5% tolerance, (brown-black-yellow-gold)
270-kΩ, $\frac{1}{4}$-W, 5% tolerance, (red-violet-yellow-gold)
330-kΩ, $\frac{1}{4}$-W, 5% tolerance, (orange-orange-yellow-gold)

POTENTIOMETERS

1-kΩ, 1-W, 10% tolerance, linear taper, Allen Bradley type JA1NO56S-102UA, $\frac{7}{8}$″ shaft or equivalent

100-kΩ, 1-W, 10% tolerance, linear taper, Allen Bradley type JA1N056S-104UA, $\frac{7}{8}$″ shaft or equivalent

Note: These potentiometers must have a 2-in. solid wire soldered to each terminal for connection to the circuit construction board.

CAPACITORS

0.1-μF, 50-V, 20% tolerance, Centralab, low voltage disc ceramic type, #UK50-104, Mallory CK05BX104K, or equivalent

10-μF, 50-V, miniature electrolytic type, Mallory #TT50X10B, CDE Type NLW 10-50, or equivalent

SEMICONDUCTORS

N-channel junction field-effect transistor, Motorola 2N4222, Heath #417-291, or equivalent

Silicon diode, Texas Instruments IN4001, Heath 56-28, or equivalent

Germanium diode, General Electric 1N4149, Heath 56-89, or equivalent

Zener diode, 5.1 V ± 10%, 500 mW, Motorola 1N5231, 1N751, Heath #56-16, or equivalent

NPN transistor, Motorola MPSA20, Heath 417-122, or equivalent

PNP transistor, Motorola MPSA70, 2N2431, Heath 417-122, or equivalent

N-channel MOSFET, Motorola 2N3796, Heath 417-935, or equivalent

Silicon-controlled rectifier, General Electric C106F, Heath 57-50, or equivalent

Triac, RCA 2N5754 or equivalent

Unijunction transistor, Motorola 2N2646, Heath 417-183, or equivalent

LED, Motorola MV5054-2, Heath #412-40, or equivalent

Phototransistor, General Electric L14F1, Heath #417-851, or equivalent

MISCELLANEOUS

12-V lamp, ALCO #2182, GTE Sylvania 12ES, Heath 412-601, or equivalent

Circuit construction board, AP Products type 923252; Continetial Specialists type UVCB; E & L Instruments #SK-10

Ac power source, filament transformer, Stancor-Essex P-8384, 12.6 V center tapped, 1A

Variable dc power supply, Heath Experimenter #ET-3100

VOM, Weston Type 6100, 660, 6000, or 4440; Beckman Tech 300, 310, or 330; B & K model 2810, 2815, 2845, or 2831; Triplett model 4000, 3400, 603, or 310; Simpson 303, 313, 265, 360-2, 460-4, 260, or 462

OPTIONAL ITEMS

Heathkit Electronic Design Experimenter ET-3100. This unit includes a variable DC source, circuit construction board, and two potentiometers.

Oscilloscope

UNIT ONE

Semiconductor Basics

UNIT INTRODUCTION

One major division of the electronics field deals with the flow of electrons through a piece of solid material. The term *solid-state electronics* is often used to describe this branch of the field. Solid-state electronics is based on the action of current carriers passing through a piece of semiconductor material. *Semiconductors* are solid materials that have electrical properties that lie somewhere between those of a conductor and an insulator. They have made possible such things as radio, television, computers, robots, and stereo amplifiers.

This unit deals with the theory of materials that are used in the construction of nearly all solid-state devices.

UNIT OBJECTIVES

Upon completion of this unit, you will be able to:

1. Explain the electrical difference between conductors, semiconductors, and insulators.
2. Describe how N-type and P-type semiconductors are produced.
3. Explain how current carriers move through semiconductors.
4. Explain the terms intrinsic, doping, holes, covalent bonding, and energy level.

IMPORTANT TERMS

Atom The smallest particle to which an element can be reduced and still retain its identity.

Compound Two or more elements combined chemically.

Conductor A material that allows electrical current to flow through it easily.

Element The basic materials that make up all other materials; they exist by themselves (such as copper, hydrogen, carbon) or in combination with other elements (water is a combination of the elements hydrogen and oxygen).

Insulator A material that offers a high resistance to electrical current flow.

Matter Anything that occupies space and has weight; can be a solid, a liquid, or a gas.

Molecule The smallest particle to which a compound can be reduced before being broken down into its basic elements.

Neutron A particle in the nucleus (center) of an atom that has no electrical charge.

Nucleus The core or center part of an atom, which contains protons having a positive charge and neutrons having no electrical charge.

Proton A particle in the center of an atom that has a positive (+) electrical charge.

Stable atom An atom that will not release electrons under normal conditions.

Semiconductor A material that has a value of electrical resistance between that of a conductor and an insulator and is used in the manufacture of solid-state devices such as diodes and transistors.

Valence electrons Electrons in the outer shell of an atom.

SEMICONDUCTOR THEORY

Nearly every study of semiconductors starts with an investigation of atomic theory. This is done purposely to acquaint the reader with some important principles that show how a material responds electronically. In this regard, we need to be familiar with such things as energy levels, ionic bonding, covalent bonding, stable atoms, and electrical conductivity. Semiconductor materials respond to these ideas. You need to be familiar with this information to see how a particular device is formed and how it responds when energy is applied.

Elements are classified according to the number of particles they possess. Atoms of an element are distinguished by the number of protons in their nucleus, which is known as the atomic number. No two elements have the same atomic number. The nucleus of every atom is composed of one or more positively charged particles called protons. In addition to the protons, in all elements except the lightest, there are one or more neutrons. Hydrogen, which is the simplest atom, has one proton and no neutrons in its nucleus. The nucleus of uranium has 92 protons and 146 neutrons. The atomic weight of an atom is the sum of protons and neutrons in the nucleus. The atomic weight of uranium is 238. The atomic weight of an element is rarely a whole number because it represents the average value of a large number of atoms. The atomic weight of a single atom is always a whole number.

In semiconductor electronics we are primarily concerned with the electron content of an atom. An electron is a negatively charged particle that revolves around the nucleus of an atom. An electron is extremely small. It weighs 1/1850 as much as a proton. Practically all of the weight of an atom is in the nucleus. Figure 1-1 shows the particle structure of an atom of copper. This structure is shown in a two-dimensional representation. Keep in mind that an atom is a three-dimensional object.

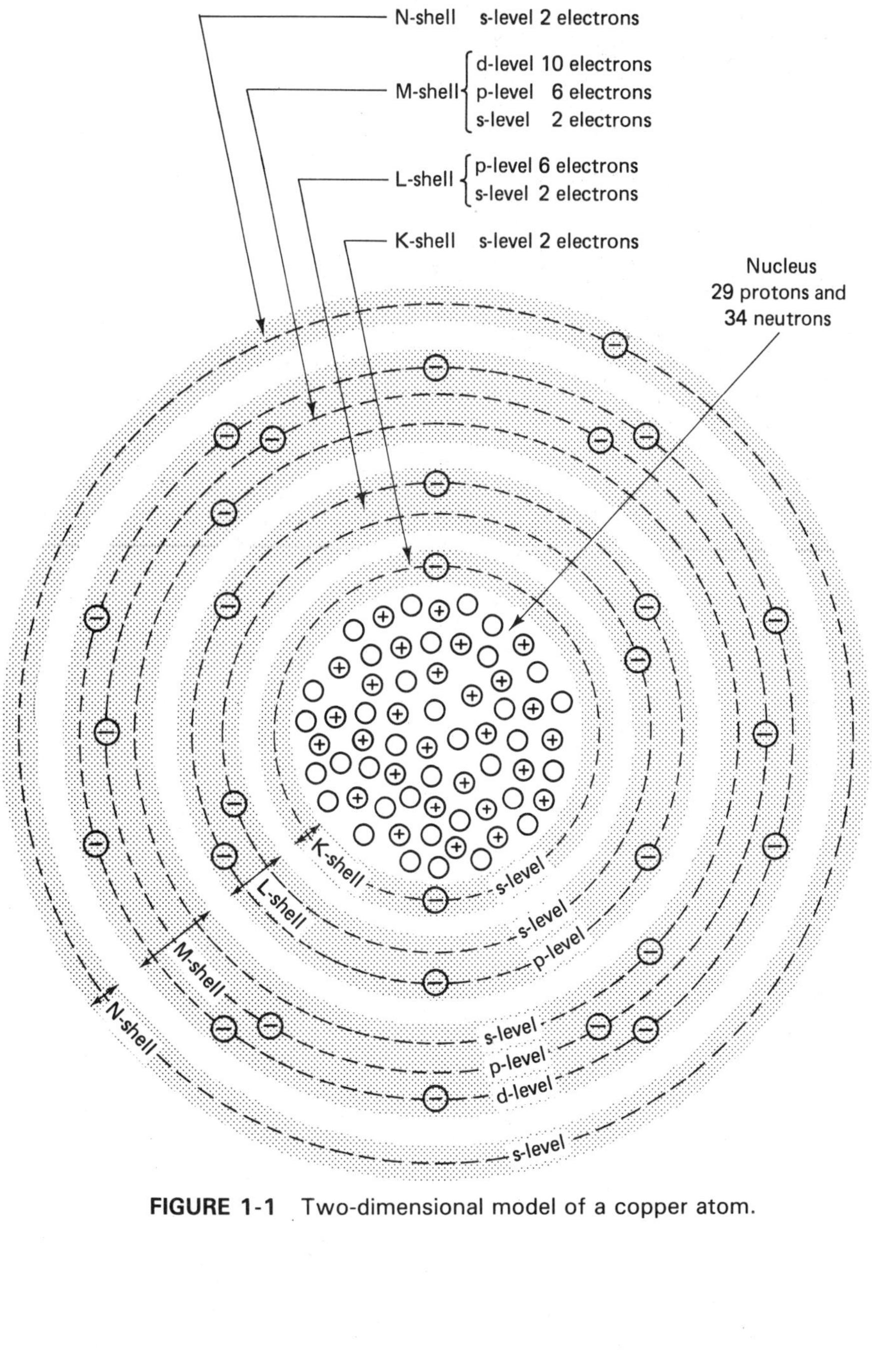

FIGURE 1-1 Two-dimensional model of a copper atom.

The electrons of an atom are not an equal distance from the nucleus. They revolve around the nucleus in shells or layers. Neils Bohr, a Danish physicist, identified these shells by the letters K, L, M, N, O, and P. The K shell is closest to the nucleus. Each shell or layer has a distinct energy level. Electrons do not exist in the space between these energy levels. The energy levels of a shell are identified by the letters s, p, d, f, g, and h. The s-level is closest to the nucleus. A shell may have from one to six distinct energy levels in its structure. Energy levels represent two things: distance from the nucleus and the amount of energy possessed by an electron. Energy levels closest to the nucleus are of the lowest value while those further away are occupied by electrons with more energy. This largely depends on the number of electrons in an atom. The number of electrons in each energy level follows a unique pattern. This pattern is 2, 6, 10, 14, 18, and 22. Table 1.1 shows the shells, the energy levels found in each shell, and the maximum number of electrons that exist in each level. Note that shell P has six energy levels. The maximum number of electrons at each level follows the 2, 6, 10, 14, 18, 22 pattern. This number sequence appears in the other shells but decreases in value with the last level. The complexity of an atom usually dictates its electron shell and energy level assignment.

The exact path an electron follows in the structure of an atom is not known. This path is generally described as an orbital. An orbital shows the mathematical probability of where an electron will appear in the structure of an atom. Today we believe there are four types of orbitals in atoms. These are identified by the letters s, p, d, and f. Figure 1-2 on page 6 shows the representation of s, p, and d orbitals. The f-type of orbital is not shown in this sequence because there is very little evidence that it is as active as the others in atomic bonding. It is more meaningful in chemical analysis.

SELF-EXAMINATION

1. The number of protons in the nucleus of an atom is an indication of the ____________ ____________.
2. The atomic weight of atom is the sum of the number of ____________ and ____________ in the nucleus.
3. A negatively charged particle that revolves around the nucleus of an atom is an ____________.
4. The (K-L-M-N-O-P-Q) shell of an atom is closest to the nucleus. ____
5. The energy levels of a shell are identified by the letters ____, ____, ____, ____, ____, and ____.
6. The ____ shell of an atom has 6 energy levels.
7. An ____________ shows the mathematical probability of where an electron will appear in the structure of an atom.

ANSWERS

1. Atomic number
2. Protons, neutrons
3. Electron
4. K
5. s, p, d, e, f, and g
6. P
7. Orbital

ATOM COMBINATIONS

Hydrogen is an atom that contains one proton in its nucleus. This proton possesses a positive charge. The number of electrons that exist in an atom tends to match the net positive charge of the

TABLE 1-1
Shells, energy levels, and maximum number of electrons for each energy level.

Shells	Energy Level	Maximum Number of Electrons
K	s	2
L	s	2
	p	6
M	s	2
	p	6
	d	10
N	s	2
	p	6
	d	10
	f	14
O	s	2
	p	6
	d	10
	f	14
	g	18
P	s	2
	p	6
	d	10
	f	14
	g	18
	h	22
Q	s	2

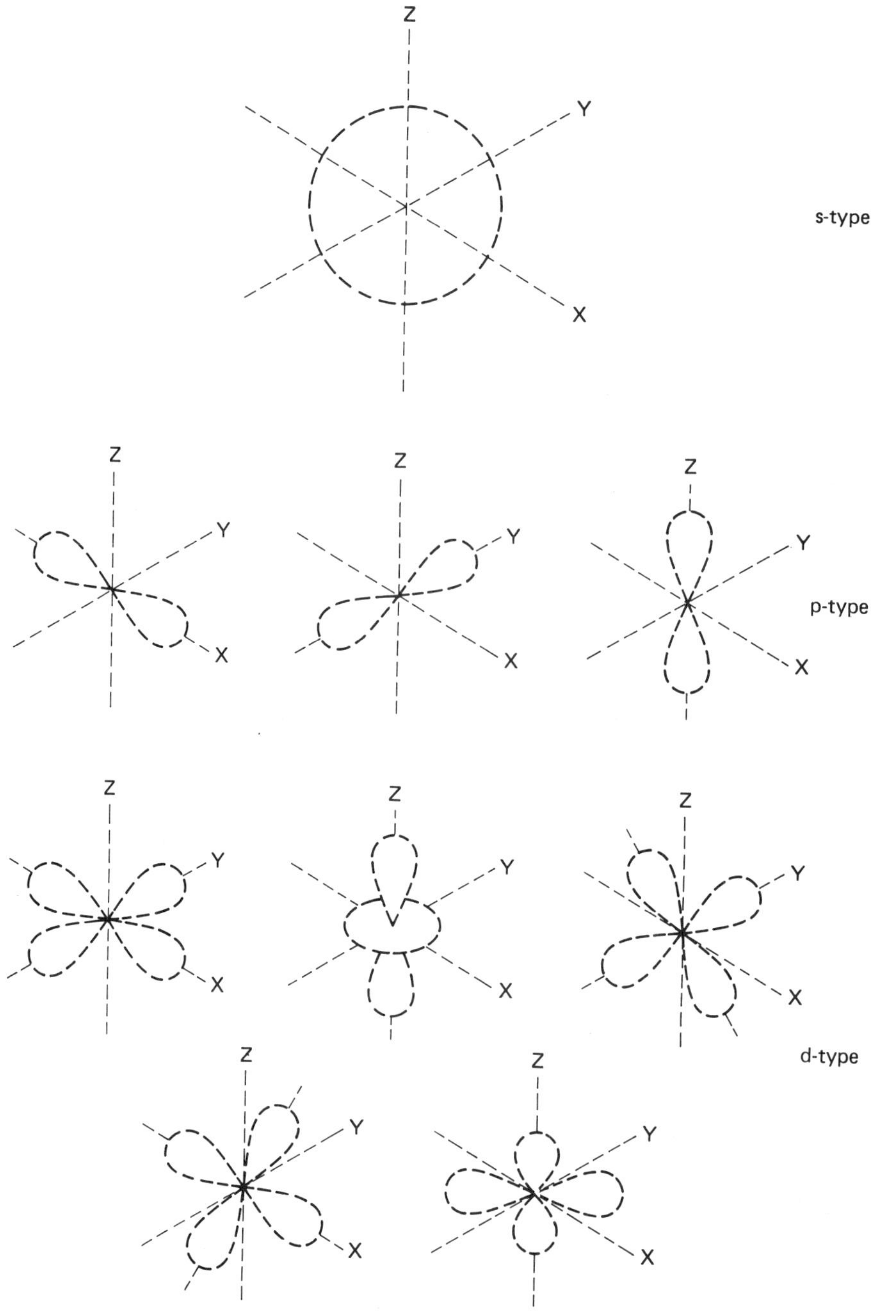

FIGURE 1-2 Orbitals of an atom.

nucleus. Hydrogen, therefore, has one electron in its structure. This electron revolves around the nucleus in the K-shell in an s-type orbital pattern. If no additional energy is added to the electron it is considered to be in its ground state or lowest energy level.

Helium is somewhat more complex than hydrogen. Helium has two protons and two neutrons in its nucleus. Two electrons revolve around the nucleus in the s-level of the K-shell. The s-energy level has a full complement of electrons in its structure. A shell with a full complement of electrons tends to be a preferred atomic condition. This is generally called a *stable state* or a *stabilized atom.*

Lithium has three protons and three neutrons in the structure of its nucleus. It also has three electrons to balance the net positive charge of the nucleus. Two of these electrons fill the s-level of the K-shell. Since the K-shell only has one energy level, the third electron of the structure must appear in the s-level of the L-shell. The L-shell has s- and p-energy levels to be filled. The s-level has a capacity of 2 and the p-level has a capacity of 6. The p-level must be filled before the next shell is formed. The electron in the s-level of the L-shell of lithium is called the valence electron. It can effectively enter into some chemical activity with another atom. The valence electron represents the highest energy level of the atom. Figure 1-3 shows a two-dimensional representation of a lithium atom.

A compound is a combination of two or more elements to form an entirely different material. The atomic structure of lithium lends itself very well to compound formation. A compound of lithium and hydrogen is called lithium hydride. In this compound the valence electron of lithium is transferred to the K-shell of a hydrogen atom. The lithium atom takes on a +1 charge because it has lost an electron. This is now considered to be a positive lithium ion. The hydrogen atom gains an electron and becomes a negative ion with a charge of −1. Since unlike charged particles have an attracting power, a single molecule of lithium hydride is formed by this combination of atoms. Compounds formed in this manner are described as electrovalent combinations or ionic bonded mixtures. The net charge of the ionized particles bonds the atoms together in a compound.

Table salt is a good example of ionic bonding that occurs naturally. Figure 1-4 shows ionic bonding between two independent

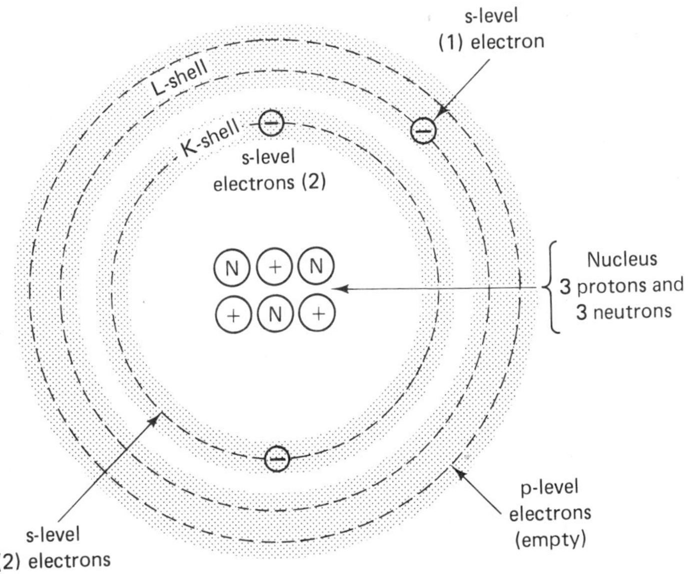

FIGURE 1-3 Two-dimensional representation of a lithium atom.

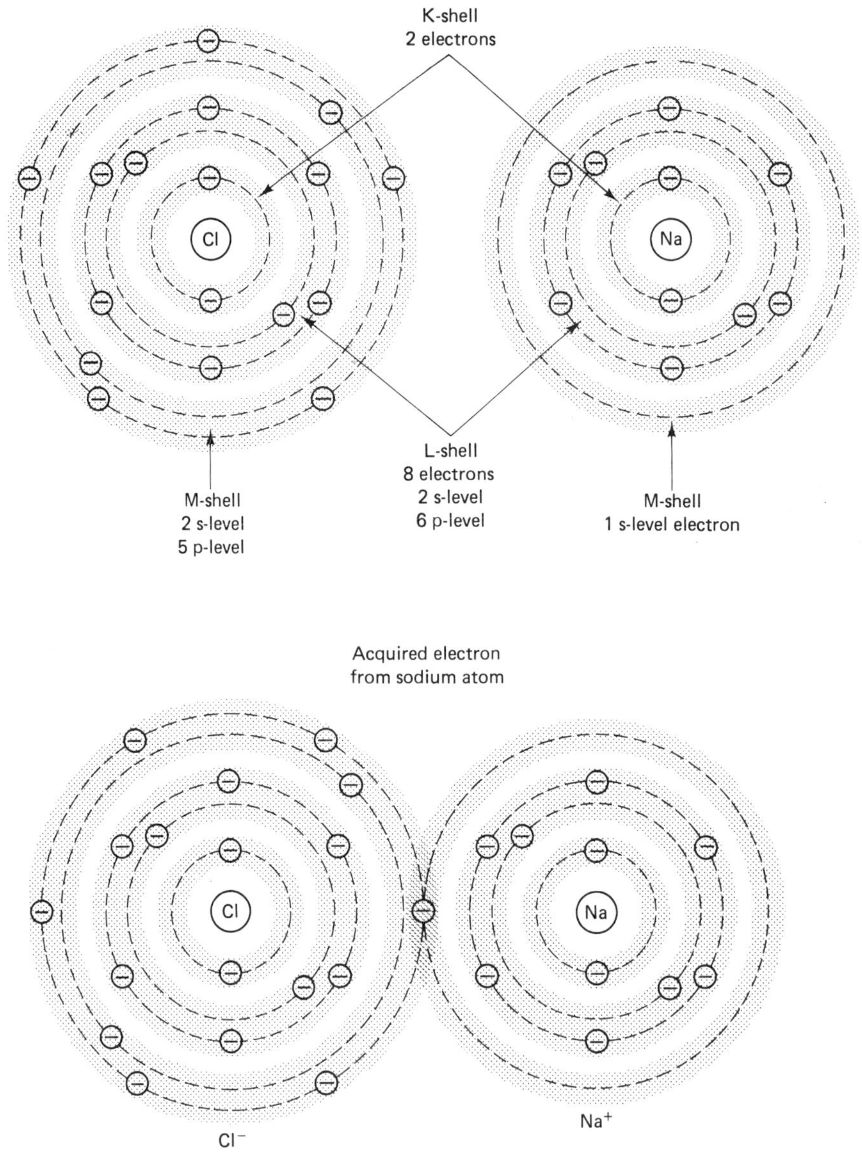

FIGURE 1-4 Ionic bonding of Cl and Na atoms to produce salt.

atoms. The sodium (Na) atom donates its one valence electron to a chlorine (Cl) atom with seven valence electrons. Upon acquiring the extra electron, Cl becomes a negatively charged ion, while the sodium atom becomes a positively charged ion. Since unlike charges attract, the Na^+ and Cl^- ions are bonded together by an electrostatic force.

Copper oxide (Cu_2O) is also an example of ionic bonding. The valence electrons of two atoms of pure copper (Cu) are transferred and combine with one atom of oxygen (O). Copper is a good electrical conductor. It has one valence electron in the s-level of the N-shell. Oxygen has six valence electrons in the L-shell, two of these are in the s-level and four are in the p-level. The combination of two copper atoms and one oxygen atom forms a stable copper oxide compound. The two copper valence electrons combine with four electrons in the p-level of the L-shell of oxygen to form a full complement of six electrons. The stability of this compound makes copper oxide a good insulating material.

Another form of bonding can take place when the outmost shell of an atom is partially filled. The atoms in this form of bonding must position themselves so that the energy levels of the valence electrons interact. This form of bonding is called *electron pair bonding* or *covalent bonding*. No ions are formed in covalent bonding. A covalent force is established between the connecting electrons. These electrons alternately shift back and forth between each atom. The covalent force bonds individual atoms together in a simulated condition of stability.

A two-dimensional illustration of covalent bonding is shown in Fig. 1-5. Two hydrogen atoms are bonded together in this drawing. Note that the K-shells of each atom overlap so that their electrons can be shared. In this form of bonding both atoms share valence electrons, and both now have the equivalent of a full complement of electrons in their s-level. The molecules of common gases such as oxygen and nitrogen are frequently held together by covalent bonding. Semiconductors such as carbon, silicon, and germanium are also bonded by this process. Covalent bonding is much stronger than ionic bonding.

Metallic bonding occurs in good electrical conductors. An electrical conductor such as copper has one electron in the s-level of its N-shell. This electron is easily influenced by outside energy and has a tendency to wander around the material between dif-

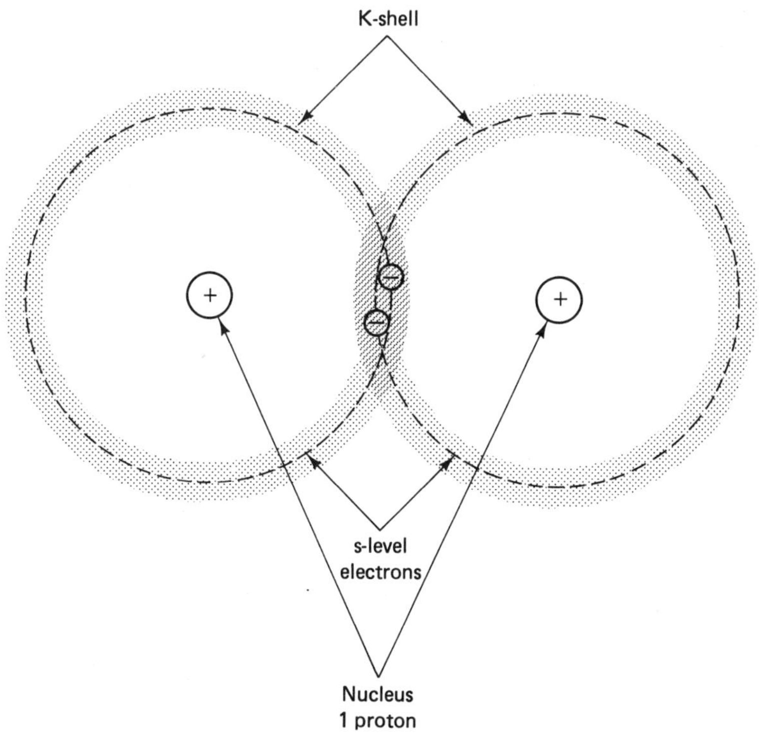

FIGURE 1-5 Covalent bonding of hydrogen atoms.

ferent atoms. Upon leaving one atom, it immediately enters the orbital of another atom. The process is repeated on a continuous basis. When an electron receives enough energy to leave an atom it causes the original atom to become a positive ion. In metal, this process occurs at temperatures of 25°C. The process takes place in a random manner. This means that one electron is always associated with an atom but is not in one particular orbital path. As a result of this condition, a large number of the structural atoms of a piece of copper tends to share valence electrons.

In metallic bonding there is a type of electrostatic force between positive ions and electrons. In a sense, electrons float around in a cloud that covers the positive ions. This floating cloud bonds the electrons randomly to the ions. Figure 1-6 shows an example of the metallic bonding of copper. An energy band discussion of electrical conductivity is introduced in the following section.

SELF-EXAMINATION

8. Hydrogen has an atomic number of ________.
9. Helium has an atomic weight of ________.
10. Lithium has an atomic weight of ________.
11. When two or more elements are used to form a different material it is called a ________.
12. Lithium hydride is a ________.
13. Table salt and copper oxide are examples of ________ bonding.

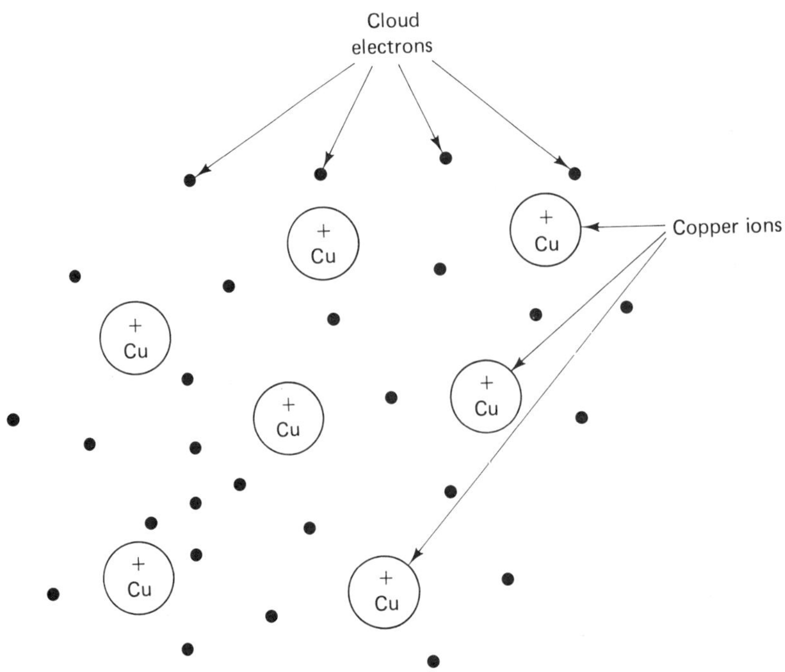

FIGURE 1-6 Metallic bonding of copper.

14. Individual atoms of silicon are frequently connected together by __________ bonding.
15. Covalently bonded silicon consists of the original atom and __________ neighboring atoms.
16. __________ bonding occurs in good electrical conductors.

ANSWERS

8.	1	9.	4
10.	6	11.	Compound
12.	Compound	13.	Ionic
14.	Covalent	15.	4
16.	Metallic		

INSULATORS, SEMICONDUCTORS, AND CONDUCTORS

Solid materials are used primarily in electronic devices to achieve electrical conductivity. These materials can be divided into insulators, semiconductors, and conductors. For discussion purposes, assume that we have a cubic centimeter block of these three materials. If the resistance is measured between opposite faces of each block, it shows some very unique differences. The insulator cube measures several million ohms. The conductor cube, by comparison, measures only 0.000001Ω. The semiconductor cube measures approximately 300Ω. In effect, this shows a rather wide range of resistance differences in the three materials.

A convenient way to evaluate the relationship of conductors, semiconductors, and insulators is through the use of energy-level diagrams. This takes into account the amount of energy needed to cause an electron to leave its valence band and go into conduction. The diagram represents a composite of all atoms within the material.

Energy-level diagrams of insulators, semiconductors, and conductors are shown in Fig. 1-7. The valence band is located at the bottom and the conduction band is at the top of each diagram. The valence band represents the highest energy level that electrons can attain and still be influenced by the nucleus. Electrons in this band normally combine with valence-band electrons of other atoms to form molecules or compounds. A number of other electrons exist below the valence energy band. As a rule, we are only concerned with the response of valence and conduction band electrons in the operation of semiconductor devices.

The valence band represents a composite energy level of the valence electrons of each atom. A specific amount of outside en-

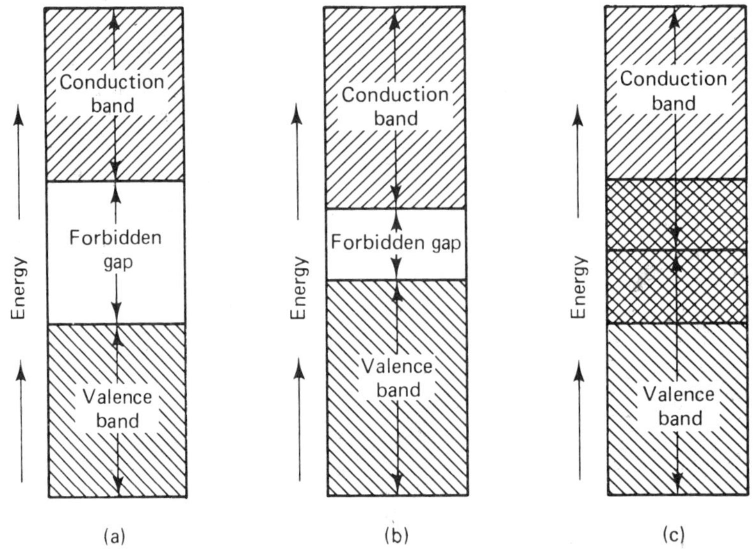

FIGURE 1-7 Energy-level diagrams for (a) insulators; (b) semiconductors; (c) conductors.

ergy must be added to valence electrons to cause them to go into conduction. The forbidden region, when it exists, separates the valence and conduction bands. A certain amount of energy is needed to cause valence electrons to cross the forbidden gap. If it is insufficient, electrons are not released for conduction. They will remain in the valence band. The width of the forbidden region indicates the conduction status of a particular material. In atomic theory the width of the forbidden gap is expressed in electron volts (eV). An electron volt is defined as the amount of energy gained or lost when an electron is subjected to a potential difference of 1 V. The atoms of each element have a specific energy-level value that permits conduction.

The energy-level diagram of an insulating material has a very wide forbidden gap. A carbon crystal or natural diamond is an excellent insulator. Crystallized carbon has a forbidden gap of approximately 6 eV. The large width of this region keeps valence electrons from crossing the forbidden gap and going into conduction. In order for valence electrons to travel through the forbidden gap, they must acquire additional energy. The amount of energy needed by good insulators is generally very high. The energy level value that causes an insulator to go into conduction is often called its breakdown voltage.

Even the best insulators will go into conduction if sufficient energy is applied. Thyrite is an example of this condition. Thyrite is commonly used as a lightning arrester. At normal voltages it is an ideal insulator. When it is subjected to high voltage, electrons cross the forbidden gap and go into conduction. This shunts the high voltage to ground, thus protecting a device from lightning. When good insulators are operated at high temperatures, the increased heat energy causes valence electrons to go into conduction.

The forbidden gap of a semiconductor is much smaller than that of an insulator. Silicon, for example, needs to gain 1.21 eV of energy at absolute zero (−273°C) to go into conduction. Energies of this magnitude are not easily acquired. As a result, the valence band remains full, the conduction band remains empty, and these materials respond as insulators. However, an increase in temperature causes the conductivity of this material to change. At normal room temperature or 25°C, the valence electrons acquire thermal energy that is greater than the normal eV value. This essentially reduces the width of the forbidden gap and causes a semiconductor to be a conductor. This particular characteristic is extremely important in solid-state electronic devices.

The energy-level diagram of a conductor is quite unusual compared with the other materials. In a sense, the valence band and conduction band are one and the same. Conductivity is explained as having an interaction of different energy levels of the valence band. Studies show that the atoms of most metals and semiconductors are in the form of a crystal lattice structure. A crystal consists of a space array of atoms or molecules built up by regular repetition in a three-dimensional pattern. The energy level of electrons in the crystal do not respond in the same manner as those of an individual atom. When atoms form crystals, it is found that the energy levels of inner-shell electrons are not affected by the presence of neighboring atoms. However, the valence electrons of individual atoms are often shared by more than one atom. The new energy level of valence electrons is found in a distinct band. The spacing between the energy levels of this band is very small compared with that of isolated atoms. Thus electrons are free to absorb energy and to move from one point to another, thus conducting heat and electricity. In good conductors, the energy-level bands of valence electrons tend to overlap. This lowers the energy level of valence electrons and increases the electrical conductivity of the material.

SELF-EXAMINATION

17. An energy level diagram shows the ____________ band at the bottom and the ____________ band at the top.
18. The ____________ band represents the highest energy level that electrons can attain and still be influenced by the nucleus.
19. The ____________ gap or region separates the valence band and conduction bands of an energy-level diagram.
20. The width of the forbidden region of an energy-level diagram is expressed in ____________ ____________.
21. The energy-level diagram of an insulator shows a ____________ forbidden gap.
22. The energy-level diagram of a semiconductor shows a ____________ forbidden gap.

23. In the energy-level diagram of a conductor the valence band and conduction band ________.

ANSWERS

17. Valence, conduction
18. Valence
19. Forbidden
20. Electron volts
21. Wide
22. Narrow or thin
23. Overlap

SEMICONDUCTOR MATERIALS

Materials such as silicon, germanium, and carbon are found naturally in crystalline form. Instead of being a random mass, these atoms are arranged in an orderly manner. A crystal of silicon or germanium forms a definite geometric pattern, namely, a cube. Figure 1-8 is a two-dimensional simplification of the silicon crystal showing only the nucleus and valence electrons. Note that the electrons of individual atoms are covalently bonded together. Germanium has a similar type of crystal structure.

Crystals of silicon and germanium can be manufactured by melting the natural elements. The process is somewhat complex and rather expensive. Manufactured crystals must be made extremely pure to be useable in semiconductor devices. A very pure semiconductor crystal is called an *intrinsic* material. Germanium is considered to be intrinsic when only 1 part of impurity exists in 10^{10} parts of germanium. Silicon is intrinsic when the impurity ratio is $1{:}10^{13}$. In more sophisticated solid-state devices, the ratio may even be higher.

An intrinsic crystal of silicon would appear as the structure in Fig. 1-8. Covalent bonding changes the electrical conductivity of the material, causing each group of atoms to have a simulated condition of stability. Thus only a limited number of free electrons are available for conduction. Silicon and germanium crystals respond to some extent as insulators.

Intrinsic silicon at $-273°C$ is considered to be a perfect insulator. The valence electrons of each atom are firmly bound together in perfect covalent bonds; no free electrons are available for conduction. In actual circuit operation, the absolute zero condition is not very meaningful because it cannot readily be attained. Any temperature above $-273°C$ causes silicon to become somewhat conductive. The insulating quality of a semiconductor material is therefore dependent on its operating temperature.

At room temperature, which is approximately 25°C, silicon atoms receive enough energy from heat to break their bondings. A number of free electrons become available for conduction. At room temperature intrinsic silicon becomes somewhat conductive.

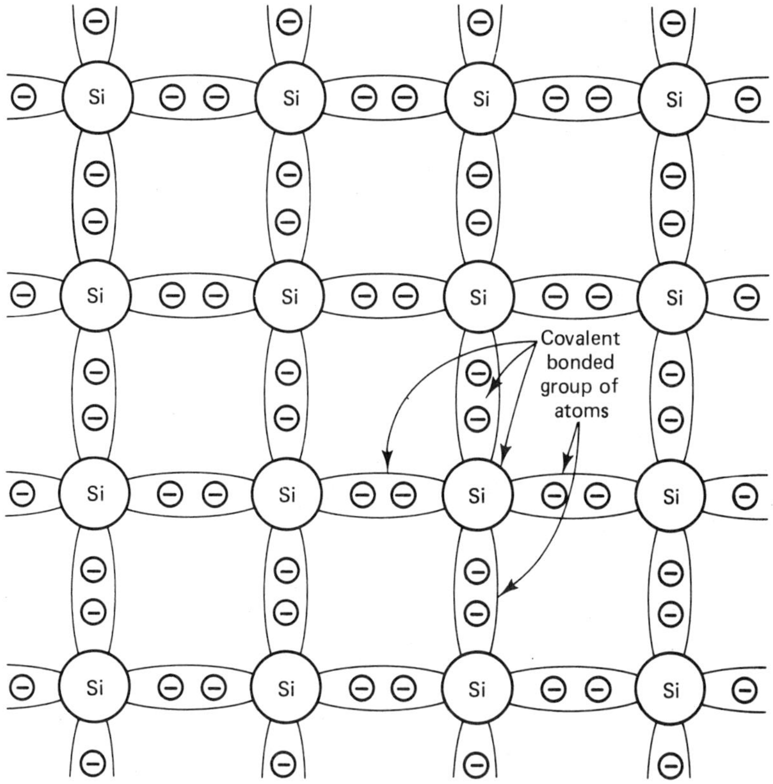

FIGURE 1-8 Intrinsic crystal of silicon.

An important solid-state event takes place when intrinsic silicon goes into conduction. For every electron that is freed from its covalent bonding a void is created. This spot, which is normally called a *hole,* represents an electron deficiency. A hole in a covalent bonding group occurs when an electron is released. An increase in the temperature of a piece of silicon causes the number of free electrons and holes to increase.

It should be remembered that when a neutral atom loses an electron, it acquires a positive charge. The atom then becomes a positive ion. Since an atom acquires a hole at the same time as the positive charge, the hole bears a positive charge. It should be noted, however, that the crystal remains electrically neutral. For every free electron there is an equivalent hole. These two balance out the overall charge of the crystal.

When a valence electron leaves its covalent bond to become a free electron, a hole appears in its place. This hole can then attract a different electron from a nearby bonded group. Upon leaving its bonded group, this electron creates a new hole. The original hole is then filled and becomes void. Each electron that leaves its bonding to fill a hole creates a new hole in its original group. In a sense, this means that electrons move in one direction and holes move in the opposite direction. Since electron movement is considered to be an electric current, holes are also representative of current flow.

Electrons are called *negative current carriers* and holes are called *positive current carriers.* These current carriers move in opposite directions in a semiconductor material. When voltage is applied, holes move toward the negative side of the source. Electron current flows toward the positive side of the source. This condition takes place only in a semiconductor material. In a conductor, such as copper, there is no covalent bonding. Current carriers are restricted to electrons.

Figure 1-9 shows how an intrinsic piece of silicon will respond at room temperature when voltage is applied. Note that the free electrons move toward the positive terminal of the battery. Electrons leaving the semiconductor flow into the copper connecting wire. Each electron that leaves the material creates a hole in its place. The holes appear to move by jumping between covalent bonded groups. Holes are attracted by the negative terminal connection. For each electron that flows out of the material a new electron enters at the negative connection point. The process of hole flow and electron flow through the material is continuous as long as energy is supplied. Current flow is the resulting carrier movement. An intrinsic semiconductor has an equal number of current carriers moving in each direction. The resulting current flow of a semiconductor is limited primarily to the applied voltage and the operating temperature of the material.

Pure silicon or germanium in its intrinsic state is rarely used as a semiconductor. Useable semiconductors must have controlled amounts of impurities added to them. The added impurities change the conduction capabilities of a semiconductor. The process of adding an impurity to an intrinsic material is called *doping.* The impurity is called a *dopant.* Doping a semiconductor causes it to be an extrinsic material. Extrinsic semiconductors are the operational basis of nearly all solid-state devices.

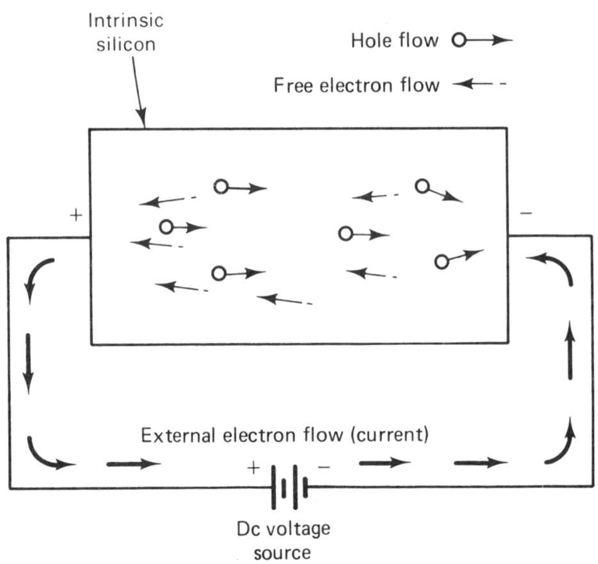

FIGURE 1-9 Current carrier movement in silicon.

N-Type Material

When an impurity is added to silicon or germanium without altering the crystal structure, an N-type of semiconductor material is produced. Atoms of arsenic (As) and antimony (Sb) have five electrons in their valence band. Locate these elements in the periodic table in Appendix A; they are in Group VB. Adding this type of impurity to silicon does not alter the crystal structure or bonding process. Each impurity atom has an extra electron that does not take part in a covalent bonding group. These electrons are loosely held together by their parent atoms. Figure 1-10 shows how a silicon crystal is altered with the addition of an impurity atom.

When arsenic is added to pure silicon, the crystal becomes an N-type material. It has extra electrons or negative (N) charges that do not take part in the covalent bonding process. Impurities that add electrons to a crystal are generally called donor atoms. An N-type material has more extra free electrons than an intrinsic piece of material. A piece of N-material is not negatively charged; its atoms are all electrically neutral. There are, however, a number of

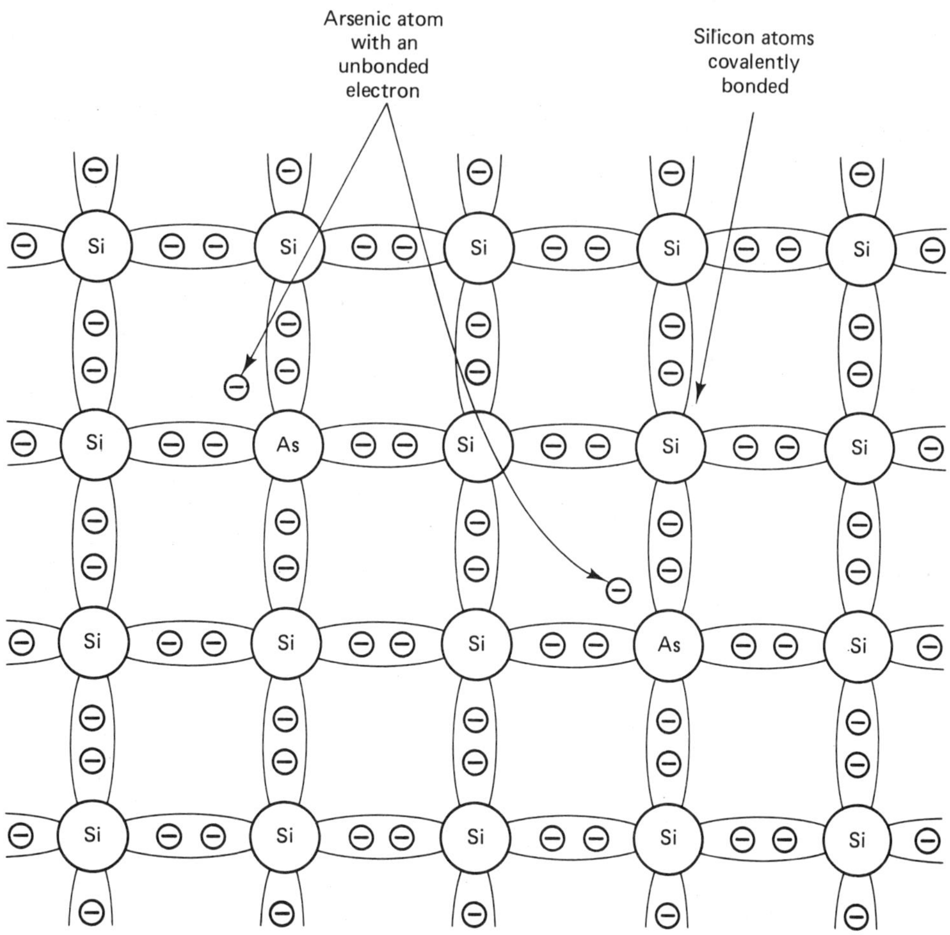

FIGURE 1-10 Extrinsic silicon crystal structure.

electrons that do not take part in the covalent bonding process. These electrons are free to move about through the crystal structure.

An extrinsic silicon crystal of the N-type will go into conduction with a very small amount of voltage applied. An intrinsic crystal (pure silicon) requires a rather substantial amount of voltage or energy for its electrons to go into conduction. Essentially this means that an N-material is a fairly good electrical conductor. In this type of crystal, electrons are considered to be the majority current carriers. Holes are the minority current carriers. The amount of donor material added to silicon determines the number of majority current carriers in its structure.

The number of electrons in a piece of N-type silicon is a million or more times greater than the number of electron-hole pairs of a piece of intrinsic silicon. At room temperature there is a decided difference in the electrical conductivity of this material. Extrinsic silicon becomes a rather good electrical conductor because there are a larger number of current carriers to take part in conduction. Current flow is achieved primarily by electrons in this material. Figure 1-11 shows how the current carriers respond in a piece of N-material. There are more electrons indicated than holes, so electrons are the majority current carriers and holes are the minority carriers.

If the voltage source of Fig. 1-11 were reversed, the current flow would reverse its direction. This means that N-type silicon will conduct equally well in either direction. The flow of current carriers is simply reversed. This is an important consideration in the operation of a device that employs N material in its construction.

P-Type Material

P-type semiconductor material is formed when indium (In) or gallium (Ga) is added to intrinsic silicon. Indium and gallium are clas-

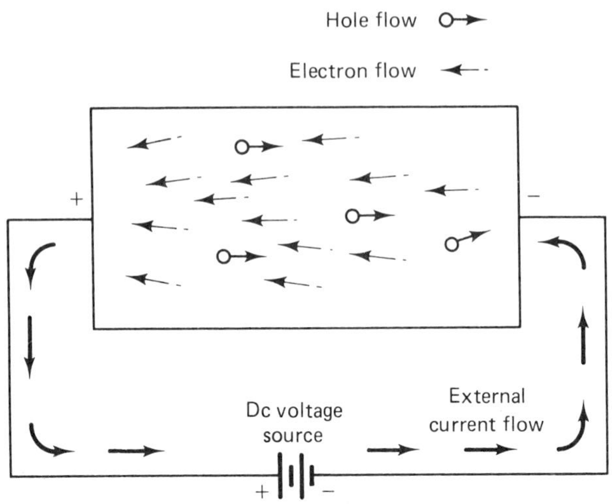

FIGURE 1-11 Current carriers in an N-type material.

sified as Group IIIB elements in the periodic table of Appendix I. These elements are often called acceptors because they readily seek a fourth electron. This type of dopant material has three valence electrons. Each covalent bond that is formed with an indium atom will have an electron deficiency or hole, which represents a positively charged area in the covalent bonding structure. Each hole in the P-material can be filled with an electron. Electrons from neighboring covalent bond groups require very little energy to move in and fill a hole. Holes in the P-material tend to wander around from one covalent bond group to another.

The ratio of doping material to silicon is typically in the range of 1 to 10^6, or 1 to 1 million. This means that P-material has a million times more holes than the heat-generated electron-hole pairs of pure silicon. At room temperature there is a very decided difference in the electrical conductivity of this material. Figure 1-12 shows how the crystal structure of silicon is altered when doped with an acceptor element. In this case the dopant is indium.

A P-material will go into conduction with only a small applied voltage. Intrinsic silicon, by comparison, requires substantially more voltage to produce conduction. Extrinsic silicon is therefore considered to be a rather good electrical conductor. In this mate-

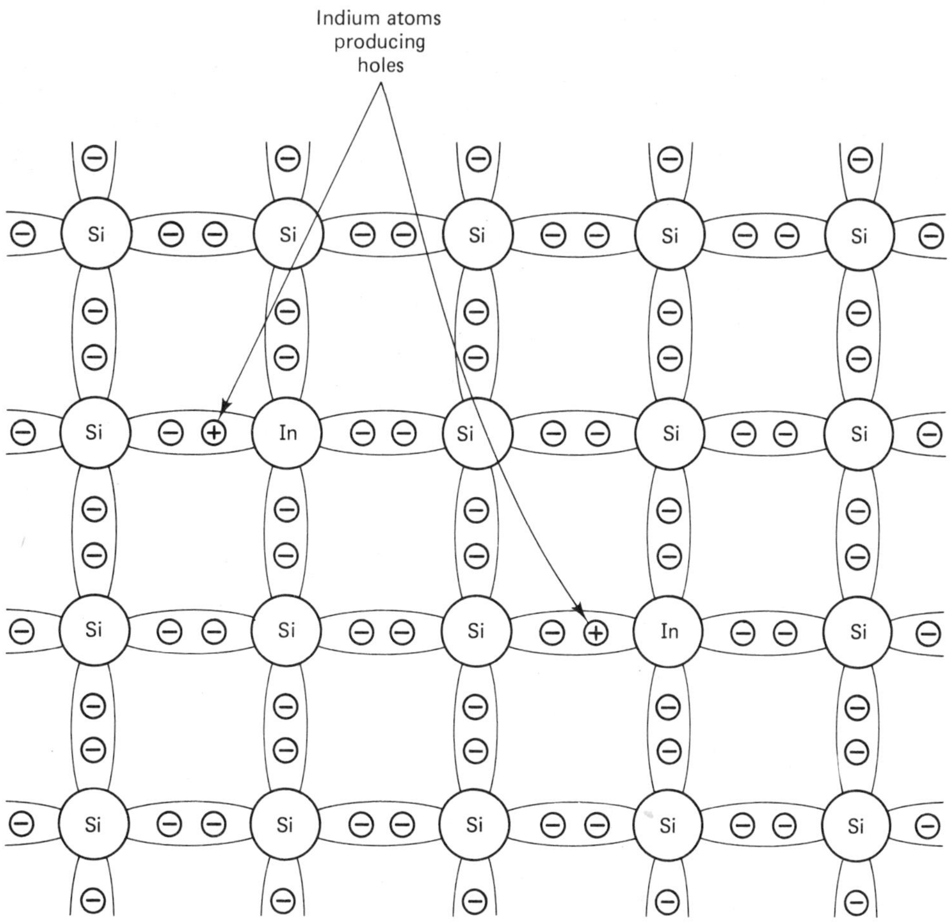

FIGURE 1-12 P-type crystal material.

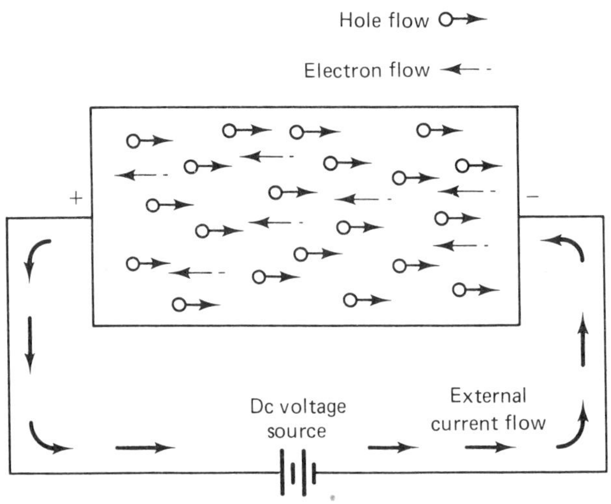

FIGURE 1-13 Current carriers in a P-type material.

rial, holes are the majority carriers and electrons are the minority carriers. The amount of acceptor material added to silicon determines the majority carrier content of the crystal.

Figure 1-13 shows how a P-type crystal will respond when connected to a voltage source. Note that there are more holes than electrons. With voltage applied, electrons are attracted to the positive battery terminal. Holes move, in a sense, toward the negative battery terminal. An electron is picked up at this point. The electron immediately fills a hole. The hole then becomes void. At the same time, an electron is pulled from the material by the positive battery terminal. Holes therefore move toward the negative terminal because electrons shift between different bonded groups. With energy applied, hole flow is continuous.

SELF-EXAMINATION

24. A very pure semiconductor material is called an __________ material.
25. At a temperature of absolute zero intrinsic silicon is an excellent insulator, at room temperature it becomes __________.
26. Electrons are called __________ current carriers, and holes are called __________ current carriers.
27. P-type material is formed by mixing an __________ type of element with intrinsic silicon.
28. An N-type material is formed by mixing a __________ type of element with intrinsic silicon.
29. How many electron-pair groups are formed around a single silicon atom in a covalent bonded structure?
    a. One b. Two c. Three d. Four e. Five

ANSWERS

24. Intrinsic	25. Conductive or a conductor
26. Negative, positive	27. Acceptor
28. Donor	29. d

SUMMARY

Semiconductor operation is directly related to atomic theory. Atoms consist of a nucleus that holds protons and neutrons and an equivalent number of electrons revolving in shells or layers. The energy level and location of electrons in the structure of an atom determines its electrical conductivity. Conductivity is the basis of material classifications such as insulators, semiconductors, and conductors.

Stabilized atoms have a full complement of electrons in the valence band. Unstable atoms try to become stable by drawing electrons away from neighboring atoms. Group IVB atoms such as silicon or germanium have four valence electrons. This type of atom is normally considered to be unstable. Silicon, for example, has four valence electrons in its M-shell. Two of these are in the s-energy level and two are in the p-level. The p-level needs four electrons to have a full complement and to be stable.

Valence electrons of different atoms are frequently bonded together to form stabilized molecules or compounds. Ionic, covalent, and metallic bonding are important in semiconductor theory. Table salt is an example of ionic bonding. A sodium atom donates one electron to chlorine that has seven valence electrons. This combination produces a form of stability and causes the two atoms to be bonded by an ionizing force. Covalent bonding occurs when the four valence electrons of one atom share four additional electrons (one each) from four neighboring atoms. Covalent bonding differs from ionic bonding in that the atoms do not form ions. Metallic bonding occurs in good electrical conductors. The valence electrons of metal tend to wander between different atoms and form a floating cloud that permits the material to be a good conductor.

Silicon and germanium are good semiconductor materials. When these Group IVB materials are to be used, they must first be made very pure. Then a doping material is added to silicon or germanium to change its level of conductivity. An intrinsic form of silicon is a rather good insulating material. When intrinsic silicon is mixed with a Group VB element, such as arsenic, it forms an N-type semiconductor material. A P-type material is formed when intrinsic silicon is mixed with Group IIIB elements such as indium or gallium.

Extrinsic silicon is considered to be a rather good conductor. When a piece of P-material is energized, holes move toward the negative side of the voltage source to gain electrons. Electrons are removed from the material where it is attached to the positive side of the source. This causes the formation of new holes. As a result, holes move toward the negative terminal and electrons move toward the positive terminal. When a piece of N-material is energized, electrons are attracted by the positive side of the source and holes are attracted by the negative side. Holes are developed when electrons leave the material. Electrons moving in the material fill holes. This action causes new holes to be created at different locations. Hole flow is opposite to that of electron flow.

EXPERIMENT 1-1
SEMICONDUCTOR MATERIAL TESTING

A semiconductor material will conduct current carriers equally in either direction—that is, it is bidirectional. When a piece of semiconductor material is connected to a source of electrical energy, it becomes conductive. The amount of current is limited by the value of the external resistor and the voltage of the source. Semiconductor current is also heat dependent, a property that varies to some extent on the type of semiconductor material used. The doped silicon of the device used in this activity is less dependent on heat than is a similar amount of germanium.

OBJECTIVES

1. To test the flow of current through a piece of semiconductor material.
2. To become familiar with the use of an electronic device.
3. To build a simple electronic circuit.
4. To connect an instrument and measure current and voltage.
5. To calculate current and compare it with a measured value.
6. To observe the influence of heat on the current flow of a semiconductor circuit.

EQUIPMENT

Circuit construction board

Volt-ohm-milliammeter (VOM), standard or digital type

N-channel junction field-effect transistor, Motorola 2N5459 or Heath 417-291

1 kΩ, $\frac{1}{4}$-W resistor

Variable dc source

PROCEDURE

1. Select an N-channel JFET from the component package. The device used in this activity is shown in Fig. 1-1A.
2. With an ohmmeter measure the resistance between any two of the three leads. If a resistance value of a few hundred ohms is measured, reverse the two leads. A

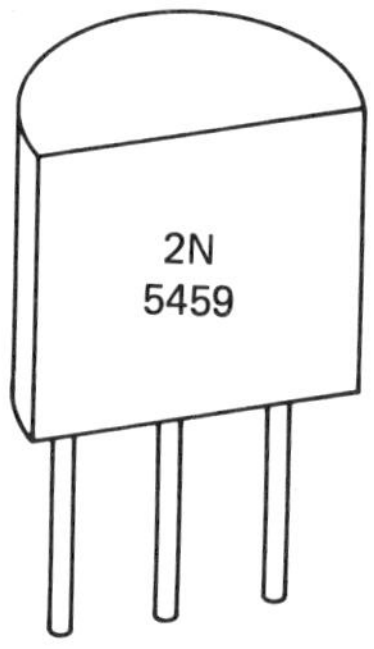

FIGURE 1-1A N-channel JFET.

reading of the same resistance in the reverse direction indicates the N-channel of the device. Find the two leads of the JFET with the same resistance value in each direction. Record the measured resistance. ________.

3. Connect the two JFET leads into the circuit as indicated in Fig. 1-1B.
4. Turn on the voltage source. We will refer to it as V_s in the remainder of the activity. Adjust V_s to produce 0 V. Measure the voltage across the resistor. This will be described as V_r. Measure the voltage across the JFET. This will be considered as the material resistance and referred to as V_m. V_r and V_m should both be equal to zero at this time.
5. Adjust the voltage source to produce 0.5 V. Again measure the voltage across the resistor and the JFET (semiconductor material).
6. Complete the chart shown in Fig. 1-1C. Record the measured values of V_r and V_m for each value of V_s. This chart is labeled forward current flow.

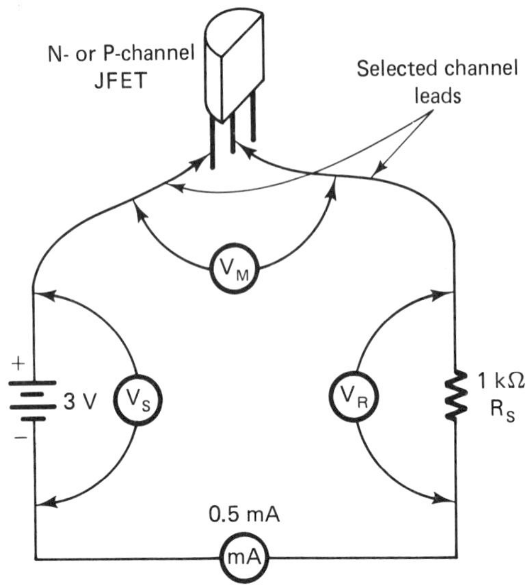

FIGURE 1-1B Semiconductor conduction circuit.

Current-flow chart

(Forward current)

V_s	V_r	V_m	I Calculated
0			
0.5			
1.0			
1.5			
2.0			
2.5			
3.0			
3.5			
4.0			
4.5			
5.0			

FIGURE 1-1C Current-flow chart.

7. Increase the voltage source in 0.5 V steps, and record the values of V_s, and V_r, and V_m up to 5.0 V.
8. Using Ohm's law, $I = V_r/R_s$, determine the current for each step of the chart. Record the current values on the chart.
9. Turn off the voltage source. Reverse the two selected leads of the JFET.
10. Repeat steps 4, 5, 6, 7, and 8 for the reversed material. The resulting chart should be labeled reverse current flow.
11. Prepare the VOM to measure current in a range that will handle the value calculated for a V_s of 5 V. What is a suitable range on the meter you are using? __________.
12. Connect the meter in series with the circuit as indicated in Fig. 1-1D.
13. With the source adjusted to 5 V, measure and record the current flow.
14. Grasp the JFET between your first finger and thumb. Hold the JFET for a minute or so while observing the current reading. Is there any change in the value of current?
15. More heat can be obtained from a soldering pencil or incandescent lamp. Do not touch the heat source directly to the JFET or apply a large amount of heat for more than 30 s. Is there any noticeable change in current when heat is applied to the JFET? ____________________.
16. Turn off the source voltage and disconnect the circuit.

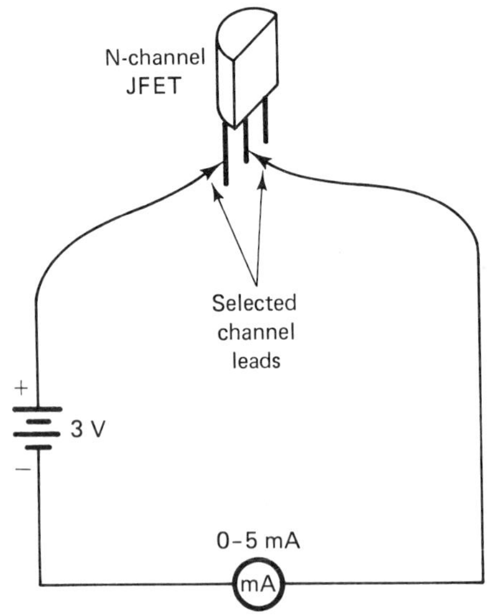

Current-flow chart

(Reverse current)

V_s	V_r	V_m	I Calculated
0			
0.5			
1.0			
1.5			
2.0			
2.5			
3.0			
3.5			
4.0			
4.5			
5.0			

FIGURE 1-1D

DISCUSSION

In this experiment you connected a piece of N-type semiconductor material in a circuit and tested its ability to conduct current. The resulting current flow was directly related to the value of the source voltage and the combined resistance of the semiconductor material and the series resistor. An increase in source voltage caused a corresponding increase in current. The voltages across the semiconductor material (V_m) and across the resistor (V_r) were measured with a voltmeter. Current through the circuit was calculated by the Ohm's law expression $I = V_r/R$. Actual circuit current was then measured and compared with the calculated value. These values should have been approximately the same.

In step 10, you reversed the direction of current flow through the semiconductor material by switching the leads of the device. You should have noted that current flow in the reverse direction was primarily the same as current flow in the forward direction by comparing the measured values of I, V_m, and V_r.

In steps 14 and 15 heat was applied to the semiconductor material. This should have caused an increase in current. A semiconductor material is considered to have a negative temperature coefficient, which means that an increase in temperature will cause a decrease in resistance. With less total resistance, there should be a corresponding increase in current. The negative temperature coefficient characteristic of a semiconductor material is an important circuit consideration. The heat of a semiconductor material must be carefully controlled or the device will be permanently damaged.

ANALYSIS

1. Explain the relationship between current (I) and source voltage (V_s) and the semiconductor material of this activity.
2. What influence does the N-type semiconductor material of this activity have on the direction of current flow?
3. How does the application of heat alter the current through a semiconductor material?

UNIT 1 EXAMINATION
SEMICONDUCTOR BASICS

Instructions: Circle the response that most correctly completes each statement.

1. A semiconductor material whose majority current carriers are holes is classified as:
   a. An N-type material
   b. A P-type material
   c. An intrinsic material
   d. An extrinsic material
   e. A doped material
2. The process by which impurities are added to a pure semiconductor crystal is called:
   a. Melting
   b. Mixing
   c. Doping
   d. Diffusing
   e. Refining
3. The type of atomic bonding most common in a semiconductor material is:
   a. Metallic
   b. Ionic
   c. Covalent
   d. Chemical
4. The impurity added to silicon to form a P-type material is considered to be:
   a. A Group IIIB element
   b. A Group VB element
   c. A donor material
   d. An ionized material
   e. A Group IVB element
5. An impurity added to silicon to form an N-type material is considered to be:
   a. An acceptor
   b. An ionized material
   c. A Group IIIB element
   d. A donor
   e. A Group IVB element

6. Electrons existing in the outermost shell of an atom are called:
   a. Ions
   b. Acceptors
   c. Valence electrons
   d. Donors
   e. Stabilized electrons
7. When an atom either gains or loses an electron, it is said to be:
   a. Ionized
   b. Bonded
   c. Excited
   d. Stabilized
   e. An acceptor
8. When two or more atoms combine to form a molecule, each atom is seeking:
   a. An ion
   b. Stability
   c. Conduction
   d. Valence
9. Table salt is a good example of:
   a. Ionic bonding
   b. Metallic bonding
   c. Covalent bonding
   d. A semiconductor
10. Ionic bonding is exemplified in molecules of:
    a. Copper oxide
    b. Metal
    c. Silicon
    d. Germanium
11. Metallic bonding is exemplified by atoms of:
    a. Germanium
    b. Silicon
    c. Copper
    d. Copper oxide
12. Covalent bonding refers to the:
    a. Sharing of electrons between atoms in a molecule
    b. Attraction of charged particles
    c. Ionization of atoms
    d. Wandering of free electrons among neighboring atoms
13. Connecting a voltage source to a piece of N-type material will cause:
    a. Current flow from $-$ to $+$

b. Current flow from + to −
c. Current flow in either direction
d. No current flow

14. How many electrons from neighboring atoms are shared by a covalently bonded silicon atom?
a. One
b. Two
c. Three
d. Four
e. Five

15. N-type silicon is considered to be a material that:
a. Is ionized
b. Has extra electrons that do not take part in covalent bonding
c. Has extra holes that upset the normal covalent bonding of atoms
d. Is stabilized

UNIT TWO

Semiconductor Diodes

UNIT INTRODUCTION

Semiconductor diodes are used rather extensively in the field of electronics. Radio, television, industrial control, computers, home entertainment equipment, and electrical appliances are a few of the applications. Diodes are probably the simplest of all electronic components because only two leads or electrodes are used in their construction. Functionally, diodes are unidirectional, which means that they conduct well in only one direction. Alternating current can easily be converted into direct current with a diode. In this application a diode is used as a rectifier. Other functions are the operational basis of more complex electronic devices. Transistors, integrated circuits, and other solid-state devices are constructed by the same techniques used to produce diodes. A person working in electronics must, therefore, be very familiar with the characteristic operation of this device.

Semiconductor diode theory is the subject of this unit. Learning this material is an essential first step in understanding how semiconductor devices operate.

UNIT OBJECTIVES

Upon completion of this unit, you will be able to:

1. Explain how a diode is constructed of N- and P-materials.
2. Connect a diode so that it is forward or reverse biased.
3. Distinguish between the forward and reverse characteristics of a diode.
4. Interpret the graphic relationship of current and voltage characteristics for solid-state diodes.
5. Identify the schematic symbol of a diode while noting the name of the electrodes and the material of its construction.
6. Identify the electrodes of a diode with an ohmmeter and be able to evaluate the condition of the device as good, shorted, or open.
7. Interpret the information from a manufacturer's data sheet for a silicon diode.
8. Identify several diodes by their package style or case structure.

IMPORTANT TERMS

Before reading Unit 2, review the following terms. These terms provide a basic understanding of terms associated with semiconductor diodes.

Anode A terminal of an electronic device that collects electrons.

Barrier potential or voltage The voltage that is developed across a P-N junction due to the diffusion of holes and electrons.

Bias forward Voltage applied across a P-N junction that causes the conduction of majority current carriers.

Bias reverse Voltage applied across a P-N junction that causes little or no current flow.

Cathode The terminal of an electronic device that is responsible for electron emission.

Covalent bond The bond formed when two atoms share a pair of electrons to form a stable molecule.

Depletion zone An area near the P-N junction void of current carriers.

Dielectric constant A number that compares the insulating ability of a material to that of air. Air has a dielectric constant of 1.

Doping The process of adding an impurity to a semiconductor material.

Electron A negatively charged subatomic particle.

Energy Something that is capable of producing work, such as heat, light, chemical, or mechanical action.

Hole A charge area or void where an electron is missing in a group of covalently bonded atoms.

Ion An atom that has either lost an electron (positive ion) or gained an electron (negative ion).

Junction The point where P and N semiconductor materials are joined.

JUNCTION DIODES

The term *junction diode* is often used to describe a two-terminal device whose crystal structure is made of P- and N-materials. One terminal is attached to the P-material and the other to the N-material. The point where these materials are joined is called a *junction.* A junction diode permits current carriers to flow readily in one direction and blocks the flow of current in the opposite direction.

Figure 2-1 shows the crystal structure of a junction diode. Notice the location of the P- and N-materials with respect to the junction. The crystal structure is continuous from one end to the other. The junction serves only as a dividing line that marks the ending of one material and the beginning of the other. This type of structure permits electrons to move readily through the entire structure.

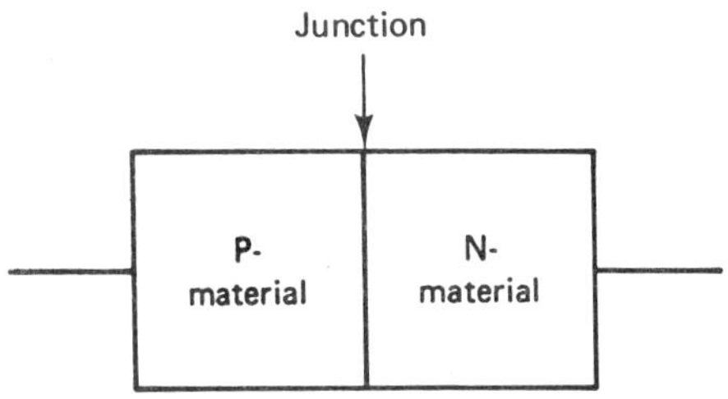

FIGURE 2-1 Crystal structure of a junction diode.

Figure 2-2 shows two pieces of semiconductor material before they are formed into a P-N junction. As indicated, each piece of material has majority and minority current carriers. The number of carrier symbols shown in each material indicates the minority or majority function. Electrons are the majority carriers in the N-material and the minority carriers in the P-material. Holes are the majority carriers in the P-material and the minority carriers in the N-material. Both holes and electrons are free to move about in their respective materials.

Depletion Zone

When a junction diode is first formed, there is a unique interaction between current carriers. Electrons from the N-material move readily across the junction to fill holes in the P-material. This action is commonly called *diffusion.* Diffusion is the result of a high concentration of carriers in one material and a lower concentration in the other. Only those current carriers near the junction take part in the diffusion process.

The diffusion of current carriers across the junction of a diode causes a change in its structure. Electrons leaving the N-material cause positive ions to be generated in their place. Upon entering the P-material to fill holes, these same electrons create negative ions. The areas on both sides of the junction then contain a large number of positive and negative ions. The number of holes and electrons in this area becomes depleted. The term *depletion zone* is used to describe this area. It represents an area that is void of majority current carriers. All P-N junctions develop a depletion zone when they are formed. Figure 2-3 shows the depletion zone of a junction diode.

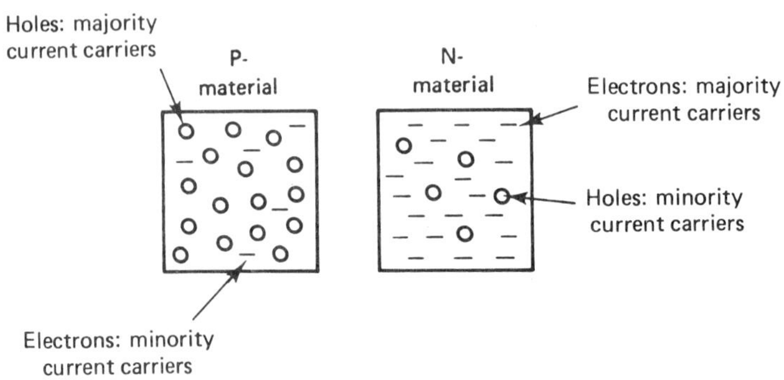

FIGURE 2-2 Semiconductor materials.

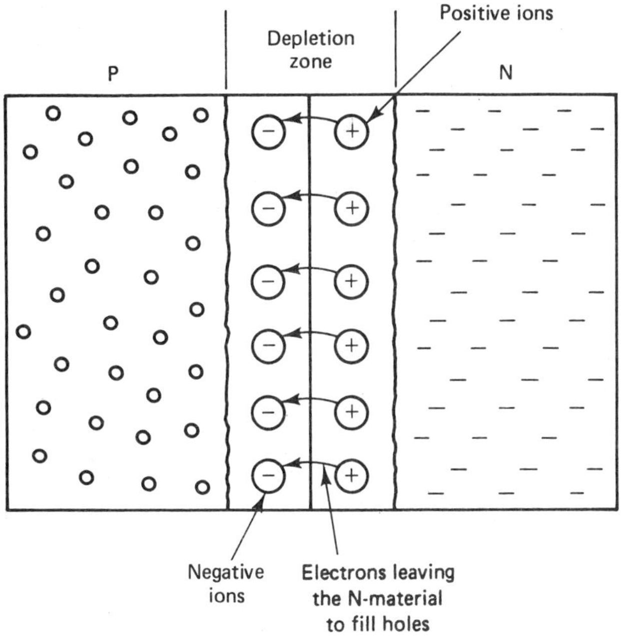

FIGURE 2-3 Depletion-zone formation.

Barrier Potential

Before N- and P-materials are joined together at a common junction, they are considered to be electrically neutral. After they are joined, however, diffusion takes place immediately. The creation of negative and positive charges on the P and N sides of the junction, respectively, tends to drive the remaining electrons and holes away from the junction. This action makes it somewhat difficult for additional charge carriers to diffuse across the junction. The end result is a charge buildup or barrier potential across the junction.

Figure 2-4 shows the resulting barrier potential as a small battery connected across the P-N junction. Note the polarity of this potential with respect to the P- and N-materials. This voltage will exist even when the crystal is not connected to an outside source of energy. The barrier potential for germanium is approximately 0.3 V, and 0.7 V for silicon; however, these voltage values cannot be measured directly. They appear only across the space-charge region of the junction. The barrier potential of a P-N junction must be overcome by an outside voltage source in order to produce current conduction.

SELF-EXAMINATION

1. The point where N- and P-materials are joined together in a diode is called a ____________.
2. A ____________ permits current carriers to flow readily in one direction and blocks the flow of current in the other direction.

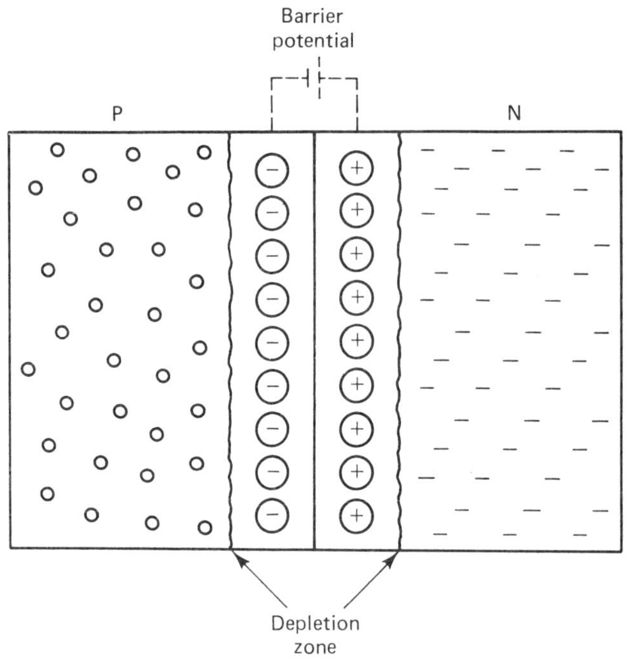

FIGURE 2-4 Barrier potential of a diode.

3. A ____________ is a two-terminal electronic device.
4. A diode is formed by joining a piece of ____________ material to a piece of ____________ material at a common junction.
5. The majority current carriers of a piece of N-material are ____________.
6. The majority current carriers of a piece of P-material are ____________.
7. ____________ are the minority current carriers of a piece of N-material.
8. ____________ are the minority current carriers of a piece of P-material.
9. When electrons move from N-material across a junction to fill holes in P-material the process is called ____________.
10. Electrons leaving the N-material to fill holes in the P-material cause a ____________ ____________ to appear across the junction.
11. The barrier potential is approximately ____________ V for germanium and ____________ V for silicon.
12. An area near the junction of a diode that is void of current carriers is called the ____________ ____________.

ANSWERS

1. Junction
2. Diode
3. Diode
4. N, P or P, N

5.	Electrons	6.	Holes
7.	Holes	8.	Electrons
9.	Diffusion	10.	Barrier potential
11.	0.3, 0.7	12.	Depletion zone

JUNCTION BIASING

An external source of energy applied to a P-N junction is called a *bias voltage* or simply a *bias*. This voltage either adds to or reduces the barrier potential of the junction. *Forward biasing* reduces the barrier potential and causes current carriers to return to the depletion zone. *Reverse biasing* adds external voltage of the same polarity to the barrier potential and causes an increase in the width of the depletion zone, which hinders current carriers from entering the depletion zone.

Reverse Biasing

When a battery is connected across a P-N junction as in Fig. 2-5 it causes reverse biasing. Note that the negative terminal of the battery is connected to the P-material, and the positive terminal is connected to the N-material. This connection causes the battery polarity to oppose the material polarity of the diode. Since unlike charges attract, the majority charge carriers of each material are pulled away from the junction. Reverse biasing of a diode normally causes it to be nonconductive.

Figure 2-6 shows how the majority current carriers are rearranged in a reverse-biased diode. As shown, electrons of the N-material are pulled toward the positive battery terminal. Each electron that moves or leaves the diode causes a positive ion to appear in its place. This causes a corresponding increase in the width of the depletion zone on the N side of the junction.

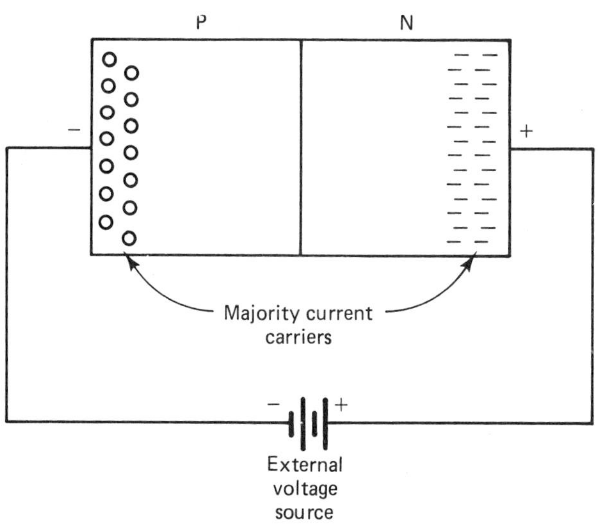

FIGURE 2-5 Reverse-biased diode.

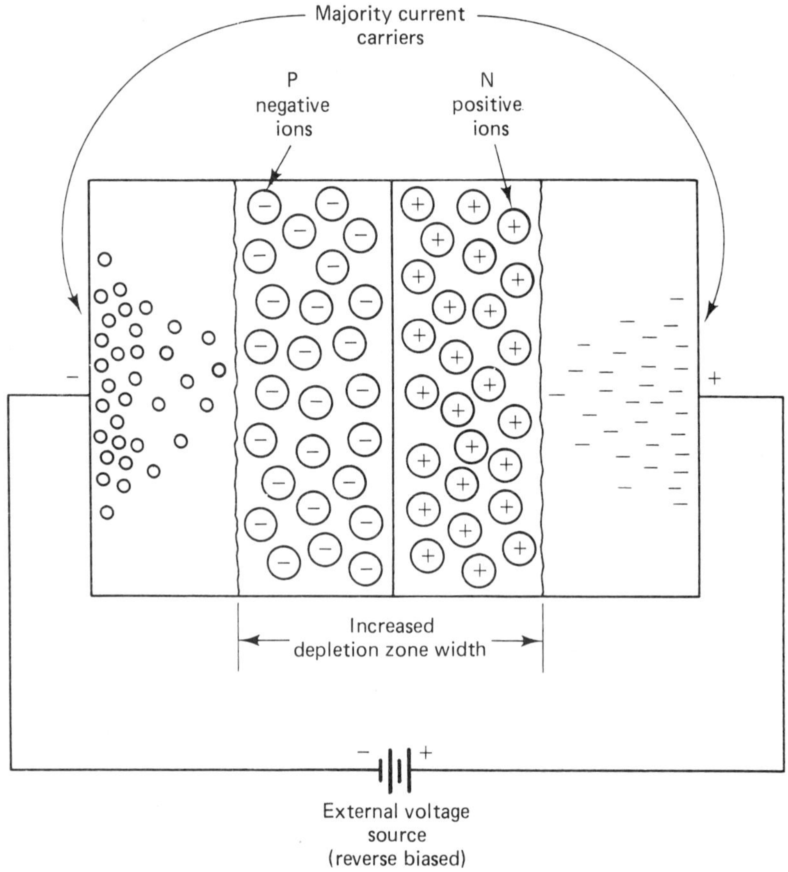

FIGURE 2-6 Reverse-biased P-N junction.

The reaction on the P side of the diode is very similar. In this case a number of electrons leave the negative battery terminal and enter the P-material. These electrons immediately move in and fill a number of holes. Each filled hole then becomes a negative ion. These ions are then repelled by the negative battery terminal and driven toward the junction. As a result, the width of the depletion zone increases on the P side of the junction.

The depletion zone width of a reverse-biased diode is directly dependent on the value of the supply voltage. With a wide depletion zone a diode cannot effectively support current flow. The charge buildup across the junction increases until the barrier voltage equals the external bias voltage. When this occurs, a diode effectively becomes a nonconductor.

Leakage Current

The wider depletion zone of a reverse-biased diode represents an area that is almost void of majority current carriers and thus will respond as an insulator. Ideally, current carriers do not pass through an insulator, but in a reverse-biased diode some current actually flows through the depletion zone. This is called *leakage current.* Leakage current is dependent on minority current carriers.

Remember that minority carriers are electrons in the P-material and holes in the N-material. Figure 2-7 shows how these carriers respond when a diode is reverse biased. It should be noted that the minority carriers of each material are pushed through the depletion zone to the junction, which causes a very small leakage current to occur. Normally, leakage current is so small that it is often considered negligible.

The number of minority current carriers in a semiconductor is primarily dependent on temperature. At normal room temperatures of 25°C (77°F) there are a rather limited number of minority carriers present in a semiconductor. However, when the surrounding temperature rises, it causes a considerable increase in minority carrier production, which creates a corresponding increase in leakage current.

Leakage current occurs to some extent in all reverse-biased diodes. In germanium diodes, leakage current is only a few microamperes. Silicon diodes normally have fewer minority current carriers, so they have less leakage current. Typical leakage current values for silicon are a few nanoamperes. The construction material of a diode is thus an important consideration. Germanium is much more sensitive to temperature than is silicon, so germanium has a higher level of leakage current. This factor is largely responsible for the widespread use of silicon in modern semiconductor devices.

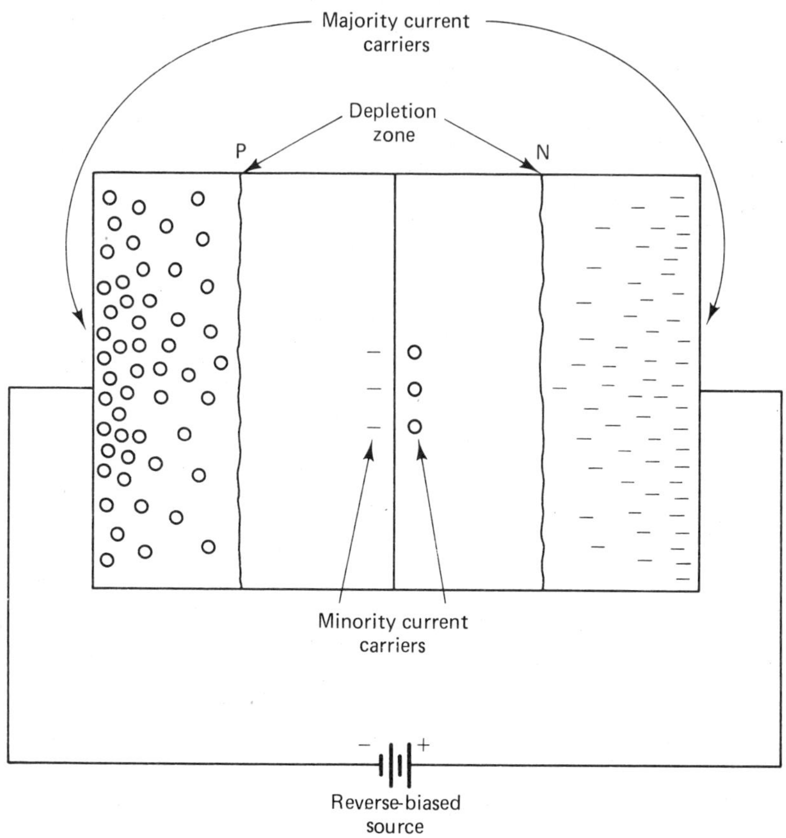

FIGURE 2-7 Minority current carriers.

Forward Biasing

When voltage is applied to a diode as in Fig. 2-8 it causes forward biasing. In this example the positive battery terminal is connected to the P-material and the negative terminal to the N-material. This voltage repels the majority current carriers of each material. A large number of holes and electrons therefore appear at the junction. On the N side of the junction, electrons move in to neutralize the positive ions in the depletion zone. In the P-material, electrons are pulled from negative ions, which causes them to become neutral again. This means that forward biasing causes the depletion zone to collapse and the barrier potential to be removed. The P-N junction therefore supports a continuous current flow when it is forward biased.

Figure 2-9 shows how the current carriers of a forward-biased diode respond. Since the diode is connected to an external voltage source, it has a constant supply of electrons. Large arrows are used in the diagram to show the direction of current flow outside the diode. Inside the diode, smaller arrows show the movement of majority current carriers. Remember that electron flow and current are synonymous.

Starting at the negative battery terminal, assume now that electrons flow through a wire to the N-material. Upon entering this material, they flow immediately to the junction. At the same time an equal number of electrons are removed from the P-material and are returned through a resistor to the positive battery terminal. This action generates new holes and causes them to move toward the junction. When these holes and electrons reach the junction, they combine and effectively disappear. At the same time new holes and electrons appear at the outer ends of the diode. These majority carriers are generated on a continuous basis. This action continues as long as the external voltage source is applied.

It is important to realize that electrons flow through the entire diode when it is forward biased. In the N-material this is quite

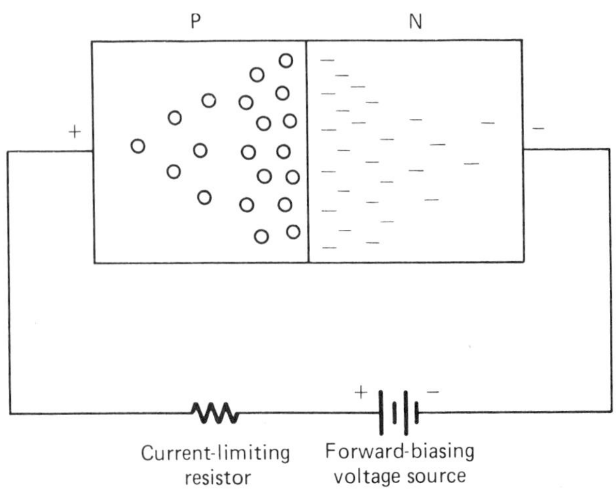

FIGURE 2-8 Forward-biased diode.

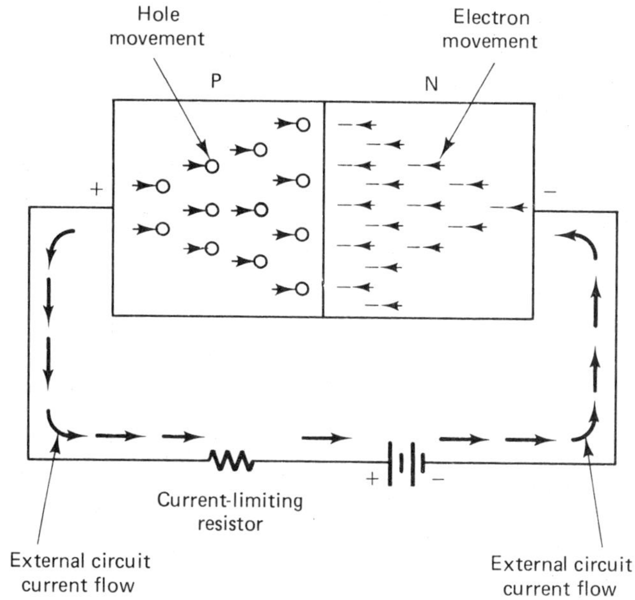

FIGURE 2-9 Current carrier flow in a forward-biased diode.

obvious. In the P-material, however, holes are the moving current carriers. Remember that hole movement in one direction must be initiated by electron movement in the opposite direction. Therefore, the combined flow of holes and electrons through a diode equals the total current flow.

The current-limiting resistor of Fig. 2-9 is essential in a forward-biasing diode circuit. This resistor is needed to keep the current flow at a safe operating level. The maximum current (I_{max}) rating of a diode represents this value. As long as diode current does not exceed this value, it can operate satisfactorily.

SELF-EXAMINATION

13. ____________ is a process of connecting a voltage source across the P-N junction of a diode.
14. ____________ biasing occurs when the negative side of the voltage source is connected to the P-material and the positive side of the source is connected to the N-material.
15. ____________ biasing occurs when the positive side of the source is connected to the P-material and the negative side of the source is connected to the N-material.
16. Since unlike charges attract, the ____________ current carriers of a reversed-biased diode are pulled away from the junction.
17. A reverse-biased diode is considered to be ____________.
18. Reverse biasing of a diode causes the width of the depletion zone to ____________.

19. ______________ current is determined by the minority current carrier content of a diode.
20. Minority current content of a diode is ______________ dependent.
21. Leakage current is greatest in diodes made of ______________.
22. ______________ biasing of a diode causes the depletion zone to collapse.
23. Current-limiting resistors are used in forward-biased diode circuits to keep the ______________ at a safe operating level.

ANSWERS

13. Biasing	14. Reverse
15. Forward	16. Majority
17. Nonconductive	18. Increase
19. Leakage	20. Temperature
21. Germanium	22. Forward
23. Current or I_m	

DIODE CHARACTERISTICS

Now that we have seen how a junction diode operates, it is time to examine some of its electrical characteristics, such as voltage, current, and temperature. Since these characteristics vary a great deal in an operating circuit, it is best to look at them graphically. This makes it possible to see how the device will respond under different operating conditions.

Forward Characteristic

When a diode is connected in the forward bias direction, it conducts forward current I_F. The value of I_F is directly dependent on the amount of forward voltage V_F. The relationship between forward voltage and forward current is called the ampere-volt, or *I-V*, characteristic of a diode. A typical diode forward *I-V* characteristic is shown in Fig. 2-10(a), with its test circuit in Fig. 2-10(b). Note in particular that V_F is measured across the diode and that I_F is a measure of what flows through the diode. The value of the source voltage V_S does not necessarily compare in value with V_F.

When the forward voltage of the diode equals 0 V, I_F equals 0 mA. This value starts at the origin (0) of the graph. If V_F is gradually increased in 0.1-V steps, I_F begins to increase. When the value of V_F is great enough to overcome the barrier potential of the P-N junction, a substantial increase in I_F occurs. The point at

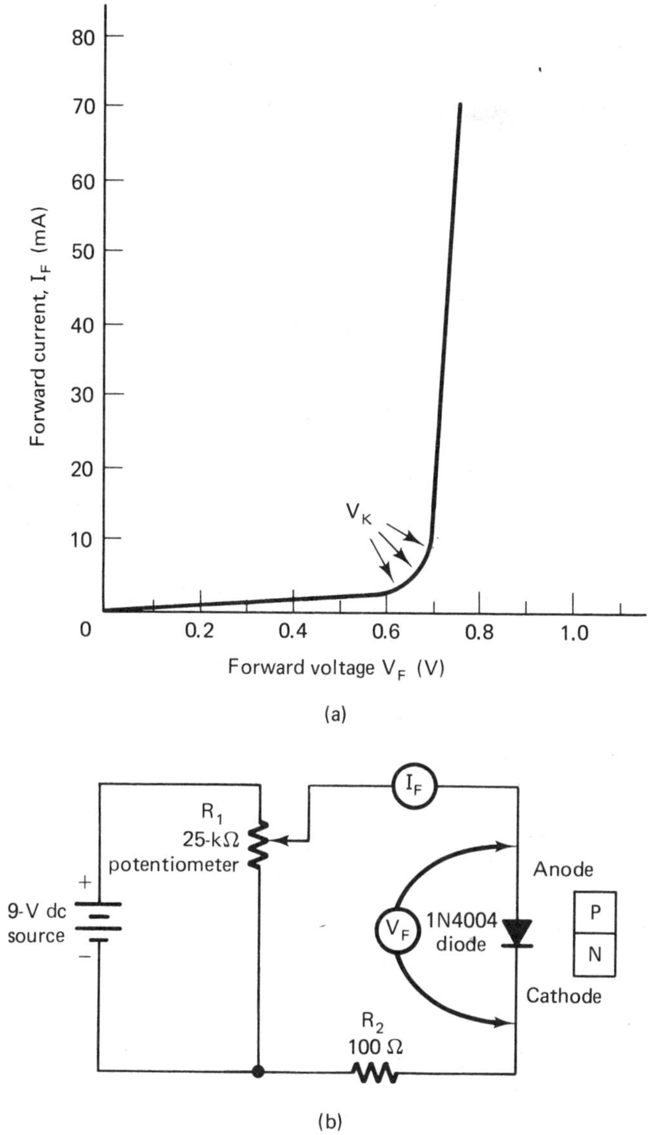

FIGURE 2-10 (a) *I-V* characteristics of a diode; (b) diode characteristic test circuit.

which this occurs is often called the *knee voltage* V_K. V_K is approximately 0.3 V for germanium diodes and 0.7 V for silicon.

If the value of V_F increases much beyond V_K, the forward current becomes quite large. This, in effect, causes heat to develop across the junction. Excessive junction heat can destroy a diode. To prevent this from happening, a protective resistor is connected in series with the diode. This resistor limits I_F to some point below the maximum current value of the diode. Diodes are rarely operated in the forward direction without a current-limiting resistor.

The operation of a silicon diode connected in the forward-conduction direction is shown by the circuit of Fig. 2-11. Note that the diode is forward biased by the source and has its current limited by a series resistor R_1. When the diode is connected in this

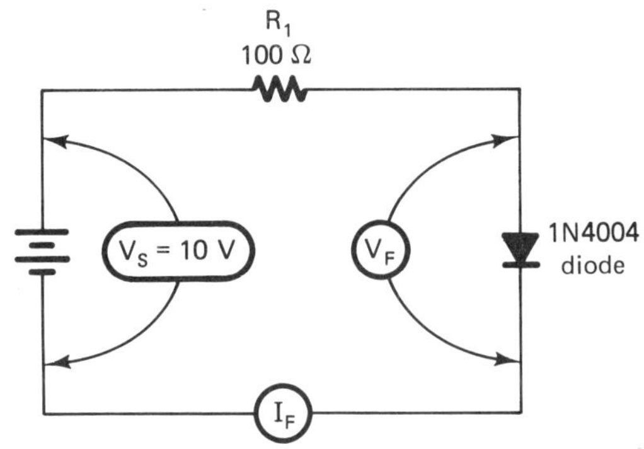

FIGURE 2-11 Silicon diode connected in forward conduction.

manner, the forward current depends on source voltage and the value of the current-limiting resistor. The resulting I_F can be determined by Ohm's law. This takes into account the voltage drop across the diode when it is forward biased. For a silicon diode this is 0.7 V. The voltage across the limiting resistor is the source voltage (V_s) minus the knee voltage (V_K) across the diode. Forward current (I_F) is determined by the expression

$$I_F = \frac{V_S - V_K}{R_1}$$

$$I_F = \frac{10 \text{ V} - 0.7 \text{ V}}{100}$$

$$I_F = 0.093 \text{ A, or } 93 \text{ mA}$$

Reverse Characteristic

When a diode is connected in the reverse bias direction it has an I_R-V_R characteristic. Figure 2-12 shows the reverse *I-V* characteristic of a diode and its test circuit. This characteristic has different values of I_R and V_R. Reverse current is usually quite small. The vertical I_R line in this graph has current values graduated in microamperes. The number of minority current carriers that take part in I_R is quite small. In general, this means that I_R remains rather constant over a large part of V_R. Note also that V_R is graduated in 100-V increments. Starting at zero when the reverse voltage of a diode is increased, there is only a slight change in I_R. At the voltage breakdown V_R point current increases very rapidly. The voltage across the diode remains fairly constant at this time. This constant-voltage characteristic leads to a number of reverse-biased diode applications. Normally, diodes are used in applications where the V_R is not reached.

The physical processes responsible for current conduction in a reverse-biased diode are called *zener breakdown* and *avalanche breakdown.* Zener breakdown occurs when electrons are pulled from their covalent bonds in a strong electric field. This occurs at a rather high value of V_R. When large numbers of covalent bonds are broken at the same time, there is a sudden increase in I_R.

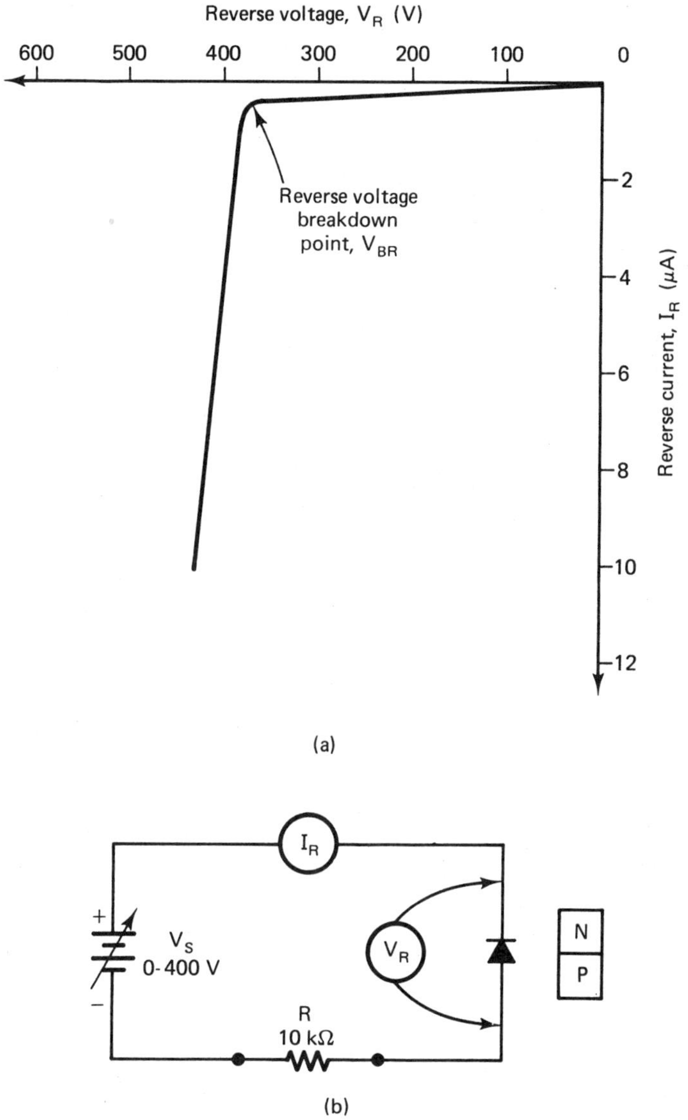

FIGURE 2-12 (a) Reverse *I-V* characteristics of a diode; (b) reverse characteristic test circuit.

Avalanche breakdown is an energy-related condition of reverse biasing. At high values of V_R minority carriers gain a great deal of energy. This gain may be great enough to drive electrons out of their covalent bonding, which creates new electron-hole pairs. These carriers then move across the junction and produce other ionizing collisions and additional electrons. The process continues to build up until an avalanche of current carriers is produced, at which point the process is irreversible.

Combined *I-V* Characteristics

The forward and reverse *I-V* characteristics of a diode are generally combined on a single characteristic curve. Figure 2-13 shows a rather standard method of displaying this curve. Forward- and

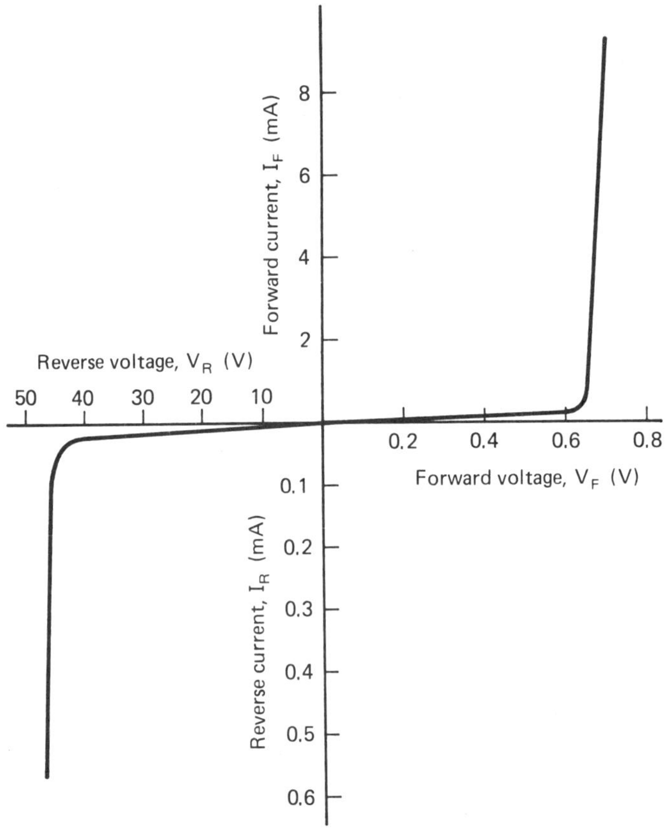

FIGURE 2-13 Diode *I-V* characteristics.

reverse-bias voltage, V_F and V_R, are usually plotted on the horizontal axis of the graph. V_F extends to the right and V_R to the left. The point of origin, or zero value, is at the center of the horizontal line. Forward and reverse current values are shown vertically on the graph. I_F extends above the horizontal axis, with I_R extending downward. The origin serves as a zero indication for all four values. This means that combined V_F and I_F values are located in the upper right part of the graph and V_R and I_R are in the lower left corner. Different scales are normally used to display forward and reverse values.

A rather interesting comparison of silicon and germanium characteristic curves is shown in Fig. 2-14. Careful examination shows that germanium requires less forward voltage to go into conduction than silicon. This characteristic is a distinct advantage in low-voltage circuits. Note also that a germanium diode requires less voltage drop across it for different values of current. This means that germanium has a lower resistance to forward current flow. Germanium therefore appears to be a better conductor than silicon. Silicon is more widely used, however, because of its low leakage current and lower production cost.

The reverse-bias characteristics of silicon and germanium diodes can also be compared in Fig. 2-14. The I_R of a silicon diode is very small compared with that of a germanium diode. Reverse current is determined primarily by the minority current content

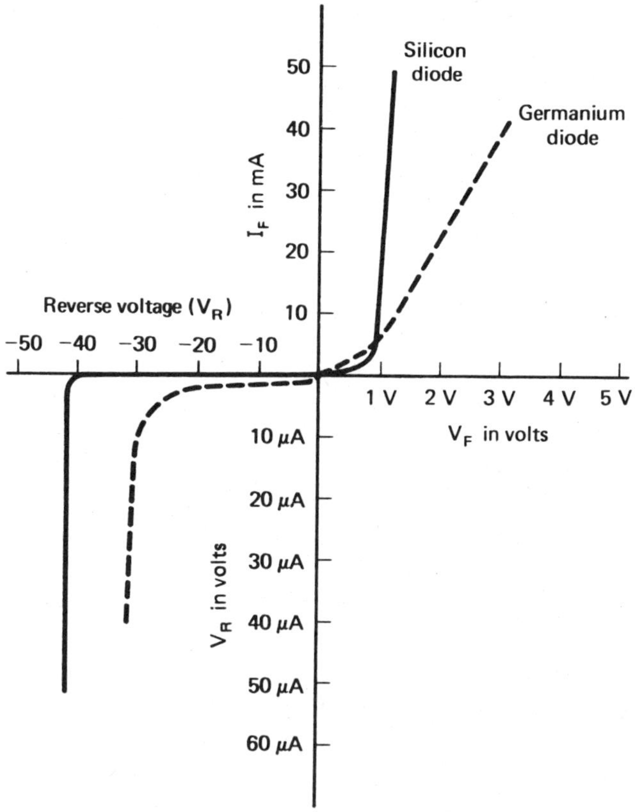

FIGURE 2-14 Silicon and germanium diode characteristics.

of the material, a condition influenced primarily by temperature. For germanium diodes, I_R doubles for each 10°C rise in temperature. In a silicon diode the change in I_R is practically negligible for the same rise in temperature. As a result, silicon diodes are preferred over germanium diodes in applications where large changes in temperature occur. Comparisons of this type are quite obvious through the study of characteristic curves.

SELF-EXAMINATION

24. In a diode, forward current (I_F) is ____________ related to the value of forward voltage (V_F).
25. Forward voltage (V_F) is measured ____________ a diode.
26. Forward current (I_F) is a measure of the current passing ____________ a diode.
27. When V_F overcomes the barrier potential of a P-N junction, there is a large increase in ____________.
28. The knee voltage (V_K) is 0.3 V for ____________ and 0.7 V for ____________.
29. ____________ breakdown occurs when the electrons of a diode are pulled from their covalent bonding in a strong reverse-biased electric field.

30. ___________ breakdown of a diode is due to an energy-related reverse-biased condition.

ANSWERS

24. Directly	25. Across
26. Through	27. Forward current, or I_F
28. Germanium; silicon	29. Zener
30. Avalanche	

DIODE SPECIFICATIONS

Selection of a diode for a specific application requires knowledge of its specifications. This type of information is usually made available through the manufacturer. Diode specifications generally include absolute maximum ratings, typical operating conditions, mechanical data, lead identification, mounting procedures, and characteristic curves. Figure 2-15 shows a representative diode data sheet. Some of the important specifications are explained in the following:

1. Maximum reverse voltage, V_{RM}: The absolute maximum or peak reverse bias voltage that can be applied to a diode. This may also be called the peak inverse voltage (P_{IV}) or peak reverse voltage (P_{RV}).
2. Reverse breakdown voltage, V_{BR}: The minimum steady-state reverse voltage at which breakdown will occur.
3. Maximum forward current, I_{FM}: The absolute maximum repetitive forward current that can pass through a diode at 25°C (77°F). This is reduced for operation at higher temperatures.
4. Maximum forward surge current, I_{FM} (surge): The maximum current that can be tolerated for a short interval of time. This current value is much greater than I_{FM}. This represents the increase in current that occurs when a circuit is first turned on.
5. Maximum reverse current I_{RM}: The absolute maximum reverse current that can be tolerated at device operating temperature.
6. Forward voltage V_{FM}: Maximum forward voltage drop for a given forward current at device operating temperature.
7. Power dissipation P_D: The maximum power that the device can safely absorb on a continuous basis in free air at 25°C (77°F).
8. Reverse recovery time T_{rr}: The maximum time that it takes the device to switch from its on to its off state.

TYPES 1N4001 THROUGH 1N4007
DIFFUSED-JUNCTION SILICON RECTIFIERS

***electrical characteristics at specified ambient† temperature**

PARAMETER		TEST CONDITIONS	MAX	UNIT
I_R	Static Reverse Current	V_R = Rated V_R, $T_A = 25°C$	10	μa
		V_R = Rated V_R, $T_A = 100°C$	50	μa
$I_{R(avg)}$	Average Reverse Current	V_{RM} = Rated V_{RM}, $I_O = 1$ a, f = 60 cps, $T_A = 75°C$	30	μa
V_F	Static Forward Voltage	$I_F = 1$ a, $T_A = 25°C$ to $75°C$	1.1	v
$V_{F(avg)}$	Average Forward Voltage	V_{RM} = Rated V_{RM}, $I_O = 1$ a, f = 60 cps, $T_A = 25°C$ to $75°C$	0.8	v
V_{FM}	Peak Forward Voltage	V_{RM} = Rated V_{RM}, $I_O = 1$ a, f = 60 cps, $T_A = 25°C$ to $75°C$	1.6	v

*Indicates JEDEC registered data.

THERMAL INFORMATION

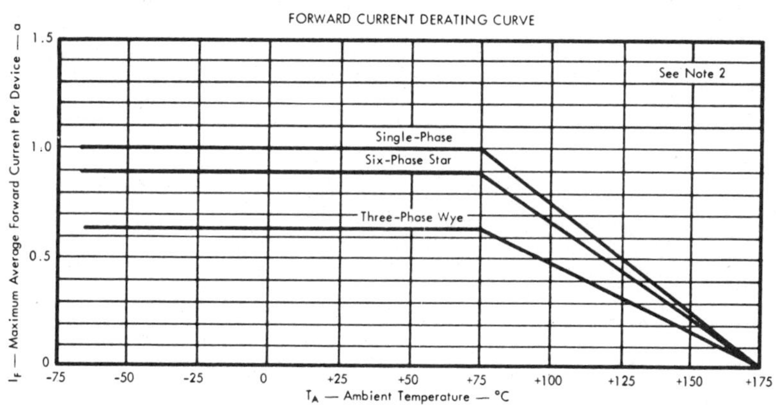

FIGURE 1

†The ambient temperature is measured at a point 2 inches below the device. Natural air cooling shall be used.

NOTE 2: This rectifier is a lead-conduction-cooled device. At (or above) ambient temperatures of 75°C, the lead temperature ⅜ inch from case must be no higher than 5°C above the ambient temperature for these ratings to apply.

TEXAS INSTRUMENTS RESERVES THE RIGHT TO MAKE CHANGES AT ANY TIME IN ORDER TO IMPROVE DESIGN AND TO SUPPLY THE BEST PRODUCT POSSIBLE.

FIGURE 2-15 Data sheet for a silicon diode. [Courtesy of Texas Instruments.]

Diode Temperature

The operation of a diode is directly related to temperature. All semiconductor materials are similar in this respect. The primary effect of temperature is the generation of additional electron-hole pairs because more valence electrons are thermally excited and

TYPES 1N4001 THROUGH 1N4007
DIFFUSED-JUNCTION SILICON RECTIFIERS

TYPES 1N4001 THROUGH 1N4007
BULLETIN NO. DL-S 657366, FEBRUARY 1965

50-1000 VOLTS • 1 AMP AVG

- **MINIATURE MOLDED PACKAGE**
- **INSULATED CASE**
- **IDEAL FOR HIGH-DENSITY CIRCUITRY**

***mechanical data**

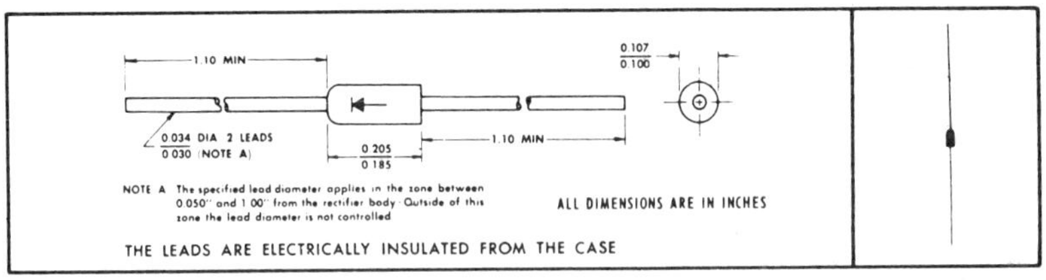

***absolute maximum ratings at specified ambient† temperature**

		1N4001	1N4002	1N4003	1N4004	1N4005	1N4006	1N4007	UNIT
V_{RM}	Peak Reverse Voltage from −65°C to +175°C (See Note 1)	50	100	200	400	600	800	1000	v
V_R	Steady State Reverse Voltage from 25°C to 75°C	50	100	200	400	600	800	1000	v
I_O	Average Rectified Forward Current from 25°C to 75°C (See Notes 1 and 2)	1							a
$I_{FM(rep)}$	Repetitive Peak Forward Current, 10 cycles, at (or below) 75°C (See Note 3)	10							a
$I_{FM(surge)}$	Peak Surge Current, One Cycle, at (or below) 75°C (See Note 3)	30							a
$T_{A(opr)}$	Operating Ambient Temperature Range	−65 to +175							°C
T_{stg}	Storage Temperature Range	−65 to +200							°C
	Lead Temperature ⅜ Inch from Case for 10 Seconds	350							°C

NOTES: 1. These values may be applied continuously under single-phase, 60-cps, half-sine-wave operation with resistive load. Above 75°C derate I_O according to Figure 1.

2. This rectifier is a lead-conduction-cooled device. At (or above) ambient temperatures of 75°C, the lead temperature ⅜ inch from case must be no higher than 5°C above the ambient temperature for these ratings to apply.

3. These values apply for 60-cps half sine waves when the device is operating at (or below) rated values of peak reverse voltage and average rectified forward current. Surge may be repeated after the device has returned to original thermal equilibrium.

* Indicates JEDEC registered data.

† The ambient temperature is measured at a point 2 inches below the device. Natural air cooling shall be used.

TEXAS INSTRUMENTS
INCORPORATED
POST OFFICE BOX 5012 • DALLAS, TEXAS 75222

FIGURE 2-15 (*continued*)

gain enough energy to move into the conduction band. These additional current carriers cause a decided change in the *I-V* characteristics of a diode.

Figure 2-16 shows how the *I-V* characteristics of a silicon diode change with temperature. These operating temperatures are indicated in degrees Celsius. Three main differences should be noted. The first of these is in the forward-bias region. This shows

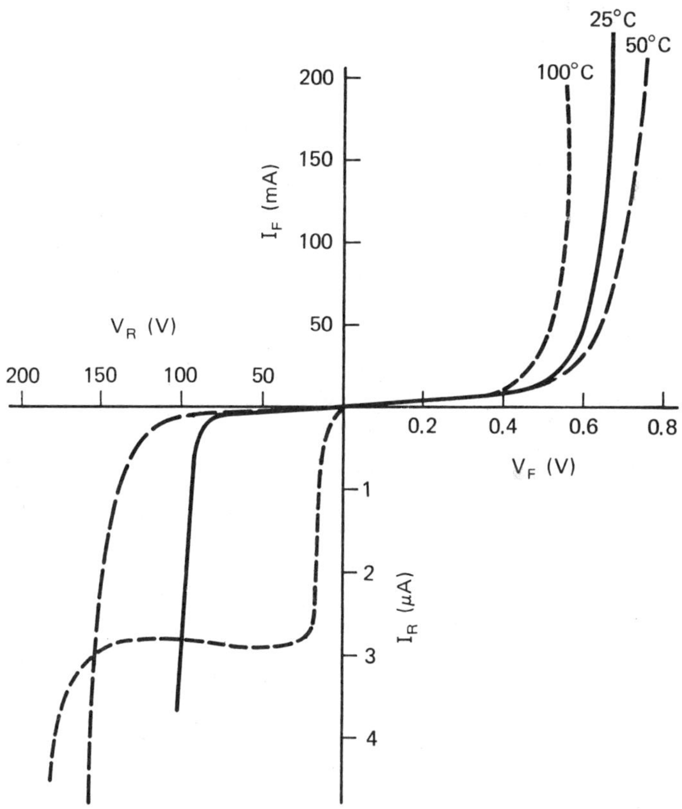

FIGURE 2-16 *I-V* characteristics of a silicon diode at 100, 50, and 25°C.

that less voltage is needed to produce conduction at elevated temperatures. The other differences are in the reverse-bias area. One of these shows that there is a great deal more leakage current at higher temperatures. The third difference is in breakdown voltage. This indicates that lower temperatures produce higher breakdown voltages. These considerations must all be taken into account when selecting a diode for a circuit application.

Electronic circuits that employ diodes are called upon to operate in a rather wide range of temperatures. Consumer-grade diodes are usually rated for a range of $-50°$ to $+100°C$. The extremes of this range are more meaningful if related to the boiling point of water (100°C, or 212°F) and the freezing point of mercury ($-39°C$, or $-40°F$). Military-grade diodes are rated at $-65°$ to 125°C. This type of diode is much more expensive than consumer-grade devices.

Junction Capacitance

A *capacitor* is defined as two or more conductive plates separated by an insulating material. A reverse-biased diode has a structure that responds as a capacitor. The two independent crystal materials serve as conductor plates, with the depletion zone acting as a dielectric material. The term *junction capacitance* is used to describe this effect.

Junction capacitance is a variable value that is dependent on bias voltage. Small amounts of reverse bias, and in some cases forward bias, produce the largest values of junction capacitance. This, in effect, reduces the width of the depletion zone. A decrease in dielectric thickness causes a corresponding increase in capacitance.

When the reverse-bias voltage of a diode is increased, there is a decrease in junction capacitance. This condition causes a corresponding increase in the width of the depletion zone. With a thicker dielectric material, there is a smaller junction capacitance.

The dielectric constant of a silicon depletion zone is approximately 12. This means that the depletion zone is 12 times better than air as a dielectric. As a result, some rather significant capacitance values appear across a junction diode. This value is extremely important in high-frequency circuit applications. A reverse-biased diode can therefore respond as a voltage-controlled capacitor. Special devices known as *varicap diodes* are designed to perform this function.

Switching Time

One rather important characteristic for some diodes is *switching time,* which refers to the time it takes for a diode to switch from one state to the other. This condition is important in computer applications that involve rapid turn-on and turn-off times.

When a junction diode is forward biased, its depletion zone begins to fill with current carriers. This action produces a large number of electron-hole recombinations near the junction. If the bias voltage is suddenly removed, the recombination process does not stop instantly. Current carriers have a type of inertia that causes them to continue moving once they are placed in motion. The nonconduction state cannot be reached until all the current carriers have cleared the depletion zone. This means that the depletion zone has a current-carrier storage function.

Changing state from reverse bias to forward bias is also time dependent in a diode. The delay involved in this characteristic is due to junction capacitance. This capacitance tends to absorb the initial forward bias current when it is being charged. After the charging operation has been completed, current carriers become available for normal conduction. Special switching diodes are now available for most high-speed control operations.

Diode Packaging

Diodes are manufactured today in a wide range of case styles and packages. A person working with these devices must be familiar with some of the common methods of packaging. Element identification and lead marking techniques are essential for proper installation and testing. A diode improperly connected in a circuit may be damaged itself or may cause damage to other parts. Figure

2-17 shows three general classes of diodes according to their current-handling capability. These represent some of the more popular package styles in use today.

The low-current diodes of Fig. 2-17(a) are the smallest of all packages; body length rarely exceeds 0.3 cm. The cathode is usually denoted by a painted color band. Glass-packaged diodes often have two or three color bands to indicate specific number types. This group of diodes is generally capable of passing forward current values of approximately 100 mA. The peak reverse voltage

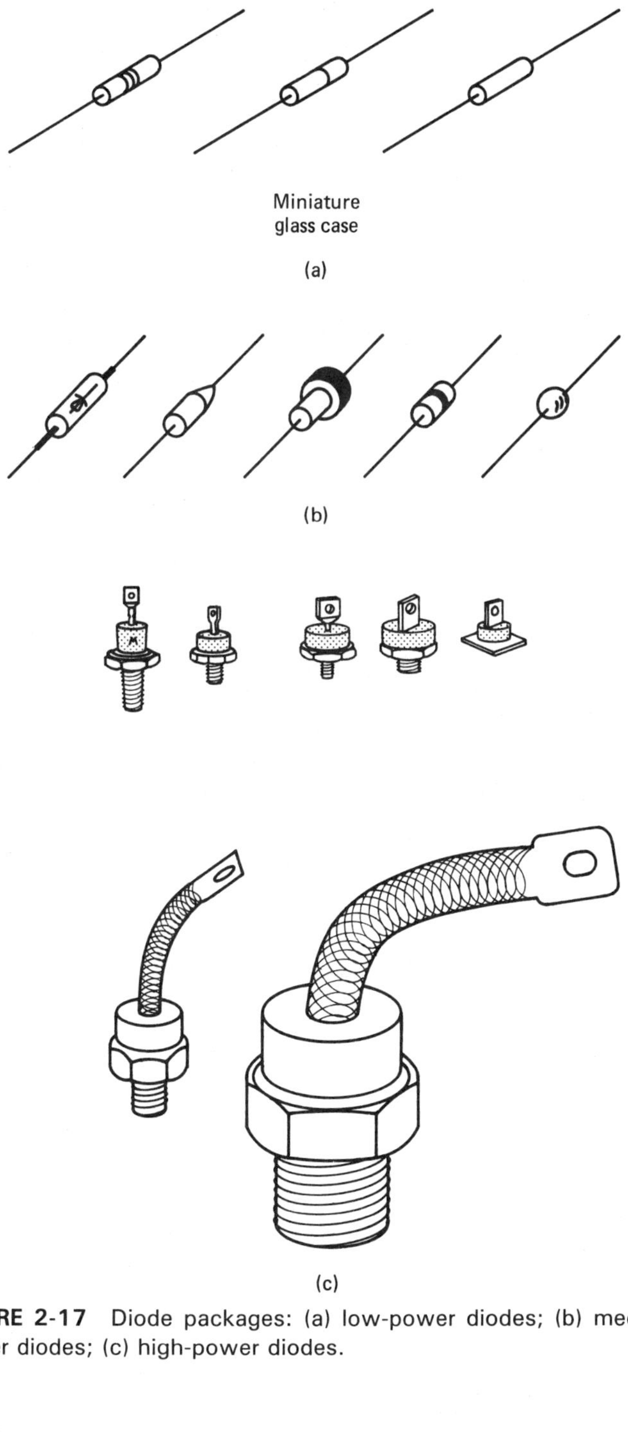

(a)

(b)

(c)

FIGURE 2-17 Diode packages: (a) low-power diodes; (b) medium-power diodes; (c) high-power diodes.

rating rarely exceeds 100 V. Low reverse current values for these devices are typically 5 μA at 25°C.

The medium-current diodes of Fig. 2-17(b) are slightly larger in size than the low-current devices. Body size is approximately 0.5 cm with larger connecting leads. Diodes in this group can pass forward current values up to 5 A. Peak reverse voltage ratings generally do not exceed 1000 V. The anode and cathode terminals may be identified by a diode symbol on the body of the device. A color band near one end of the case is also used to identify the cathode. Low- and medium-current diodes are usually mounted by soldering. Any heat generated during operation is carried away by air or lead conduction.

High-current, or power, diodes are the largest of all diode types shown in Fig. 2-17(c). These devices normally generate a great deal of heat. Air convection of heat is generally not adequate for most installations. These devices are designed to be mounted on metal heat sinks, which conduct heat away from the diode. Diodes of this classification can pass hundreds of amperes of forward current. Peak reverse voltage ratings are in the 1000-V range. Numerous packaging types and styles are used to house power diodes. As a rule, the rating of the device and method of installation usually dictate its package type.

SELF-EXAMINATION

31. The absolute maximum reverse bias voltage that can be applied to a diode is called the ______ ______ ______.
32. I_{FM} refers to the absolute maximum ______ ______ that can pass through a diode.
33. The *I-V* characteristics of a silicon diode change with ______.
34. The crystal structure of a reverse-biased diode responds as a ______.
35. The depletion zone of a reverse-biased diode serves as the ______ material of a capacitor.
36. A ______ diode responds as a voltage-controlled capacitor.
37. ______ ______ of a diode refers to its ability to change from one state to another.
38. Diodes are generally packaged according to their ______ rating.

ANSWERS

31. Peak inverse voltage or peak reverse voltage
32. Forward current

33. Temperature	34. Capacitor
35. Dielectric	36. Varicap
37. Switching time	38. Power

DIODE TROUBLESHOOTING

Diode operation is determined by its connection in a circuit. Current conduction is directly dependent on the polarity of the device. Heavy conduction occurs when it is forward biased, and very little current flows when it is reverse biased. Connecting a diode backward in a circuit reverses its conduction. This can destroy the diode and possibly damage several other circuit parts. A person working with diodes must be absolutely certain that they are connected properly in a circuit.

The schematic symbol of a diode is commonly used to identify its polarity. Figure 2-18 shows a diode symbol, element names, and the polarity of the crystal material. The term *anode* is commonly used to describe the electrode that attracts or gathers in electrons. The P-material of the crystal serves as the anode in a diode. The term *cathode* is used to denote the electrode that gives off or emits electrons. The N-material of the crystal serves as the cathode. Note that forward current passes through a diode from the cathode to the anode. A diode symbol indicates this movement from the bar to the point of the arrow.

Because diodes are available in many different package types and styles, there is often some confusion about lead identification. A person working with these devices is frequently called on to test diodes and identify leads. Volt-ohm-milliammeters (VOMs) are used to perform this operation. The ohmmeter section of the instrument is used to perform this test.

Figure 2-19 shows an analog or hand-deflection meter being used to test a diode. In preparation for this test the meter should be placed in the R × 100 or an equivalent 1-kΩ range. The two meter leads are then connected to the diode. Exact resistance readings are not particularly important. Only high- or low-resistance indications are used. Figure 2-19(a) shows how an ohmmeter is connected to forward bias a diode. This test normally causes the meter to show a low resistance. Reverse biasing of a diode is shown

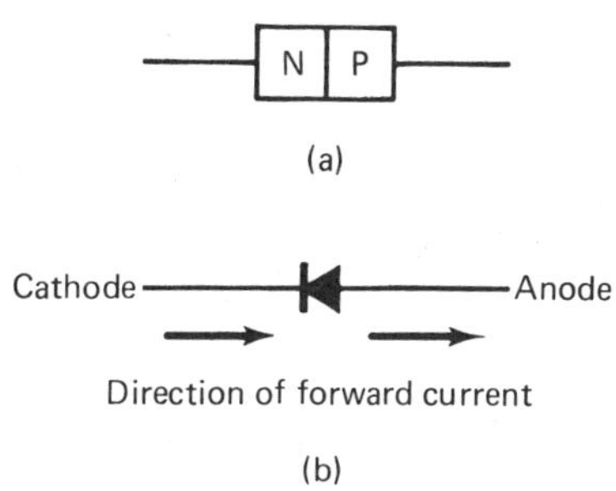

FIGURE 2-18 Diode: (a) crystal structure; (b) symbol and element names.

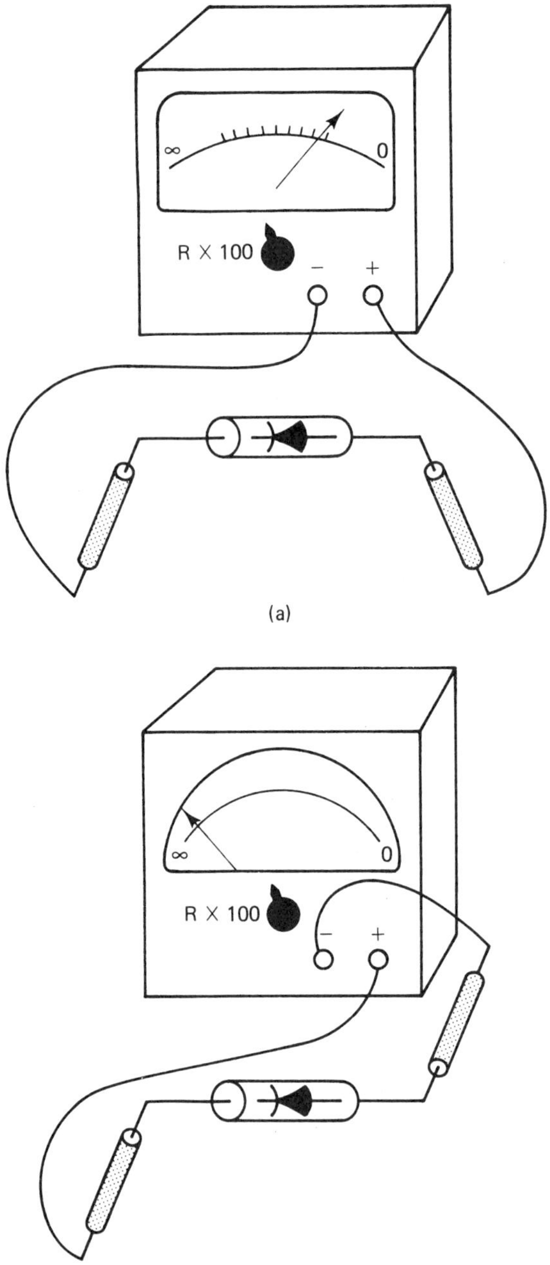

FIGURE 2-19 Diode testing with an ohmmeter: (a) forward-bias test connection; (b) reverse-bias test connection.

in Fig. 2-19(b). An infinite resistance is indicated by the ohmmeter during this test.

A good diode normally shows low resistance when forward biased and high resistance when reverse biased. If a diode does not respond in this manner, it is probably defective. A diode that shows low resistance in both directions is considered to be shorted. Shorting usually occurs when the maximum current or voltage ratings of the device are exceeded. Shorted diodes often cause damage to other circuit components. Blown fuses, tripped circuit breakers, or overheated resistors are generally good indicators of a shorted diode. It is a good practice to test other circuit components before replacing a shorted diode.

When a diode shows infinite resistance in both directions, it is considered to be open. Open diodes generally cause a circuit to be nonconductive. This condition rarely causes damage to other circuit parts. Operation simply stops, and there is no current flow through the diode. Open diodes do not occur very often in electronic circuits. A quick ohmmeter check will readily detect this type of fault should it occur.

An ohmmeter can also be used to identify the leads of a diode. The polarity of the ohmmeter voltage source must be known for this test to work. Straight-polarity ohmmeters have the black or common lead negative and the red lead positive. A diode connected to the ohmmeter will show resistance when forward biased and infinite resistance when reverse biased. Forward biasing occurs when the polarity of the ohmmeter matches the crystal polarity of the diode. The red lead is therefore connected to the anode and the black lead to the cathode. Lead identification is simply a matter of finding the forward bias direction of a diode and noting the polarity of the ohmmeter leads.

Ohmmeters with reverse polarity have the black lead positive and the red lead negative. Leads are identified with this instrument by connecting the ohmmeter to the diode and determining the direction of forward biasing. Forward biasing will again cause the ohmmeter to read low resistance. The lead polarity of the diode, however, is a reverse of the ohmmeter lead polarity. The positive lead must therefore be connected to the cathode and the negative lead to the anode. Lead identification is simply a matter of finding the forward-bias direction of a diode and then noting the reverse polarity of the ohmmeter.

It is essential that the polarity of an ohmmeter be known before it can effectively be used to identify diode leads. An unfamiliar meter can easily be checked with a separate dc voltmeter. Simply connect the voltmeter to the ohmmeter probes and observe the voltage indication. An up-scale voltage indication shows that the ohmmeter voltage matches the voltmeter polarity. A straight-polarity ohmmeter will respond in this manner. Reverse-polarity ohmmeters will cause an off-scale voltage indication to occur when the probe polarity is matched. As a rule, this polarity test should be made on all ranges of the ohmmeter. In some instruments there may be a reverse polarity only on higher resistance ranges.

The value of an ohmmeter's voltage supply is also an important consideration when testing diodes. If the supply voltage is less than 0.2 V, it will not be large enough to forward bias all diodes. This could cause a good diode to appear to be open. The ohmmeter would therefore show an infinite resistance in both directions. You may recall that it takes at least 0.3 V to cause conduction in germanium diodes and 0.7 V in silicon. Many electronic VOMs and all digital VOMs respond in this way. The voltage supply of this type of instrument must be evaluated to see if it can be used to test a diode.

The supply voltage of some ohmmeters may also be extremely large. This type of meter could actually develop enough current or voltage to damage certain diodes. As a rule, high-frequency detection diodes are very susceptible to being damaged in this way. To avoid this kind of problem, do not attempt to test high-frequency diodes with high-voltage ohmmeters.

SELF-EXAMINATION

39. When a diode is tested with an ohmmeter, ____________ conduction occurs when it is forward biased.
40. A good diode normally shows ____________ resistance when forward biased and ____________ resistance when reverse biased.
41. A diode that shows low resistance in either direction of biasing is ____________.
42. A diode that shows infinite resistance in either direction of biasing is considered to be ____________.
43. The ____________ of the voltage source of an ohmmeter must be known to properly identify the leads of a diode.

ANSWERS

39. Heavy
40. Low, infinite
41. Shorted
42. Open
43. Polarity

DIODE APPLICATIONS

Junction diodes are used rather extensively in electronic circuits. In general, diodes are used to control the conduction of electric current. When a diode is forward biased it becomes conductive and responds as a closed circuit. Reverse biasing causes nonconduction and produces the effect of an open circuit. Diodes are therefore designed to block or pass current according to their biasing. This permits the diode to be used as a switch. Diodes can also be used to alter the direction of current passing through other electronic parts. In this application the diode is used to change ac to dc, a process called *rectification.* Rectification is one of the most common applications of the diode.

Rectification

The term *rectification* is used to describe something that is being changed or altered. Electronically, rectification refers to the process of changing alternating current to direct current. Figure 2-20 shows the circuitry of a diode used as a rectifier. In this application only one alternation of the ac input appears in the output. This is called *half-wave rectification.* The resulting output is called *pulsating dc.* A half-wave rectifier can remove either the positive or negative alternation, depending on the polarity of the diode. When the output has a positive alternation, the resulting dc is positive with respect to the common point or ground. Figure 2-20(a) shows this condition of operation. A negative output occurs when the positive alternation is removed. Figure 2-20(b) shows an example of a rectifier with a negative output.

A simplification of the rectification process is shown in Fig. 2-21. Part (a) of this diagram shows how the diode will respond during the positive alternation. This alternation causes the diode to be forward biased. Note that the anode is positive and the cathode is negative. The resulting current flow is shown by arrows. The output of the rectifier appears across the resistor (R_L). The voltage across R_L is positive at the top and negative at the bottom.

Figure 2-21(b) shows how the diode will respond during the negative alternation. This alternation causes the diode to be re-

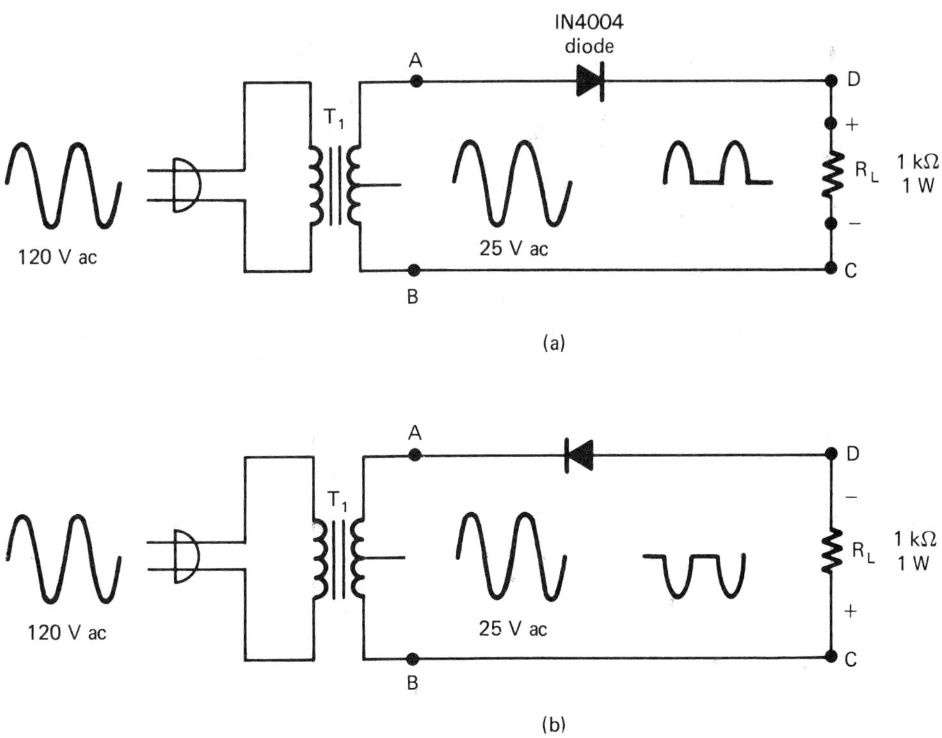

FIGURE 2-20 Half-wave power supplies: (a) positive output; (b) negative output.

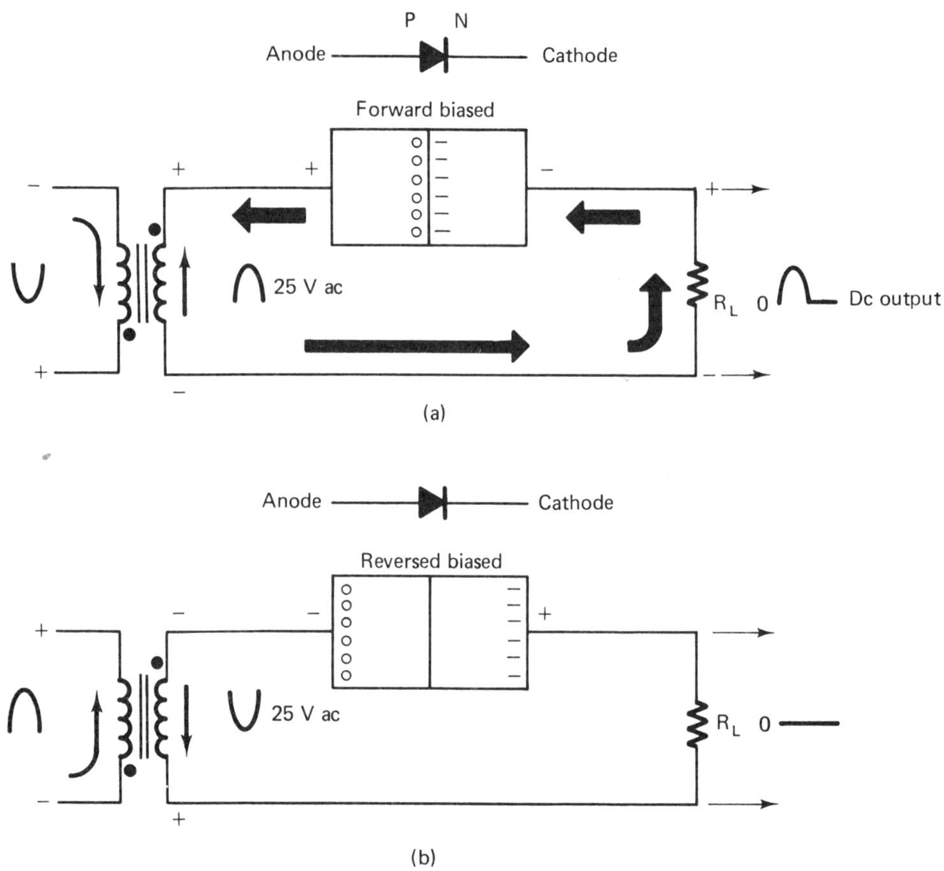

FIGURE 2-21 Simplification of rectification: (a) positive alternation; (b) negative alternation.

verse biased. The anode is now negative and the cathode is positive. No current will flow during this alternation. The output across resistor R_L is zero for this alternation.

The dc output of a half-wave rectifier appears as a series of pulses across R_L. A dc voltmeter or ammeter would therefore indicate the average value of these pulses over a period of time. Figure 2-22(a) shows an example of the pulsating dc output. The average value of one pulse is 0.637 of the peak value. Since the second alternation of the output is zero, the composite average value must take into account the time of both alternations. The composite output is therefore 0.637/2, or 0.318 of the peak value. Figure 2-22(b) shows a graphic example of this value.

The equivalent dc output of a half-wave rectifier without filtering can be calculated when the value of the ac input is known. The peak value of one alternation is determined by first multiplying the root-mean-square, or rms, value by 1.414. The composite dc output is then determined by multiplying the peak value by 0.318.

A simplified method of calculating the dc output of a half-wave rectifier is to combine different values. This means that 1.414 × 0.637/2 equals 0.45. The composite dc output of a half-wave rectifier is therefore 45%, or 0.45, of the rms input voltage. In an

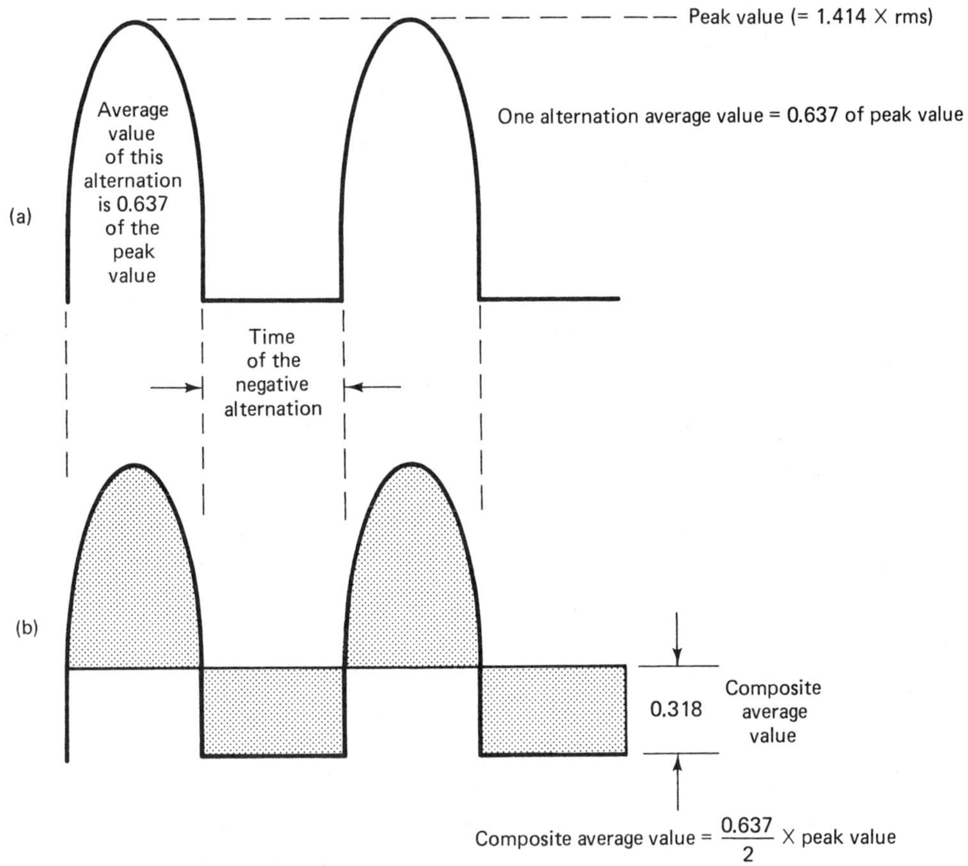

FIGURE 2-22 Half-wave rectifier output: (a) pulsating dc output; (b) composite output.

actual circuit we must also take into account the voltage drop across the diode. When a silicon diode is used, this value is 0.6 V. The dc output of an unfiltered half-wave rectifier is, therefore, 0.45 × rms − 0.6 V. Using this procedure, let us determine the dc voltage of the rectifier of Fig. 2-21.

Note that the applied ac voltage to the rectifier of Fig. 2-21 is 25 V. This is considered to be the rms, or effective, value of the applied ac. To determine the resulting dc output, we must multiply the rms value by 0.45 and subtract 0.6 V:

$$\begin{aligned}\text{dc output voltage} &= 0.45 \times \text{rms} - 0.6 \text{ V}\\ &= 0.45 \times 25 - 0.6 \text{ V}\\ &= 10.65 \text{ V}\end{aligned}$$

The output pulse frequency of a half-wave rectifier will occur at the same rate as the applied ac input. With 60 Hz input, the pulse rate or ripple frequency of a half-wave rectifier is 60 Hz. Only one pulse appears in the output for each complete sine-wave input. Potentially, this means that only 50% of the ac input is transformed into a useable dc output. Half-wave rectification is therefore only 50% efficient. Due to this low efficiency rating half-wave rectification is not used very extensively in electronic circuits today. Other, more efficient forms of rectification tend to be used in place of half-wave circuits. Full-wave rectifiers that employ two or four diodes are more efficient. This type of rectification converts both alternations of the ac input into dc output.

SELF-EXAMINATION

44. The process of changing ac to dc is called ________.
45. The output of a half-wave rectifier is ________ ________.
46. Conduction of a half-wave rectifier occurs only when a diode is ________ biased.
47. The dc output of an unfiltered half-wave rectifier appears as a series of ________.
48. An ammeter or voltmeter connected across the load resistor of a half-wave rectifier indicates the ________ value of the dc pulses.
49. The average value of one dc pulse is ________ of the peak value.
50. If an unfiltered half-wave rectifier is energized by 12.6 V rms, the dc output that appears across the load resistor is ________ V.

ANSWERS

44. Rectification
45. Direct current, or dc
46. Forward
47. Pulses
48. Average
49. 0.637
50. 5.07

SUMMARY

The term junction diode is used to describe the crystal structure of a two-element electronic device made of N- and P-materials. The P-N junction formed when these materials are joined responds as a continuous crystal. Current carriers diffuse across the junction when it is formed. The diffusion process causes a barrier potential to be established across the junction. For germanium the barrier potential is approximately 0.3 V, and for silicon it is 0.7 V.

When an external source of energy is applied to the P-N material of a diode, it either adds to or reduces the barrier potential of the junction. Forward biasing reduces the barrier potential of a junction and causes the diode to be conductive. Forward biasing is achieved by connecting the positive side of the source to the P-material and the negative side of the source to the N-material. Reverse biasing adds to the barrier potential of a diode and causes it to be nonconductive. Reverse biasing occurs when the negative side of the source is connected to the P-material and the positive side of the source to the N-material.

The characteristics of a diode shows how it will respond when connected in a circuit that is forward biased or reverse biased. In the forward-biased direction, conduction occurs at a few tenths of a volt. This causes a very rapid increase in forward current. In the reverse-biased direction there is very little current with an extremely large value of reverse voltage. Reverse current is generally temperature dependent.

When selecting a diode for an application, one must consider its specifications, which include maximum reverse voltage, reverse breakdown voltage, forward current, surge current, maximum reverse current, power dissipation, and reverse recovery time. Diode operation is related to temperature. Most of its characteristics change with temperature. Junction capacitance is a characteristic that changes with bias voltage. Switching time is a variable that refers to the time it takes a diode to change from one state to another.

Diodes can be tested and the leads can be identified with an ohmmeter. When the positive lead of an ohmmeter is connected to the P-material and the negative lead to the N-material, a good diode will show a low resistance reading. Reversing the ohmmeter leads will show high resistance. A shorted diode will indicate low resistance in both directions. An open diode will show infinite resistance in both directions. The anode and cathode are identified by matching the polarity of the ohmmeter leads with the P- and N-materials. The anode is P, and the cathode is N.

A common diode application is rectification of alternating current to direct current by alternately changing the bias voltage between the forward and reverse direction. The output of an unfiltered half-wave rectifier is pulsating dc. A voltmeter or ammeter can measure the average value of these pulses. A half-wave rectifier is 50% efficient because it produces output for only one alternation of the ac input.

EXPERIMENT 2-1
DIODE TESTING WITH AN OHMMETER

The internal circuitry of an ohmmeter has a voltage source connected to its leads when it measures resistance. This voltage can be used to bias a diode without connecting it to a circuit. When the positive side of the ohmmeter voltage source is connected to the P-material and the negative side to the N-material, the diode is forward biased. This condition causes the diode to have low resistance. The ohmmeter will indicate some resistance on an appropriate scale. Reversing the two ohmmeter leads will cause the diode to be reverse biased. No conduction occurs when a diode is reverse biased. As a result, the ohmmeter will not register resistance of any magnitude. This procedure serves as a good-bad test for a diode.

Resistance testing varies between different diodes. A good rectifier diode should have a forward to reverse resistance ratio of at least 10:1. Most other diodes must have a 100:1 ratio to be acceptable. An open diode will show infinite resistance in both directions. A shorted diode will indicate low resistance in both directions.

Lead identification of a diode is achieved by knowing the polarity of the voltage source of the ohmmeter leads. Most ohmmeters have straight polarity. This means that the red probe is positive and the black or common probe is negative. If this is known about the ohmmeter, the anode and cathode of a diode can be identified. Simply connect the ohmmeter leads to the diode and find a low resistance indication. The positive lead must be connected to the anode and the negative lead to the cathode. The reverse is true for an ohmmeter with reverse polarity. The polarity of the ohmmeter can be tested with a voltmeter.

OBJECTIVES

1. To demonstrate the testing of a diode with an ohmmeter.
2. To evaluate the status of a diode with an ohmmeter.
3. To identify the leads of a diode with an ohmmeter.

EQUIPMENT

Volt-ohm-milliammeter (VOM), standard or digital type

Silicon diode, Texas Instruments 1N4001 or Heath 56-28 or equivalent.

Germanium diode, General Electric 1N4149, Heath 56-89, or equivalent.

Assortment of diodes (optional)

PROCEDURE

1. Prepare the VOM to measure resistance on its R × 1, or lowest resistance range. For digital VOMs use the 2-kΩ range or its nearest equivalent.
2. Refer to Fig. 2-1A(a). Connect the ohmmeter leads across the diode as indicated. The ideal forward resistance would be 0 Ω.
3. Record the measured forward resistance (R_F). ________ R_F.
4. Refer to Fig. 2-1A(b). Connect the ohmmeter leads across the diode as indicated. The ideal reverse resistance would be infinite or ∞.
5. Record the measured reverse resistance (R_R). ________ R_R.
6. Calculate the forward and reverse resistance ratio from the formula R_R/R_F.
7. The calculated resistance ratio is 1 : ________.
8. Repeat steps 1 through 7 for the germanium diode.
9. If additional diodes are available, repeat steps 1 through 7 for one or more diodes.
10. If some unmarked diodes are available, identify the anode and cathode by using the same test procedure. Keep in mind that the polarity of the ohmmeter indicates the correct assignment of the anode and cathode.
11. Make a drawing of the identified leads of each diode.

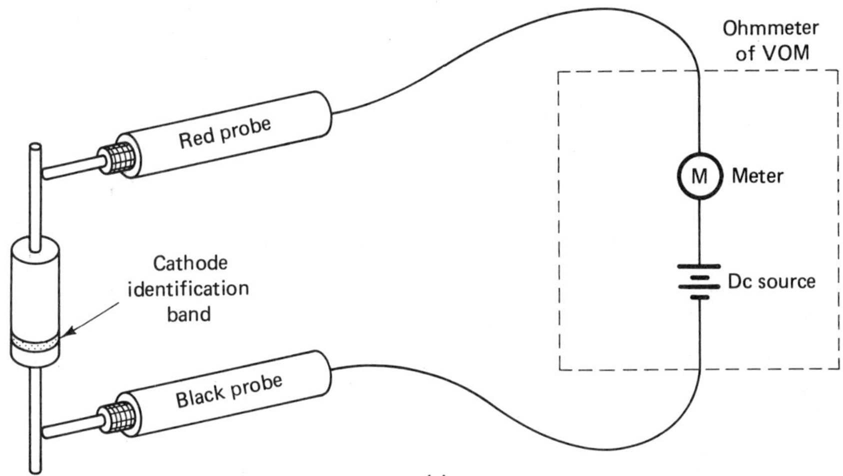

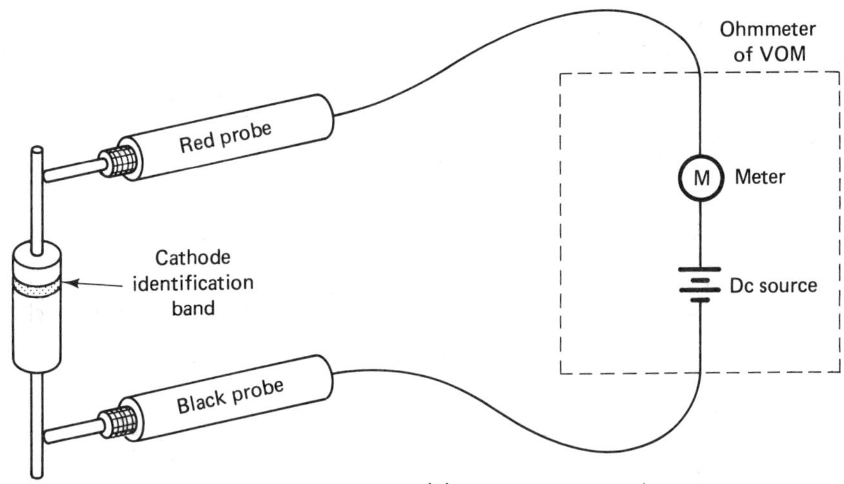

FIGURE 2-1A Ohmmeter test connections for a diode.

DISCUSSION

This experiment shows that a good diode has low forward resistance and high reverse resistance. The forward-to-reverse-resistance ratio is a good indicator of the status of a diode. The actual resistance values are not significant. If the resistance ratio is 1:10 for a germanium diode or 1:100 for a silicon diode, it is considered good. Silicon diodes are primarily used for rectification, while germanium diodes are used for signal detection.

A rather unusual observation may have been encountered when applying the ohmmeter test procedure to diodes. If the range of the ohmmeter is changed during the test procedure it usually indicates a different resistance value for each range. Changing the range of an ohmmeter alters the internal resistance of the meter circuit. This generally causes the probe voltage and current to be reduced. The resulting resistance indication of a diode will therefore change when the ohmmeter range is switched.

ANALYSIS

1. Explain why an ohmmeter can be used to determine the status of a diode.
2. Explain how the anode and cathode of an unmarked diode can be identified with an ohmmeter.

EXPERIMENT 2-2
CURRENT-VOLTAGE CHARACTERISTICS OF A DIODE

The *I-V* characteristics of a diode tell a great deal about its performance in an operating circuit. The data for an *I-V* curve can be determined by voltage and current measurements obtained from an operating circuit. These data can then be plotted on a graph and used to see how a diode will respond when it is placed in operation.

A number of important diode characteristics can be obtained from the data of an *I-V* curve, namely, the point where forward voltage causes an increase in forward current, the effect that reverse voltage has on reverse current, the influence that temperature has on forward and reverse conduction, and the conduction differences between germanium and silicon. The forward and reverse resistance of a diode can also be determined from *I-V* data. This information is very useful in predicting diode performance in a circuit.

OBJECTIVES

1. To determine the forward voltage, current, and resistance of a silicon and a germanium diode.
2. To determine the reverse voltage, current, and resistance of a silicon and a germanium diode.
3. To plot an *I-V* curve for a silicon and a germanium diode.

EQUIPMENT

Variable dc source
Volt-ohm-milliammeter (VOM), standard or digital type
Silicon diode, Texas Instruments 1N4001 or Heath 56-28
Germanium diode, General Electric 1N60, Heath 56-28
1-kΩ, $\frac{1}{4}$-W resistor (brown-black-red-gold)
Circuit construction board

PROCEDURE

1. Construct the circuit shown in Fig. 2-2A. The VOM is used to measure different circuit voltages. It must be moved to different locations to observe these voltages.

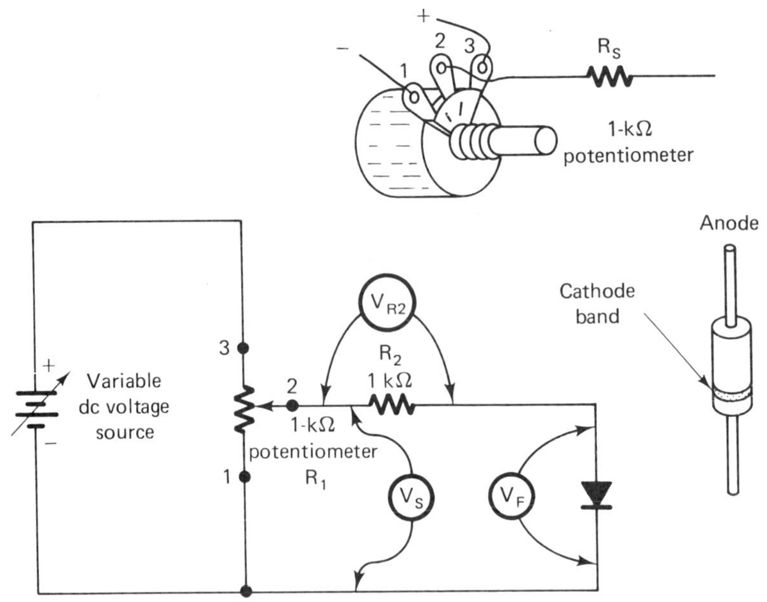

FIGURE 2-2A Diode *I-V* test circuit.

2. Turn on the dc voltage source (V_{ps}) and adjust it to produce 10 V, as measured across the 1-kΩ potentiometer. This voltage will appear across the two outside terminals (1 and 3) of the potentiometer. The source voltage (V_s) for the circuit is a variable value that is adjusted by the potentiometer. V_s will appear at terminals 1 and 2 of the potentiometer. In this activity, you are directed to adjust potentiometer R_1 to produce a specific voltage across the diode (V_F). Then measure the voltage across the series resistor (R_2) and circuit source voltage across terminals 1 and 2 of the potentiometer. Record these values in Data Table 1 of Fig. 2-2B.
3. The first part of this experiment is designed to measure the forward *I-V* characteristics of a diode. Note that only voltage values are called for in the procedure. Forward current (I_F) is determined by calculation. Ohm's law is used to determine current by the formula $I_F = V_{R2}/R_2$. Forward resistance R_F is determined from the expression $R_F = V_F/I_F$.
4. Adjust V_s to produce a V_F of 0 V across the diode. Measure and record the values of V_s and V_{R2}.
5. Adjust the value of V_s to produce a V_F of 0.1 V across the diode. Measure the values of V_s and V_{R2}. Record the measured values of V_s and V_{R2} in Data Table 1.
6. Repeat step 3 for each of the *V* values of Data Table 1. *Note:* The V_F voltage of a silicon diode is generally considered to be 0.7 V. In an operating circuit this may vary from 0.5 to 0.7 V according to the amount of current flowing through the diode. As a rule, a larger V_F will appear across a diode only when the value of I_F has been

1N4001				
Data table 1				
Measured			Calculated	
V_F	V_{R2}	V_S	I_F	R_F
0				
0.1				
0.2				
0.3				
0.4				
0.5				
0.6				
0.7				

1N4001				
Data table 2				
Measured			Calculated	
V_R	V_{R2}	V_S	I_F	R_F
0				
2				
5				
7				
10				
12				
15				
20				

1N60				
Data table 3				
Measured			Calculated	
V_F	V_{R2}	V_S	I_F	R_F
0				
0.1				
0.2				
0.3				
0.4				

1N60				
Data table 4				
Measured			Calculated	
V_R	V_{R2}	V_S	I_F	R_F
0				
2				
5				
7				
10				
12				
15				
20				

FIGURE 2-2B Data tables.

increased significantly. In this circuit the value of I_F is kept at a rather low value. With some diodes the upper values of V_F may not be obtained. If this occurs, use only the values that can be achieved by the circuit. Do not try to reach the upper values by increasing the power source voltage.

7. Calculate the I_F and R_F values for each of the measured V_F values.
8. Turn off the power source and remove the 1N4001 diode from the circuit. Reverse the two leads and return it to the circuit so that it is connected in the reverse-bias direction. The voltmeter will be used to measure reverse voltage values across the diode (V_R), resistor voltage (V_{R2}), and the voltage source (V_s). The reverse current I_R

and reverse resistance R_R are calculated from these values.

9. Turn on the variable dc power source and adjust it to 20 V measured across terminals 1 and 3 of the potentiometer. Repeat steps 2 through 5 for the reverse-connected diode. The values of V_R are indicated in Data Table 2 of Fig. 2-2B. Note that these values are much larger than the previous V_F values.
10. Turn off the power source.
11. Using the V_F and I_F data from Table 1, plot a forward conduction curve on the graph of Fig. 2-2C. The forward curve is located in the upper right quadrant of the graph.
12. Using the V_R and I_R data from Table 2, plot a reverse conduction curve on the graph of Fig. 2-2C. The reverse curve is located in the lower left quadrant of the graph.
13. Repeat steps 1 through 12 for the 1N60 germanium diode. In this procedure the V_F of the 1N60 diode should not be adjusted to values in excess of 0.4 V. When plotting the forward and reverse curves on the graph of Fig. 2-2C, use dots or dashed lines to distinguish 1N60 data from the 1N4001.

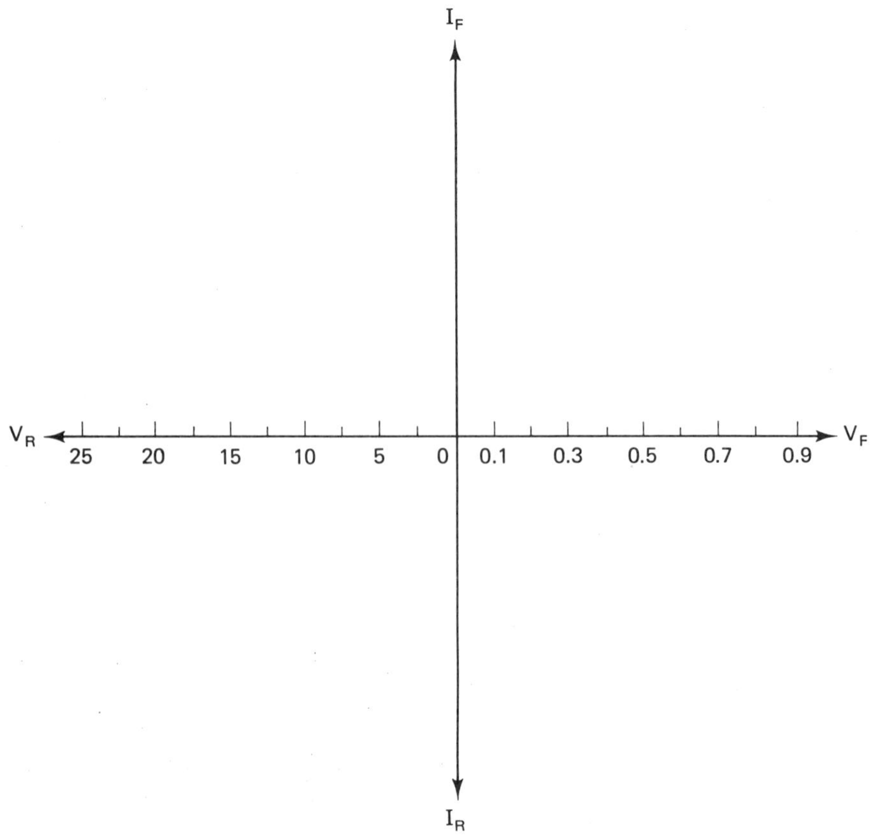

FIGURE 2-2C *I-V* graph.

DISCUSSION

In this experiment you were asked to measure the forward and reverse voltage values of a diode to develop data for plotting an *I-V* curve. The data for a silicon diode and a germanium diode were both obtained and used to develop curves.

An investigation of the curves developed in this activity shows a great deal about the characteristic differences of silicon and germanium diodes. It should be apparent that germanium goes into conduction in the forward direction at a lower voltage than silicon. These values are approximately 0.3 to 0.4 V for germanium and 0.6 to 0.7 V for silicon. The forward knee voltage of each diode shows where a very pronounced increase in forward current occurs. This voltage is generally considered to be of a constant value after the knee has been reached.

The reverse characteristics of the diodes used in this activity are very important. This characteristic shows the maximum reverse voltage that can be applied to a diode without its being damaged. You were unable to reach this voltage value with the two diodes. Reverse current flow is very small in most diodes. In the silicon diode practically no I_R could be observed. The germanium diode has a more significant I_R. Reverse current is also described as leakage current. Leakage current is very sensitive to temperature.

ANALYSIS

1. Explain what the forward characteristics of a silicon diode tell about its operation.
2. Explain what the reverse characteristics of a silicon diode tell about its operation.

UNIT 2 EXAMINATION
SEMICONDUCTOR DIODES

Instructions: Circle the response that most correctly completes each statement.

1. The construction of a junction diode consists of:
   a. A piece of N-material
   b. A piece of P-material
   c. A combination of N- and P-materials
   d. Electrically conductive materials
2. When electrons leave the N-material to fill holes in the P-material the process is called:
   a. Depletion
   b. Diffusion
   c. Mixing
   d. Doping
3. Diffision current in a diode is caused by:
   a. Chemical energy
   b. Heat energy
   c. Voltage
   d. Crystal formation
4. The depletion zone or region of a diode is:
   a. Due to reverse biasing
   b. Due to forward biasing
   c. An area created by crystal doping
   d. An area void of current carriers
5. When the P-N materials of a junction diode are formed, diffusion current causes:
   a. A barrier potential to be formed across the junction
   b. Covalent bonding
   c. The mixing of current carriers
   d. Forward biasing
   e. Reverse biasing
6. The forward voltage (V_F) of a fully conductive silicon diode is:
   a. Approximately 0.3 V
   b. Approximately 0.7 V
   c. Equal to the source voltage (V_s)
   d. Equal to the voltage drop across the series resistor

7. When a diode is externally forward biased:
   a. The width of its depletion zone increases.
   b. The width of its depletion zone decreases.
   c. Its barrier potential increases.
   d. Its depletion zone or region increases.
8. When a diode is externally reverse biased:
   a. The width of the depletion zone increases.
   b. The width of the depletion zone decreases.
   c. The barrier potential of the junction is reduced.
   d. The depletion zone or region of the junction is reduced.
9. The leakage current of a diode is caused by:
   a. Heat energy
   b. Chemical energy
   c. Internal barrier voltage
   d. The crystal doping material
10. The arrowhead or pointed end of a diode symbol represents the:
    a. Anode
    b. Cathode
    c. N-material
    d. Direction of electron flow
11. A shorted diode tested with an ohmmeter will show:
    a. Low resistance in either direction
    b. High resistance in only one direction
    c. Low resistance in only one direction
    d. High resistance in both directions
12. An open diode tested with an ohmmeter will show:
    a. Low resistance in both directions
    b. High resistance in only the forward direction
    c. High resistance in both directions
    d. Low resistance in one direction and high resistance in the other
13. The diode of Fig. 2-2A is:
    a. Forward biased
    b. Reverse biased
    c. Shorted
    d. Open
14. The V_F meter of Fig. 2-2A will show a voltage value in the range of:
    a. 0.0 to 0.4 V
    b. 0.41 to 0.7 V
    c. 0.71 to 1.0 V
    d. 1.01 to 10 V

15. The I_F meter of Fig. 2-2A will show the current to be in the range of:
    a. 0.0 to 5.0 mA
    b. 5.1 to 10 mA
    c. 10.1 to 100 mA
    d. 101 mA to 10 A
16. The V_{R2} meter of Fig. 2-2A will show a voltage value in the range of:
    a. 0.0 to 0.4 V
    b. 0.41 to 0.7 V
    c. 0.71 to 1.0 V
    d. 1.1 to 10 V
17. If the diode of Fig. 2-2A is reversed in the circuit, the V_F meter will show a voltage in the range of:
    a. 0.0 to 0.4 V
    b. 0.41 to 0.7 V
    c. 0.71 to 1.0 V
    d. 1.1 to 10 V
18. If the diode of Fig. 2-2A is reversed in the circuit, the V_{R2} meter will show a voltage in the range of:
    a. 0.0 to 0.4 V
    b. 0.41 to 0.7 V
    c. 0.71 to 1.0 V
    d. 1.1 to 10 V
19. If the diode of Fig. 2-2A is reversed in the circuit, the I_F will show current in the range of:
    a. 0.0 to 5.0 mA
    b. 5.1 to 10. mA
    c. 10.1 to 100 mA
    d. 101 mA to 10 A

UNIT THREE

Zener Diodes

UNIT INTRODUCTION

In the previous unit we were concerned with the diode as a forward conducting device. As a rule, reverse biasing of a diode will not produce conduction unless the peak reverse voltage is exceeded. When this occurs, a junction diode is usually destroyed. Operation of a conventional diode is generally related to forward conduction applications.

In this unit we examine a group of diodes referred to as avalanche, breakdown, or zener diodes. This type of diode operates in the reverse direction. The reverse breakdown voltage characteristic of the device is used to a unique advantage: Voltage can be regulated with these devices. Reverse breakdown of a semiconductor is an important solid-state operational characteristic.

UNIT OBJECTIVES

Upon completion of this unit, you will be able to:

1. Analyze the forward and reverse characteristics of a zener diode.
2. Explain the meaning of reverse breakdown in the operation of a zener diode.

3. Determine the safe operating characteristics of a zener diode.
4. Determine operating characteristics such as power dissipation, zener impedance, zener current, and zener voltage.
5. Explain how voltage regulation is achieved.
6. Develop an experimental voltage regulator circuit and test its operation.
7. Design a voltage regulator circuit with a zener diode.

IMPORTANT TERMS

Before reading Unit 3, review the following terms. These terms provide a basic understanding of terms associated with zener diodes.

Avalanche breakdown P-N junction breakdown that results when thermally generated holes and electrons gain enough energy from a reverse-biased source to produce new current carriers.

Dissipation The giving off, expending, using, or dispersing of something such as heat or power.

Impedance The total opposition to current flow in an ac circuit that is a combination of resistance, inductive reactance, and capacitive reactance.

Power dissipation The ability of a device to change energy from one form to another, give off heat, or use power.

Regulation Holding constant some conditions, like voltage, current, or power.

Zener breakdown Current carrier production due to the influence of a strong electric field that causes large numbers of electrons in the depletion region to form covalent bonds across a reverse-biased P-N junction. This occurs in silicon P-N junctions at values less than 5 V.

CRYSTAL STRUCTURE AND SYMBOL

The symbol and crystal structure of a zener diode are shown in Fig. 3-1. Notice that the crystal structure is very similar to that of a conventional junction diode. The zener diode symbol is, however, uniquely different. The cathode is drawn with a bent line to distinguish it from the regular diode symbol. The bent line of the cathode represents the letter Z. It only appears as a Z, however, when the symbol is oriented horizontally in a diagram. The bent line of the cathode is very useful in distinguishing the zener diode from a conventional diode. If a diode symbol is printed on the housing of a device, it will be a standard diode symbol. The zener diode symbol is rarely printed on the case of a device for identification.

Zener diodes are normally connected in a circuit in the reverse bias direction. This means that the anode must be connected to the negative side of the voltage source and the cathode to the positive side. This action causes the depletion zone of the junction to

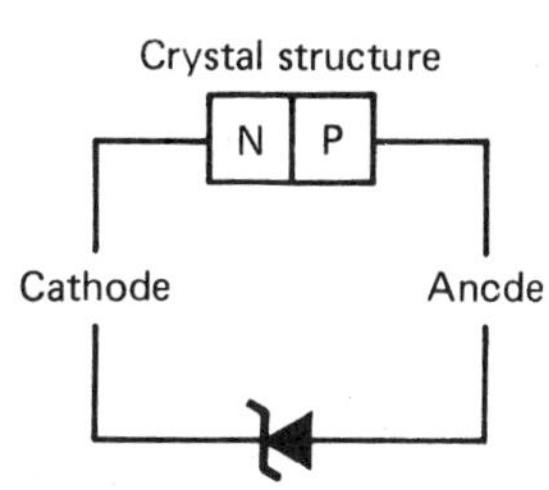

FIGURE 3-1 Zener diode symbol and crystal structure.

become wider. Under normal circumstances, we would expect a reverse-biased junction to be nonconductive. In a zener diode, two conditions are responsible for causing this device to become conductive in the reverse-biased direction, namely, *avalanche breakdown* and *zener breakdown.*

Avalanche breakdown occurs when thermally generated holes and electrons gain enough energy from the reverse-biased source to produce new carriers. These current carriers are the result of removing valence electrons from their normal bonding. These new carriers, in turn, produce additional carriers that disrupt the bonds of other atoms. The process continues with the number of carriers building up in increasing numbers. The end result is a large current flow through the reverse-biased junction. Avalanche breakdown is primarily responsible for reverse current conduction above 8 V. Diodes of this type were once called avalanche diodes.

Zener breakdown is the result of a strong electric field that appears across the P-N junction. This field causes covalent bonds near the junction to break apart. As a result, a large quantity of new holes and electrons are produced. These newly generated holes and electrons represent a substantial increase in current. Zener breakdown is primarily responsible for conduction of a reverse-biased junction below 6 V. Diodes that conduct in this range of voltage are called *zener diodes.* Today the term zener diode is primarily used to describe all diodes that operate in the reverse bias direction.

ZENER CHARACTERISTICS

The characteristics of a junction diode show that it is designed for operation in the forward direction. Forward biasing produces a large I_F for a rather small value of V_F. Reverse biasing generally does not cause current conduction until higher values of reverse voltage are reached. If V_R is great enough, however, breakdown occurs and causes a reverse current flow. Junction diodes are usually damaged when this occurs. Avalanche diodes, breakdown diodes, and zener diodes are all designed to operate in the reverse direction without being damaged. The term zener diode is commonly used to describe all these devices.

Figure 3-2 shows the volt-ampere characteristic of a typical zener diode. In the forward-bias direction, this device behaves like an ordinary silicon diode. In the reverse-bias direction, there is practically no reverse current (labeled I_z) until the breakdown voltage (labeled V_z) is reached. When this occurs a very large change in zener current is accompanied by only a small change in zener voltage. Varying amounts of I_z can then pass through a zener diode without causing damage. The reversed zener voltage across the diode remains relatively constant even when the source voltage is increased. The zener current has a safe operating limit or maxi-

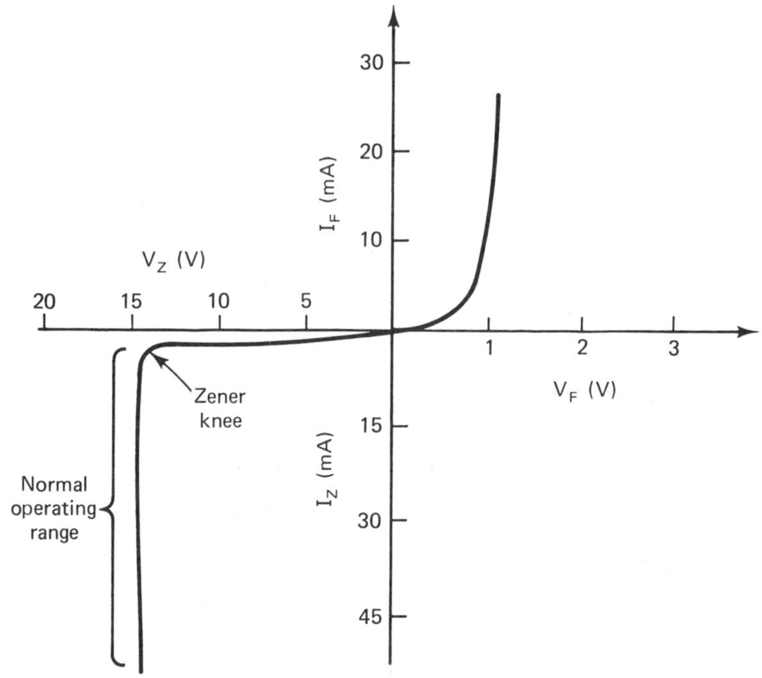

FIGURE 3-2 *I-V* characteristics of a zener diode.

mum value that must be observed. It is determined by the wattage rating of the diode.

Zener Voltage

Zener diodes are available in a wide range of breakdown voltages. The zener breakdown voltage, or simply zener voltage, is usually identified by the letters V_z. Typical V_z values range from 1.4 to 200 V. Some popular low-voltage values are 3.0, 3.6, 4.7, 5.1, 6.2, 6.8, 7.5, 9.1, 10, 12, 15, 18, and 20 V_z. A number of other V_z values are also available. These values vary to some extent among different manufacturers.

The accuracy of the zener voltage of a diode is a very important selection consideration and is largely responsible for the cost of a zener diode. The accuracy or tolerance rating of a zener diode is a range of values within which the actual V_z will appear. Some of the more popular tolerance ratings are 20%, 10%, 5%, and 1%. Low-cost zeners have the highest tolerance percentage, while 1% tolerance devices are the most expensive. A zener diode rated at 9.1 V with 10% tolerance will have a V_z value that is ± 0.91 V of 9.1 V. Its actual value could be somewhere between 8.19 and 10.01 V. As a rule, diodes with a 10% tolerance rating will work for most circuit applications. Zeners with a 1% tolerance are used only in special circuit applications that require critical voltage values.

Zener voltages are generally identified at a minimum zener current value. Figure 3-2 shows that a certain amount of I_z is needed to enter the knee of the I_z curve. In this case approximately

1 mA of I_z is needed to produce conduction. After this value has been reached, the V_z remains fairly constant over a wide range of I_z. The normal operating range of I_z in this region is determined by the power dissipation rating.

If the V_z region of Fig. 3-2 is examined very carefully, it shows that breakdown does not occur at a precise location. The curve is actually rounded in this area. This is called the V_z knee of the curve. Ideally, the knee should have a very sharp edge. When it is rounded, V_z occurs more gradually. Zeners with a low tolerance rating generally have a sharp V_z knee.

SELF-EXAMINATION

1. A zener diode is normally connected in the ________-biased direction.
2. When a zener diode is reverse biased it will go into ________ at a prescribed voltage value.
3. ________ refers to the zener voltage accuracy of a zener diode.
4. The actual zener voltage of a 6.8-V_z diode with 10% tolerance is somewhere between ________ and ________ V.
5. The characteristics of a zener diode show that after a minimum value of I_z is reached the ________ ________ remains at a constant value.
6. ________ breakdown is the result of thermally generated holes and electrons gaining enough energy from the source to move valence electrons out of their bonding.
7. ________ breakdown is the result of covalent bonds breaking up due to a strong electric field across a reverse-biased P-N junction.
8. The ________ of a zener diode symbol is drawn as a bent line or bar.

ANSWERS

1. Reverse	2. Conduction
3. Tolerance	4. 7.48, 6.12 V
5. Zener voltage or V_z	6. Avalanche
7. Zener	8. Cathode

ZENER CURRENT RATINGS

Manufacturers rate zener diodes according to their zener voltage value and the maximum power dissipation (P_D) at 25°C. P_D refers to the ability of the zener's junction to give off heat. In the process of dissipating heat a certain amount of power is used or consumed by the zener. This characteristic gives an indication of the maximum zener current (I_z) that a diode can safely conduct. A 1.0-W, 15-V_z zener diode, for example, can conduct an I_z of 0.066 A or 66 mA. This is determined by the basic power equation $P = IV$, by solving for I. We do, however, substitute P_D for P, I_z for I, and V_z for V in the formula. The modified formula then becomes

$$\text{maximum } I_z = \frac{P_D}{V_z}$$

$$= \frac{1 \text{ W}}{15\text{V}_z}$$

$$= 0.066 \text{ A, or } 66 \text{ mA}$$

The I-V characteristics of a IN1775, 1-W zener are shown in the graph of Fig. 3-3. This zener has a maximum I_z of 66 mA. The minimum zener current is also indicated by the curve. This value is approximately 1 mA. The region between these two extreme conditions is labeled the safe operating range. The zener can operate anywhere between these two extreme points without being damaged. If the current exceeds the maximum I_z value, additional heat will be developed. This heat may become so intense that it can cause permanent damage to the junction. As a rule, the area beyond the maximum I_z point should be avoided in normal diode applications.

Power Dissipation

A principal cause of solid-state device destruction is heat, because when power is applied to a device, the device heats up. If too much power is applied, the device will overheat and be destroyed. The *power dissipation rating* of a zener diode is an indication of how much heat the device can give off or dissipate. It is rated in watts at a certain temperature. Typically this rating is expressed as 1 W at 25°C. The temperature value is an indication of device operation at ambient temperature. *Ambient* refers to something that goes around or surrounds an object. In this case ambient refers to the air surrounding the operating device. A temperature of 25°C (77°F) is considered to be the temperature of an inside room. Some manufacturers indicate the power dissipation rating at temperatures such as 50°C (122°F) and 75°C (169°F). These are generally considered to be working temperatures. The actual power that can be dissipated by a zener diode decreases as the temperature increases and increases when the temperature decreases in value. In other words, the P_D of a zener diode has a negative temperature coefficient.

The P_D rating of a zener diode can be altered to some extent

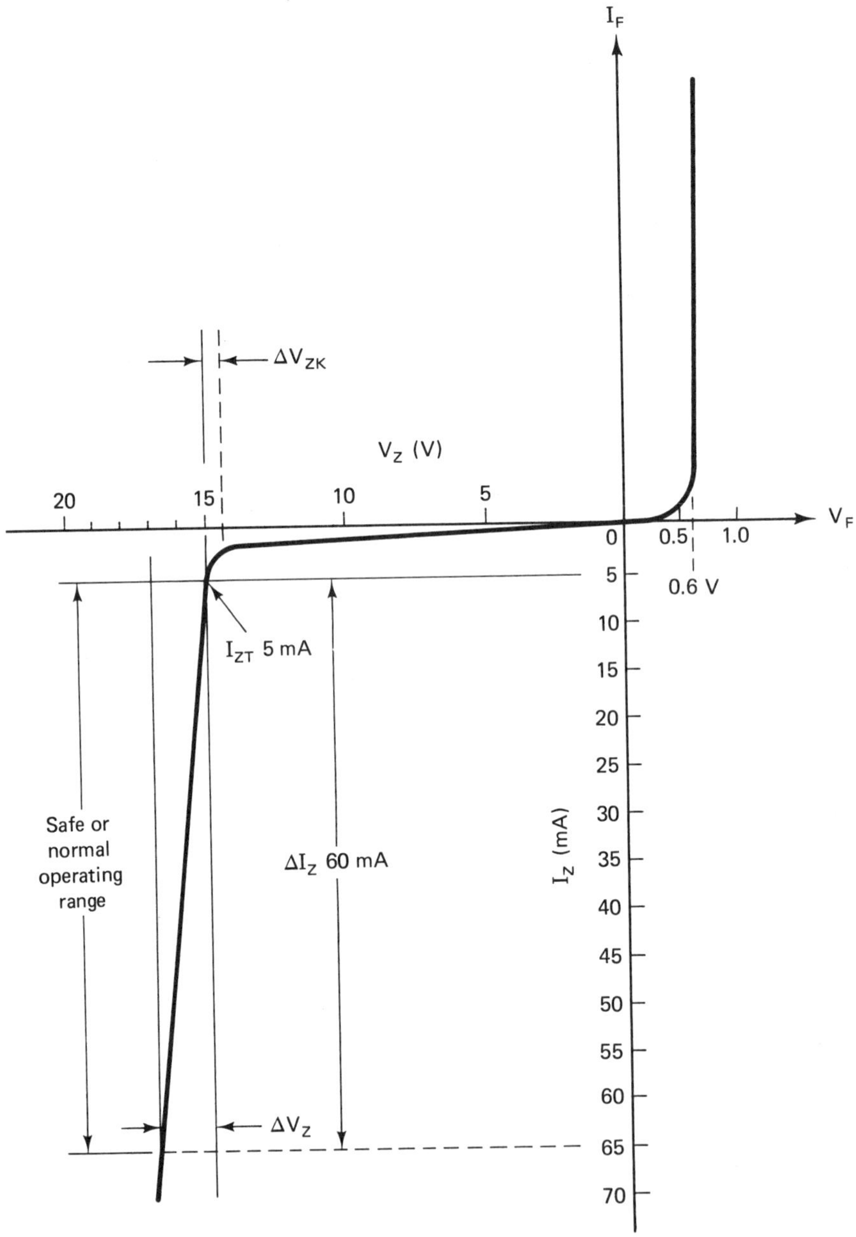

FIGURE 3-3 I_z-V_z characteristic of a 1N1775 zener diode.

by its lead length and its mounting in a circuit. The P_D value can be derated when these factors are considered. *Derating* permits a device to handle more power than its normal P_D value indicates. This is achieved by reducing lead length or mounting the device on a piece of metal, which serves as a heat sink. Manufacturers usually provide data about device derating in a chart that correlates lead length and the area of the heat sink. Many zener diodes are housed in packages that permit them to be attached to a heat sink. In some installations the metal chassis that houses a circuit may be used as the heat sink. Derating is an important consideration when a zener diode operates close to its maximum power rating.

Zener Impedance

Another matter of importance in zener diode selection is zener impedance (Z_z). This characteristic deals with the slope of the I_z-V_z curve after conduction has been achieved. A zener diode actually shows a slight increase in the value of its V_z when the I_z increases. This increase is very noticeable because it means that the reverse characteristic is not entirely vertical. Zener impedance is responsible for this characteristic and is determined by the expression

$$Z_z = \frac{\Delta V_z}{\Delta I_z}$$

where the Greek letter delta (Δ) denotes a change in value.

Figure 3-3 shows an I_z-V_z curve that indicates the location of the items being discussed. The ΔI_z value is derived from the safe or normal operating range of the zener. Some manufacturers select the range of values from a prescribed test area near the middle of the curve. The range of V_z must be equated to corresponding values of I_z. Note the location of the ΔV_z and ΔI_z values on the chart. The Z_z for this range of operation is

$$\begin{aligned} Z_z &= \frac{\Delta V_z}{\Delta I_z} \\ &= \frac{17 - 15 \text{ V}}{65 - 5 \text{ mA}} \\ &= \frac{2 \text{ V}}{60 \text{ mA}} \\ &= \frac{2 \text{ V}}{0.06 \text{ A}} \\ &= 33.33\ \Omega \end{aligned}$$

The reverse knee impedance (Z_{zk}) of a zener diode can be determined by the same procedure. This shows the impedance of the device near the breakdown point. Zener knee impedance is a good indication of the slope or sharpness of the knee area of the curve. Determine the Z_{zk} of the zener in Fig. 3-3. Use the indicated values of ΔV_{zk} and ΔI_{zk} in your calculation.

In practice, a zener diode with a sharp knee will produce a lower value of zener impedance than a rounded knee. Ideally, the zener knee should be very sharp. Manufacturers usually include knee impedance and zener impedance in their specifications of a device.

SELF-EXAMINATION

9. Zener diodes are rated according to their ____________ ____________ and ____________ ____________.
10. A zener diode rated at 1 W, 12 V_z has a maximum current or I_z capability of ____________ A or ____________ mA.
11. The space between the minimum and maximum I_z values of a zener diode is called the ____________ operating range.
12. When the maximum I_z value of a zener diode is exceeded, damage will likely occur due to excessive ____________.
13. The principal cause of zener diode destruction is ____________.
14. Power dissipation of a zener diode refers to the rated ____________ at a particular ____________.
15. When the junction temperature of a zener diode goes up, the P_D rating goes ____________.
16. Zener ____________ is a characteristic based on the value of V_z/I_z.
17. A low value of zener impedance is generally a good indication of the sharpness of the ____________ ____________.

ANSWERS

9. Power dissipation, zener voltage
10. 0.083 A, or 83 mA
11. Safe or normal
12. Heat
13. Heat
14. Watts, temperature
15. Down
16. Impedance or Z
17. Zener knee

ZENER DIODE VOLTAGE REGULATORS

The most widely used application of zener diodes is in voltage regulation. The constant V_z characteristic of a reverse-biased zener is utilized in this operation. A large change in zener current will cause only a small change in V_z, so a zener can be used as an alternate path for the output current of a dc power supply. The constant V_z developed across the zener can then be connected to a load device. The load voltage thus remains at a constant value by altering the I_z flowing through the zener diode.

The dc output of an unregulated power supply has a tendency to change value under normal operating conditions. Changes in ac input voltage and variations in the resistance of the load are pri-

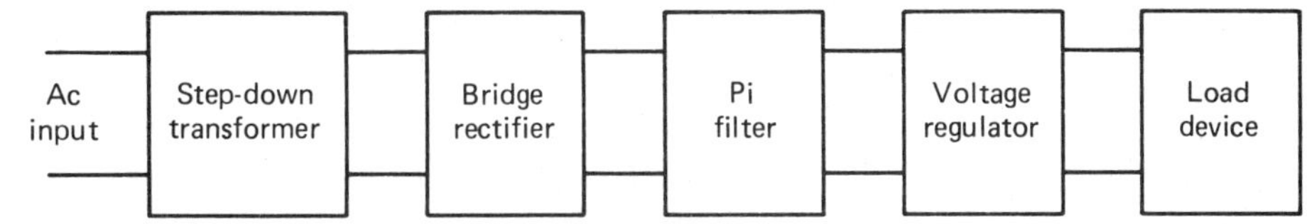

FIGURE 3-4 Block diagram of a power supply.

marily responsible for these fluctuations. In some power-supply applications, voltage changes do not represent a serious problem, but in many electronic circuits voltage changes may cause improper operation. When a stable dc voltage is required, power supplies must employ a voltage regulator. Figure 3-4 shows the location of a regulator in a power supply. If the regulator block of the power supply is removed, the load device will be connected to an unregulated dc output. With the regulator block in the circuit, the dc output voltage will be of a constant value. The V_z of the zener diode selected for use in the regulator determines the voltage value of the dc output.

The Zener Diode Voltage Regulator

A number of voltage regulator circuits have been developed for power supplies. Zener diodes are used exclusively in power supplies that have a fixed or single-value voltage output. Figure 3-5 shows a basic zener diode voltage regulator. This regulator is located between the filter circuit and the load device. The dc input voltage for operation of the regulator comes from the filter circuit. The value of this voltage has a tendency to change according to variations in the ac input voltage. The dc output of the regulator is applied to the load device of the power supply. The load device is represented by a resistor. The zener diode is connected in shunt or parallel with the load resistor. The V_z of the zener diode keeps the voltage of the load device at a constant value.

The voltage regulator of Fig. 3-5 consists of a zener diode and a series resistor. The series resistor (R_s) connects the unregulated dc input to the load resistor. The zener diode is connected in parallel with the load. It is reverse biased by the dc source voltage. When V_z is reached, the zener becomes conductive. The zener diode then serves as a bypass route through which varying amounts of

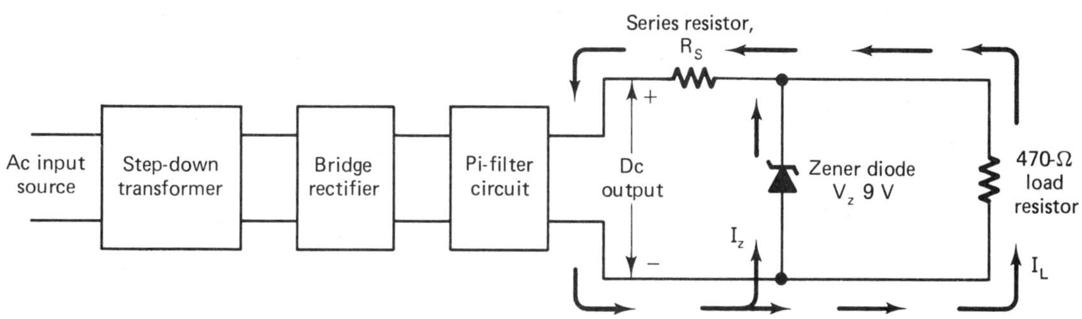

FIGURE 3-5 Zener diode regulator.

current will flow. The amount of I_z conduction depends on the value of the dc input voltage. The value of V_z across the diode, however, remains constant over a wide range of dc input. Regulation is achieved by altering the conduction of I_z while maintaining V_z at a constant value. The combined I_z and load current I_L must both pass through the series resistor. This current value determines the voltage drop across R_s. Variations of current through R_s are used to keep the output voltage at a constant value.

Voltage Regulator Operation

The operation of a voltage regulator is best explained by using an actual circuit. In this case we refer to the regulator circuit of Fig. 3-6. The dc input of this circuit is 16.6 V. The zener diode has a rating of 9.1 V at 1 W. This means that the output of the regulator is 9.1 V. The series resistor (R_s) is therefore responsible for lowering the dc input voltage to the V_z value of the zener diode. The voltage drop of R_s is symbolized by V_{RS} and can be determined by the formula

$$\begin{aligned} V_{RS} &= V_{in} - V_z \\ &= 16.6 \text{ V} - 9.1 \text{ V} \\ &= 7.5 \text{ V} \end{aligned}$$

The value of V_{RS} is also an indication of the total current (I_T) passing through the series resistor. I_T can be determined by Ohm's law. In this case the expression is $I_T = V_{RS}/R_s$. For the regulator this would be

$$\begin{aligned} I_T &= \frac{V_{RS}}{R_s} \\ &= \frac{7.5 \text{ V}}{100 \ \Omega} \\ &= 0.075 \text{ A, or } 75 \text{ mA} \end{aligned}$$

This current must all pass through the zener diode as I_z when the regulator is operating without a load connected to the output; 9.1 V will then appear at the two output terminals of the power supply. This represents the no-load (NL) condition of operation.

When an external load is connected to a power supply, more current is required. Ideally, this current should be available with 9.1 V of output. Assume now that the power supply is connected to a 270-Ω external load resistor. The total resistance or composite load of the power supply is now 270 Ω. With 9.1 V applied to the load, there is a specific value of load current (I_L). I_L can be determined from the formula $I_L = V_z/R_L$. For the regulator circuit this is

$$\begin{aligned} I_L &= V_Z/R_L \\ &= 9.1 \text{ V}/270\Omega \\ &= 0.0337 \text{ A, or } 33.7 \text{ mA} \end{aligned}$$

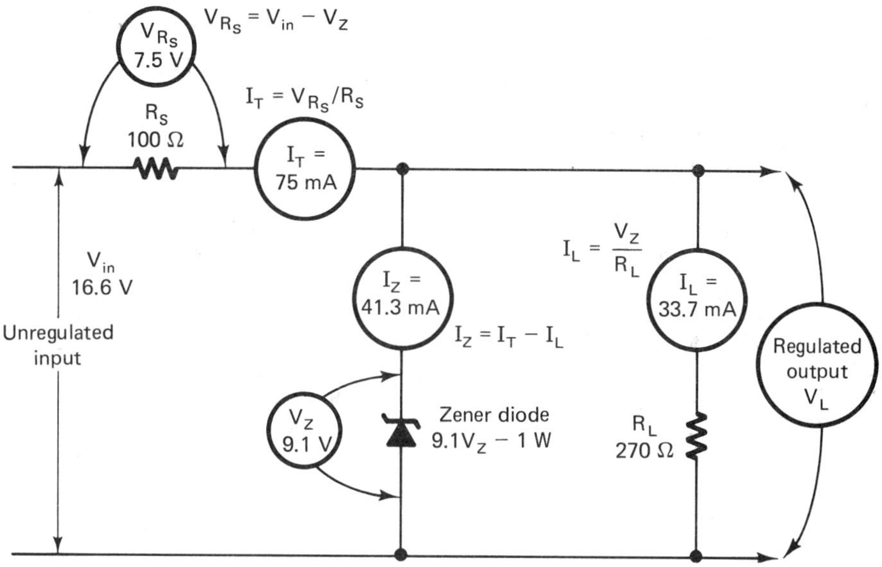

FIGURE 3-6 A voltage regulator circuit.

The total output current available for the power supply is that which passes through the series resistor. This was determined to be 0.075 A, or 75 mA. With a 270-Ω load connected to the power supply, I_L is 9.1 V/270Ω = 0.0337 A, or 33.7 mA. The current passing through the zener diode is determined by subtracting the load current from the total current. This is 75 mA of I_T – 33.7 mA of I_L or 41.3 mA of I_z. The output voltage of the power supply remains at 9.1 V with the increase in load. This represents the full load (FL) condition of the power supply.

With a slightly smaller external load resistor, the power supply will be out of its regulation range. Should this occur, no I_z will flow, and the value of I_L alone will determine the current passing through R_s. An I_T value in excess of 75 mA will cause a greater voltage drop across R_s. This, in effect, will cause the output voltage to be less than 9.1 V. All voltage regulator circuits of this type have a maximum load limitation. Exceeding this limit will cause the output voltage to go out of regulation.

If the resistance of the external load is increased, it causes a reduction in load current. This condition causes a reverse in the operation of the regulator. With a larger value of R_L there is less I_L. As a result, the zener diode must conduct more heavily. The maximum current-handling rating of the zener diode determines this condition. An infinite load, for example, demands no I_L. The zener diode therefore conducts the full current passing through R_s. This value was calculated to be 75 mA for our circuit. For a 1-W, 9.1-V zener diode the maximum current rating is 1 W/9.1 V = 0.10989, or 109.89 mA. In this case, the diode is capable of handling the maximum possible I_L that will occur for an infinite load. In effect, the regulating range of our power supply is from an infinite value to approximately 120 Ω. The output voltage remains at 9 V over this entire range of load values.

A regulator must also be responsive to changes in input voltage. If the input voltage, for example, were to increase by 20%, it would cause the regulator circuit to see approximately 20 V instead of 16.6 V. The voltage drop across R_s would then be 20.0 − 9.1 V, or 10.9 V. This, in turn, would cause an increase in total current through R_s to a value of 10.9 V/100 Ω, or 109 mA. The current flow of the zener diode would therefore increase to 109 mA with no external load resistor. Since the zener is capable of handling up to 109.89 mA, it could respond to this change and maintain the output voltage at 9 V. An input of 20 V, however, approximates the maximum input that is possible for the zener diode being used.

The response of a regulated power supply to a decrease in input voltage must also be taken into account. A 20% decrease in input voltage would cause the supply to be 20% × 16.6, or 3.32 V less. The resulting input would then be 16.6 V − 3.32 V, or 13.28 V. The voltage drop across R_s would change to 13.28 − 9.1 V, or 4.18 V. With this reduced voltage, the total current would drop in value to 4.18 V/100Ω = 0.0418 A, or 41.8 mA. This value could certainly be handled by the zener diode. The maximum low resistance of the load would, however, increase to some extent. Under this condition the load might drop to only 160Ω before going out of regulation. A 20% decrease in input voltage, however, represents a very significant change in the control capabilities of a regulator.

Zener Diode Regulator Design

When working with power supplies for electronic circuits, one is often called upon to design a voltage regulator for a specific application. This calls for a general understanding of the circuit and the component ratings. The components selected for the regulator consist of a zener diode and a series resistor. The zener must have a V_z value and a power dissipation that will accommodate the circuit ratings. The series resistor must be selected to accommodate the range of voltages being used. It must also have a wattage rating that will accommodate the total current. Component selection uses some of the same procedures outlined in the voltage regulator part of this unit.

Regulator design requires some knowledge of the circuit being controlled and the values of voltage and current being accommodated. Let us assume that the unregulated output of a power supply ranges from 20 to 15 V. The desired regulated voltage is 12 V, and the load current will range from 0 to 40 mA. What value of series resistor and zener diode are needed to achieve regulation in a circuit attached to the power supply?

Twelve volts is a standard voltage value for a zener diode. This particular device can be selected from a tolerance range of 1 to 20%. The power dissipation rating of the zener will be made after the value of R_s has been determined by calculation.

A series resistor must also be selected that will permit the zener diode to operate within its prescribed range. The value of R_s is equal to the minimum value of the input voltage (V_{in}) minus the zener voltage (V_z) divided by the maximum value of the load current (I_L). The voltage difference between V_{in} and V_z represents the voltage that will appear across the series resistor. This voltage value divided by the maximum load current equals the value of R_s. For our circuit we therefore find the value of R_s to be

$$R_s = \frac{V_{in}\,(\text{min}) - V_z}{I_L(\text{max})}$$

$$= \frac{15\text{ V} - 12\text{ V}}{0.04\text{ A}}$$

$$= 75\Omega$$

The wattage rating to the resistor must also be determined from the basic power formula, $P = IV$. For this particular application the wattage rating of the series resistor is determined from the values of the maximum load current (I_L) and the series resistor voltage (V_{RS}), or

$$P = I_L\,(\text{max}) \times V_{RS}$$

$$= 0.04 \times 3\text{ V}$$

$$= 0.012\text{ W}$$

Since resistors do not have a standard wattage value of 0.012 W, a value of 0.25, or $\frac{1}{4}$ W, will be selected for use in this circuit. This resistor can withstand significantly more heat than the calculated value.

The power dissipation rating, or P_D, of the zener diode is a calculated value. In this case power dissipation is determined by the basic $P_D = I_z V_z$ formula. The maximum I_z value of the formula is first determined by finding the I_T passing through R_s then subtracting the minimum load current (I_L) from I_T. The power dissipation rating of the zener diode is thus

$$P_D = V_z \left[\frac{V_{in}\,(\text{max}) - V_z}{R_s - I_L\,(\text{min})}\right]$$

$$= 12\text{ V} \times \frac{(20\text{ V} - 12\text{ V}}{75\Omega - 0)}$$

$$= 12\text{ V} \times \left(\frac{8\text{ V}}{75\Omega}\right)$$

$$= 12\text{ V} \times 0.1067\text{A}$$

$$= 1.28\text{ W}$$

Zener diodes are not available with a P_D rating of 1.28 W. A 5-W zener would be selected for this application. This particular zener could withstand a great deal more current than the calculated value.

Our calculations of the zener diode and series resistor of the regulator are based on the assumption that the value of V_z remains fairly constant under normal circuit applications. This is done to simplify the calculations. As a rule, this is a good practical assumption. If the actual value of V_z does change in the operation of a regulator, the zener impedance (Z_z) of the diode could be used to determine changes in I_z values. Z_z is generally given in the manufacturer's specifications of a diode.

SELF-EXAMINATION

18. The most widely used application of a zener diode is in ____________ ____________.
19. The output ____________ of regulator varies in value because of a change in the input voltage and changes in the value of the load resistor.
20. Zener diodes are used exclusively in regulated power supplies that have a ____________ value of output voltage.
21. A voltage regulator circuit consists of a ____________ ____________ and a ____________ resistor.
22. The zener diode of a voltage regulator is connected in shunt or parallel with the ____________ ____________.
23. The ____________ ____________ of a voltage regulator is used to connect the unregulated dc input of the power supply to the load resistor.
24. The ____________ ____________ passing through the series resistor of a voltage regulator is the sum of I_z and I_L.
25. The ____________ across the series resistor (R_s) of a voltage regulator is determined by $V_{in} - V_z$.
26. The total ____________ passing through the series resistor of a voltage regulator is determined by the expression V_{RS}/R_s.
27. The ____________ ____________ of a voltage regulator is determined by V_z/R_L.
28. Design of a zener diode voltage regulator calls for a 10-V regulated output with a fixed load current of 30 mA. The unregulated input varies from 15 to 20 V dc. A suitable value of series resistor is ____________ Ω, and that of the zener diode should be ____________ V_z at ____________ W.

ANSWERS

18. Voltage regulation
19. Voltage
20. Single
21 Zener diode, series

22. Load resistor
23. Series resistor
24. Total current
25. Voltage
26. Current
27. Load current
28. $R_s = V_{in}\text{ (min)} - V_z/I_L\text{ (max)}$
$= (15\text{ V} - 10\text{ V})/(0.03\text{ A})$
$= (5\text{ V})/(0.03\text{ A})$
$= 166.67\ \Omega$; a standard value of 180 Ω would be used.
$P = V_z\,[(V_{in}\text{ (max)} - V_z/R_s - I_L\text{ (min)}]$
$= (10\text{ V})[(20\text{ V} - 10\text{ V})/(180\Omega) - 0\text{ mA}]$
$= (10\text{ V})(10\text{ V}/180)$
$= (10\text{ V})(0.055\text{ A})$
$= 0.59$ W; a 1-W zener with a 10-V_z rating would be used.

SUMMARY

Zener diodes are normally connected in a circuit in the reverse-bias direction. This means that the anode must be connected to the negative side of the voltage source and the cathode to the positive side. This action causes the depletion zone of the diode to become wider. Under normal circumstances, we would expect this junction to be nonconductive. In zener diode use the junction becomes conductive at a specific V_z value. The V_z of the diode remains fairly constant over a large range of I_z. This characteristic of the zener is used in applications that call for a stable voltage value. Voltage regulation is an important application of the zener diode.

A zener diode has a number of characteristics that must be taken into account when selecting a device for use in an electronic circuit. The V_z characteristic is used to show the accuracy that a zener has in its operation. V_z has a tolerance rating that indicates the value of its accuracy. Low-tolerance V_z diodes are more expensive than high-tolerance devices. Power dissipation is another characteristic that must be taken into account when selecting a zener for a specific application. P_D refers to the ability of a zener to dissipate heat. This rating is also used to determine the I_z operating range of a zener. $I_z = P_D/V_z$. The zener impedance is also important. Z_z deals with the slope of the I_z-V_z curve. A low-tolerance zener has a sharp knee and very little angle in its V_z.

A very common application of the zener diode is in voltage regulation. A simple voltage regulator is made up of a series resistor and a zener diode. The regulator is placed between the filter output of a power supply and the load device. The zener diode is placed in parallel with the load device. It is reverse biased by the dc input. When the V_z is reached, the zener becomes conductive and provides an alternate path through which current will flow. The combined I_z and I_L current flows through the series resistor. This causes the input voltage to drop in the value to the V_z. The voltage drop across the series resistor (R_s) can be determined by the expression $V_{RS} = V_{in} - V_z$. The value of the series resistor can be determined by the formula $R_s = V_{in}$ (min) $- V_z/I_L$ (max). The P_D rating of the zener is determined by the formula $P_D = V_z[V_{in}$ (max) $- V_z/R_s - I_L$ (min)].

EXPERIMENT 3-1
ZENER DIODE TESTING

An ohmmeter can be used to identify the leads of a zener diode and to detect whether it is shorted or open. The ohmmeter, however, does not determine how a zener diode will respond when voltage is applied. A simple test procedure can be performed to evaluate this condition of operation using a variable dc source, a series resistor, and a zener diode. Its zener voltage can be determined by altering the variable source voltage while observing the voltage across the zener. The voltage will remain constant across the zener until the V_z value is reached. This will show the tolerance, the V_z, and the knee voltage of the device. This is a significant characteristic when the zener is reverse biased.

OBJECTIVES

1. To demonstrate a practical method of testing a zener diode with a voltage source.
2. To evaluate the V_z of a zener diode.
3. To connect a zener diode in a circuit for normal operation.

EQUIPMENT

Variable dc source
Volt-ohm-milliammeter (VOM), standard or digital type
Zener diode, 5.1 V ± 10%, 500 mW, Motorola 1N5231, 1N751, or Heath 56-16
100-Ω, $\frac{1}{4}$-W, 5% resistor (brown-black-brown-gold)
1-kΩ, 1-W linear potentiometer
Circuit construction board

PROCEDURE

1. Prepare the VOM meter to measure resistance on its R × 1 or lowest resistance range. For a digital VOM, use the 2 k or an equivalent range.
2. Connect the two leads of the ohmmeter across the diode. A good zener will show low resistance when it is forward biased and infinite resistance when reverse biased. Identify the anode and cathode of the diode with the ohmmeter.

3. Make a sketch of the identified leads of the device.

4. Construct the circuit shown in Fig. 3-1A. The 1-kΩ potentiometer is adjusted to produce different values of input voltage. Adjust the dc power source to produce 12 V. Connect this voltage across the potentiometer as indicated.
5. Adjust the potentiometer to produce 0 V of input voltage for the circuit. This voltage is labeled dc input in the chart of Fig. 3-1A(b).
6. Then measure the voltage that appears across the zener diode. Record this value in the chart.
7. Adjust the potentiometer to produce 1 V of input voltage. Measure the V_z and record its value in the chart.
8. Repeat this procedure for each of the indicated input voltage values.
9. The voltage across the series resistor (V_{RS}) of the circuit is the difference between the dc input voltage and V_z. For a few of the input voltage values around 5 V, calculate the V_{RS}. These voltage values can also be measured with the VOM for the appropriate input voltage value.
10. The zener current of the circuit can be calculated by the expression $I_z = V_{RS}/R_s$. Calculate a few of the I_z values of the diode around the 5 V_z part of the chart.
11. Reverse the zener diode in the circuit of Fig. 3-1A(a). Adjust the potentiometer to produce an input voltage of 0 V. Measure and record the diode voltage V_F and the series resistor voltage V_{RS}. Adjust the input voltage across the potentiometer to 0.1 V. Again measure and record the V_F and V_{RS}.
12. Repeat the procedure for each of the dc input values in the chart of Fig. 3-1A(c). Note that the input voltage

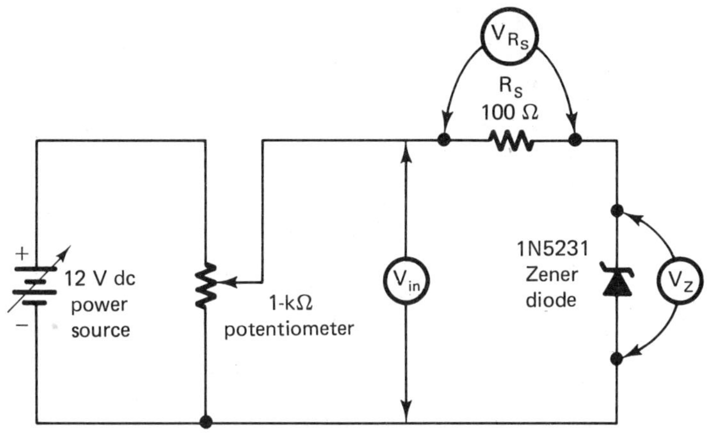

(a)

Dc V_{in}	V_Z	V_{R_S}	I_Z
0			
1			
2			
3			
4			
5			
6			
7			
8			
9			
10			
11			
12			

(b)

Dc V_{in}	V_F	V_{R_S}	I_F
0			
0.1			
0.2			
0.3			
0.4			
0.5			
0.6			
0.7			
0.8			
0.9			
1.0			

(c)

FIGURE 3-1A Voltage regulator test circuit.

does not exceed a value in excess of 1 V. This diode conducts very heavily in the forward direction. This should be noted by the observed voltage values. An increase in the value of V_F indicated a corresponding increase in I_F.

13. Calculate the forward current for the zener diode for the measured values of the chart. $I_F = V_{RS}/R_s$. Record the calculated values in the chart.
14. Turn off the power source and disconnect the circuit.

DISCUSSION

This experiment showed that a good zener diode has a low forward resistance and an infinite reverse resistance when tested with a

VOM. The leads can be identified by the ohmmeter when the diode is forward biased. The positive lead is connected to the P-material or anode and the negative lead is connected to the N-material or cathode. This is based on the assumption that the VOM has straight polarity.

The next part of the activity dealt with a zener diode connected in the reverse direction. The V_z and V_{RS} voltages were measured or calculated. The I_z of the circuit was then calculated for these values. It should be noted that reverse connection of a zener diode causes conduction to take place at the V_z point. This voltage remains very constant after V_z has been reached. The I_z, which is calculated using V_{RS}/R_s, increases in value when the dc input voltage is increased. This characteristic of the zener diode is used to control current that is used to regulate a dc voltage. The V_z value of this zener is in the range of 5.1 V. The measured V_z of the zener should be very close to this value. Any variation in V_z is due to the tolerance of the zener. This particular diode has a 10% tolerance rating. The V_z can vary ± 0.51 of 5.1 V.

The last part of the experiment had a zener diode connected in the forward direction. This showed that the zener responds as a standard silicon diode. As the forward voltage V_F increases, the diode goes into conduction at approximately 0.6 V. This shows a corresponding increase in the value of I_F. As a rule, zener diodes are not used in the forward direction.

ANALYSIS

1. Explain what causes a zener diode to maintain its V_z at a constant value when the source voltage is increased.
2. Explain why the power dissipation rating of a zener diode is an important selection characteristic.

EXPERIMENT 3-2
ZENER DIODE VOLTAGE REGULATION

A zener diode is used primarily to control or regulate the output voltage of a power supply. This type of circuit responds in such a way that a change in the input and a change in the value of the load resistor can be altered to produce a rather constant dc output voltage. The regulator circuit is located between the filter and the load resistor. A simple shunt regulator employing a zener diode is used in this circuit application. The I_z of a zener diode varies with changes in input voltage and load resistance. Heavy conduction of the zener added to the load current cause an increase in the total current passing through the series resistor. This causes an increase in the voltage drop across R_s and lowers the output voltage to a constant value. The output voltage of a regulator is maintained at a constant value because of the V_z characteristic of the zener diode.

OBJECTIVES

1. To construct a zener diode voltage regulator circuit that controls the voltage across the output terminals of a power supply.
2. To observe how a change in dc input voltage is regulated by a zener diode.
3. To observe how a change in the value of a load resistor is regulated by a zener diode.

EQUIPMENT

Variable dc voltage source
Volt-ohm-milliammeter (VOM), standard or digital type
Zener diode, 5.1 V ± 10%, 500 mW, Motorola 1N5231, 1N751, or Heath 56-16
1000-Ω, $\frac{1}{4}$-W resistor (brown-black-red-gold)
470-Ω, $\frac{1}{4}$-W resistor (yellow-violet-brown-gold)
220-Ω, $\frac{1}{4}$-W resistor (red-red-brown-gold)
100-Ω, $\frac{1}{4}$-W resistor (brown-black-brown-gold)
47-Ω, $\frac{1}{4}$-W resistor (yellow-violet-black-gold)
1000-Ω, 1-W, linear potentiometer
Circuit construction board

PROCEDURE

1. Construct the regulator circuit of Fig. 3-2A(a). The load resistor is not used for the first part of this activity. Connected in this way, the regulator does not have a load.
2. The source voltage is adjusted to 12 V as measured by a VOM. The input voltage or V_{in} is measured across the variable arm and one side of the potentiometer. In the circuit this is indicated by the meter symbol labeled V_{in}. Other voltages measured by the VOM are labeled V_{RS} and V_z. The V_{in} voltage is adjusted to each value indicated on the chart of Fig. 3-2A(b). The corresponding V_z and V_{RS} values are then measured and recorded in the chart. The I_T flowing through R_s is then calculated for each V_{in} value. The formula is $I_T = V_{RS}/R_s$. The I_z of the circuit is equal to I_T.
3. Adjust V_{in} to each value of the chart and then measure and record V_{RS} and V_z for each setting.

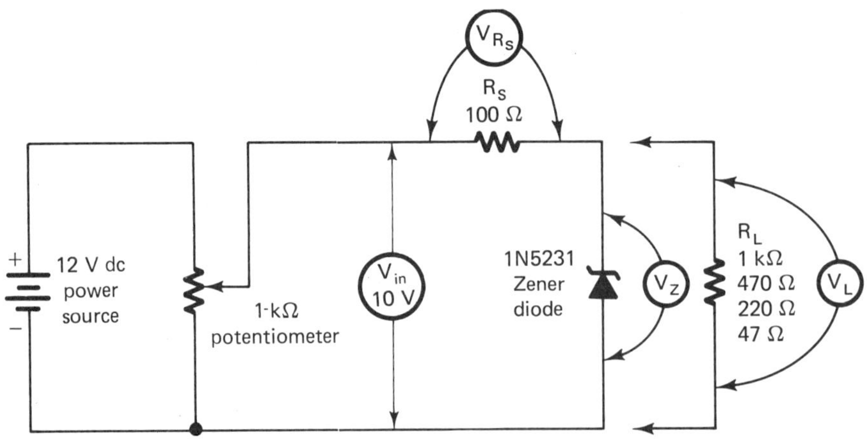

(a)

V_{in}	V_Z	V_{R_S}	I_Z
12			
10			
8			
7			
6			
5			
4			
3			

(b)

R_L	V_{in}	V_{R_S}	I_T	V_L	I_L	I_Z
1 kΩ						
470 Ω						
220 Ω						
47 Ω						

(c)

FIGURE 3-2A Voltage regulator test circuit and data charts.

4. Calculate the I_z or I_T for each V_{in} setting. Record the calculated values in the chart.
5. Adjust V_{in} to 10 V for the remaining steps. Do not adjust V_{in} unless it changes value during the operation of the circuit. It should be checked periodically for each change in circuit load.
6. Refer to the chart of Fig. 3-2A(c) for the remaining operations.
7. The load resistance of this circuit is changed to see how the regulator will respond. A specific value of R_L is connected to the output of the regulator and the indicated voltage values are then measured with the VOM. These values are recorded in the chart under the appropriate heading.
8. Connect the 1-kΩ load to the circuit across the zener diode. This is designated as the R_L connection points in the diagram. Measure V_{RS} and V_L. Record the measured values in the provided space of the chart.
9. Remove the 1-kΩ resistor and replace it with the 470-Ω load resistor. Measure V_{RS} and V_L. Record these values in the correct space of the chart.
10. Repeat step 9 for the 220-Ω and 47-Ω resistors.
11. Calculate the values of I_T, I_z, and I_L, $I_T = V_{RS}/R_s$, $I_L = V_z/R_L$, $I_z = I_L$.

DISCUSSION

In this experiment you had an opportunity to build a shunt voltage regulator with a zener diode and test its response to input voltage changes and load changes. The first part of the activity showed how the regulator responded to changes in input voltage. In this part of the experiment the zener diode was used as a regulator without a load. All the current flowing through the series resistor had to pass through the zener as I_z. The zener has the ability to conduct varying amounts of I_z while maintaining its V_z at a constant value. In this case the V_{in} was varied from a high value of 12 V to a low value of 3 V. The measured zener voltage should have remained very close to the rated V_z value of the diode. Any variation in voltage was due to the tolerance of the zener. The measured series resistor voltage V_{RS} changed according to the value of the input voltage. The sum of V_z and V_{RS} equals V_{in}. The value of V_{RS} is current dependent. When V_{in} is high, V_{RS} is also high. The value of V_z developed across the zener is dependent on the conduction status of the diode. When it is conductive, V_z appears across the zener. When V_{in} drops below 6 V, the zener becomes nonconductive. This causes the same effect as an open circuit and $V_z = V_{in}$. There was very little V_{RS} appearing across R_s for the lower V_{in} values. This showed that the regulator has a rather wide operating range. It cannot achieve regulation when V_{in} is less than 1 V higher than V_z.

The last part of the experiment showed how a regulator responds to changes in the load. V_{in} was adjusted to 10 V. Different values of R_L were connected across the zener or to the circuit's output. The values of V_{RS} and V_z were measured with a VOM. The I_T, I_L, and I_z of the circuit were then calculated. It should be noted that when the load is high resistant there is a rather large amount of I_z. When the value of the load resistor is reduced, the load current increases and I_z is reduced. I_L and I_z added together equal the I_T. When the resistance of R_L decreases, there is a corresponding increase in I_L and a decrease in I_z. For this activity, an R_L of 220 Ω still provided regulation. For R_L values less than this the circuit went out of its regulating range. The current demand of the load became so great that it caused an excessive amount of I_T. This action increased the voltage drop across R_s and lowered the diode's voltage to a value below the V_z. The zener dropped out of conduction. This in turn caused the circuit to respond as a voltage divider consisting of R_s and R_L. The voltage drop across each resistor was then dependent on the I_T. With R_s being larger than R_L, more voltage appeared across it. The output voltage or V_L dropped to a rather low value. In general, the operating range of this regulator is from an infinite value to some value below 220 Ω.

ANALYSIS

1. What determines the range of regulated voltage for Fig. 3-2A?
2. When the value of R_s is reduced, the load current ________ ________ and zener current ________________.

UNIT 3 EXAMINATION
ZENER DIODES

Instructions: Circle the response that most correctly completes each statement.

1. A zener diode achieves voltage regulation when it is:
   a. Forward biased across R_L
   b. Connected in series with R_s
   c. Reverse biased across R_L
   d. Connected in parallel with R_s
2. When thermally generated holes and electrons gain enough energy from the reverse-biased source of a zener diode to produce new current carriers, it is called:
   a. Avalanche breakdown
   b. Reverse breakdown
   c. Zener breakdown
   d. Low-voltage breakdown
3. A strong electric field across a P-N junction that causes covalent bonds to break apart:
   a. Is called avalanche breakdown
   b. Is called reverse breakdown
   c. Is called zener breakdown
   d. Results in breakdown above 8 V
4. The reverse current that flows through a zener diode is called:
   a. I_T
   b. I_L
   c. I_z
   d. I_F
5. The voltage across a zener diode that is responsible for its ability to achieve regulation is called:
   a. V_{RS}
   b. V_L
   c. V_z
   d. V_{in}
6. The percentage of accuracy that a zener diode has when it goes into conduction is called:
   a. Tolerance
   b. Knee voltage

c. Minimum V_z
d. V_z safe operating range

7. The P_D rating of a zener diode refers to its ability to:
   a. Regulate voltage
   b. Dissipate heat
   c. Control I_z
   d. Change impedance
8. A 1N1775A zener diode is rated 15 V at 1 W. In what range does the maximum I_z of this device fall?
   a. 0 to 0.05 A
   b. 1 to 50 mA
   c. 51 to 75 mA
   d. 76 to 100 mA
   e. 101 mA or above
9. A zener diode is rated at 9.1 V and has a 10% tolerance. What is the operating range where the V_z of this device is expected to fall?
   a. 1 to 9.1 V
   b. 9.1 to 20 V
   c. 9.1 to 10.01 V
   d. 8.09 to 9.1 V
   e. 8.09 to 10.01 V
10. In a zener diode voltage regulator the value of V_z is determined by:
    a. $V_{in} - V_{RS}$
    b. $V_{RS} - V_{in}$
    c. $V_L - V_{RS}$
    d. $V_{RS} + V_{in}$
11. In a zener diode voltage regulator a small increase in the value of the input voltage (V_{in}) will cause:
    a. An increase in V_z
    b. A decrease in V_{RS}
    c. A decrease in I_z
    d. An increase in I_z
12. In a zener diode regulator, reducing the value of R_L will cause
    a. An increase in V_z
    b. A decrease in V_z
    c. An increase in I_z
    d. A decrease in I_z
13. Applied input voltage beyond the V_z point of a zener diode causes:
    a. An increase in I_z
    b. A decrease in I_z

c. The I_z to level off
d. The I_z to cut off
e. Only a trace of I_z

14. Design of a zener diode voltage regulator calls for a 10-V output with a fixed load of 50 mA. The unregulated input voltage varies from 12 to 15 V. What is a suitable value of series resistor for the circuit. Indicate the resistance range where this value will fall.
    a. 1 to 50Ω
    b. 51 to 75Ω
    c. 76 to 100Ω
    d. 101Ω or higher
15. What would be a suitable wattage rating for the series resistor selected for question 14?
    a. $\frac{1}{8}$ W
    b. $\frac{1}{4}$ W
    c. $\frac{1}{2}$ W
    d. 1 W
    e. 2 W

UNIT FOUR

Special Semiconductor Diodes

UNIT INTRODUCTION

In the previous three units we were concerned with semiconductor materials, junction diodes, and zener diodes. In this unit we examine some semiconductor diodes that have unique characteristics that distinguish them from other diodes. Applications are rather specialized and in some cases may represent a very small part of diode technology. As a rule, these devices may require some unusual construction technique or a specialized circuit application.

This unit is included in the study of semiconductor devices to expand your knowledge of diode technology. Through it you will become familiar with some of the unusual characteristics of a diode. This will provide a better understanding of the diode and some of its unusual applications. The diode is indeed the building block of all semiconductor devices.

UNIT OBJECTIVES

Upon completion of this unit, you will be able to:

1. Interpret the graphic data of a tunnel diode.
2. Locate the peak current, valley current, peak voltage, and valley voltage of a tunnel diode's *I-V* curve.

3. Identify the crystal structure and schematic symbols of a tunnel diode.
4. Describe how capacitance is produced across the junction of a varactor diode.
5. Describe how the internal capacitance of a P-N junction is changed with bias voltage.
6. Identify the crystal structure and schematic symbols of a varactor diode.
7. Describe the distinguishing characteristics of Schottky, PIN, IMPATT, and Gunn-effect diodes.

IMPORTANT TERMS

Before reading Unit 4, review the following terms. These terms provide a basic understanding of terms associated with semiconductor diodes.

Ambient temperature The surrounding temperature of an object or device.

Barrier An area in a semiconductor PN junction when it is formed where electron and hole movement establishes a voltage or difference in potential.

Bistable Two stable or stationary operating conditions of an electronic switching circuit.

Breakdown A semiconductor condition where current carriers pass through a PN junction due to a high reverse-biased field voltage or due to impact ionization of the current carriers.

Capacitance An inherent electrical property of a device that opposes a change in voltage. This property permits energy to be stored in an electrostatic field. A capacitor is composed of two or more conductors separated by an insulating material.

IMPATT diode—impact avalanche and transit time diode An avalanche diode used as a high-frequency oscillator or amplifier. The negative resistance of this device depends on the transient time of current carriers through the depletion layer.

Intrinsic A semiconductor material that is relatively free of impurities.

Negative resistance An electronic condition in which an increase in voltage across a resistive element causes a reduction in current and vice versa. Certain semiconductors have a negative resistance region in their operation.

Oscillator An electronic device and its circuit that generates or changes dc into ac.

Resonance An electrical condition in which the components of a circuit oscillate at a frequency that is dependent on its inductance and capacitance.

Schottky barrier A barrier formed between metal and a semiconductor material. This barrier achieves rectification but avoids slowing down the current carriers, so that high-frequency ac can be rectified.

Temperature coefficient A temperature value that serves as a measure of some specific property of a device.

THE TUNNEL DIODE

A *tunnel diode* is a two-element semiconductor device with construction similar to that of a conventional junction diode. This device has an anode connected to the P-material and a cathode connected to the N-material. The two materials are joined together at a common point or junction. The tunnel diode is, however, quite different beyond this point. The P- and N-materials, for example, are rather heavily doped. This means that the material has a rather high concentration of impurities. Typical doping levels may be one hundred to several thousand times that of a regular diode.

Heavy doping of the semiconductor material of a tunnel diode causes the N-material to have a large number of free electrons and the P-material to have a large number of holes. The high current carrier content of the crystal material causes the width of the depletion region to be very thin. In general, depletion zone width is only $\frac{1}{100}$ of that of a regular diode. This permits current carriers to "tunnel" through the barrier rather than move across it by reducing the bias voltage. Normally, electrons must have sufficient external energy to cross the surface barrier of a junction. Due to the increased number of current carriers and the thin barrier that exists, electrons tend to pass through the barrier with an extremely small amount of energy. In many cases, ambient temperature may be sufficient to cause some conduction. This can occur in some cases when very little bias voltage has been applied. Current carriers tend to move across the barrier as if it did not exist. Figure 4-1 shows an example of electrons tunneling through the barrier of a P-N junction with a small amount of external voltage.

Tunnel Diode *I-V* Characteristics

The current-voltage characteristic of a tunnel diode is quite different from that of other diodes. The tunnel diode has three distinct characteristics that makes it electrically different. Each characteristic depends on a value of bias voltage applied to the device. Within a certain range of bias voltage, a tunnel diode is conductive in both directions. See the *I-V* characteristics of Fig. 4-2. Note the area on each side of the zero crossover point. In the

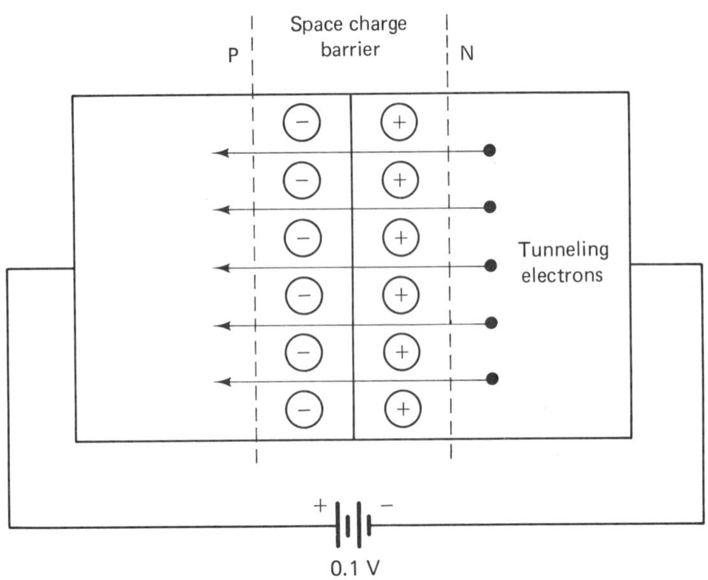

FIGURE 4-1 Tunneling electrons.

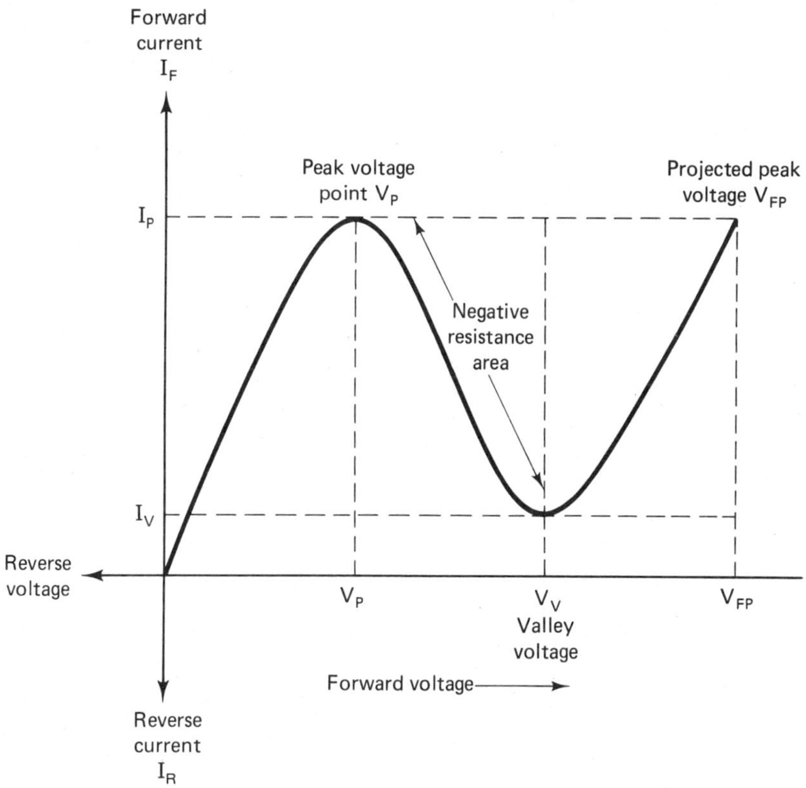

FIGURE 4-2 *I-V* characteristics of a tunnel diode.

second range of bias voltage, the tunnel diode has a negative resistance characteristic. This is located between the forward voltage peak (V_p) and the valley voltage (V_v) point. Negative resistance refers to an area where an increase in voltage causes a decrease in current. The third characteristic of bias voltage deals with an area beyond the valley voltage point. Increasing the forward voltage beyond the V_v point of a tunnel diode causes it to respond as a conventional diode. This part of the curve shows that an increase in forward voltage causes a corresponding increase in forward current.

The operation of a tunnel diode can be changed according to the value and polarity of the applied bias voltage. The three bias conditions discussed all occur within a few millivolts. The general area of operation is somewhere between 0 and 600 mV. This is all dependent on the material used in the construction of the diode. Germanium or gallium arsenide is used in the construction of most tunnel diodes. The doping level of the P- and N-materials has a great deal to do with other characteristics. The peak current (I_p) for example, can vary from a few microamperes to 100 A. The peak voltage, however, is limited to a maximum value in the range of 600 mV. For this reason a tunnel diode can be easily damaged. A VOM with a 1.5-V cell energizing the ohmmeter could damage a tunnel diode by simply measuring its resistance. When using the tunnel diode, one must use extreme care in connecting it into a circuit and altering the circuit so that it will produce a desired operating condition.

Tunnel Diode Specifications

Three common symbols for the tunnel diode are shown in Fig. 4-3. Two of these symbols resemble the conventional diode symbol with a slight modification. The third symbol, which is widely used, consists of a line and a half-circle. Notice the parts of the symbol that are used to identify the anode and cathode. For the standard diode symbol the anode and cathode remain unchanged. For the line-half-circle symbol the anode is represented as a line, and the half-circle denotes the cathode. These symbols may all be drawn within a circle or with the circle omitted.

Tunnel diodes are usually packaged in a special housing. Figure 4-4 shows a representative package. Note the dimensions of this device. It is extremely small in comparison with a conventional diode. As a rule, the enclosure is usually metal. The anode is insulated from the metal housing, and the cathode is attached to it. Since tunnel diodes are used primarily in high-frequency signal applications, metal enclosures are purposely used to isolate the internal diode structure from stray electromagnetic fields.

The data from a tunnel diode specification sheet is shown in Fig. 4-5. The tunnel diode represented in the sheet is a low-power device. Note that the data is listed for three different conditions of operation, namely, minimum, typical, and maximum operation. This particular diode is used for oscillators, high-frequency amplification, and high-speed switching applications. The specification sheet generally lists a number of characteristics that influence the high-frequency response of the diode, such as capacitance of the junction (C), terminal lead inductance (L_s), negative resistance ($-R$), and lead resistance (R_s). The other listings are defined in Fig. 4-5.

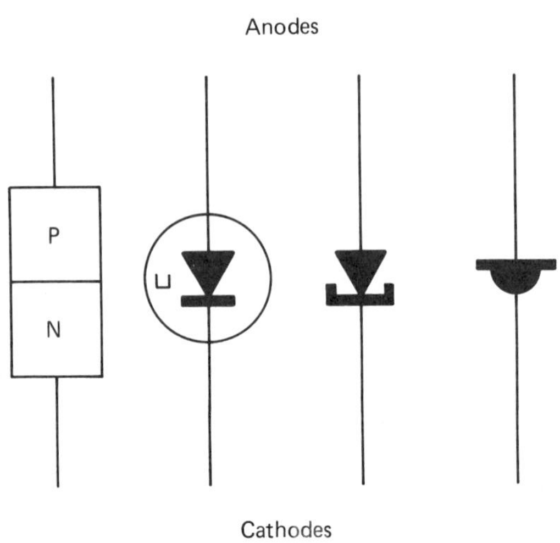

FIGURE 4-3 Tunnel diode symbols and crystal structure.

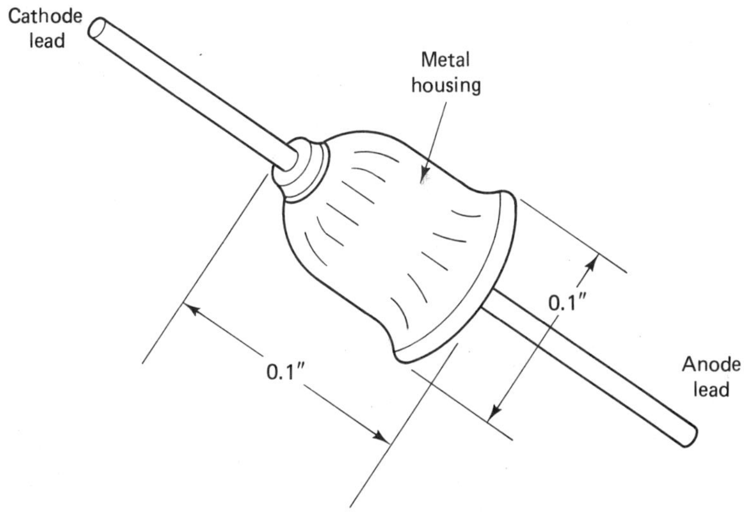

FIGURE 4-4 Tunnel diode package.

TUNNEL DIODE APPLICATIONS

When the tunneling principle of a highly doped P-N junction was first discovered, the tunnel diode was recognized as an important high-speed switching device. State switching could essentially take place at the speed of light. The response time of the switching action is primarily limited by diode capacitance. As a rule, this is in the order of 1 to 10 pF. This means that switching can occur from the zero point to the peak point with a rise time as small as 22 ps.

Specifications: Ge 1N2939

ABSOLUTE MAXIMUM RATINGS (25°C)

Current				
Forward (−55 to +100°C)		5 mA		
Reverse (−55 to +100°C)		10 mA		

ELECTRICAL CHARACTERISTICS (25°C)(1/8 in. Leads)

	Min.	Typ.	Max.	
I_P	0.9	1.0	1.1	mA
I_V		0.1	0.14	mA
V_P	50	60	65	mV
V_V		350		mV
REVERSE VOLTAGE (I_R = 1.0 mA)			30	mV
FORWARD PEAK POINT CURRENT VOLTAGE V_{fp}	450	500	600	mV
I_p/I_v		10		
−R		−152		Ω
C		5	15	pF
L_s		6		nH
R_S		1.5	4.0	Ω

FIGURE 4-5 Tunnel diode data.

The negative resistance characteristic of a tunnel diode has also opened the door for a number of other applications. High-frequency signal generation or oscillation can now be achieved by using this characteristic. The negative resistance characteristic also permits the tunnel diode to be used as a high-frequency amplifying device. High-frequency signal control is probably the most widely used application of the tunnel diode today.

High-Frequency Oscillators

The negative resistance characteristic of a tunnel diode can be used to a unique advantage when it is connected across an *LC* tank circuit. If an inductor and capacitor are connected in parallel it forms an *LC* tank circuit. Figure 4-6 shows a tank circuit connected to a dc source. When the switch of the circuit is closed momentarily, it causes the tank circuit to be shocked into oscillation. The resulting damped oscillatory waveform is shown in Fig. 4-6. Note that the amplitude of the wave decreases with each succeeding wave. This is caused by the effective resistance of the tank circuit. The resistance essentially dissipates the power so that it causes the waves to die out after a few cycles of operation. If the tank circuit had only pure inductance and capacitance, the oscillations

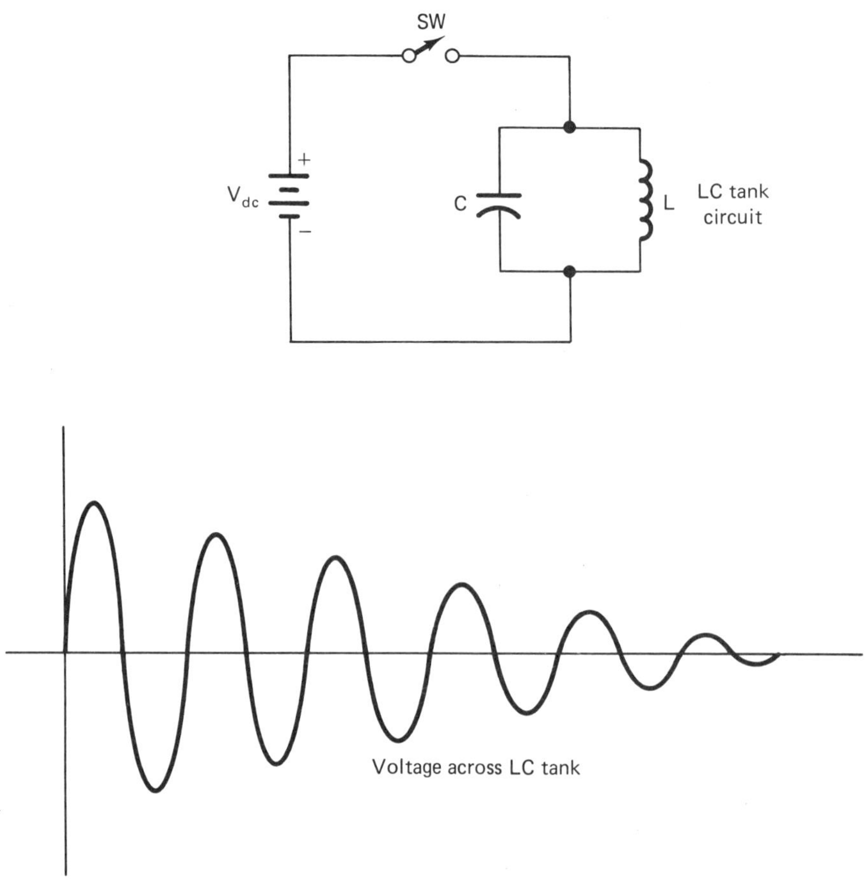

FIGURE 4-6 *LC* tank circuit and damped oscillatory waveform.

would be continuous after the initial voltage was applied. In practice, the circuit will always have some resistance. This means that the resulting oscillations will die out after a few cycles of operation.

The negative resistance effect of a tunnel diode can be used to cancel or overcome the effective resistance of a tank circuit. This will permit the tank circuit to behave as if it has no resistance. As a result, when the tank is shock-excited with voltage, oscillations will occur without a change in amplitude. The frequency of the oscillating wave is determined by the capacitance and inductance of the tank circuit. This oscillator is very simple in operational theory and construction.

In order to use a tunnel diode for the tank circuit of an oscillator, the diode must be forward biased. The amount of forward bias voltage needed to reach the negative resistance area is very critical. Ideally this operating point is located between the peak and valley voltage points. This part of the characteristic curve is in the order of 60 to 350 mV. A representative operating voltage would be somewhere near the center of this voltage range, or approximately 145 mV.

Figure 4-7 shows a tunnel diode oscillator with a variable capacitor that permits frequency changes. Biasing of the tunnel diode is achieved by a voltage-divider network consisting of resistors R_1 and R_2. The generated output signal developed by the circuit appears across the tank. An oscillator of this type has good stability and can generate signals in the microwave range. It is, however, rather sensitive to changes in temperature and bias voltage.

Tunnel Diode Switching

A very common application of the tunnel diode today is in high-speed switching circuits. A tunnel diode can perform logic functions and memory. The tunnel diode offers the advantages of small size, low operating power, high speed, low cost, and high reliability.

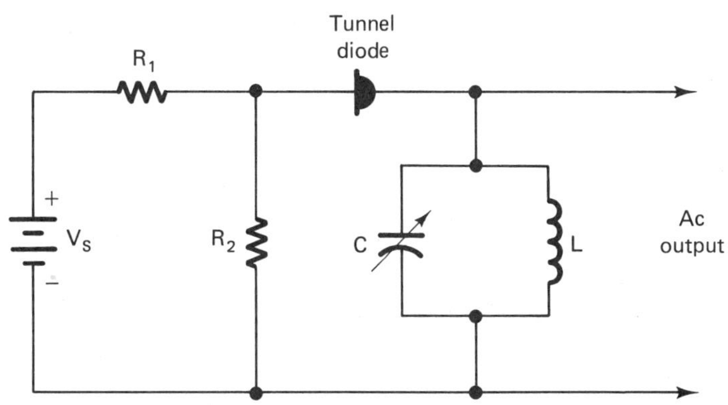

FIGURE 4-7 A tunnel diode oscillator.

It is possible to form a simple two-state, or *bistable,* switching circuit by connecting a tunnel diode in series with a voltage source and a single resistor. Figure 4-8 shows this type of circuit and an *I-V* curve for the diode. In order to see how the circuit responds as a switch, the operating range of the circuit must be plotted. A solid line extending across the *I-V* curve shows the range of anticipated operation. If a source voltage of 0.5 V is applied to a 100-Ω resistor, it will cause a forward current of 5.0 mA. Note the location of these two points on the solid line of the *I-V* curve. Note also that the operating line crosses the *I-V* curve in three places. Point *A* occurs near the peak current point, or I_p. Point *B* is in the negative resistance region of the curve. This is considered to be an unstable operating point in a switching circuit. Point *C* is located slightly above the valley voltage point (V_v). Points *A* and *C* are considered to be the stable operating points of the switching circuit.

In order to see how a tunnel diode is used as a bistable switch, we must consider some specific operating conditions for the circuit in Fig. 4-8. Point *A* is located at an I_F of 4.75 mA and a V_F of 25 mV. Point *C* is located at an I_F of 1.0 mA and a V_F of 400 mV. The circuit can be switched to either of these positions, and it will remain there until a state change occurs. The operational state of the circuit is determined by the value of the source voltage. Before the circuit is energized, both I_F and V_F are zero. When the circuit is energized by the 500-mV source, its operation will change to

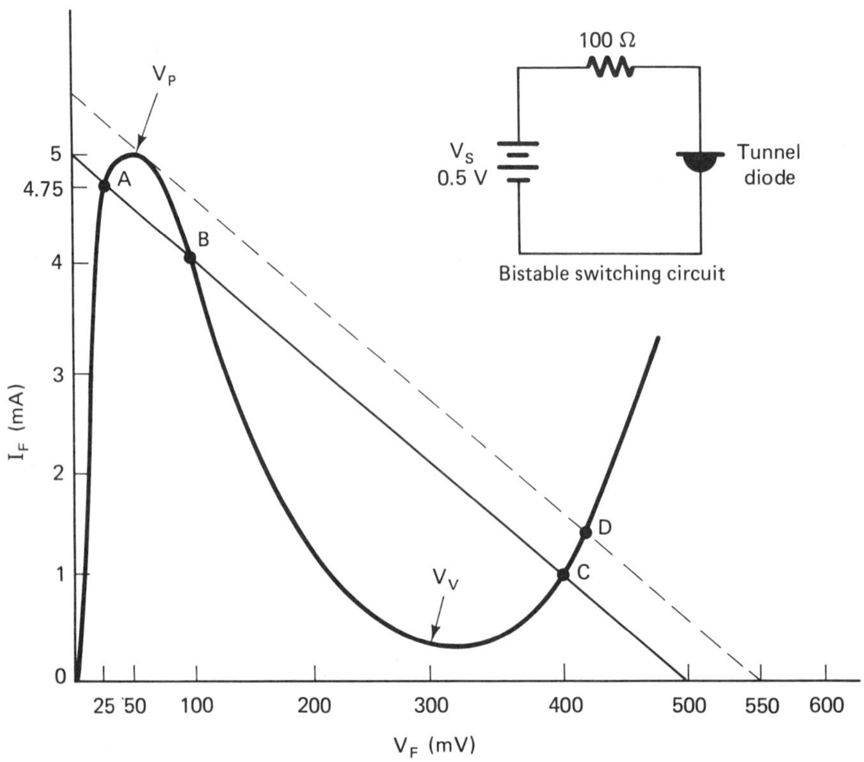

FIGURE 4-8 Bistable switching circuit and operating characteristics.

point *A*. In order for the circuit to operate at point *C*, the source voltage must be increased to 550 mV. This will cause the operation of the circuit to shift to the dotted line where it intersects the *I-V* curve at stable point *D*. The circuit will remain at this operating point as long as the source remains at 550 mV. However, reducing the source to a value of 500 mV will cause the circuit to remain in its stable condition by moving to point *C* on the curve. To change the state of operation back to point *A*, the source voltage would have to drop in value below 25 mV. When the source voltage is changed back to 500 mV, the circuit will remain in its stable state at point *A*. The circuit has two stable states of operation depending on the value of the source voltage. This type of circuit is called a *bistable voltage switch.*

SELF-EXAMINATION

1. Heavy doping of the P- and N-materials of a tunnel diode causes the depletion zone to be very __________.
2. The __________ effect is due to the heaving doping of P- and N-materials and a thin depletion zone.
3. __________ __________ refers to an area of a tunnel diode where an increase in voltage causes a decrease in current.
4. An increase in forward voltage beyond the __________ __________ point causes a tunnel diode to respond as a regular diode.
5. When a tunnel diode is used as a switch it has _____ stabilized conditions of operation.
6. When a tunnel diode is used as an oscillator, it is biased to operate in the __________ __________ region.
7. A __________ __________ will conduct in either the forward or reverse direction when properly biased.
8. A tunnel diode generally operates in the (forward-reverse) direction for most of its applications.
9. When a tunnel diode is used as an oscillator, the __________ __________ cancels the effective resistance of a tank circuit.
10. The two stable operating points of a bistable tunnel diode switch are to the left of the __________ __________ point and to the right of the __________ __________ point.

ANSWERS

1. Thin	2. Tunneling
3. Negative resistance	4. Valley voltage
5. Two	6. Negative resistance

7. Tunnel diode	8. Forward
9. Negative resistance	10. Voltage peak (V_p), valley voltage (V_v)

VARACTOR DIODES

In an ordinary junction diode the depletion region is an area that separates the P- and N-materials on each side of the junction. This area develops when the junction is initially formed. It represents a unique part of the diode that is essentially void of current carriers. In this regard, the depletion region serves as a dielectric medium or insulator. When a diode has bias voltage applied, its depletion region will change in width. In a sense, this means that a diode responds as a voltage-variable capacitor. By definition, a capacitor is two or more conductors separated by an insulating material. The P- and N-materials, being semiconductors, are separated by a depletion region insulator. Devices designed to respond to the capacitance effect are called varactors, varicap diodes, or voltage-variable capacitors.

Varactor Characteristics

A varactor diode is a specially manufactured P-N junction with a variable concentration of impurities in its P- and N-materials. In a conventional diode, doping impurities are usually distributed equally throughout the material. Varactors have a very light dose of impurities near the junction, but away from the junction the impurity level increases. This type of construction produces a much steeper voltage-capacitance relationship. Figure 4-9 shows a comparison of the capacitance between a conventional silicon diode and a varactor diode. It can be seen that an ordinary diode possesses only a small value of internal capacitance. In general, this capacitance is too small to be of practical value.

Varactor diodes are normally operated in the reverse bias direction. With an increase in reverse biasing, the depletion region increases its width. This means less resulting capacitance. A decrease in reverse bias voltage causes a corresponding increase in capacitance. In effect, the capacitance of a diode varies inversely with its bias voltage. This relationship, however, does not change linearly. Notice the nonlinear area between 0 and 0.2 V of the varactor diode of Fig. 4-9. The value of capacitance increases rather significantly when the reverse bias voltage decreases. The capacitance will continue to increase even when the diode is forward biased. In fact, the greatest capacitance is produced just before the forward bias barrier voltage is reached. This is approximately 0.6 V for a silicon diode. However, when the barrier is reached, the diode becomes conductive. This causes the diode to respond as a shorted capacitor. Because of this condition, varactor diodes

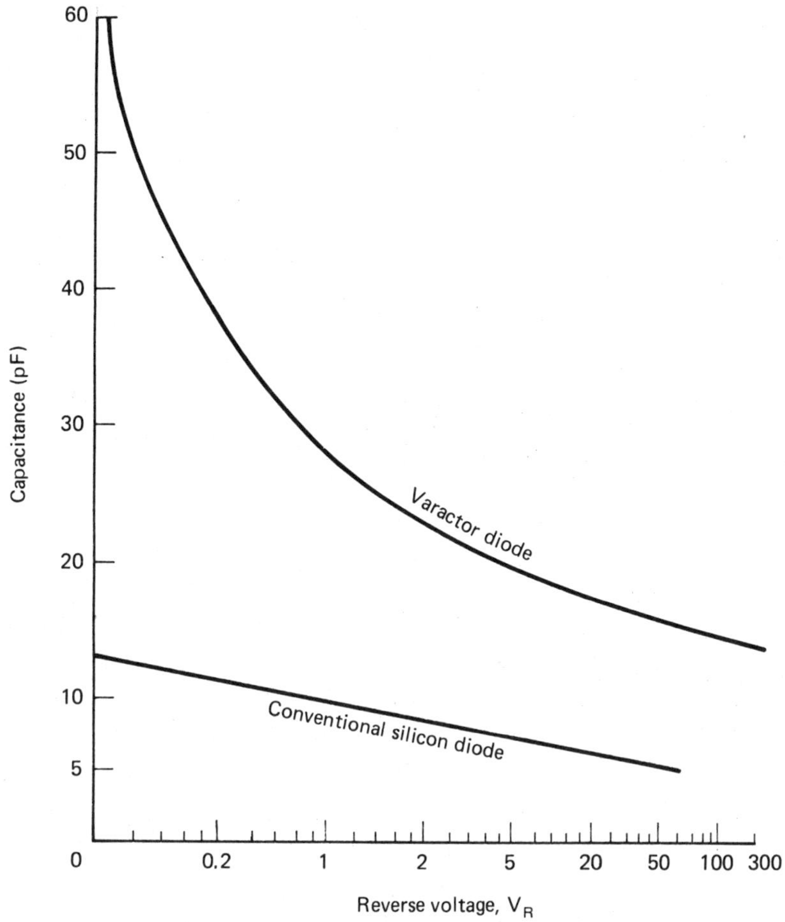

FIGURE 4-9 Diode capacitance values, T_A - 25°C.

should not be used in the forward bias region. Their normal range of operation is between zero and the reverse breakdown voltage.

The internal capacitance of a varactor diode will change to some extent with temperature. Manufacturers usually take this into account by rating the operational characteristics of the device at some temperature value. The capacitance-voltage characteristic of the diode in Fig. 4-9 is rated at 25°C. This shows that these C-V characteristics will occur only when the device is operating at or near its rated temperature value. The amount of capacitance change that takes place for a given change in temperature is usually expressed as the temperature coefficient of capacitance. The letters TCc denote this characteristic of the diode. As a rule, the TCc value of a varactor diode has a positive temperature characteristic. This shows that as the temperature increases, the depletion region decreases in width. A decrease in the depletion region will cause a corresponding increase in capacitance. A positive TCc value therefore shows that an increase in temperature causes a corresponding increase in capacitance.

The electrical characteristics of a representative varicap diode are shown in Fig. 4-10. This data sheet shows the absolute maximum ratings, a housing outline, and electrical characteristics for

BB139

VHF/FM VARACTOR DIODE

DIFFUSED SILICON PLANAR

- C_3/C_{25} . . . 5.0–6.5
- MATCHED SETS (Note 2)

ABSOLUTE MAXIMUM RATINGS (Note 1)

Temperatures	
Storage Temperature Range	−55°C to +150°C
Maximum Junction Operating Temperature	+150°C
Lead Temperature	+260°C
Maximum Voltage	
WIV Working Inverse Voltage	30 V

DO-35 OUTLINE

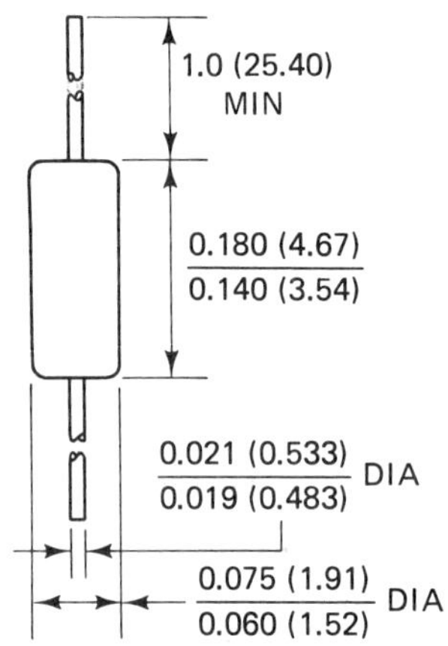

NOTES:
Copper clad steel leads, tin plated
Gold plated leads available
Hermetically sealed glass package
Package weight is 0.14 gram

ELECTRICAL CHARACTERISTICS (25°C Ambient Temperature unless otherwise noted)

SYMBOL	CHARACTERISTIC	MIN	TYP	MAX	UNITS	TEST CONDITIONS
BV	Breakdown Voltage	30			V	I_R = 100 μA
I_R	Reverse Current		10 0.1	50 0.5	nA μA	V_R = 28 V V_R = 28 V, T_A = 60°C
C	Capacitance	 4.3	29 5.1	 6.0	pF pF	V_R = 3.0 V, f = 1 MHz V_R = 25 V, f = 1 MHz
C_3/C_{25}	Capacitance Ratio	5.0	5.7	6.5		V_R = 3 V/25 V, f = 1 MHz
Q	Figure of Merit		150			V_R = 3.0 V, f = 100 MHz
R_S	Series Resistance		0.35		Ω	C = 10 pF, f = 600 MHz
L_S	Series Inductance		2.5		nH	1.5 mm from case
f_O	Series Resonant Frequency		1.4		GHz	V_R = 25 V

NOTES
1. These ratings are limiting values above which the serviceability of the diode may be impaired
2. The capacitance difference between any two diodes in one set is less than 3% over the reverse voltage range of 0.5 V to 28 V

FIGURE 4-10 Electrical characteristics of a varactor diode. [Courtesy of Fairchild Camera and Instrument Corp.]

operation at 25°C. The data listed are for minimum, typical, and maximum operation. Most of these characteristics have some influence on the operational frequency of the device. These are tested at a representative operating frequency. Note in particular that this diode has a series resonant frequency (f_o) of 1 GHz. A gigahertz is 1,000,000,000 Hz. This particular varactor diode is used in adjustable tuning circuits and for automatic frequency control applications.

Operating Efficiency

When the varactor diode is used as a capacitor in a resonant circuit, its operational efficiency becomes very important. The efficiency rating of a capacitor is a ratio of the amount of energy stored compared with the actual amount of energy used by the capacitor in the storing process. This relationship is called the *quality factor* or Q of a capacitor. A varactor diode has a Q rating very similar to that of a capacitor. The Q of a varactor diode is an expression of the capacitive reactance (X_c) divided by the series resistance (R_s), or

$$Q = \frac{X_c}{R_s}$$

Since the X_c of this expression is frequency dependent, the Q will also change with frequency. The X_c of a capacitor is $1/2\pi fC$. X_c is measured in ohms. For a varactor diode the X_c and R_s values of the Q expression are combined in a more workable formula. This is expressed as

$$Q = \frac{1}{2\pi fCR_s}$$

Note in this formula that Q is inversely related to the values of f, C, and R_s.

The Q of a varactor diode takes into account such things as internal capacitance, series resistance, and frequency. The series resistance is generally due to lead length and the bulk resistance of the semiconductor material. In conventional operation the reverse-biased voltage of the diode blocks dc from passing through the device. The internal resistance of a diode is a form of opposition to an ac signal. Since ac passes through the capacitance of a reverse-biased diode, its value is influenced by the R_s that it sees. As a rule, the R_s of a varactor diode is quite small at high frequencies. The data sheet of Fig. 4-10 shows R_s to be 0.35 at a frequency of 600 MHz. This means that a varactor diode offers some measurable amount of opposition to an ac signal. This opposition is frequency dependent. The Q of a varactor diode is often expressed as a number value at an operational frequency. On the data sheet the Q is 150 at 100 MHz.

Typically, these devices have a rather high value of Q. In general, the Q increases at lower frequencies and decreases in value at higher frequencies. This means that the device works very efficiently at some designated range of ac frequency. If the varactor is subjected to frequencies higher than the designated operating frequency, the Q will drop in value. If the normal operating frequency is exceeded, the Q can be reduced to a value of 1. The frequency at which this occurs is called the cutoff frequency (f_o). The cutoff frequency of a varactor diode can be calculated by the formula

$$f_o = \frac{1}{2\pi C R_s}$$

This operating condition is an important selection characteristic for the varactor diode.

Varactor Construction

Varactor diodes are a rather broad general class of semiconductor devices that are designed for a variety of different applications. As might be expected, the construction of the device has a great deal to do with its use in different applications. The material used in its construction, for example, dictates the frequency response. Silicon is used largely for devices that operate on frequencies up to 1.0 GHz. Gallium arsenide is generally used for devices that operate on frequencies in excess of 1.0 GHz. The type of housing or packaging of a device is another important selection factor; a variety of different packages are available. The DO-35 package of Fig. 4-10 is preferred for most low-power and low-frequency applications. This type of package is glass enclosed with axial leads that can be soldered into a circuit. The power rating of the device is in the range of 500 mW. Power rating is another important selection consideration. Power ratings range from 500 mW to 50 W for stud-mounted packages. The power rating refers to the ability of a varactor diode to dissipate heat. Generally this is based on the series resistance (R_s) value of the device. The power being dissipated is in the form of ac energy that is being manipulated. Ac can pass through a reverse-biased diode that responds as a capacitor.

A number of schematic symbols are used to represent the varactor diode. Figure 4-11 shows three popular methods of representation. The symbol on the left has a small capacitor inside the

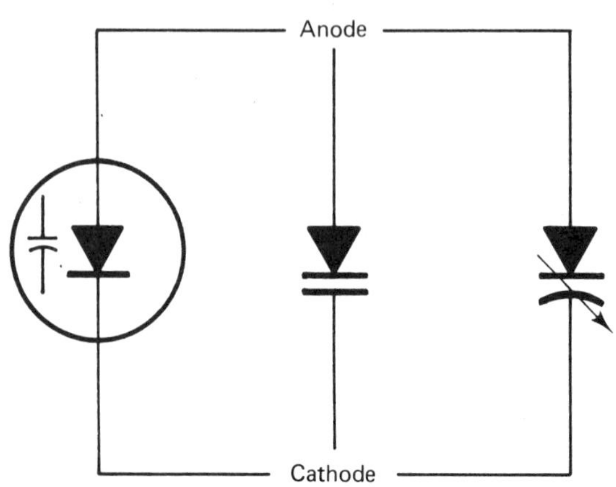

FIGURE 4-11 Varactor diode symbols.

circle surrounding the diode. This symbol tends to be used more frequently than the other two. Industrial electronic circuits and specialized engineering schematics often use the other two symbols.

Applications

Varactor diodes are used primarily in resonant circuits where some level of tuning or frequency control is desired. As a rule, this device is commonly used as a variable capacitor in a tuned circuit. The resonant frequency of the circuit is controlled by altering the reverse bias voltage of the varactor diode. Essentially, the bias voltage of the diode must not interfere with the operation of the tuned circuit. A large capacitor generally isolates the tuned circuit from the diode. Ac will pass through the capacitor, while dc is blocked.

Figure 4-12 shows a representative varactor diode–tuned *LC* circuit. The resonant frequency of this circuit is determined by the inductor (L) and the series combination of C_D and C_1. C_1 normally has a larger capacitance value than C_D. C_1 is used to prevent the dc bias voltage of C_D from going into the inductor. R_1, R_2, and R_3 serve as the bias voltage control network for C_D. Adjusting R_3 will change the amount of reverse bias voltage to C_D. Moving the wiper arm of R_3 up causes more reverse bias voltage to C_D. This increases the capacitance of C_D. Moving the wiper arm down reduces the bias voltage, which causes a reduction in capacitance. Used in this manner the varactor becomes a tuning component for resonant frequency. Other applications of the varactor diode are automatic frequency control (AFC) in FM radio and TV receivers. Varactor tuning applications are widespread in new solid-state circuitry.

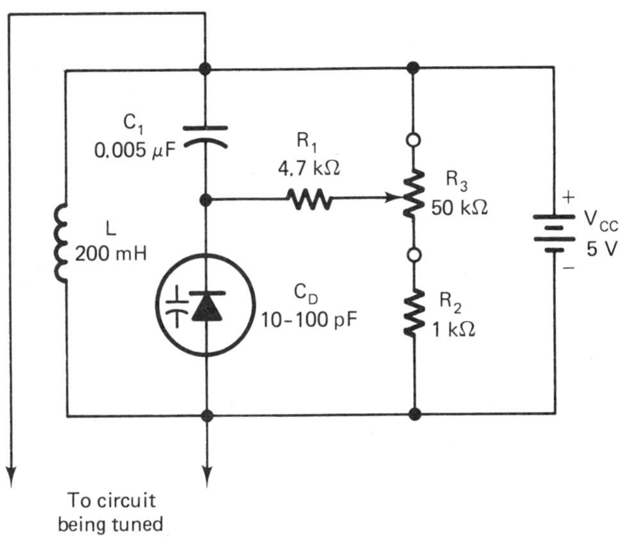

FIGURE 4-12 Varactor *LC* circuit.

SELF-EXAMINATION

11. The depletion region of a varactor diode serves as the ____________ material of a capacitor.
12. The P- and N-materials of a varactor diode serve as the ____________ of a capacitor.
13. The ____________ of a varactor diode is changed by altering the reverse bias voltage of the P-N junction.
14. The junction of a varactor diode is normally ________ ____________ when operated as a voltage-variable capacitor.
15. An increase in the reverse bias voltage of a varactor diode causes the junction capacitance to (increase, decrease).
16. A decrease in the reverse bias of a varactor diode causes the junction capacitance to (increase, decrease).
17. The *C-V* characteristics of a varactor diode will change to some extent with ____________.
18. ____________ is commonly used in the construction of a varactor diode that will control frequencies below 1 GHz.
19. The ____________ rating of a varactor diode refers to its ability to give off or dissipate heat.
20. Varactor diodes are commonly used as the ____________ component of an *LC* resonant circuit.
21. The *Q* factor of a varactor diode is an expression of ____________ ____________.
22. When the *Q* of a varactor diode drops to a value of 1 this is called the ____________ frequency.
23. Varactor diodes generally have a rather (high, low) *Q* rating.

ANSWERS

11. Dielectric	12. Conductors
13. Capacitance	14. Reverse biased
15. Decrease	16. Increase
17. Temperature	18. Silicon
19. Power	20. Tuning
21. Operating efficiency	22. Cutoff
23. High	

MISCELLANEOUS DIODES

There are a number of two-terminal devices that have a single P-N junction or its equivalent that respond in some unconventional way. These devices have unique characteristics that distinguish them from conventional junction diodes. As a rule, these devices may have a different method of operation, terminals, or construction. High-frequency oscillators and amplifiers, power control, and switching are included in this group of diodes. Some of the devices found in this classification are Schottky, IMPATT, Gunn-effect, PIN, and switching diodes.

Schottky-Barrier (Hot-Carrier) Diodes

The *Schottky-barrier, surface-barrier,* or *hot-carrier* diode is a two-terminal device that has a number of important applications in electronics. Its areas of application were first limited to the very high frequency (VHF) range of operation. In recent years this device has been found in low-voltage/high-current switching power supplies. It is also used in computer integrated circuits, radar systems, communication circuits, and instrumentation.

The construction of a Schottky diode is very different from the conventional P-N junction diode. The junction formed by this device is considered to be a metal-semiconductor structure. Figure 4-13 shows the structure of a Schottky diode and its schematic symbol. The semiconductor material is a piece of N-type silicon. The metal can be a variety of different materials, such as gold, silver, platinum, chrome, or tungsten. The construction techniques and materials of this device each result in a different frequency response and biasing voltage. In general, these characteristics are very similar in many respects. The device can be

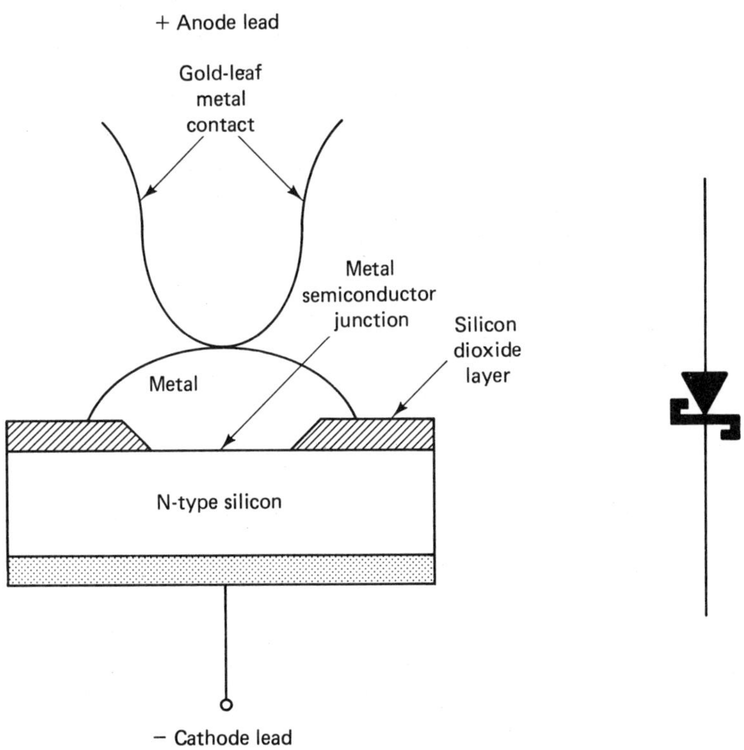

FIGURE 4-13 Schottky-barrier diode construction and symbol.

manufactured so that it will fit into a variety of different applications.

In the two materials of a Schottky diode, electrons are considered to be the majority current carriers. Metal naturally contains an abundance of electrons. The N-type of semiconductor material is purposely doped so that it has a large number of electrons that do not take part in the covalent bonding process. When the two materials are joined during the forming process, electrons from the N material immediately flow into the adjoining metal. These electrons possess a rather high level of energy compared with those of the metal piece. The injected electrons are commonly called *hot carriers*. In a conventional P-N junction, the current carriers are represented as electrons and holes. The energy level of the two current carriers is primarily the same. Schottky diodes are unique in that conduction is entirely by majority carriers. The heavy flow of electrons from N-material to metal causes a depletion of current carriers near the material junction. This depletion area is similar to that of a conventional P-N junction.

When voltage is applied to a Schottky diode, the negative side of the source is connected to the cathode, and the positive side is connected to the metal anode. This action reduces the barrier and causes current carriers to return to the depletion region. The barrier is less than that of a conventional P-N junction in both the forward and reverse directions. This means that a higher current flow can be achieved with a smaller bias voltage. This is a very desirable feature in the forward bias direction but is rather undesirable in the reverse bias direction. Schottky diodes generally have a very low peak inverse voltage (PIV) rating. A 50-A Schottky diode, for example, will have a PIV of only 50 V. A conventional diode of the same type would have a PIV of several hundred volts. The *I-V* characteristics of a Schottky diode are very similar to those of a conventional junction diode.

The absence of minority current carriers in a Schottky diode makes it an attractive high-speed switching device. The minority current carrier content of a diode normally slows down the reverse recovery time of a solid-state device. With the minority carrier content at a minimum, the Schottky diode can be used to change states very quickly. Switching frequencies approaching 20 GHz are very common for the Schottky diode.

PIN Diodes

A new development in the miscellaneous group of diodes is the *PIN* diode. The abbreviation PIN refers to the structure of the semiconductor material, which is quite different from that of a conventional junction diode. The letter I denotes a layer of undoped, or intrinsic, semiconductor material between layers of heavily doped P and N materials. Figure 4-14 shows the structure of a PIN diode and some of the schematic symbols that represent this device.

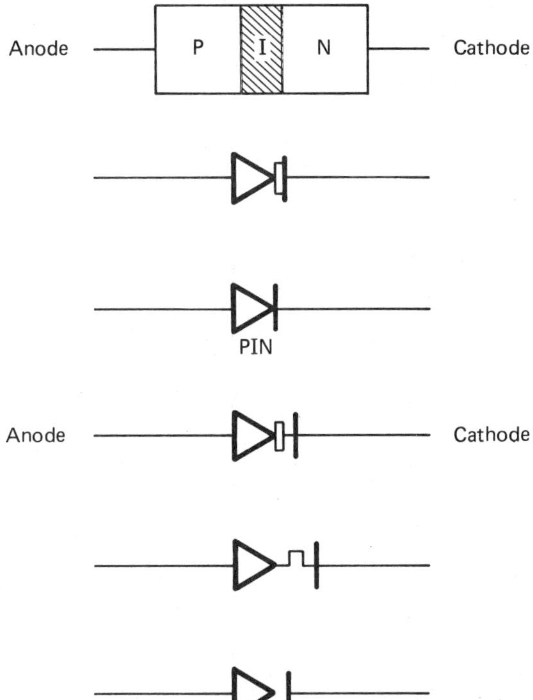

FIGURE 4-14 PIN diode crystal structure and symbols.

Probably the most important feature of a PIN diode is its ability to respond as almost a pure resistor at high radio frequencies. Resistance can be varied from 10,000 Ω to less than 1 Ω by the control of current passing through the diode. Most diodes have this characteristic to some degree. The PIN diode however, is designed to achieve a relatively wide range of resistance with good linearity, low distortion, and low current drive. The PIN diode is widely used in high-speed switching applications, high-frequency control, and microwave circuits.

At radio frequencies a forward-biased PIN diode behaves as a pure resistance. The resistance of this device is determined by the bias voltage, thickness of the intrinsic layer, and the properties of the current carriers. The resistance is inversely proportional to the forward bias current. Typically, only the on and off resistance are of any major concern in the operation of this device as a switch.

The housing or packaging of a PIN diode is similar in many respects to that of other high-frequency diodes. Figure 4-15 shows some representative high-frequency diode packages. The application of the device, its operating frequency range, power dissipation, and chip structure are some of the factors that dictate the packaging for a diode.

Gunn-Effect Diodes

Another semiconductor device that is used to control high-frequency ac is the *Gunn-effect* diode. This particular device is primarily designed to operate in the microwave region. Typical op-

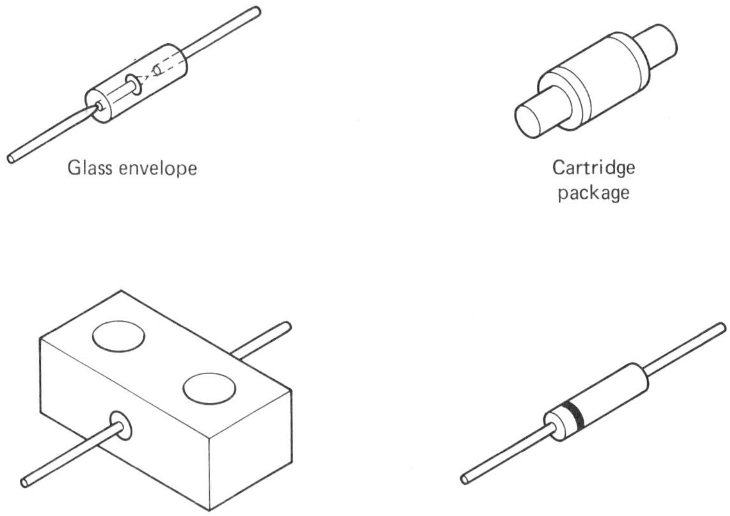

FIGURE 4-15 PIN diode package types.

erating frequencies are in the range of 5 to 100 GHz. This device is capable of controlling high-frequency ac with a minimum of parts and a low-voltage dc energizing source.

The Gunn-effect diode differs from other semiconductor devices in its construction. It, for example, does not have a distinct P-N junction. Like the Schottky diode, the Gunn-effect device has a piece of semiconductor material connected between two metal connections. This material is unevenly doped, so the crystal will break into different conduction regions with fields of different intensities across them. With the application of a specific dc voltage value, the device will go into conduction and have a negative resistance effect. This effect can be used to generate or amplify *RF* signals. Operation depends on the amount or bulk of material involved in the structure. Gallium arsenide is used to form the semiconductor material of the device.

The Gunn-effect diode must have a specific voltage polarity applied to its material in order to produce the negative resistance effect, that is, it must be properly biased to produce this characteristic. Most devices will be damaged if the polarity of the source voltage is reversed. Typical Gunn-effect devices respond to voltages of 8 to 12 V_{dc}. The operational current for a 10-mW device is 500 to 850 mA. In most microwave applications the generated signal is radiated directly from the semiconductor material. A common application of this device is in portable radar systems.

The IMPATT Diode

The *IMPATT* diode is similar in many respects to a conventional diode in crystal construction and operation. The term IMPATT stands for IMPact Avalanche and Transit Time. The term avalanche indicates the area of operation, which is the reverse-bias area near the avalanche region of conduction. A small change in reverse voltage will cause the diode to produce a negative resistance characteristic. This effect can be used to amplify or generate high-frequency ac signals. RF signal control is in the range of 2 to 10 GHz. The operational efficiency of this device is rather low. At 5 GHz a working device would have an efficiency rating of 10%. Efficiency, in this case, is the relationship of power output to power input. Although an efficiency rating of 10% seems to be exceedingly low, in practice this figure represents one of the more efficient methods of generating microwave power.

The housing or packaging of an IMPATT diode is very similar to that of other high-frequency diodes. Most of the packages used for the PIN diode could be used for the IMPATT diode. One very popular package not shown in Fig. 4-15 has the diode mounted inside a small threaded bolt structure. This type of package permits the diode to be placed in a metal cavity that responds to microwave signals and also permits heat to be conducted away from the small PN junction. Packaging is primarily dictated by the application of the device and the frequency being controlled. The IMPATT diode is very similar in operation to the Gunn-effect diode, but the IMPATT diode has a PN junction, while the Gunn-effect diode has a metal–N material structure.

SELF-EXAMINATION

24. A ____________ diode has a metal-semiconductor type of construction.
25. In a Schottky diode ____________ are the majority current carriers.
26. The absence of minority current carriers in a Schottky diode improves its ability to perform high-speed ____________ operations.
27. The high-energy electrons of a Schottky diode are generally called ____________ ____________.
28. A device that has a piece of intrinsic material separating the N and P semiconductors is called a ____________ diode.
29. An important feature of the PIN diode is that it responds as a ____________ at high RF.
30. A Gunn-effect diode has a piece of ____________ material connected between two metal terminals.
31. An IMPATT diode operates in reverse bias near the ____________ region.

ANSWERS

24. Schottky
25. Electrons
26. Switching
27. Hot carriers

28. PIN	29. Resistor
30. Semiconductor	31. Avalanche

SUMMARY

A tunnel diode is a two-element semiconductor with construction similar to that of a conventional diode. The PN materials of this device are rather heavily doped, which causes the depletion zone to be very thin. Due to the increased number of current carriers and the thin depletion zone, carriers tend to tunnel through the depletion zone with a very small amount of applied energy. This condition causes the tunnel diode to have some rather unusual characteristics. Within a certain bias voltage range a tunnel diode is conductive in both directions. In the forward direction a tunnel diode has a negative resistance region within the first 500 mV. This region is located between the forward peak voltage point and the valley voltage point. Beyond the valley voltage point, conduction in the forward direction rises very quickly, similar to that of a conventional diode.

Applications of the tunnel diode are primarily restricted to high-frequency signal control and switching. Switching is achieved by altering the forward voltage so that one stable point is to the left of V_p, and the other stable operating point is to the right of V_v. Stable operation is achieved by changing the bias voltage between these two operating points. The negative resistance region of the tunnel diode is also used to control frequency. In this way the negative resistance region can be used to overcome the effective resistance of a tank circuit. This will cause the oscillations of a tank to be continuous without a change in amplitude. The frequency effected by this operation is dependent on the *LC* values of the tank circuit.

Varactor diodes are specially manufactured devices that produce a depletion zone across a P-N junction when it is reverse biased. The reverse-bias voltage causes the depletion zone to change according to its value. The depletion zone serves as the dielectric material, and the semiconductors respond as the plates of a capacitor. The value of capacitance can be changed according to the bias voltage.

Applications of the varactor diode are primarily restricted to frequency control. In this use a varactor responds as a voltage-variable capacitor. The diode is connected across an inductor and serves as the capacitor of an *LC* tank circuit. By increasing the reverse bias voltage, the varactor causes a decrease in its capacitance. The capacitance of this device is inversely related to the reverse bias voltage. Television, FM radio, and automatic frequency control circuits use varactor diodes.

Schottky diodes are two-terminal devices that are constructed of metal and a piece of semiconductor material. The electrons of the N-material possess a rather high level of energy compared with the electrons of the metal. When the two materials are joined during the forming process, injected electrons from the N-material become hot carriers. This type of construction causes the Schottky diode to respond entirely to majority current carriers. Switching frequencies approaching 20 GHz are common with this device.

A PIN diode is constructed of an intrinsic layer of semiconductor material placed between the P- and N-materials of the junction. Diodes of this type have the ability to respond as resistors to high radio frequencies. Resistance can be varied from 10,000Ω to less than 1Ω by control of the current passing through the device. PIN diodes are widely used as high-speed switching devices in microwave control circuits.

The Gunn-effect device is rather unusual compared with other diodes because it does not have a distinct PN junction in its construction. Rather, it has a piece of semiconductor material connected between two metal terminals. When a particular dc voltage is applied to the device, it responds by producing a negative resistance. This effect can be used to amplify or generate microwave signals. Operating frequencies are in the range of 5 to 100 GHz.

IMPATT diodes are rather unusual high-frequency devices. The term IMPATT refers to impact avalanche and transit time. These diodes operate in the reverse bias region near the avalanche point of conduction. A small change in reverse voltage will cause this device to have a negative resistance characteristic. This can be used to control RF signals in the 2- to 10-GHz range. Although these diodes have a rather low efficiency rating, under normal circumstances this efficiency rating is much better than that of other high-frequency generators.

UNIT 4 EXAMINATION
SPECIAL SEMICONDUCTOR DIODES

Instructions: Circle the response that most correctly completes each statement.

1. A semiconductor device that resembles a voltage variable capacitor is the:
   - **a.** Tunnel diode
   - **b.** Varactor diode
   - **c.** Schottky diode
   - **d.** IMPATT diode
   - **e.** PIN diode
2. A semiconductor device that has high-energy electrons or hot carriers is the:
   - **a.** Tunnel diode
   - **b.** Varactor diode
   - **c.** Schottky diode
   - **d.** IMPATT diode
   - **e.** PIN diode
3. A diode that has an intrinsic layer of semiconductor material separating the materials of its P-N junction is the:
   - **a.** Tunnel diode
   - **b.** Varactor diode
   - **c.** Schottky diode
   - **d.** IMPATT diode
   - **e.** PIN diode
4. A diode that has a negative resistance region when forward biased is the:
   - **a.** Tunnel diode
   - **b.** Varactor diode
   - **c.** Schottky diode
   - **d.** IMPATT diode
   - **e.** PIN diode
5. A semiconductor device that operates in the avalanche region is the:
   - **a.** Tunnel diode
   - **b.** Varactor diode
   - **c.** Schottky diode
   - **d.** IMPATT diode
   - **e.** Gunn-effect diode

6. A semiconductor device that has heavily doped P- and N-materials, which produce a thin depletion region is the:
   a. Tunnel diode
   b. Varactor diode
   c. Schottky diode
   d. IMPATT diode
   e. PIN diode
7. When the reverse bias voltage of a varactor diode increases, the:
   a. Leakage current increases
   b. Capacitance decreases
   c. Depletion zone decreases
   d. Majority current carrier content increases
   e. Negative resistance increases
8. When the Q factor of a varactor diode drops to a value of 1, it shows that the device has reached:
   a. The cutoff frequency
   b. Avalanche breakdown
   c. Full conduction
   d. Heavy leakage current
   e. Thermal breakdown
9. A semiconductor device that operates with a forward-biased metal–semiconductor junction is a:
   a. Tunnel diode
   b. Varactor diode
   c. Schottky diode
   d. IMPATT diode
   e. PIN diode
10. A device that responds as a pure resistance to RF signals is a:
    a. Tunnel diode
    b. Varactor diode
    c. Schottky diode
    d. IMPATT diode
    e. PIN diode
11. The operational efficiency of a varactor diode is based on:
    a. Capacitance
    b. Capacitance and series resistance
    c. Capacitance and inductance
    d. Capacitance and temperature
    e. Leakage current
12. A Schottky diode has the same type of construction as a:

a. Tunnel diode
b. Varactor diode
c. IMPATT diode
d. Gunn-effect diode
e. PIN diode

UNIT FIVE

Bipolar Transistors

UNIT INTRODUCTION

Invention of the transistor in 1948 signaled the start of a whole new era in electronics. Circuits could be built exceedingly small, operate without heating power, amplify signals with a low-voltage source, and be extremely rugged. Integrated circuits, microprocessors, pocket calculators, home computers, and communication electronics have all been made possible through this development. The impact of this device on the electronics field has still not been fully realized.

Bipolar transistors are three-element devices made of semiconductor materials. These devices have two P-N junctions in their construction. Current flow is controlled by altering the voltage applied to the two junctions. Through this procedure it is possible for the device to achieve amplification, switching, or signal generation.

Bipolar transistor operation is an extremely important electronic principle. It is probably used more today than any other single electronic device. Some very sophisticated integrated circuits, for example, may employ thousands of these devices in their operation. Each device responds in the same basic manner.

In this unit we will study the operation of a bipolar transistor. You will become familiar with basic transistor principles, device test procedures, and lead identification. A person working with these devices must be well versed in these procedures. This is necessary in order to truly understand the device and its basic operation.

Study the material of this unit very carefully. Examine the objectives in order to see what you are expected to accomplish in this study.

UNIT OBJECTIVES

Upon completion of this unit, you will be able to:

1. Describe the physical construction of NPN and PNP bipolar transistors.
2. Understand the fundamental operation of a bipolar transistor.
3. Explain how the emitter, base, and collector of a bipolar transistor is biased to make the device operational.
4. Understand the relationship between emitter-, base-, and collector-current flow in the operation of a bipolar transistor.
5. Understand some of the basic terms associated with transistor characteristics such as beta, active region, cutoff, and saturation region.
6. Inspect the characteristic curves of a transistor and understand the relationship of collector current, base current, and collector-emitter voltage.
7. Inspect the characteristics of a transistor and explain how it will respond in different regions of operation.
8. Test a bipolar transistor with a VOM to determine if it is good or defective and to identify its leads.

IMPORTANT TERMS

Before reading Unit 5, review the following terms. These terms provide a basic understanding of terms associated with bipolar transistors.

Active region An area of transistor operation between cutoff and saturation.

Alloy A substance composed of two or more metals or a metal and a nonmetal substance united by being fused together and dissolving in each other when in a molten state; a transistor junction fabrication process.

Base A thin layer of semiconductor material between the emitter and collector of a bipolar transistor.

Beta A designation of transistor current gain determined by I_C/I_B.

Bipolar A type of transistor with two P-N junctions. Current carriers have two polarities, including holes and electrons.

Collector A section of a bipolar transistor that collects majority current carriers.

Current carriers Electrons or holes that support current flow in solid-state devices.

Cutoff A condition or region of transistor operation where current carriers cease to flow through the device.

Depletion region The area of a P-N junction where the number of current carriers has diminished.

Emitter A semiconductor section that is responsible for the release of majority current carriers.

Epitaxial A transistor fabrication procedure in which atoms are formed on a surface so that they are an extension of the original crystal structure.

Gain A ratio of voltage or current output to voltage or current input.

Saturation region A condition of operation in which a device is conducting to its full capacity.

BIPOLAR TRANSISTORS

Transistor types are usually identified by letters. NPN and PNP are two types of bipolar transistors. The letters are used to denote the polarity of the semiconductor material used in its construction. Material polarity is determined by the type of impurity added to silicon or germanium when it is manufactured.

A PNP transistor has a thin layer of N-material placed between two pieces of P-material. The schematic symbol of the PNP transistor and its crystal structure are shown in Fig. 5-1. Leads attached to each piece of material are identified as the emitter, base, and collector. In the symbol, when the arrowhead of the emitter lead "Points iN" toward the base, it indicates that the device

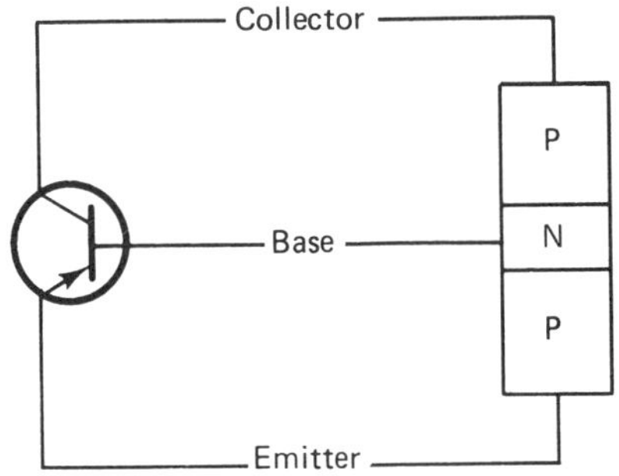

FIGURE 5-1 PNP transistor symbol and crystal structure.

is of the PNP type. Lead identification is extremely important when working with transistors.

An NPN transistor has a thin layer of P-material placed between two pieces of N-material. The schematic symbol, crystal structure, and lead designations of the transistor are shown in Fig. 5-2. The schematic symbol of this device is the same as the PNP except for the direction of the arrowhead. An NPN transistor has the arrow "not pointing" toward the base. The polarity of the source voltage applied to each element is determined by the material used in its construction.

PNP and NPN transistors are bipolar devices. The term *bipolar* refers to conduction by both holes and electrons. Electrons are the majority current carriers of the N-material, whereas holes are the majority current carriers in the P-material. Current conduction in the transistor is based on the movement of both current carriers.

TRANSISTOR CONSTRUCTION

Since the invention of the first transistor in 1948, a variety of construction processes have been used to manufacture this device. The first transistors manufactured were of the point-contact type. Semiconductor materials were connected together by pointed wires that were fused to the material. Today transistors constructed by this method are primarily used in high-frequency applications. Other modern manufacturing methods are alloy junction and diffusion.

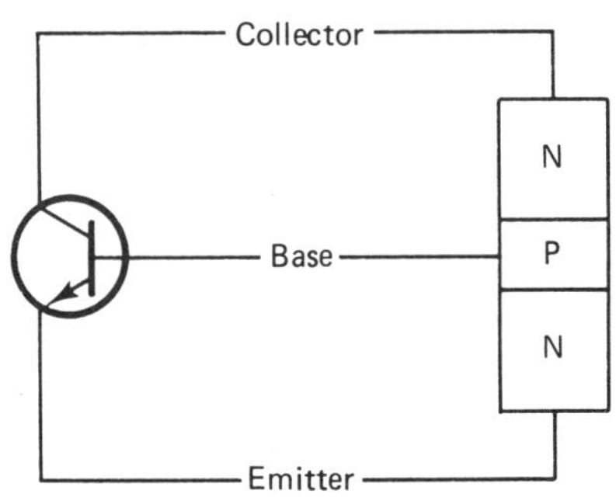

FIGURE 5-2 NPN transistor symbol and crystal structure.

The following discussion of each technique is very brief. Our primary concern is with the fundamental steps of fabrication to familiarize you with the basic fabrication terms.

Alloy-Junction Transistors

The alloy-junction method of bipolar transistor construction is an outgrowth of a procedure that is used to form diodes. In transistor construction this procedure is achieved by attaching two small pieces of metal on opposite sides of a thin piece of semiconductor material. The metal pieces serve as an impurity or dopant for the semiconductor. Figure 5-3 shows a structure with this formation. The entire structure is then heated until the impurity melts into the semiconductor material. The melting process causes an alloy of a different type to be formed on each side of the semiconductor. In this case, a PNP type of device is formed by the alloying procedure.

The three semiconductor materials formed by the alloying process represent the emitter, base, and collector of a bipolar transistor. Leads are attached to each material area. The entire structure is then placed in a housing with the three leads serving as external connection points to the respective material areas.

Notice in Fig. 5-3 that the collector-base area of the transistor is made larger than the emitter-base area. This is purposely done to permit the collector to have greater power dissipation. The alloy-junction technique of transistor fabrication is used rather extensively today in the manufacturing of high-power transistors.

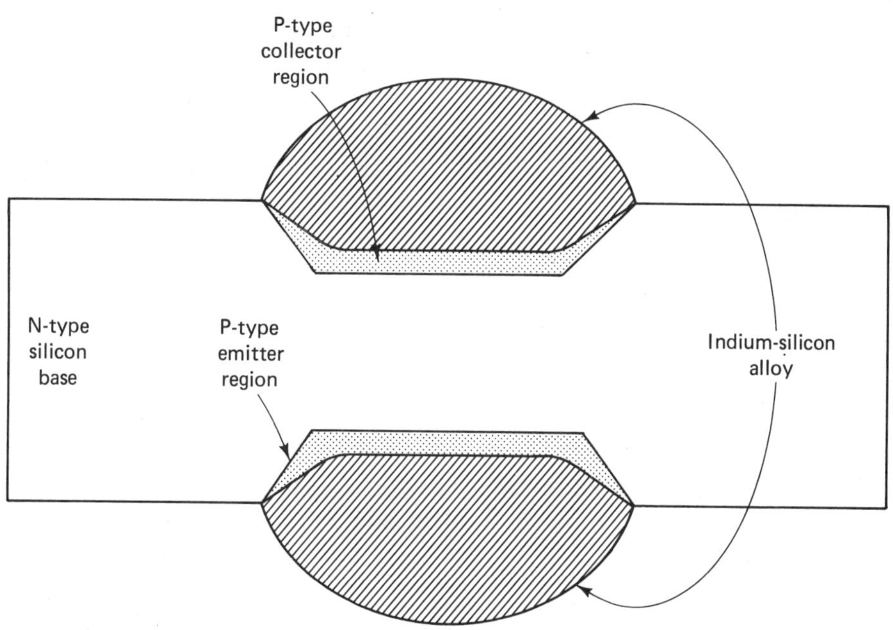

FIGURE 5-3 PNP alloy-junction transistor formation.

Diffusion Transistors

Bipolar transistors are largely manufactured today by the diffusion technique. This process involves the movement of N- or P-type impurity atoms into a piece of silicon. To achieve this movement it is necessary to heat a piece of silicon to 1250°C in the presence of a controlled impurity vapor. The movement process is very slow. Typical formation rates are 0.00001 in. per hour.

The diffusion process is carried out in two steps. The first step consists of heating a piece of silicon in an impurity dopant vapor to form a high concentration of the dopant on its surface. This is called the *deposition* function. The silicon piece then has its temperature elevated. This causes the dopant atoms to be absorbed or diffuse into the silicon. This is called the *diffusion* step. Manufacturing involves repeated deposition and diffusion steps.

Figure 5-4 shows transistor formation by the diffusion process. The first step involves an N-type semiconductor that has a P-type vapor deposited on its surface. In the next step, increased heat causes the P-material to diffuse into the N-material. This forms an N-type of collector and a P-type base. The next step is a repeat of the initial process. In it, however, the dopant vapor is a pentavalent material. The deposition area is on top of the base. This causes an area of N-type material to be formed on the base. Diffusion causes this material to move into the base. The structure now has an N-type emitter. The completed transistor is an NPN device.

A refined version of the diffusion process is called *epitaxial growth.* In this fabrication technique, N- or P-materials are grown or formed on the surface of another material. The general shape of specific transistor elements can be aletered through this procedure. The term epitaxial is derived from the Greek words "epi," meaning upon, and "taxi," meaning arrange. This describes the process in which atoms are formed on a surface so that they are an extension of the original crystal structure. When the epitaxial process is used, and the completed structure has a flat top or level plane surface, it is called a *planar transistor.* When the completed structure rises above the primary surface, forming a plateau, it is called a *mesa transistor.* Mesa construction is used to expose the emitter and base regions so that electrical connections can easily be made.

Transistor Packaging

Once a transistor is constructed, the entire crystal assembly must be placed in an enclosure or package. The unit is then sealed to protect it from dust, moisture, or outside contaminants. Electrical connections are made to each element through leads attached to the package. When metal packages are used, the housing also serves as a heat sink, which carries heat away from the crystal. A common practice with this kind of unit is to attach the collector

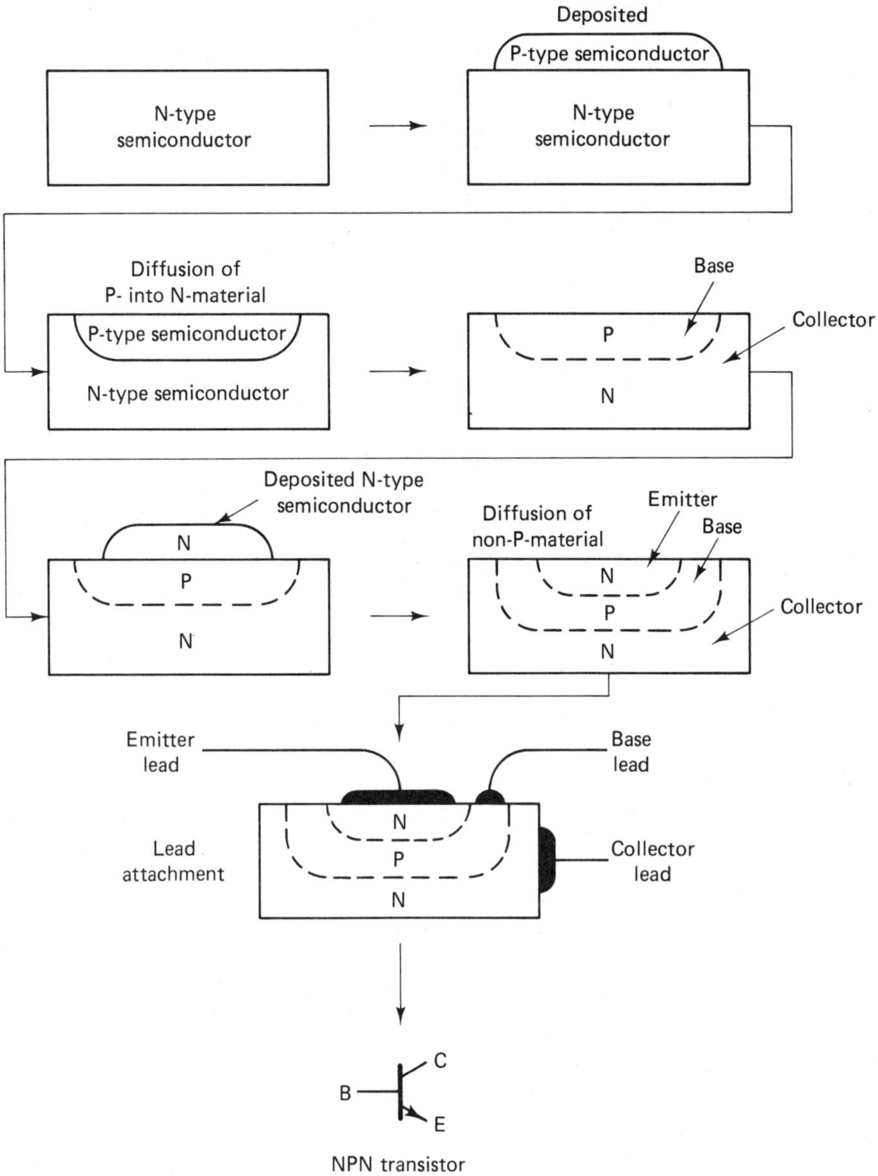

FIGURE 5-4 Transistor formed by the diffusion process.

to the metal case. This helps the collector to dissipate heat. Packages of this type should not be permitted to touch other circuit components when they are in operation.

Ceramic- and epoxy-packaged transistors are also available. As a rule, this type of packaging is somewhat cheaper to produce. These devices are usually not as rugged as the metal-packaged devices. Most small-signal transistors are now being housed in epoxy packages. The outside case of this device is insulated from all transistor elements. It is specifically designed for printed-circuit board construction and close placement with other circuit components.

Several of the more popular transistor package styles are shown in Fig. 5-5. Part (a) shows devices designed for small-signal

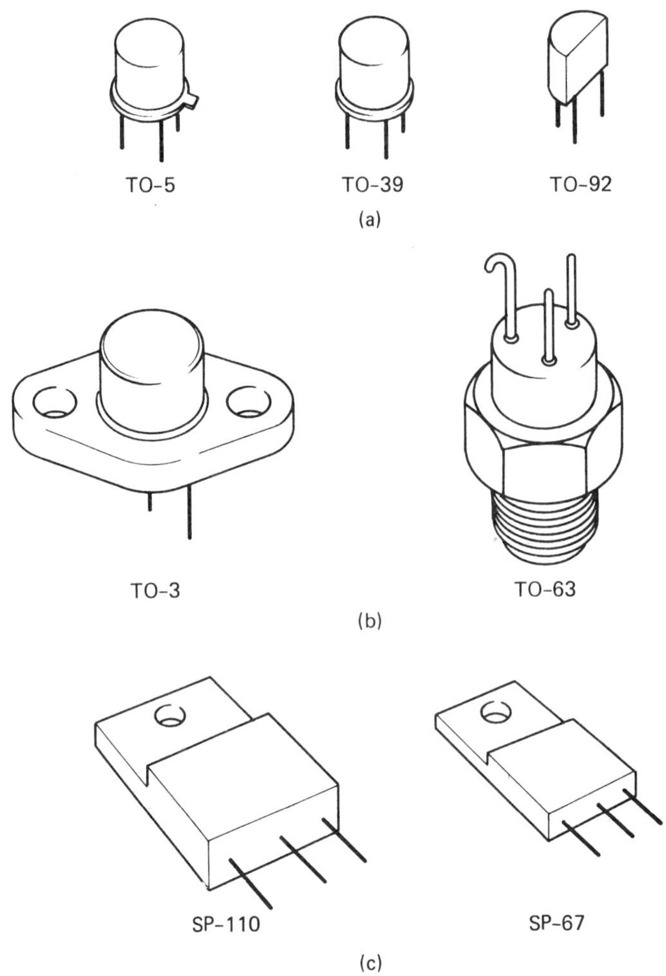

FIGURE 5-5 Typical transistor packages.

applications. The letter designation "TO" stands for transistor outline. These same packages are also used to house other solid-state devices. Part (b) shows some of the packages used to house large-signal or power transistors. Heavy metal cases of this type are used to dissipate heat generated by the transistor. The collector is usually attached to the metal case. Part (c) shows a few of the common epoxy packages. The letter designation "SP" stands for semiconductor package. This type of package is used to house a number of other semiconductor devices.

With the wide range of package styles available today, it is difficult to have a standard method of identifying transistor leads. Each package generally has a unique lead location arrangement. It is usually a good idea to consult the manufacturer's specifications for proper lead identification.

SELF-EXAMINATION

1. The center piece of semiconductor material of a transistor serves as the ____________ electrode or terminal.

2. The term ____________ refers to the conduction of holes and electrons in the operation of a transistor.
3. The arrowhead of a transistor symbol refers to the ____________ electrode.
4. When the arrowhead of a transistor symbol points toward the base it shows that the device is (NPN, PNP).
5. ____________ ____________ transistors are formed by melting two small metal pieces into opposite sides of a thin piece of semiconductor material.
6. ____________ takes place when a thin layer of dopant material is formed on the surface of silicon.
7. ____________ occurs when a concentration of dopant material is absorbed into a piece of silicon.
8. A fabrication procedure that causes a flat surface structure to be formed produces a ____________ transistor.
9. A fabrication procedure that causes a plateau area to rise on the surface of a structure produces a ________ transistor.
10. A bipolar transistor has ____________ junctions.

ANSWERS

1. Base	2. Bipolar
3. Emitter	4. PNP
5. Alloy-junction	6. Deposition
7. Diffusion	8. Planar
9. Mesa	10. Two

BASIC TRANSISTOR OPERATION

In order for a transistor to function properly, it must have electrical energy applied to all its electrodes. In semiconductor circuits the source voltage is called *bias voltage* or simply *bias*. Bipolar transistors must have both junctions biased in order to function. In an earlier unit we discussed diode biasing. Forward biasing of a P-N junction takes place when the polarity of the source voltage is positive to the P-material and negative, to the N-material. This condition causes a current flow. The depletion zone of the device is reduced, and majority current carriers are driven toward the junction. One junction of a transistor must be forward biased when it operates.

Reverse biasing of a P-N junction is achieved when the polarity of the source is negative to the P-material and positive to the N-material. This condition does not ordinarily permit current flow. The width of the depletion zone is increased by this action, and minority current carriers move toward the junction. In normal circuit operation reverse bias current is extremely small. One junction of a transistor must be reverse biased in order for the device to be operational.

NPN Transistor Biasing

Consider now the biasing of the NPN transistor of Fig. 5-6. In this diagram an external voltage source has been applied only to the emitter-base regions of the transistor. In an actual circuit this source is called the emitter-base voltage or V_{BE}. Note that the polarity of V_{BE} causes the emitter-base junction to be forward biased. This condition causes majority current carriers to be forced together at the emitter-base junction. The resulting current flow in this case would be quite large. In our example, note that the same amount of current flows into the emitter and out of the base. Emitter current (I_E) and base current (I_B) are used to denote these values. Under normal circuit conditions a V_{BE} source would not be used independently.

Figure 5-7 shows an external voltage source connected across the base-collector junction of an NPN transistor. The base-collector voltage, or V_{CB}, is used to reverse bias the base-collector junction. In this case there is no indication of base current I_B or collector current I_C. In an actual reverse-biased junction there could be a very minute amount of current. This would be supported primarily by minority current carriers. As a general rule, this is called *leakage current.* In a silicon transistor leakage current is usually considered to be negligible. In an actual circuit V_{CB} is not normally applied to a transistor without the V_{BE} voltage source.

For a transistor to function properly the emitter-base junction must be forward biased and the base-collector junction reverse biased. Both junctions must have bias voltage applied at the same time. In some circuits this voltage may be achieved by separate V_{BE} and V_{CB} sources. Other circuits may utilize a single bat-

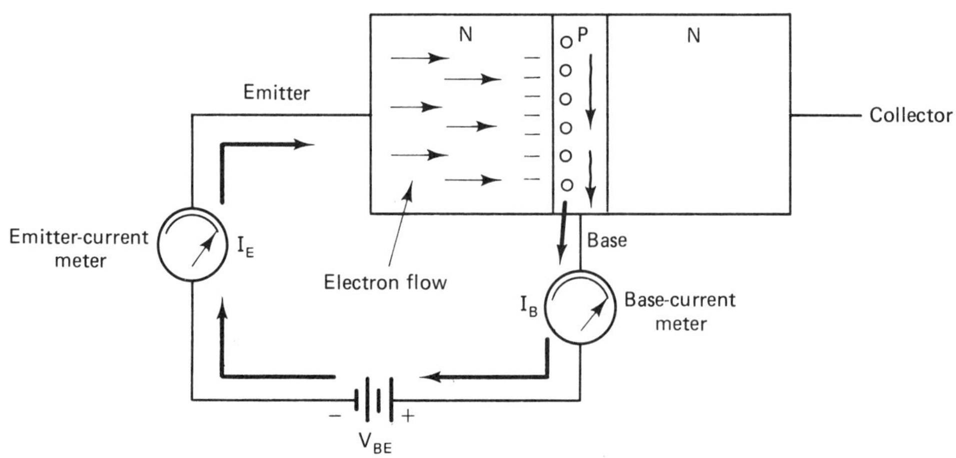

FIGURE 5-6 Emitter-base biasing.

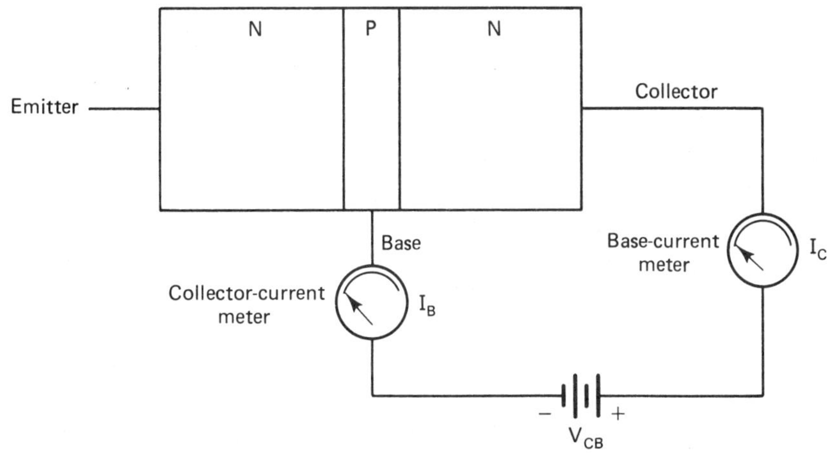

FIGURE 5-7 Base-collector biasing.

tery with specially connected bias resistors. In either case, the transistor responds quite differently when all its terminals are biased.

Consider now the action of a properly biased NPN transistor. Figure 5-8 shows separate V_{BE} and V_{CB} sources connected to the transistor. V_{BE} provides forward bias for the emitter-base junction, whereas V_{CB} reverse biases the collector junction. Connected in this manner, the two junctions do not respond as independent diodes.

Figure 5-9 shows how current carriers pass through a properly biased NPN transistor. The emitter-base junction being forward biased causes a large amount of I_E to move into the emitter-base junction. Upon arriving at the junction, a large number of the electrons do not effectively combine with holes in the base. The base is usually made very thin (0.0025 cm, or 0.001 in.) and it is

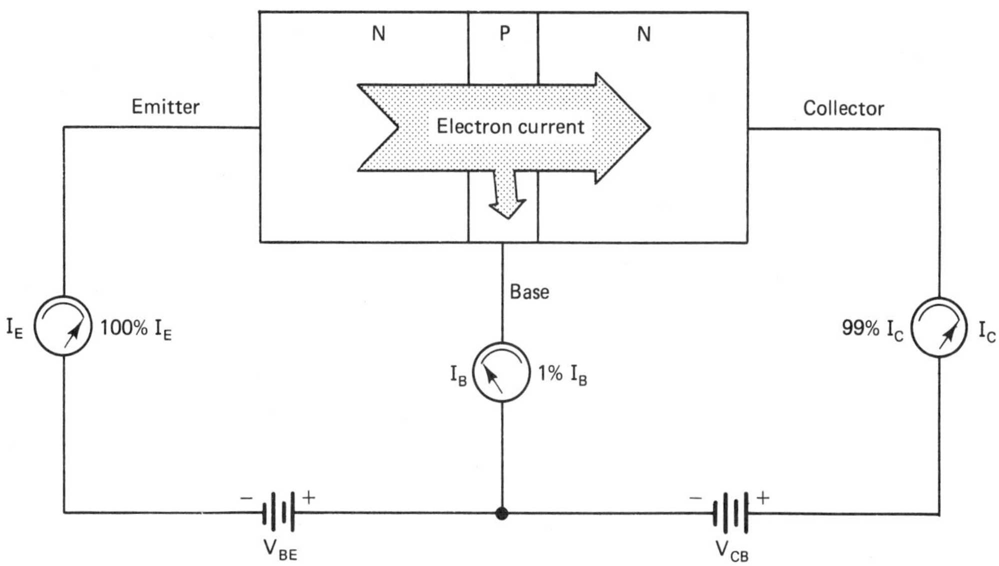

FIGURE 5-8 NPN transistor biasing.

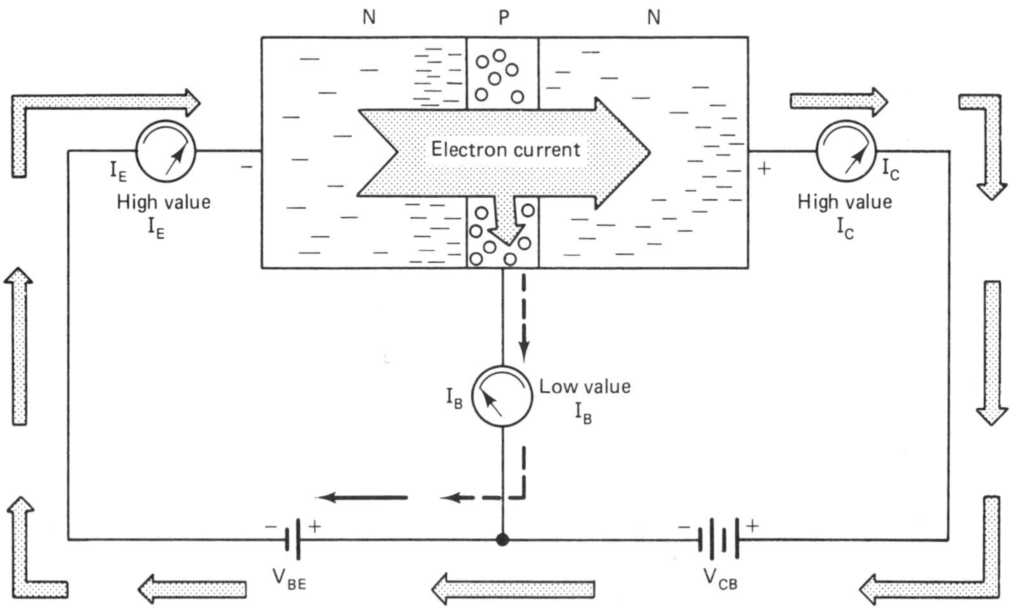

FIGURE 5-9 Current carriers passing through an NPN transistor.

lightly doped. This means that the majority current carriers of the emitter exceed the majority carriers of the base. Most of the electrons that cross the junction do not combine with holes. They are, however, immediately influenced by the positive V_{CB} voltage applied to the collector. A very high percentage of the original emitter current enters the collector. Typically, 95 to 99% of I_E flows into the collector junction and becomes collector current. After passing through the collector region, I_C returns to V_{CB}, V_{BE}, and ultimately to I_E. The current flow inside the transistor is indicated by a large arrow. Outside the transistor current flow is indicated by small arrows.

The difference between the amount of I_E and I_C of Fig. 5-9 is base current. Essentially, base current is due to the combining of a small number of electrons and holes in the emitter-base junction. In a typical circuit, I_B is approximately 1 to 5% of I_E. With the base region being very narrow, it cannot support a large number of current carriers. A small amount of base current is needed, however, to make the transistor operational. Note the direction of I_B and its flow path in the diagram.

The relationship of I_E, I_B, and I_C in a transistor can be expressed by the equation

$$\text{emitter current} = \text{base current} + \text{collector current}$$

$$I_E = I_B + I_C$$

The largest current flow in a transistor takes place in the emitter. I_C is slightly less than I_E. This means that the current flow through a reverse-biased collector-base junction is nearly equal to that of the forward-biased emitter-base junction. The difference in I_E and I_C is I_B.

If the base of a transistor were not made extremely thin and low doped, it would not respond as just described. The thin base region makes it possible for large amounts of I_E to pass through the base and into the collector region. A thicker base would cause more I_E to combine with holes in the base. Low doping of the base means that the majority current carriers of the emitter cannot effectively combine in the base. Instead of combining in the base, they pass through it and enter the collector. This improves the effectiveness of the base as a control element.

The amount of base current that flows in a transistor is very small but extremely important. Suppose, for example, that the base lead of the transistor in Fig. 5-9 is momentarily disconnected. With this element open it should be obvious that there will be no base current. Closer examination also shows that V_{BE} and V_{CB} are now connected in series. This means that their voltages are added together. As a result of this action, the collector becomes more positive and the emitter more negative. One would immediately think that this condition would cause the transistor to conduct very heavily. It does not, however, permit any conduction at all. In effect, this means that base current has a direct influence on emitter and collector current. The base-emitter junction of a transistor must be forward biased in order for it to produce collector current.

Transistor Beta

A small change in transistor base current is capable of producing a very large change in collector current. This, in effect, shows that a transistor is capable of achieving current gain. The term *beta* is commonly used to express current gain. Beta can be determined by dividing the collector current by the base current. This relationship is expressed by the equation beta equals collector current divided by base current or

$$\beta = \frac{I_C}{I_B}$$

For most transistors beta is usually quite large. In a previous circuit, we described I_C as being 95 to 99% of I_E, with I_B being from 1 to 5%. For this example, the beta of a transistor with this ratio would be

$$\beta = \frac{95\%}{5\%} = 19 \quad \text{or} \quad \beta = \frac{99\%}{1\%} = 99$$

In both examples note that beta is expressed as a pure number. The percentage signs in the numerator and denominator of the equation cancel each other. If actual current values in milliamperes or amperes were used, they would also cancel. Beta is simply expressed as a whole-number value.

Transistors are manufactured today with a very wide range of beta capabilities. Power transistors, which respond to large current values, usually have a rather low beta. Typical values are in the range of 20. Small-signal transistors may be capable of beta in the range of 400. A good average beta for all transistors is in the range of 100. Beta is a very important transistor characteristic. It is not very meaningful unless it is applied to a specific transistor and its circuit components.

PNP Transistor Biasing

Biasing of a PNP transistor is very similar to that of the NPN transistor. The emitter-base junction must, for example, still be forward biased and the collector-base reverse biased. The polarity of the bias voltage is, however, reversed for each transistor. The majority current carriers of a PNP transistor are holes instead of electrons as in the NPN device. Except for these two differences, operation is primarily the same.

Figure 5-10 shows how current carriers pass through a properly biased PNP transistor. The emitter-base junction being forward biased causes a large number of holes to move through the junction. Upon arriving in the base region, a very small number of these holes combine with electrons. This is representative of the base current Ninety-five to 99% of the holes move through the base region and enter the collector junction. These holes do not find electrons to combine with in the thin base region. They are, however, immediately influenced by the negative V_{CB} voltage of the collector. This action causes hole current to flow through the reverse-biased base-collector junction. The end result of this is I_C passing through the collector region. It then leaves the collector and flows into V_{CB}, V_{BE}, and returns to the emitter. Hole current

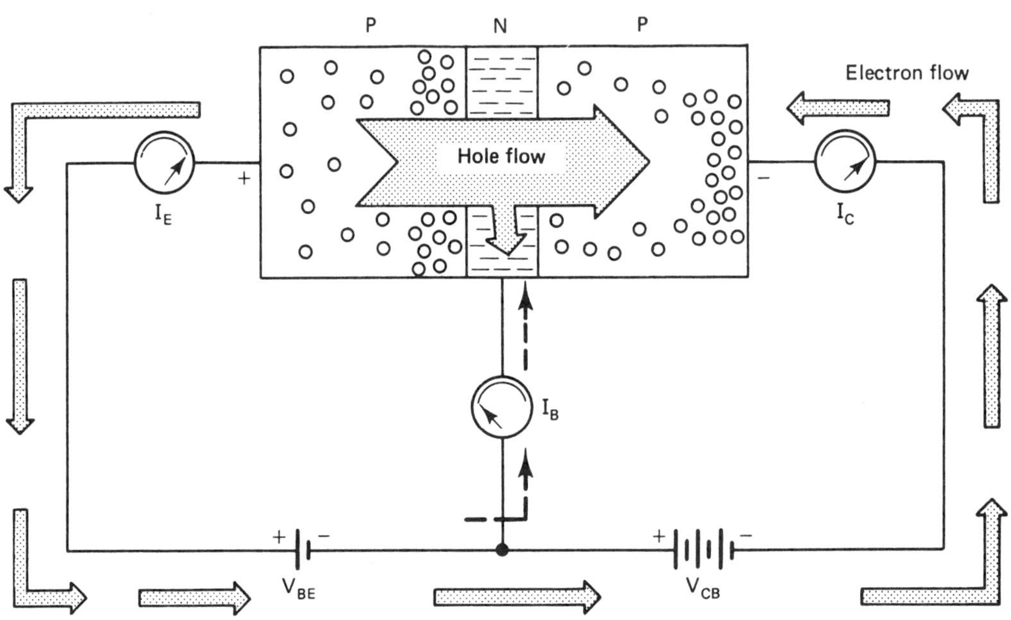

FIGURE 5-10 Current carriers of a PNP transistor.

flow is indicated by the large arrow inside the transistor. Outside the transistor, current flow is achieved by electrons. Small arrows in the diagram show current outside the transistor.

It is rather interesting to compare the transistors of Figs. 5-9 and 5-10. Specific attention should be directed to the polarity of V_{BE} and V_{CB}. This shows the primary difference in device biasing. Note also the type of majority current flow through each transistor. This is represented by the large arrow of each diagram. It is different for each device. This means that the two transistor types are not directly interchangeable. Substituting one for the other would necessitate a complete reversal of bias voltage.

PNP transistors are not as widely used today as the NPN type. Electron current flow of the NPN device has much better mobility than the hole flow of a PNP type. This means that electrons have a tendency to move more quickly through the crystal material than do holes. Because of this characteristic, NPN transistors tend to respond better at high frequencies. In general, this means that the NPN device has a wider range of applications. Manufacturers usually have a larger selection of these transistors in their lines. With a better selection of transistors available, circuit designers find it easier to select devices of a desired characteristic. As a result, the NPN device is much more popular than its PNP counterpart.

SELF-EXAMINATION

11. Transistor operation depends on the emitter-base region being ____________ biased and the base-collector region being ____________ biased.
12. When the emitter-base junction of a bipolar transistor is biased as in Fig. 5-6, the resulting I_E and I_B are ____________.
13. When the emitter-base junction of a bipolar transistor is forward biased, there is a ____________ value of I_E and I_B.
14. A reverse-biased NP or PN junction normally causes (zero, medium, maximum) current flow.
15. When the base-collector junction of a bipolar transistor is biased as in Fig. 5-7, the resulting current flow is ____________.
16. When the emitter-base junction of a transistor is forward biased, and the base-collector junction is reverse biased, the values of ____________ ____________ and ____________ ____________ are nearly the same value.
17. (I_E, I_B, I_C) is the largest value of current that flows in a bipolar transistor.
18. The difference between I_E and I_C is ____________ ____________.

19. ____________ is I_C/I_B.
20. The ____________ current of a bipolar transistor is usually only 1 to 5% of the emitter current.

ANSWERS

11. Forward, reverse	12. Equal
13. Large	14. Zero
15. Zero	16. Emitter current or I_E, collector current of I_C
17. I_E	18. Base current or I_B
19. Beta	20. Base

TRANSISTOR CHARACTERISTICS

When a transistor is manufactured it has characteristic data that distinguish it from other devices. These data are extremely important in predicting how a transistor will perform in a circuit. One very common way of showing this information is with the use of graphs. A collector family of characteristic curves is widely used to show transistor data.

A collector family of characteristic curves for an NPN transistor is shown in Fig. 5-11. The vertical part of this graph shows different collector current values in milliamperes. The horizontal part shows the collector-emitter voltage. Individual lines of the graph represent different values of base current. The zero-base-current line is normally omitted from the graph. This condition shows when the transistor is not conducting.

A family of collector curves tells a great deal about the operation of a transistor. Take for example, point A on the 60-μA base-current line. If the transistor has 60 μA of I_B and a V_{CE} of 6 V, there is a collector current of 3 mA. This is determined by projecting a line to the left of the intersection of 60 μA and 6 V. Any combination of I_C, I_B, and V_{CE} can be quickly determined from the display of these curves.

If the values of I_B and I_C can be determined from a family of curves, they can be used to predict other values. The emitter current (I_E) of a transistor can be determined from the values of I_B and I_C. Remember that $I_E = I_B + I_C$. For point A, the I_E is 60 μA + 3 mA, or 3.06 mA.

The current gain or beta of a transistor can also be determined from a family of collector curves. Since beta is I_C/I_B, it is possible to determine this condition of operation. For point A on the curves, I_C is 3 mA, and I_B is 60 μA. The current gain of a transistor operating at this point would be determined by the equation

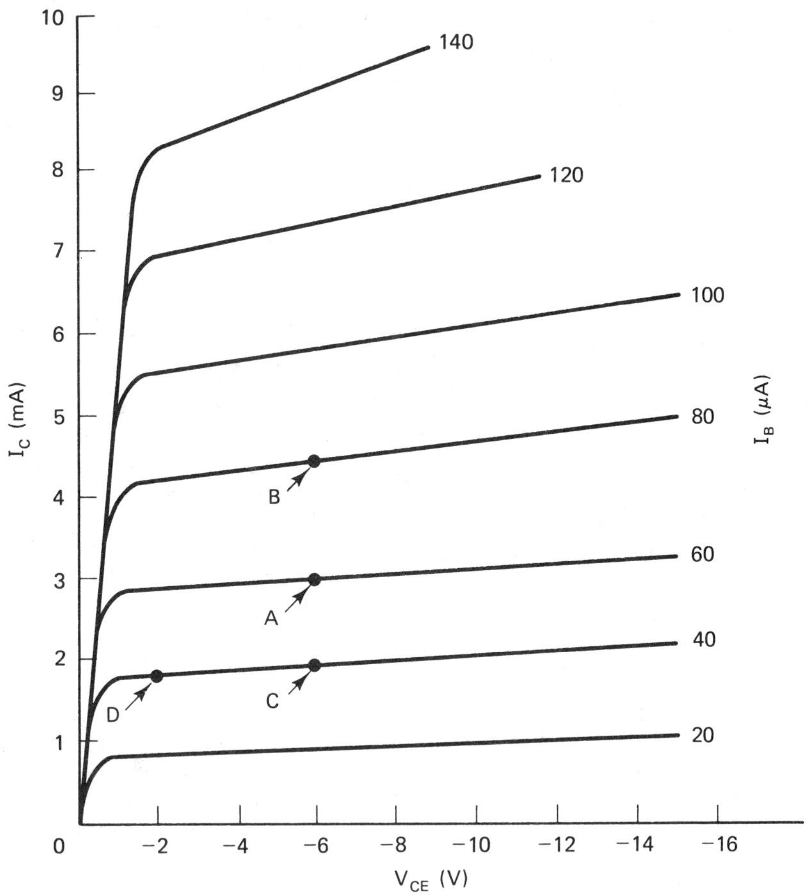

FIGURE 5-11 Collector family of characteristic curves for an NPN transistor.

$$\text{Beta} = \frac{I_C}{I_B} = \frac{3\ \text{mA}}{60\ \mu\text{A}} = \frac{0.003}{0.00006} = 50$$

This condition is called the direct-current beta. It shows how the transistor will respond when dc voltage is applied. Dc beta is an important consideration when a transistor is operated as a dc amplifier.

In many applications a transistor must amplify ac signals. Dc voltage values are applied to the device to make it operational. An ac signal voltage applied to the base would then cause a change in the value of I_B. In Fig. 5-11 assume that the input voltage causes I_B to change between points B and C. Point A is considered to be the operating point. The change in I_B is from 80 μA to 40 μA. This is normally called ΔI_B. The Greek letter delta (Δ) is used to denote a changing value in I_B. A corresponding change in I_C is determined by projecting a line to the left of each I_B value. ΔI_C is therefore 4.5 mA to 2 mA.

The ac beta of a transistor is determined by dividing ΔI_C by ΔI_B. For points B and C this is

$$\text{Ac beta} = \frac{\Delta I_C}{\Delta I_B} = \frac{4.5 \text{ mA} - 2 \text{ mA}}{80\ \mu\text{A} - 40\ \mu\text{A}} = \frac{2.5 \text{ mA}}{40\ \mu\text{A}}$$

$$= \frac{0.0025}{0.00004} = 62.5$$

The ac beta in this case is somewhat different from that of the dc beta. As a general rule, ac beta is larger than dc beta. To amplify an ac voltage, a transistor must operate over a range of different values. This usually accounts for the difference in beta values.

The same family of collector curves can also be used to show the effectiveness of a transistor in controlling collector current. Note points D and C on the 40-μA curve. A change in V_{CE} from 2 to 6 V occurs between these two points. Projecting points D and C to the left indicates a change in I_C from 1.8 to 0.2 mA. A 4-V change in V_{CE}, therefore, causes only a 0.2-mA change in I_C. Using the ac beta for comparison, a 40-μA change in I_B causes a ΔI_C of 2.5 mA. This indicates that I_C is more effectively controlled by I_B changes than by V_{CE} changes.

Operating Regions

A collector family of characteristic curves is also used to show the desirable operating regions of a transistor. The center area is called the *active region.* This area is located anywhere between the two shaded areas of Fig. 5-12. Amplification is achieved when the transistor operates in this region.

A second possible region of operation is called *cutoff.* On the family of curves this is where $I_C = 0$ mA. When a transistor is cut off no I_B or I_C flows. Any current flow that occurs in this region is due to leakage current. As a general rule, a transistor operating in the cutoff region responds as a circuit with an open switch. When a transistor is cut off, there is infinite resistance between the emitter and collector.

The shaded area on the left side of the family of curves is called the *saturation region.* This area of operation is where maximum I_C flows. A transistor operating in this region responds as a closed switch. A transistor is considered to be fully conductive in the saturation region. The resistance between emitter and collector is extremely small when a transistor is saturated.

Figure 5-13 shows how a transistor responds in the three regions of operation. In part (a) the transistor is operating in the active area. The value of V_{CE} is somewhat less than that of the source voltage V_{CC}. Collector current through the transistor causes a voltage drop across R_C. The base current of a transistor operating in this region would be some moderate value between zero and saturation. Transistor amplifiers usually operate in this area.

Figure 5-13(b) shows how a transistor responds when it is in the cutoff region. With no current flow through the transistor, there is no voltage drop across R_C. V_{CE} of this circuit is therefore

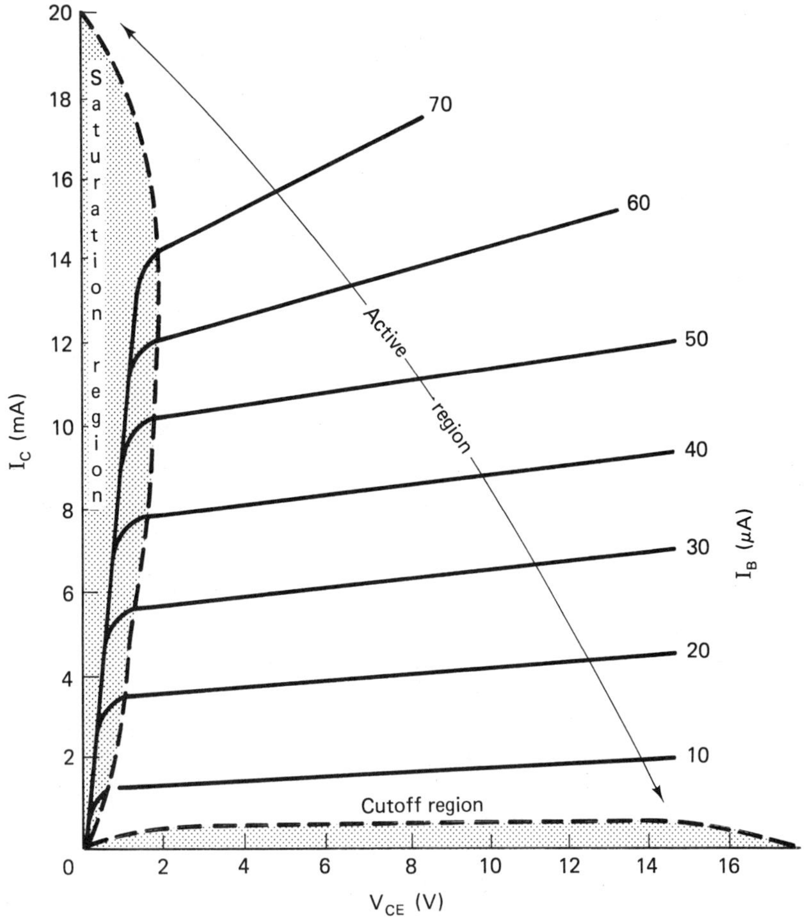

FIGURE 5-12 Operating regions of a transistor.

the same as the source voltage. A transistor is cut off when no I_C occurs.

Figure 5-13(c) shows how a transistor operates in the saturation region. In this condition of operation the transistor responds as a closed switch. V_{CE} is approximately equal to 0 V. Heavy I_C values cause nearly all V_{CC} to appear across R_C. In a sense, V_{CE} is considered to be at ground. Current passing through the transistor is limited by the value of V_{CC} and R_C. A large value of I_B is needed to cause a transistor to saturate.

Characteristic Curves

Special circuits are used to obtain the data for a family of characteristic curves. Figure 5-14 shows a test circuit that can be used to find the data points of a family of collector curves. Three meters are used to monitor this information. Base current and collector current are observed by meters connected in series with the base and collector. Collector-emitter voltage is measured with a voltmeter connected across these two terminals. V_{CE} is adjusted to different values by a variable power supply.

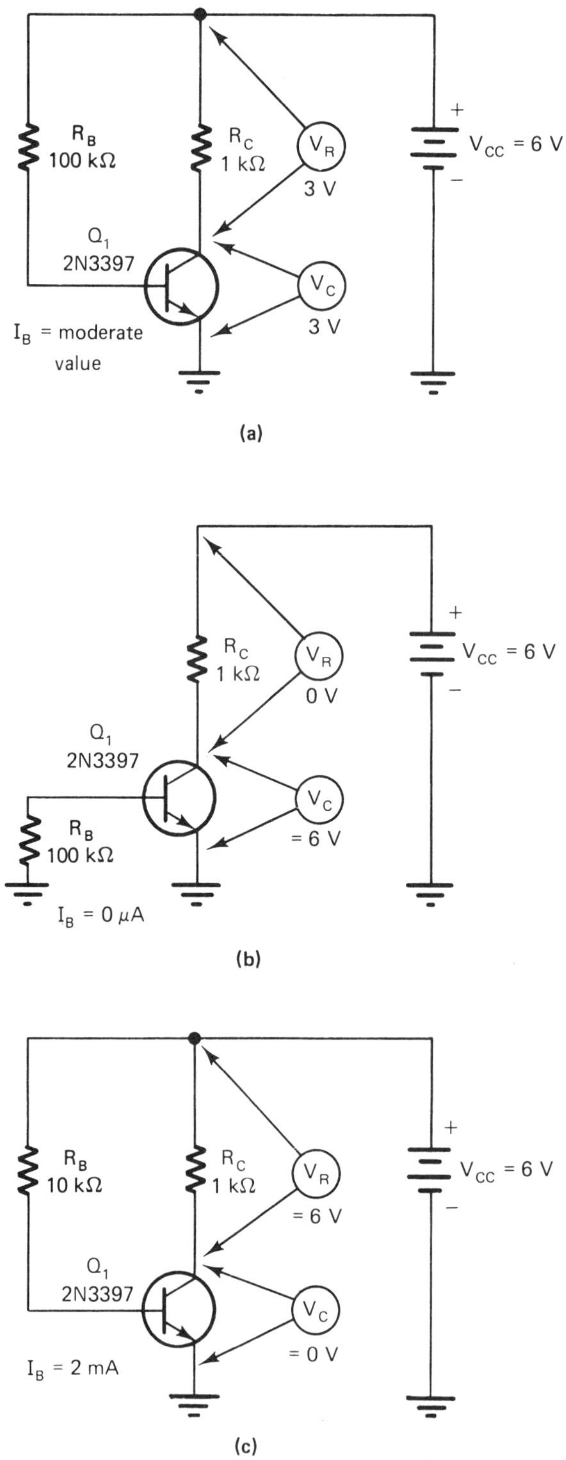

FIGURE 5-13 Transistor operation by regions: (a) active; (b) cutoff region; (c) saturation region.

The data points for a single curve are developed by first adjusting I_B to a constant value. A representative value for a small-signal transistor is 10 μA. Resistor R_B is used to adjust the value of I_B. The V_{CE} is then adjusted through its range, starting at 0 V. As V_{CE} is increased in 0.1-V steps up to 1 V, the I_C increases very

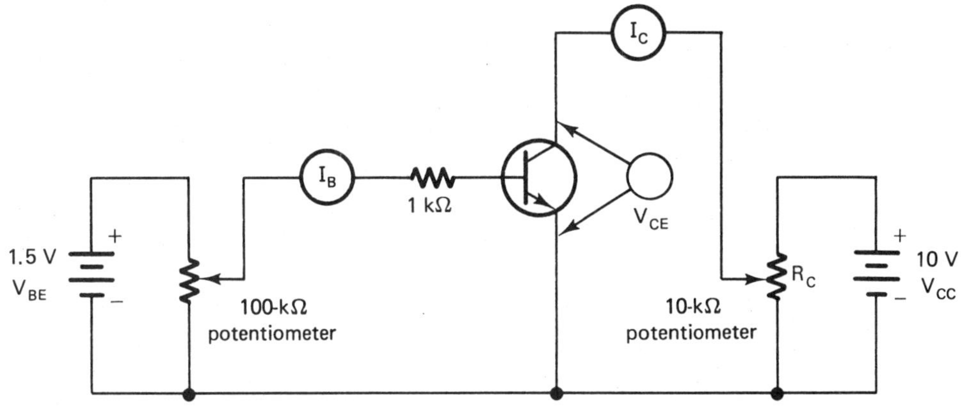

FIGURE 5-14 Transistor characteristic curve circuit.

quickly. Corresponding I_C and V_{CE} values are plotted on the graph. V_{CE} is then increased in 1-V steps while I_C is monitored. The resulting collector current usually levels off to a fairly constant value. The corresponding I_C and V_{CE} values are then added to the first data points. All points are then connected by a continuous line. The completed curve would be labeled 10 μA of I_B.

To develop the second curve, V_{CE} is first returned to zero. I_B is then increased to a new value. A common value for a small-signal transistor would be 20 μA. The second curve is then plotted by recording the I_C values for each V_{CE} value.

To obtain a complete family of curves, the process would be repeated for several other I_B values. A representative family of curves may have 8 to 10 different I_B values. The step values of V_{CE} and the increments of I_B and I_C will change with different transistors. Small-signal transistors may use 10-μA I_B steps with I_C values ranging from 0 to 20 mA. Large-signal transistors may use 1-mA I_B steps with I_C values going to 100 mA or more. Each transistor may require different values of I_B, I_C, and V_{CE} to obtain a suitable curve.

SELF-EXAMINATION

21. A collector family of characteristic curves for a bipolar transistor has 8 to 10 individual lines that represent different values of ____________ current.
22. Current gain or ____________ of a bipolar transistor is I_C/I_B.
23. An I_C of 10 mA and an I_B of 10 μA would produce a beta of ____________.
24. An I_C that changes from 10 mA to 5 mA for an I_B of 100 to 50 μA would cause an ac beta of ____________.
25. The center area of a family of bipolar characteristic curves is called the ____________ region.

26. The area of a family of bipolar characteristic curves that shows where maximum I_C flows represents the ____________ region.
27. The area of a family of bipolar characteristic curves that shows where no I_C occurs is called the ____________ region.
28. When a transistor operates in the active region with a moderate value of I_B, the V_{CE} will be ____________ than the source voltage V_{CC}.
29. When a transistor operates in the cutoff region, the V_{CE} will be equal to the ____________ voltage.
30. When a transistor operates in the saturation region, the V_{CE} will be approximately ____________.

ANSWERS

21. Base	22. Beta
23. 100	24. 100
25. Active	26. Saturation
27. Cutoff	28. Less
29. Source voltage	30. Zero

BIPOLAR TRANSISTOR TESTING

Transistor testing is a rather common procedure that must be performed periodically when working with semiconductor devices. Lead identification, gain, open leads, shorted conditions, and leakage are some of the tests commonly performed on a transistor. Curve tracers, gain testers, and sound-producing instruments can effectively be used for these tests. These instruments, as a general rule, are not always available. An ohmmeter, however, can be used to perform many of these tests. An ohmmeter is one of the functions of a volt-ohm-milliammeter (VOM) or digital volt-ohm-meter (DVOM).

Transistor Junction Testing

A good transistor has two P-N junctions in its construction. Both of these junctions can be tested with an ohmmeter. Figure 5-15 shows the junctions of PNP and NPN transistors. Each junction responds as a diode when it is tested. Forward biasing occurs when the material polarity of the junction matches the voltage polarity of the ohmmeter. When the ohmmeter polarity is reversed, the same junction will be reverse biased. Forward biasing is indicted by a low resistance reading, while reverse bias will show an extremely high resistance value. Both junctions of a transistor will respond to this test in the same way if the transistor is good.

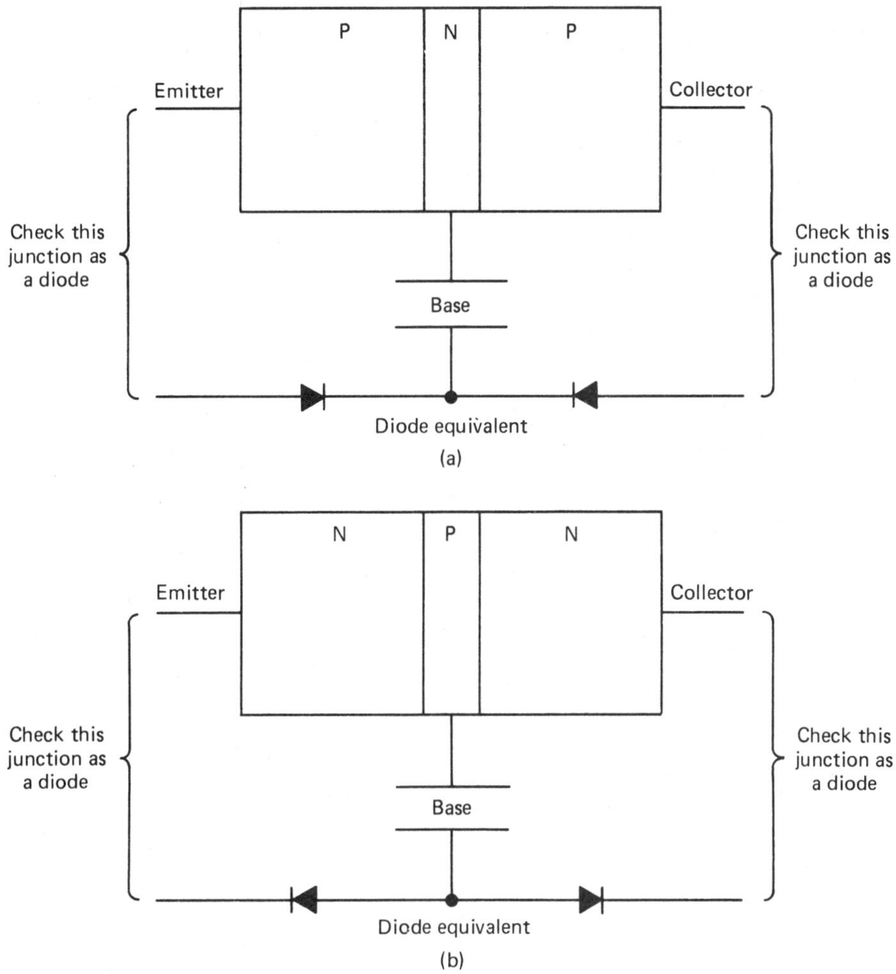

FIGURE 5-15 Junction polarity of (a) PNP and (b) NPN transistors.

A faulty P-N junction does not show a difference in its forward and reverse resistance connections. Low resistance in both directions indicates that a junction is shorted. Excessive current generally causes this type of failure in a transistor. Infinite or exceedingly high resistance in both directions indicates an open condition. An open junction is normally caused by a broken internal connection. This may be the result of a current overload or excessive shock. Open and shorted transistor failures, as a general rule, occur suddenly and cannot be corrected.

A rather quick test of both junctions can be made simultaneously with the ohmmeter. Connect one ohmmeter lead to the emitter and the other to the collector. Ohmmeter lead polarity in this case is not important. This connection should cause an extremely high resistance reading. Then reverse the two ohmmeter leads. High resistance should also occur in this direction. A good silicon transistor will show an infinite resistance in either direction. Any measurable resistance in either direction indicates leakage. Germanium transistors will show some leakage in this test. Unless the leakage is excessive, it can be tolerated in a germanium transistor. A perfect transistor would have no indication of emitter-collector leakage.

Lead Identification

An ohmmeter can also be used to identify the leads of a transistor. The polarity of the ohmmeter voltage source must be known in order for this test to be meaningful. Straight-polarity ohmmeters have the black or common lead negative and the red or probe lead positive. Reverse polarity ohmmeters are connected in the opposite direction. The polarity of an ohmmeter can be tested with a separate dc volt meter if it is unknown. The ohmmeter used in this explanation has straight polarity.

To identify transistor leads, inspect the lead location of the device under test. Pick out the center lead of the transistor. Assume that it is the base lead. Connect the negative lead of the ohmmeter to it as shown in Fig. 5-16. Then alternately touch the positive ohmmeter lead to the two outside transistor leads. If a low resistance indication occurs for each lead, the center lead is actually the base. The test also indicates that the transistor is the PNP type. If the resistance is high between the center lead and the two outside leads, reverse the meter polarity. The positive lead should now be connected to the assumed base. Once again alternately switch the negative ohmmeter lead between the two outside transistor leads. If a low resistance reading is obtained, the transistor is an NPN type.

If the center lead does not produce low resistance in either of the two conditions, it is not the base. Then select one of the outside leads as the assumed base. Try the same procedure again with the newly assumed base. If this does not produce results, try the other outside lead. One of the three leads must respond as the base if the transistor is good. If a response cannot be obtained, the transistor must be open or shorted.

Thus far we have identified the base lead and determined the polarity of the transistor to be NPN or PNP. It is now possible to test the gain and to identify the remaining two leads. A 100,000-Ω resistor and an ohmmeter are needed for this part of the test. The resistor is used to provide base current from the ohmmeter to the transistor. If a power transistor is being tested, use the R × 1 meter range and a 1000-Ω base resistor.

The process of troubleshooting transistor gain should be more meaningful if we know how it works. Figure 5-17 shows how PNP and NPN transistors will respond to the ohmmeter test. The energy source of the ohmmeter is used to supply the bias voltage to each of the transistor elements. For the PNP transistor of part (a) the positive ohmmeter lead is connected to the emitter and the negative lead to the collector. When a resistor is connected between the collector and the base, it causes base current. With the emitter and base forward biased and the collector reverse biased, the transistor becomes low resistant. The ohmmeter will respond

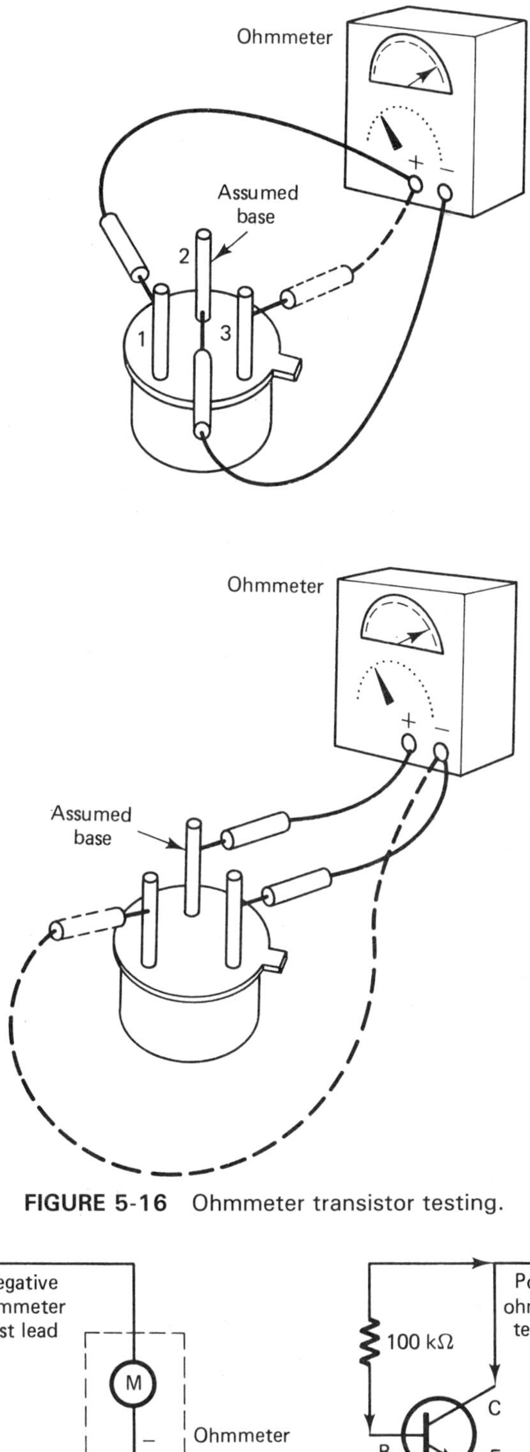

FIGURE 5-16 Ohmmeter transistor testing.

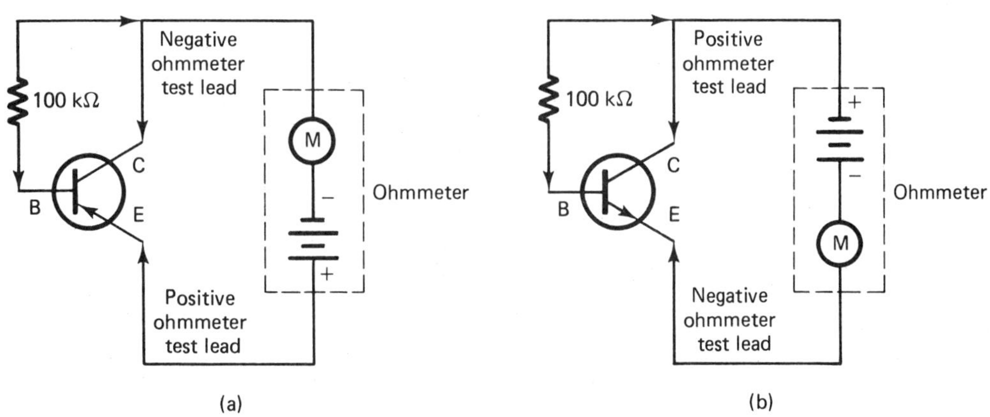

FIGURE 5-17 Transistor gain test; (a) PNP test circuit; (b) NPN test circuit.

to this condition by showing a change in resistance. The resistance of the transistor will decrease in value.

The ohmmeter test of an NPN transistor is shown in Fig. 5-17(b). For this type of transistor the emitter is connected to the negative ohmmeter lead and the collector to the positive lead. When a resistor is connected between the collector-base, it will cause base current to flow. With the emitter-base forward biased and base-collector reverse biased, the transistor becomes low resistant. An indication of this type on the ohmmeter shows that the transistor has gain and correct lead selection.

The procedure for checking an NPN transistor is shown in Fig. 5-18. In this case the ohmmeter is connected to the two outside leads. We have previously identified the center lead as the base. If another lead were found to be the base, it would be used in place of the center lead. Note that one end of the resistor is connected to the base with the other lead open. We have in this case assumed that the negative ohmmeter lead is connected to the emitter and the positive lead to the collector. If the assumed leads are correct, the emitter will be forward biased and the collector reverse biased. Touching the open end of the base resistor to the positive lead will cause a base current flow. If it does, the ohmmeter will indicate a low resistance. In effect, the emitter-base junction is forward biased, and the base-collector junction is reverse biased. If no base current flows, the assumed emitter-collector leads are reversed. Simply reverse the ohmmeter leads and again touch the resistor lead to the positive ohmmeter lead. If the transistor is good, and the assumed leads are correct, the ohmmeter will show a low resistance. If it does not, the transistor has a low gain. In some transistors, low gain is permissible. As a rule, low gain is an indication of some type of transistor problem. The ohmmeter test of transistor gain is only a close approximation of the status of the device.

Figure 5-19 shows the procedure for testing a PNP transistor. In this case the ohmmeter is connected to the two outside leads.

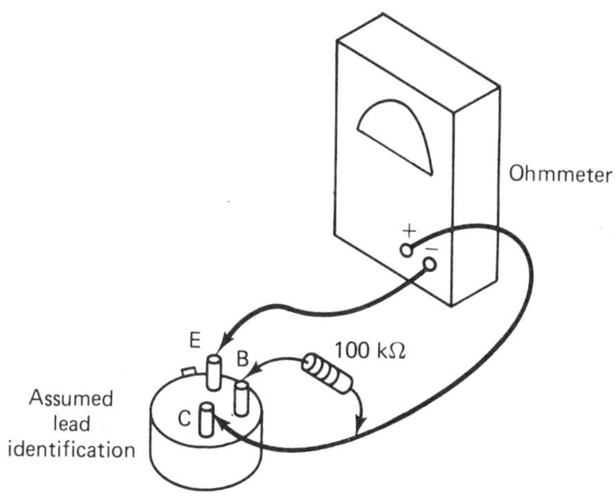

FIGURE 5-18 NPN gain test.

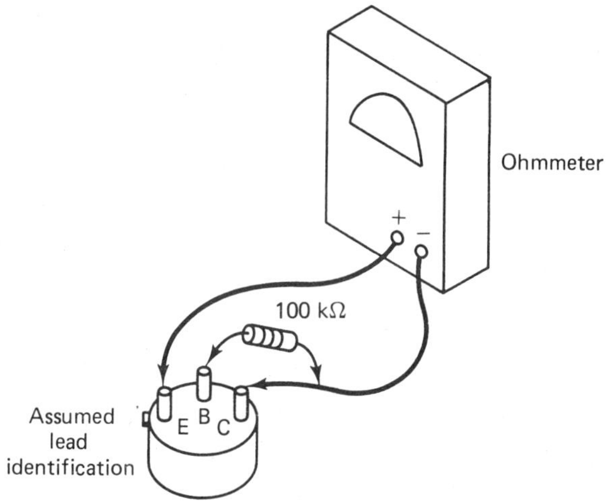

FIGURE 5-19 PNP gain test.

We have identified the center lead as the base. The base resistor is now connected to this lead. For a PNP transistor the positive ohmmeter lead goes to the emitter and the negative lead to the collector. If the assumed leads are correct, touching the open end of the base resistor to the negative lead will cause base current to flow. If it does, the ohmmeter will show a low resistance reading. This indicates that the emitter-base is forward biased and the base-collector reverse biased as in the circuit. If no current flows, the assumed emitter-collector leads are reversed. Reverse the two ohmmeter leads, and again touch the base resistor to the negative ohmmeter lead. A good transistor with correct lead identification will indicate low resistance. A transistor with low gain will not cause much of a change in resistance.

SELF-EXAMINATION

31. Each P-N junction of a good transistor will respond as a ____________ when tested with an ohmmeter.
32. A(n) ____________ P-N junction will not show a difference in its forward and reverse resistance connections.
33. When an ohmmeter is connected between the emitter-collector leads of a good bipolar transistor it will show (high, low) resistance in either polarity.
34. A good bipolar transistor is connected so that the positive ohmmeter lead goes to the base and the negative lead is switched alternately between the other two leads. If this test shows low resistance between each of the alternate leads, the device is (NPN, PNP).
35. If a bipolar transistor shows low resistance between any two leads in either direction of ohmmeter polarity, it indicates that the device is ____________.
36. Infinite or exceedingly high resistance in both directions

across either P-N junction of a bipolar transistor indicates that the junction is ____________.

ANSWERS

31. Diode	32. Faulty, or bad
33. High	34. NPN
35. Shorted	36. Open

SUMMARY

Bipolar transistors are three-element devices made of semiconductor materials. These devices have two P-N junctions in their construction. Current flow is controlled by altering the voltage applied to the two junctions.

A bipolar transistor has a thin layer of doped semiconductor material placed between two layers of doped semiconductor material of the opposite polarity. The designations NPN or PNP denote the polarity of the semiconductor material used in its construction.

A bipolar transistor is represented by a schematic symbol that has three leads. These are the emitter, base, and collector. The emitter is represented by an arrow point. When the arrow points toward the base (a straight line), it indicates that the transistor is a PNP type. When the arrow points away from the base, the symbol designates an NPN device. The collector is represented by a slanted line attached to the base.

Transistors are constructed in a variety of ways. The point-contact method of construction has semiconductor materials connected together by pointed wires that are fused to the material. Alloy-junction transistors are formed by attaching two small pieces of metal on opposite sides of a thin piece of semiconductor material. The diffusion process is used to form a semiconductor material on another surface when the material is placed in a vapor of the dopant material, and the dopant is deposited on the surface. Diffusion occurs when the temperature of the material increased, causing the dopant atoms to be absorbed by the material. Planar and mesa transistors are constructed by this process.

In order for a transistor to function it must have electrical energy applied to its electrodes. The emitter-base junction is forward biased, and the collector-base junction is reverse biased. A forward biased emitter-base junction causes a large amount of current to flow into the base region. Upon entering this region, a few current carriers combine to form base current. The other current carriers pass through the base and enter the collector region. This current is a rather substantial increase over the normal reverse-biased collector current. A very high percentage of all emitter current flows through the base and enters the collector. As a result of this action a bipolar transistor has current gain capabilities. The term beta is an expression of current gain. Beta is collector current divided by base current.

Transistor testing, lead identification, and construction material polarity can be determined with an ohmmeter. The two junctions of a transistor are tested as diodes. The emitter-collector will indicate infinite resistance in either direction of ohmmeter polarity. A good transistor will show very little resistance between the emitter-collector leads. An open transistor will show infinite resistance in each direction across the emitter-base and base-collector junctions. A shorted transistor will show low resistance in both directions across the emitter-base and base-collector junctions.

Lead and material polarity identification of a transistor is achieved by knowing the polarity of the ohmmeter used in its evaluation. The ohmmeter's voltage source is used to bias the transistor into operation. A base lead is assumed, and one ohmmeter lead is attached to this lead. The other lead is switched between the two remaining leads. If the assumed base is positive, and a low resistance occurs between the other two leads, the transistor is an NPN device. Reversing the polarity of the ohmmeter will cause an infinite resistance to appear between the assumed base and the other two leads. Reversing the ohmmeter so that the assumed base is negative and the positive lead shows low resistance between the other two leads identifies the transistor as a PNP device.

EXPERIMENT 5-1
BIPOLAR TRANSISTOR TESTING

The internal circuitry of an ohmmeter has a source of voltage connected to its leads when measuring resistance. This voltage can be used to bias a transistor when it is connected to the ohmmeter. A bipolar transistor has two PN or NP junctions in its construction. When these junctions are connected together, an NPN or PNP type of transistor is formed. A very important step in transistor testing is evaluation of the two junctions. They must respond as a diode in order for a transistor to be good. This means that forward biasing will cause low resistance, and reverse biasing will show infinite or high resistance. The two outside leads of a transistor will show infinite or high resistance to either method of biasing.

A good bipolar transistor must show low resistance when its emitter-base and base-collector are forward biased. Reversing the polarity of the ohmmeter will cause the same two junctions to have an extremely high resistance. The emitter-collector junctions must show high resistance to either ohmmeter polarity if the device is good. If it does not, the transistor has leakage current. Silicon transistors generally have no leakage, while germanium transistors show some leakage. Leakage current is temperature dependent in bipolar transistors.

OBJECTIVES

1. To test the status of a bipolar transistor with an ohmmeter.
2. To identify the polarity of the construction material of a bipolar transistor with an ohmmeter.
3. To identify the leads of a bipolar transistor with an ohmmeter.
4. To test the gain capabilities of a bipolar transistor with an ohmmeter.

EQUIPMENT

Volt-ohm-milliammeter (VOM), standard or digital type
NPN transistor, Motorola MPSA20 or Heath 417-801
PNP transistor, Motorola MPSA70, 2N2431, or Heath 417-122
10-kΩ, $\frac{1}{4}$-W resistor

PROCEDURE

1. Determine the lead polarity of the ohmmeter by placing it in the ohms position and measuring the voltage with a second voltmeter. If a second voltmeter is not available, determine the polarity of the ohmmeter from the instrument operational manual. The polarity of the ohmmeter is (straight, reversed).
2. In the first testing operation we will determine if the transistor is NPN or PNP. The polarity of the ohmmeter is very important in this procedure. Set the range switch of the ohmmeter to the R × 100 or R × 1000 position. If a digital meter is used, set the range-select switch to the 2-kΩ position or an equivalent range.
3. Select the MPAS70 transistor for this test procedure. Look at the three leads of the transistor. Notice if there are any identification marks or unusual lead placement for this transistor. Make a sketch of the lead placement of the transistor in the outer margin. Show the bottom view.
4. Pick out the center lead of the transistor and assume that it is the base. Connect the negative lead of the ohmmeter to it. If the lead is actually the base of the transistor, it will show a low resistance reading between each of the other two leads for a PNP transistor.
5. If the resistance in step 4 is high in both directions, reverse the meter leads so that the positive lead goes to the assumed base. Once again test the transistor by connecting the negative lead alternately between the two outermost leads. If a low resistance reading is obtained between the center and outside leads, the transistor is an NPN type.
6. If using the center lead does not produce a low resistance between the two outside leads, it is not the base terminal. Select one of the outside leads and assume that it is the base. Try the same test procedure again. Note that a shorted transistor will more than likely show low resistance in either polarity direction. It is rather difficult to determine the polarity of a shorted transistor.
7. After you have determined the polarity of a transistor to be either NPN or PNP by this procedure, determine if the transistor has any leakage. This test is applied to the two outside leads or to the two leads that do not include the assumed base. These leads will be identified as the emitter and collector. In general, there should be no leakage current or resistance in either direction that the ohmmeter is connected. Test the emitter and collector leads for leakage. Record the leakage resistance. Leakage resistance is ____________________.

8. We have assigned the names emitter, base, and collector to the leads of the transistor. We must now verify if the lead assignment is correct. To perform this test, assume that the base has been identified and that the transistor is of the PNP type. Select one lead and assume that it is the emitter and that the other lead is the collector. Connect the positive lead of the ohmmeter to the assumed emitter and the negative lead to the assumed collector. Then connect a 10-kΩ resistor between the collector and the base. If the lead assignments are correct, the ohmmeter will show a substantial change in resistance. This is an indication of transistor gain or beta. If there is no change in resistance, the lead assignments for the emitter and collector have been reversed. Switch the ohmmeter leads and again connect the resistor between the newly assigned collector (negative ohmmeter lead) and the base. This should produce a change in resistance indicating current gain or beta. If the ohmmeter responds in this manner, the lead assignments of the emitter, base, and collector are correct.
9. Make a bottom-view drawing of the transistor in the outer margin and identify the location of the leads.
10. Repeat the test procedure for the MPAS20 transistor. The drawing, type of transistor, and lead identification should be placed in the outer margin near item 10.

DISCUSSION

This experiment showed how to evaluate the status of a bipolar transistor with an ohmmeter. In the first part of the activity, the ohmmeter was used to identify a transistor as PNP or NPN. This used the polarity of the ohmmeter's voltage source to bias a junction. Since a transistor has two junctions, evaluation is based on junction testing. This is the same as testing a diode. The junction will indicate a low resistance when forward biased and high resistance when reverse biased. The base lead of the transistor will show low resistance between the emitter and the collector when the ohmmeter is alternately switched between these two leads. For a PNP transistor the negative ohmmeter lead goes to the base, and the positive lead will show low resistance when it is switched between the emitter and collector. For an NPN transistor the positive ohmmeter lead goes to the base and the negative lead will show low resistance when it is switched between the emitter and collector. This procedure was used to identify the polarity of the transistor's material.

The test procedure of a transistor also involves evaluation of the emitter-collector for leakage current. The ohmmeter was connected between the emitter and collector. Normally there should be no indication of resistance in either direction of ohmmeter polarity. A silicon transistor generally shows no indication of resistance (leakage current), while a germanium transistor may show some resistance. Transistor leakage is temperature dependent.

In addition to the leakage test, polarity determination, and lead identification, you had an opportunity to verify the lead assignments and test current gain. This was done by forward biasing the emitter and reverse biasing the collector with an ohmmeter and connecting a resistor to the base to forward bias it. If the lead assignments were properly made, a transistor would show a change in resistance. This indicates the ability of the transistor to achieve current gain.

ANALYSIS

1. Explain how an ohmmeter is used to determine if a transistor is an NPN or a PNP type.
2. How is the leakage of a transistor evaluated with an ohmmeter?

EXPERIMENT 5-2
TRANSISTOR OPERATING REGIONS

A bipolar transistor has three regions in which it can operate, the saturation, active, and cutoff regions. Applications of the transistor dictate the operating region being utilized. Amplification is achieved by operation in the active region. Switching applications call for transistor operation in the cutoff and saturation regions. The transistor switches back and forth between these two regions of operation.

A person working with electronic circuits must be familiar with the general response of a transistor in its three operating regions. When a transistor operates in the saturation region, it responds as a closed switch. This means that the collector-emitter voltage (V_{CE}) drops to a very small value. A transistor that is cut off has no collector current, and the V_{CE} approximates the source voltage. A transistor operating in the active region can have varying values of collector current and V_{CE} depending on the range of conduction. This is determined by the amount of base current. The operating region of a transistor is primarily determined by the value of its base current.

OBJECTIVES

1. To construct a transistor circuit and measure its bias voltage values.
2. To evaluate the operation of a transistor circuit in the active, cutoff, and saturation regions.

EQUIPMENT

Variable dc source
Volt-ohm-milliammeter (VOM), standard or digital
NPN silicon transistor, Motorola MPSA20, or Heath 417-801
1-kΩ, $\frac{1}{4}$-W resistor
10-kΩ, $\frac{1}{4}$-W resistor
100-kΩ, $\frac{1}{4}$-W resistor

PROCEDURE

1. Construct the bipolar transistor circuit of Fig. 5-13(a). The VOM is used to measure different circuit voltage values. V_{CE} is measured across the emitter-collector of

the transistor, and V_{RC} is measured across resistor R_C. The collector current is calculated by V_{RC}/R_C. The base voltage is measured across resistor R_B. Base current is equal to the base voltage divided by the value of the base resistor R_B. The value of V_{CE}, I_C, and I_B are used to determine the transistor's region of operation.

2. Apply voltage to the transistor circuit. Measure the V_{CE}, V_{RC}, and V_B with the VOM. Record these values in Table 5-2A for circuit 1.

TABLE 5-2A

Circuit	Measured Values			Calculated Values		Operating Region
	V_{CE}	V_{RC}	V_B	I_C	I_B	
5-13 (a)						
5-13 (b)						
5-13 (c)						

3. Calculate the I_C and I_B for the transistor circuit. $I_C = V_{RC}/R_C$, and $I_B = V_B/R_B$. Indicate the operating region for this circuit in Table 5-2A.
4. Turn off the voltage source and alter the circuit of Fig. 5-13(a) to conform with the circuit of Fig. 5-13(b). The base resistor is the only circuit modification.
5. Apply source voltage to the circuit. Measure the V_{CE}, V_{RC}, and V_B with the VOM. Record these values in Table 5-2A for circuit 2.
6. Calculate the I_C and I_B for the transistor circuit. Indicate the operating region for this circuit in Table 5-2A.
7. Turn off the voltage source and alter the circuit of Fig. 5-13(b) to conform with the circuit of Fig. 5-13(c). The value of the base resistor and its connection to the source have been changed.
8. Apply source voltage to the circuit. Measure the V_{CE}, V_{RC}, and V_B with the VOM. Record these values in Table 5-2A for circuit 3.
9. Calculate the I_C and I_B for the transistor circuit. Indicate the operating region of this circuit in Table 5-2A.
10. Turn off the source voltage and disconnect the circuit.

DISCUSSION

In this experiment you were asked to construct three transistor circuits. Each circuit represented a different region of operation. The first circuit operated in the active region. This condition of operation should have been evident because the values of V_{CE} and

V_{RC} were nearly the same. This meant that operation was near or in the approximate center of the operating range. Transistor amplifiers generally operate in this region.

The second circuit operated in the cutoff region. This was shown by a drastic reduction of collector current and an increase in the value of the collector-emitter voltage. In order for this condition of operation to be achieved, the emitter-base must be reverse biased. This is accomplished by connecting resistor R_B to the negative side of the source. With the base reverse biased there will be no base current and no collector current. All the source voltage appears across the transistor as V_{CE}. No voltage appears across R_C because there is no collector current.

Circuit 3 had the transistor operating in the saturation region. When this occurs the transistor responds as a closed switch. The internal resistance of the transistor becomes small and there is very little V_{CE}. A rather large value of collector current will occur in this circuit. The value of R_C and the source voltage determine the value of I_C that will occur. In order for a transistor to be saturated, the base current must rise to a rather large value. This should have been noted by the value of I_B when compared with the other two circuits.

ANALYSIS

1. Explain what conditions of operation must be met by a bipolar transistor in order for it to operate in the saturation region.
2. Explain what conditions of operation must be met by a bipolar transistor in order for it to operate in the cutoff region.
3. What are the conditions of operation that indicate when a bipolar transistor is operating in the active region?

UNIT 5 EXAMINATION
BIPOLAR TRANSISTORS

Instructions: Circle the response that most correctly completes each statement.

1. The largest current flow of a bipolar transistor occurs:
   a. In the emitter
   b. In the base
   c. In the collector
   d. Through the collector-base
   e. Through the emitter-collector
2. Conventional biasing of a bipolar transistor has the:
   a. Emitter-base forward biased and the collector-base forward biased.
   b. Emitter-base reverse biased and the collector-base forward biased.
   c. Emitter-base forward biased and the collector-base reverse biased.
   d. Emitter-base reverse biased and the collector-base reverse biased.
3. A bipolar transistor has a collector current of 9 mA and a base current of 150 μA. In which of the following ranges will the emitter current be found?
   a. 1 to 150 μA
   b. 151 μA to 915 μA
   c. 916 μA to 1.5 mA
   d. 1.51 mA to 9.15 mA
   e. 9.16 mA to 91.5 mA
4. The beta of a bipolar transistor is an expression of:
   a. Base current/collector current
   b. Collector current/base current
   c. Collector current/emitter current
   d. Emitter current/base current
   e. Base current/emitter current
5. When a bipolar transistor is saturated:
   a. V_{CE} is very low, and I_C is zero.
   b. V_{CE} is low, and I_C is high.
   c. V_{CE} equals the source voltage, and I_C is zero.
   d. V_{CE} equals the source voltage, and I_C is high.
6. When a bipolar transistor is cut off:
   a. V_{CE} equals the source voltage and I_C is high.

b. V_{CE} is low and I_C is high.
c. V_{CE} equals the source voltage and I_C is zero.
d. V_{CE} is high and I_C is low.

7. When an ohmmeter is connected across the emitter-collector junctions of a good bipolar transistor, it will show:
   a. Low resistance in both directions of ohmmeter polarity
   b. Low resistance in one direction and high resistance in the other
   c. Some resistance in both directions of ohmmeter polarity
   d. High resistance in both directions of ohmmeter polarity
8. In a properly connected bipolar transistor, an increase in base current will cause an increase in:
   a. Collector current only
   b. Emitter current only
   c. Emitter current and collector current
   d. Leakage current
9. The polarity of the collector voltage of an NPN bipolar transistor is:
   a. Positive
   b. Negative
   c. The same polarity as the emitter
   d. The same polarity as the base
10. When a bipolar transistor responds as an amplifier it:
   a. Operates in the cutoff region
   b. Operates in the saturation region
   c. Operates in the active region
   d. Switches between the saturation and cutoff regions
11. If a bipolar transistor shows low resistance between any two leads in either direction of ohmmeter polarity, it indicates that the device is:
   a. Shorted
   b. Open
   c. Leaky
   d. Normal
12. If a bipolar transistor shows an infinite or exceedingly high resistance across either P-N junction it indicates that the junction is:
   a. Shorted
   b. Open
   c. Leaky
   d. Normal

UNIT SIX

Bipolar Transistor Amplification

UNIT INTRODUCTION

Electronic systems must perform a variety of basic functions in order to accomplish a particular operation. An understanding of these functions is an essential part of system operational theory. In this unit we investigate the amplification function. Amplification is achieved by devices that will produce a change in signal amplitude. Bipolar transistors are frequently used for this purpose.

In the first part of the unit, we look at amplification in general. Next, we discuss how the bipolar transistor achieves amplification. Circuit operation, biasing methods, circuit configurations, classes of amplification, and operational conditions are considered. As a rule, this unit is directed toward small-signal amplification. Power amplification is achieved in the same manner, with the only difference being amount of current flow being controlled.

UNIT OBJECTIVES

Upon completion of this unit, you will be able to:

1. Explain the meaning of amplification with respect to voltage, current, and power.

2. Explain how a bipolar transistor achieves amplification of an ac signal.
3. Determine how a bipolar transistor amplifier is biased.
4. Analyze the operation of a bipolar transistor by the load-line method.
5. Define bipolar transistor characteristics according to linear and nonlinear operation.
6. Explain the characteristic differences in common-base, common-emitter, and common-collector transistor circuit configurations.
7. Explain the meaning of the amplifier terms alpha, beta, dynamic and static characteristics, thermal stability, and biasing.

IMPORTANT TERMS

Before reading Unit 6, review the following terms. These terms provide a basic understanding of terms associated with bipolar transistor amplification.

Alpha Current gain of a common-base amplifier. Alpha is equal to the collector current divided by the emitter current.

Beta Current gain of a common-emitter amplifier. Beta is equal to the collector current divided by the base current.

Beta-dependent biasing A method of biasing that is related to the amplifying capabilities of a transistor.

Bypass capacitor A capacitor that provides an alternate path around a component or to ground.

Circuit configuration A method of connecting a circuit, such as common-emitter, common-collector, or common-base.

Dynamic characteristics Electronic device operational characteristics that show how a device responds to ac or changing signal voltages.

Impedance A form of opposition to alternating current measured in ohms. Electronic devices have input and output impedance values.

Linear amplifiers Amplifying circuits that increase the amplitude of a signal that is a duplicate of the input. These devices operate in the active region of the characteristic curves.

Load line A line drawn on a family of characteristic curves that shows how a device will respond in a circuit with a specific value of load resistor.

Power dissipation An electronic device characteristic that indicates the ability of the device to give off heat.

Q point An operating point for an electronic device that indicates its dc or static operation with no signal applied.

Static state A dc operating condition of an electronic device with operating energy but no signal applied.

Thermal stability The condition of an electronic device that indicates its ability to remain at an operating point without variation due to temperature.

AMPLIFICATION PRINCIPLES

Amplification is achieved by an electronic device and its associated components. As a general rule, the amplifying device is placed in a circuit. The components of the circuit usually have about as much influence on amplification as the device itself. The circuit and the device must be supplied electrical energy for it to function. Typically, amplifiers are energized by direct current. This may be derived from a battery or a rectified ac power supply. The amplifier then processes a signal of some type when it is placed in operation. The signal may be either ac or dc, depending on the application.

Reproduction and Amplification

The signal to be processed by an amplifier is first applied to the input part of the circuit. After being processed, the signal appears in the output circuitry. The output signal may be reproduced in its exact form, amplified, or both amplified and reproduced. The value and type of input signal, operating source energy, device characteristics, and circuit components all have some influence on the output signal.

Figure 6-1 illustrates the process of reproduction, amplification, and the combined amplification-reproduction functions of an amplifier. In part a the amplifier performs only the reproduction function. Note that the input and output signals have the same size and shape. Part b shows only the amplification function. In this case the input signal is increased in amplitude. The output signal is amplified but does not necessarily resemble the input signal. In many applications amplification and reproduction must be achieved simultaneously. Part c shows this function of an amplifier. Depending on the application, an amplifier must be capable of developing any of these three output signals.

Voltage Amplification

A system designed to develop an output voltage that is greater than its input voltage is called a *voltage amplifier.* A voltage amplifier has voltage gain, which is defined as a ratio of the output

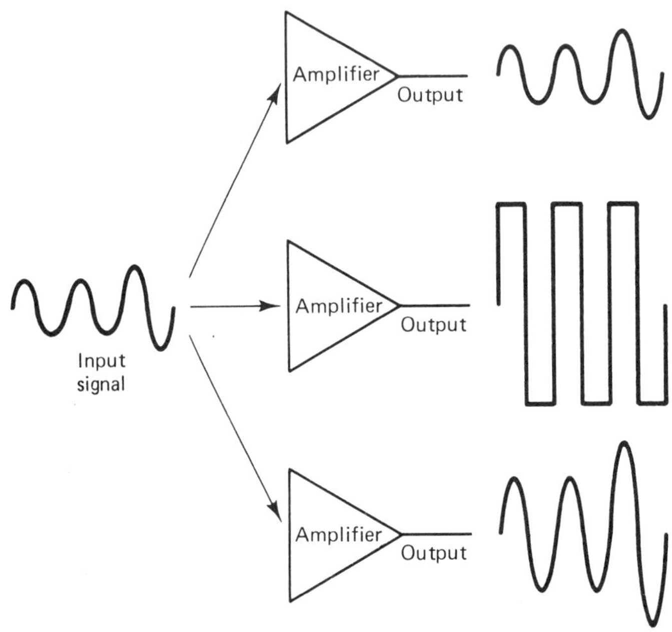

FIGURE 6-1 Amplification and reproduction: (top) reproduction; (middle) amplification; (bottom) reproduction.

signal voltage to the input signal voltage. In equation form voltage amplification is expressed as

$$\text{Amplification voltage} = \frac{\text{output voltage}}{\text{input voltage}}$$

or

$$A_v = \frac{V_{\text{out}}}{V_{\text{in}}} \quad \text{or} \quad \frac{\Delta v_{\text{out}}}{\Delta v_{\text{in}}}$$

The uppercase letters in V_{out} and V_{in} denote dc voltage values. Ac voltages are usually expressed by lowercase letters. The Greek capital letter delta (Δ) indicates a changing value. Voltage amplifiers are capable of a wide range of amplification values.

Current Amplification

An amplifying system designed to develop output current that is greater than the input current is called a *current amplifier.* A current amplifier has current gain, which is defined as the ratio of output current to input current. In equation form, it is expressed as

$$\text{Current amplification} = \frac{\text{output current}}{\text{input current}}$$

or

$$A_i = \frac{I_{\text{out}}}{I_{\text{in}}} \quad \text{or} \quad \frac{\Delta i_{\text{out}}}{\Delta i_{\text{in}}}$$

Current gain takes into account the beta of a transistor and all the associated components that make the device operational.

Power Amplification

Power amplification is a ratio of the developed output signal power to the input signal power. Note that this refers only to the power gain of the signal. All amplifiers consume a certain amount of power from the energy source during operation. This is usually not included in an expression of power gain. An equation of power gain shows that

$$\text{power amplification} = \frac{\text{output signal power}}{\text{input signal power}}$$

or

$$A_p = P_{\text{out}}/P_{\text{in}}$$

Power signal gain can also be expressed as the product of voltage gain and current gain. In this regard, the equation is:

Power amplification
$$= \text{voltage amplification} \times \text{current amplification}$$

or

$$A_p = A_v \times A_i$$

It should be apparent from the A_p equation that A_v and A_i do not both need to be large to have power gain. Typical power amplifiers may have a large current gain and a voltage gain less than 1. Power gain must take into account both signal voltage and current gain in order to be meaningful.

SELF-EXAMINATION

1. ____________ is achieved by an electronic device and its associated components.
2. The signal to be processed by an amplifier is first applied to the ____________ part of the circuit.
3. After an input signal has been processed by an amplifier, it appears in the ____________ circuitry.
4. ____________ occurs when the input signal is increased in amplitude in the output circuitry.
5. ____________ occurs when the input signal is increased in amplitude in the output circuitry.
6. ____________ ____________ is an expression of output voltage divided by input voltage.
7. ____________ ____________ is an expression of output current divided by input current.
8. ____________ ____________ is an expression of output signal power divided by input signal power.

ANSWERS

1. Amplification	2. Input
3. Output	4. Reproduction
5. Amplification	6. Voltage amplification, or A_v
7. Current amplification, or A_i	8. Power amplification, or A_p

BIPOLAR TRANSISTOR AMPLIFIERS

The primary function of an amplifier is to provide voltage gain, current gain, or power gain. The resulting gain depends a great deal on the device being used and its circuit components. Normally, the circuit is energized by dc voltage. An ac signal is then applied to the input of the amplifier. After passing through the transistor, an amplified version of the signal appears in the output. Operating conditions of the device and the circuit determine the level of amplification to be achieved.

In order for a bipolar transistor to respond as an amplifier, the emitter-base junction must be forward biased and the collector-base junction must be reverse biased. Specific operating voltage values must then be selected that will permit amplification. If both reproduction and amplification are to be achieved, the device must operate in the center of its active region. Remember that this is between the saturation and cutoff regions of the collector family of characteristic curves. Proper circuit component selection permits a transistor to operate in this characteristic region.

Basic Amplifiers

Let us look at the circuitry of a basic amplifier. Figure 6-2 shows a schematic symbol of this type of circuit. Notice that a schematic symbol of the NPN transistor is used in this diagram. Schematic symbols are normally used in all circuit diagrams. Keep in mind the crystal structure of the device represented by the symbol.

The basic amplifier has a number of parts that are needed to make it operational. V_{CC} is the dc energy source. The negative side of V_{CC} is connected to ground. The emitter is also connected to ground. This type of circuit configuration is called a grounded-emitter, or common-emitter, amplifier. The emitter, one side of the input, and one side of the output are all commonly connected together. In an actual circuit these points are usually connected to ground.

Resistor R_B of the basic amplifier is connected to the positive side of V_{CC}. This connection makes the base positive with respect to the emitter. The emitter-base junction is forward biased by R_B. Resistor R_L connects the positive side of V_{CC} to the collector. This connection reverse biases the collector. Through the connection of R_B and R_L, the transistor is properly biased for operation. A transistor connected in this manner is considered to be in its *static,* or

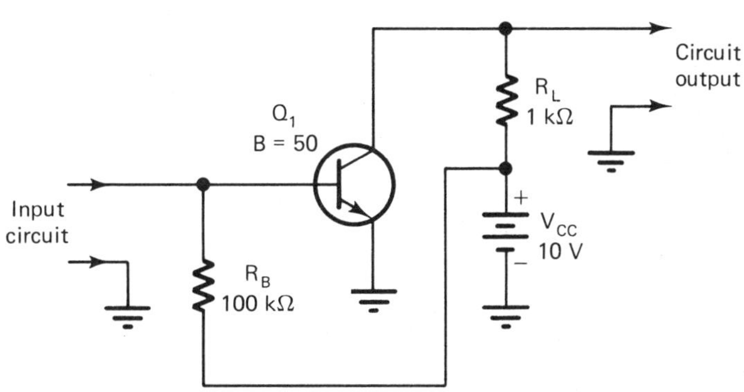

FIGURE 6-2 Basic bipolar amplifiers.

dc, operating state. It has the necessary dc energy applied to be operational. No signal is applied to the input for amplification.

Let us now consider the operation of the basic amplifier in its static state. With the given values of Fig. 6-2 base current (I_B) can be calculated. I_B, in this case, is limited by the value of R_B and the base-emitter junction resistance. When the base-emitter junction is forward biased, its resistance becomes very small. The value of R_B is therefore the primary limiting factor of I_B. I_B can be determined by an application of Ohm's law. The equation is

$$\text{base current} = \frac{\text{source voltage}}{\text{base resistance}}$$

or

$$I_B = \frac{V_{CC}}{R_B}$$

For our basic transistor amplifier this is determined to be

$$\begin{aligned} I_B &= \frac{V_{CC}}{R_B} \\ &= \frac{10 \text{ V}}{100{,}000 \ \Omega} \\ &= \frac{10 \text{ V}}{100 \times 10^3 \ \Omega} \\ &= 100 \times 10^{-6}\text{A, or } 100 \ \mu\text{A} \end{aligned}$$

The value of base current, in this case, is extremely small. Remember that only a small amount of I_B is needed to produce I_C.

The beta of the transistor used in Fig. 6-2 has a given value. With beta and the calculated value of I_B, it is possible to determine the collector current of the circuit. Beta is the current gain of a common-emitter amplifier. In Unit 5 beta was described by the formula:

$$\text{beta} = \frac{\text{collector current}}{\text{base current}}$$

or

$$\beta = \frac{I_C}{I_B}$$

In our transistor amplifier I_B has been determined and beta has a given value of 50. By transposing the beta formula, I_C can be determined by the equation

$$\text{collector current} = \text{beta} \times \text{base current}$$

or

$$I_C = \beta \times I_B$$

For the amplifier circuit, collector current is determined to be

$$\begin{aligned} I_C &= \beta \times I_B \\ &= 50 \times 100 \times 10^{-6} \\ &= 5 \times 10^{-3}\ \text{A}, \quad \text{or} \quad 5\ \text{mA} \end{aligned}$$

This means that 5 mA of I_C will flow through R_L when an I_B of 100 μA flows.

In any electric circuit, we know that current flow through a resistor will cause a corresponding voltage drop. In a basic transistor amplifier, I_C will cause a voltage drop across R_L. In the preceding step I_C was calculated to be 5 mA. By using Ohm's law again, we can determine the voltage drop across R_L from the equation

$$R_L \text{ voltage} = \text{collector current} \times \text{collector resistance}$$

or

$$V_{RL} = I_C \times R_L$$

For the amplifier circuit, the voltage drop across R_L is

$$\begin{aligned} V_{RL} &= I_C \times R_L \\ &= 5 \times 10^{-3} \times 1 \times 10^{3} \\ &= 5\ \text{V} \end{aligned}$$

This means that half or 5 V of V_{CC} will appear across R_L. With a V_{CC} value of 10 V, the other 5 V will appear across the collector-emitter of the transistor. A V_{CE} voltage of 5 V from a V_{CC} of 10 V means that the transistor is operating near the center of its active region. Ideally, the transistor should respond as a linear amplifier. When a suitable signal is applied to the input, it should amplify and reproduce the signal in the output.

The operational voltage and current values of the basic transistor amplifier in its static state are summarized in Fig. 6-3. Note that an I_B of 100 μA will cause an I_C of 5 mA. With a beta of 50, the amplifier has the capability of rather substantial amount of output current. The collector current passing through R_L causes V_{CC} to be divided. This establishes operation in the approximate center of the collector family of characteristic curves. Ideally, this

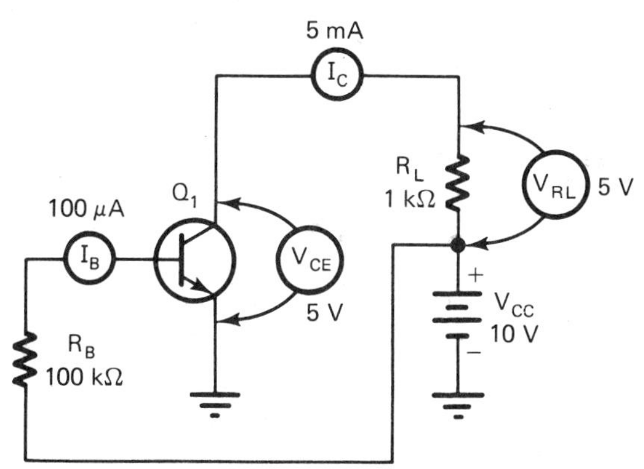

FIGURE 6-3 Static operating condition of a basic amplifier.

static condition should cause the amplifier to respond as a linear device when a signal is applied.

Signal Amplification

In order for the basic transistor circuit of Fig. 6-3 to respond as an amplifier, it must have a signal applied. The signal may be either voltage or current. An applied signal causes the transistor to change from its static to a dynamic condition. Dynamic conditions involve changing values. All ac amplifiers respond in the dynamic state. The output of this should develop an ac signal.

Figure 6-4 shows a basic transistor amplifier with an ac signal applied. A capacitor is used in this case to couple the ac signal source to the amplifier. Remember that ac passes easily through a capacitor, whereas dc is blocked. As a result, the ac signal is injected into the base-emitter junction. Dc does not flow back into the signal source. The ac signal is therefore added to the dc operating voltage. The emitter-base voltage is a dc value that changes at an ac rate.

Consider now how the applied ac signal alters the emitter-base junction voltage. Figure 6-5 shows some representative voltage and current values that appear at the emitter-base junction of the transistor. They can be measured with a voltmeter or observed with an oscilloscope. Part a shows the dc operating voltage and base current. This occurs when the amplifier is in its steady or static state. Part b shows the ac signal that is applied to the emitter-base junction. Note that the amplitude change of this signal is a very small value. This also shows how the resulting base current and voltage change with ac applied. The ac signal is essentially riding on the dc voltage and current.

Refer now to the schematic diagram of the basic amplifier in Fig. 6-6. Note in particular the waveform inserts that appear in the diagram. These show how the current and voltage values respond when an ac signal is applied.

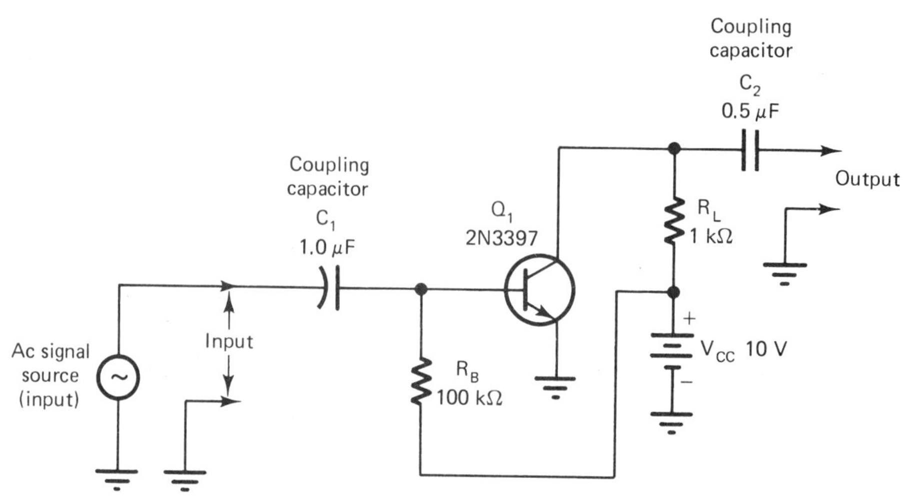

FIGURE 6-4 Amplifier with ac signal applied.

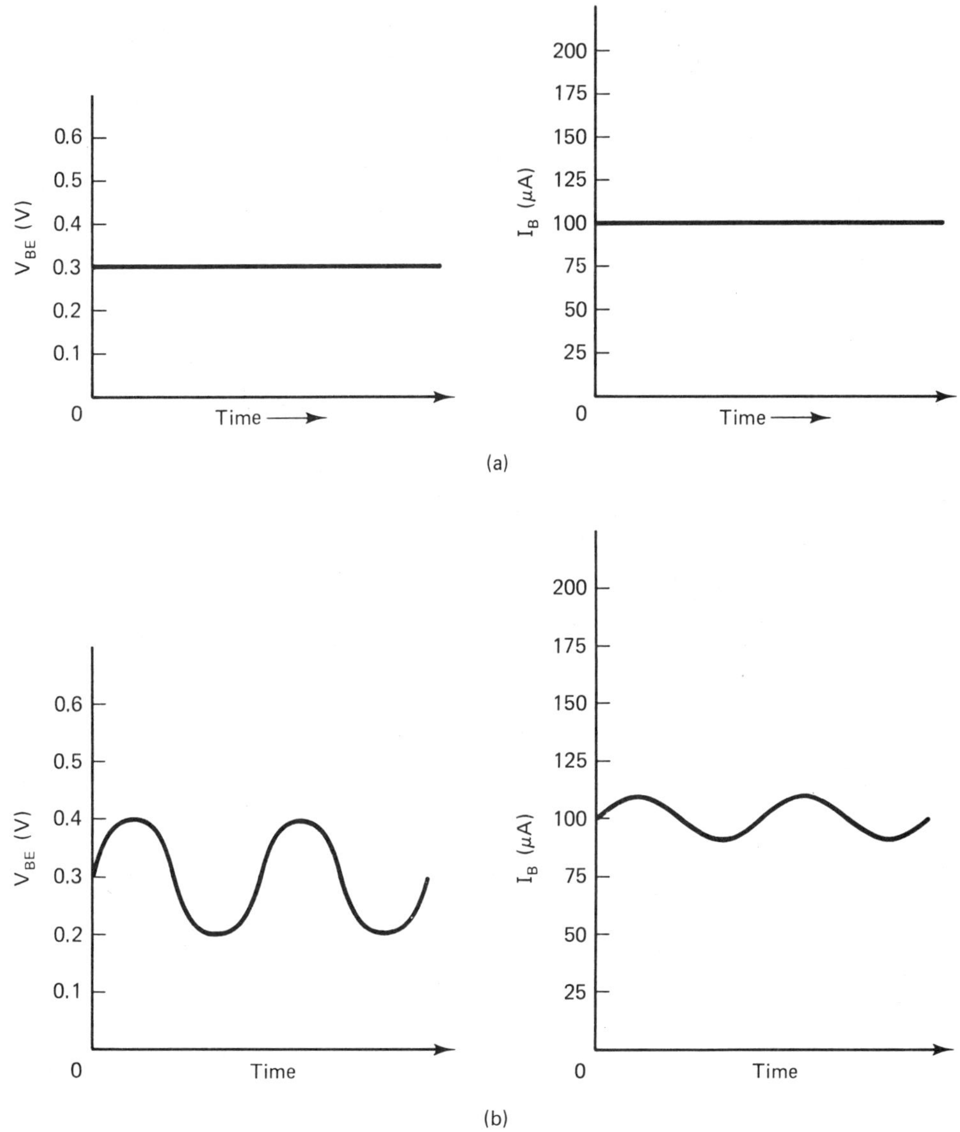

FIGURE 6-5 I_B and V_{BE} conditions: (a) static; (b) dynamic.

The ac signal applied to the input of our basic amplifier rises in value during the positive alternation and falls during the negative alternation. Initially, this causes an increase and a decrease in the value of V_{BE}. This voltage has a dc level with an ac signal riding on it. The indicated I_B is developed as a result of this voltage. The changing value of I_B causes a corresponding change in I_C. Note that I_B and I_C both appear to be the same. There is, however, a very noticeable difference in values. I_B is in microamperes, while I_C is in milliamperes. The resulting I_C passing through R_L causes a corresponding voltage drop (V_{RL}) across R_L. V_{CE} appearing across the transistor is the reverse of V_{RL}. These signals are both ac values riding upon a dc level. The output, V_{out}, is changed to an ac value. Capacitor C_2 blocks the dc component and passes only the ac signal.

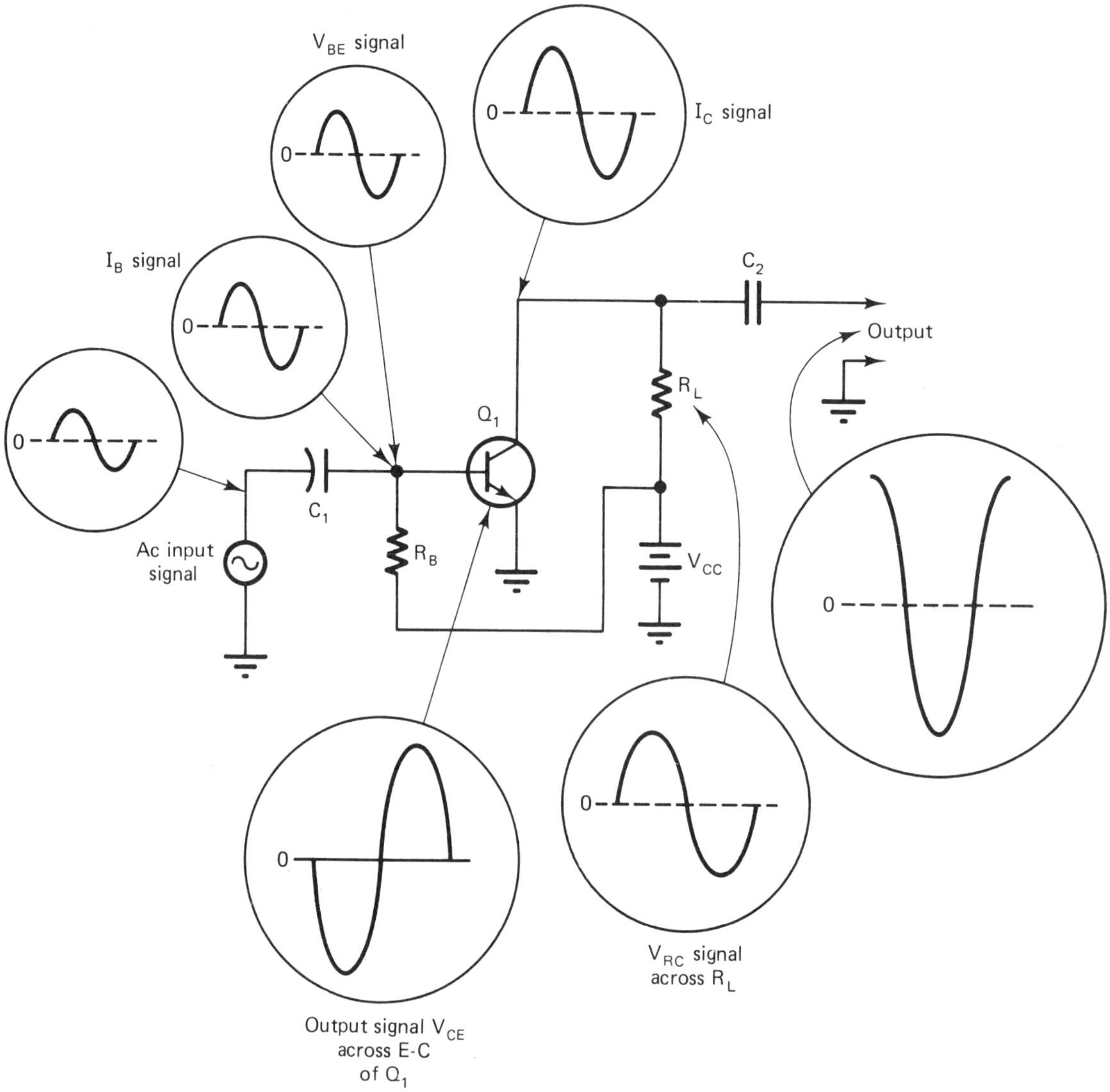

FIGURE 6-6 Ac amplifier operation.

It is interesting to note in Fig. 6-6 that input and output signals of the amplifier are reversed. When the ac input signal rises in value, it causes the output to fall in value. The negative alternation of the input caused the output to rise in value. This condition of operation is called *phase inversion.* Phase inversion is a distinguishing characteristic of the common-emitter amplifier.

Amplifier Bias

If a transistor is to operate as a linear amplifier it must be properly biased and have a suitable operating point. Its steady state of operation depends a great deal on base current, collector voltage, and collector current. Establishing a desired operating point requires proper selection of bias resistors and a load resistor to provide proper input current and collector voltage.

Stability of operation is a very important consideration. If the operation point of a transistor is permitted to shift with temperature changes, unwanted distortion may be introduced. In addition to distortion, operating point changes may also cause damage to a transistor. Excessive collector current, for example, may cause the device to develop too much heat.

The method of biasing a transistor has a great deal to do with its thermal stability. Several different methods of achieving bias are in common use today. One method of biasing is considered to be beta dependent. Bias voltages are largely dependent on transistor beta. A problem with this method of biasing is transistor response. Transistor beta is rarely ever the same for a specific device. Biasing set up for one transistor will not necessarily be the same for another transistor.

Biasing that is independent of beta is very important. This type of biasing responds to fixed voltage values. Beta does not alter these voltage values. As a general rule, this form of biasing is more reliable. The input and output of a transistor are very stable and the results are very predictable.

Beta-Dependent Biasing

Four common methods of beta-dependent biasing are shown in Fig. 6-7. The steady-state or static conditions of operation are a function of the transistor's beta. These methods are rather easy to achieve with a minimum of parts. They are not very widely used today.

The circuit in Fig. 6-7(a) is a very simple form of biasing for a common-emitter amplifier. This method of biasing was used for our basic transistor amplifier. The base is simply connected to V_{CC} through R_B. This causes the emitter-base junction to be forward biased. The resulting collector current is beta times the value of I_B. The value of I_B is V_{CC}/R_B. As a general rule, biasing of this type is very sensitive to changes in temperature. The resulting output of the circuit is rather difficult to predict. This method of biasing is often called *fixed biasing.* It is primarily used because of its simplicity.

The circuit in Fig. 6-7(b) was developed to compensate for the temperature sensitivity of circuit a. Bias current is used to counteract changes in temperature. In a sense, this method of biasing has negative feedback. R_B is connected to the collector rather than V_{CC}. The voltage available for base biasing is that leftover after voltage drop across the load resistor. If the temperature rises, it causes an increase in I_B and I_C. With more I_C, there is more voltage drop across R_L. Reduced V_C voltage will cause a corresponding drop in I_B. This, in turn, brings I_C back to normal. The opposite reaction will occur when transistor temperature becomes less. This method of biasing is called *self-biasing.*

The circuit in Fig. 6-7(c) is an example of *emitter biasing.* Thermal stability is improved with this type of construction. I_B is

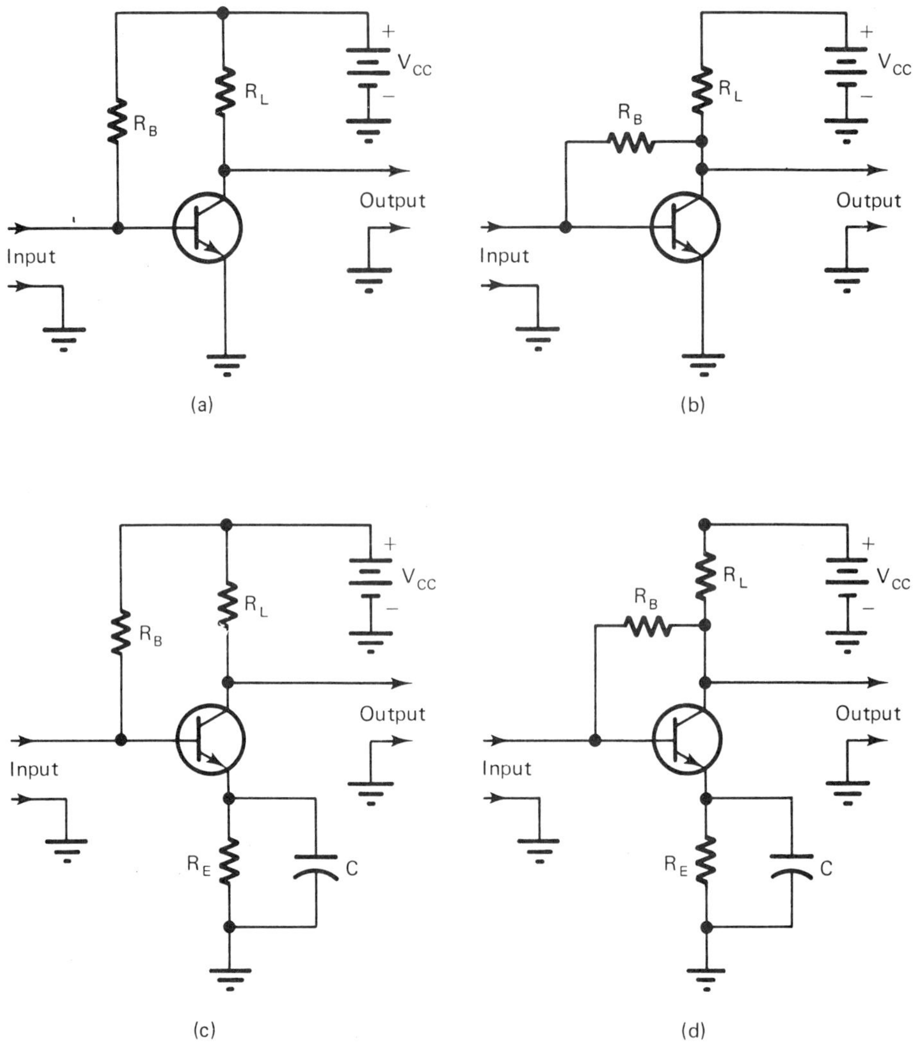

FIGURE 6-7 Methods of beta-dependent biasing: (a) fixed; (b) self-biasing; (c) fixed-emitter biasing; (d) self-emitter biasing.

again determined by the value of R_B and V_{CC}. An additional resistor is placed in series with the emitter of this circuit. Emitter current passing through R_E produces emitter voltage (V_E). This voltage opposes the base voltage developed by R_B. Proper values of R_B and R_E are selected so that I_B and I_E will flow under ordinary operating conditions. If a change in temperature should occur, V_E will increase in value. This action will oppose the base bias. As a result, collector current will drop to its normal value. The capacitor connected across R_E is called an *emitter bypass capacitor.* It provides an ac path for signal voltages around R_E. With C in the circuit, amplifier gain would be reduced. The value of C is dependent on the frequencies being amplified. At the lowest possible frequency being amplified, the capacitive reactance (X_C) must be 10 times smaller than the resistance of R_E.

Circuit (d) of Fig. 6-7 is a combination of circuits (b) and (c). It is often called *self-emitter bias.* In the same regard, circuit (c) could be called *fixed-emitter* bias. As a general rule, emitter biasing is not very effective when used independently. Circuit (d) has good thermal stability. The output has reduced gain because of the base resistor connection.

Independent Beta Biasing

Two methods of biasing a transistor that are independent of beta are shown in Fig. 6-8. These circuits are extremely important because they do not change operation with beta. As a general rule, these circuits have very reliable operating characteristics. The output is very predictable, and stability is excellent.

Circuit (a) is described as the *divider method* of biasing. It is used widely today. The base voltage (V_B) is developed by a voltage-divider network made up of R_B and R_1. This network makes the circuit independent of beta changes. Voltage at the base, emitter, and collector all depend on external circuit values. By proper selection of components, the emitter-base junction is forward biased, with the collector being reverse biased. Normally, these bias volt-

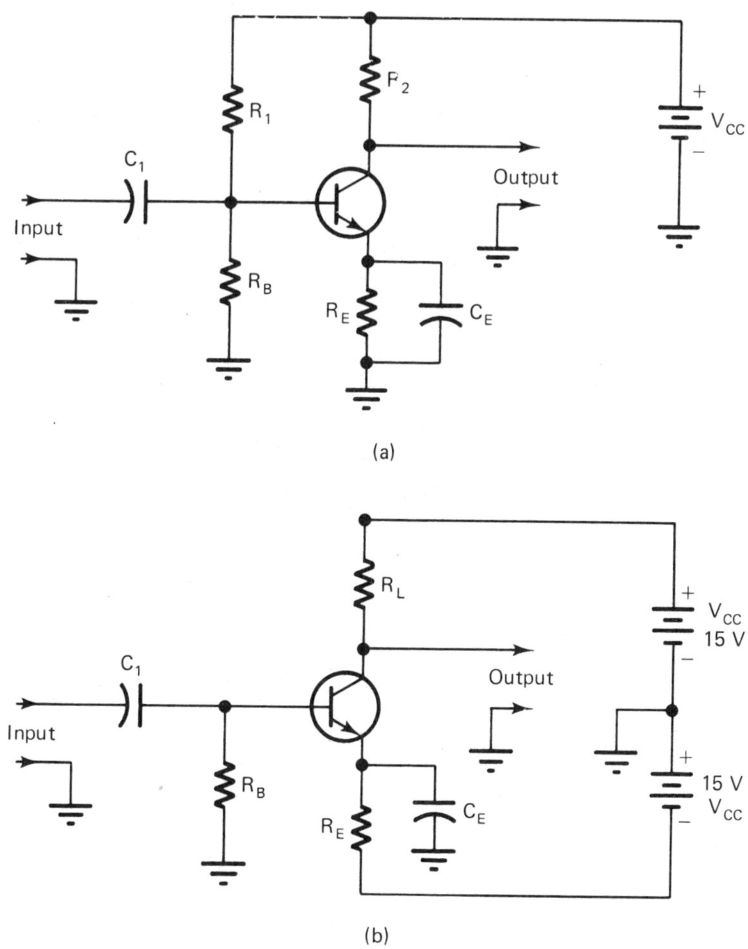

FIGURE 6-8 Independent beta biasing: (a) divider bias; (b) divider bias with split supply.

ages are all referenced to ground. The base, in this case, is made slightly positive with respect to ground. This voltage is somewhat critical. A voltage that is too positive, for example, will drive the transistor into saturation. With proper selection of bias voltage, however, the transistor can be made to operate in any part of the active region. The temperature stability of the circuit is excellent. With proper R_E bypass capacitor selection, this method of biasing produces very high gain. The divider method of biasing is often a universal biasing circuit. It can be used to bias all transistor amplifier circuit configurations.

The circuit in (b) is very similar in construction and operation that of (a). One fewer resistor is used. The power supply requires two voltage values with reference to ground. A split power supply is used as an energy source for this circuit. Note the indication of $+V_{CC}$ and $-V_{CC}$. R_B is connected to ground. In this case, the value of R_B would determine the value of I_B with only half of the total supply voltage. The values of R_L and R_E are usually made larger to accommodate the increased supply voltage. If R_E is properly bypassed, the gain of this circuit is very high. Thermal stability is excellent.

SELF-EXAMINATION

9. In order for a bipolar transistor to respond as an amplifier the emitter-base junction must be (forward, reverse) biased and the collector-base must be (forward, reverse) biased.
10. The ____________ ____________ of a bipolar transistor is equal to the beta times the base current.
11. If a bipolar transistor has a beta of 100 and a base current of 80 μA, the collector current is ____________ mA.
12. When a transistor has dc voltage supplied to its elements and no signal applied, it is considered to be in its ____________ state of operation.
13. When a bipolar transistor is energized by dc and an ac signal is applied for amplification, it is considered to be in a ____________ state of operation.
14. In a common-emitter transistor circuit configuration one of the leads must be common to both the ____________ and the ____________ of the circuit.
15. The method of biasing a bipolar transistor has a great deal to do with its ____________ ____________ during operation.
16. In bipolar transistor operation when base current increases, ____________ ____________ increases.
17. In bipolar transistor operation when collector current increases, ____________ ____________ decreases.

ANSWERS

9. Forward, reverse	10. Collector current, or I_C
11. 8 mA	12. Static
13. Dynamic	14. Input, output
15. Thermal stability	16. Collector current, or I_C
17. Collector voltage, or V_C	

LOAD-LINE ANALYSIS

Earlier in this unit we looked at the operation of an amplifier with respect to its beta. Current and voltage were calculated and operation was related to these values. This method of analysis is very important. Amplifier operation can also be determined graphically. This method employs a collector family of characteristic curves. A load-line is developed for the graph. With the load-line method of analysis. it is possible to predict how the circuit will respond graphically. A great deal about the operation of an amplifier can be observed through this method.

The load-line method of circuit analysis is widely used today. Engineers use this method in designing new circuits. The operation of a specific circuit can also be visualized by this method. In circuit design a specific transistor is selected for an amplifier. Source voltage, load resistance, and input signal levels may be given values in the design of the circuit. The transistor is made to fit the limitation of the circuit.

For our application of the load line, assume that an amplifier circuit is to be analyzed. Figure 6-9(a) shows the circuit being analyzed. A collector family of characteristic curves for the transistor is shown in Fig. 6-9(b). Note that the power dissipation rating of the transistor is included in the diagram.

Power Dissipation Curve

A common practice in load-line analysis is first to develop a power dissipation curve. This gives some indication of the maximum operating limits of the transistor. Power dissipation (P_D) refers to maximum heat that can be given off by the base-collector junction. Usually, this value is rated at 25°C. P_D is the product of I_C and V_{CE}. In our circuit the P_D rating for the transistor is 300 mW.

To develop a P_D curve, each value of V_{CE} is used with the P_D rating to determine an I_C value. The formula is

$$\text{collector current} = \frac{\text{power dissipation}}{\text{collector-emitter voltage}}$$

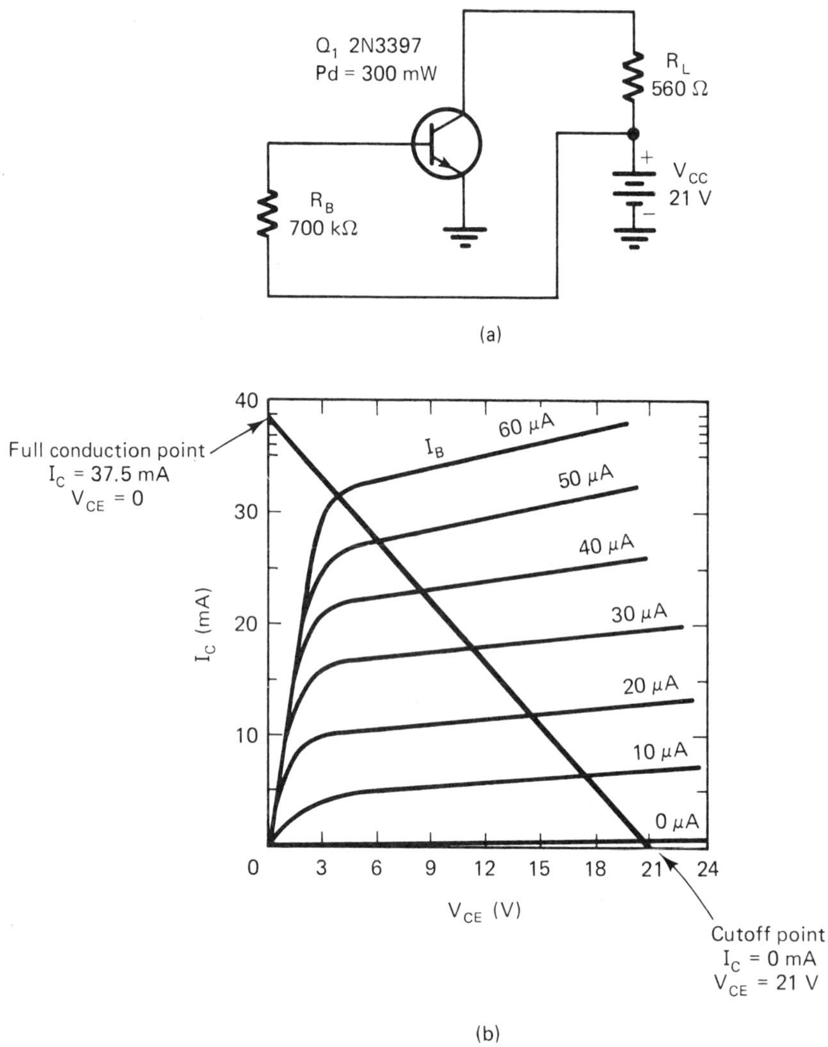

FIGURE 6-9 (a) Circuit and (b) characteristic curves of a circuit to be analyzed.

or

$$I_C = \frac{P_D}{V_{CE}}$$

Using this formula, calculate the I_C value for each of the V_{CE} values of the family of curves. Using the V_{CE} value and the corresponding calculated I_C value, note the location on the curve. These points connected together are representative of the 300-mW power dissipation curve. In practice, the load line must be located to the left of the established P_D curve. Satisfactory operation without excessive heat generation can be assured in that area of the curve.

Static Load-Line Analysis

The load line of a transistor amplifier represents two extreme conditions of operation. One of these is in the cutoff region. When the transistor is cut off, there is no I_C flowing through the device. V_{CE}

equals the source voltage with zero I_C. The second load-line point is in the saturation region. This point assumes full conduction of I_C. Ideally, when a transistor is fully conductive, $V_{CE} = 0$, and $I_C = V_{CC}/R_L$.

The two load-line construction points for the analysis circuit are shown on the curves of Fig. 6-9. The cutoff point is located at the zero I_C, 21 V_{CE} point. The value of V_{CC} determines this point. At cutoff, $V_{CE} = V_{CC}$. The saturation, or full-conduction, point is located at 37.5 mA of I_C at 0 V_{CE}. The I_C value is calculated using V_{CC} and the value of R_L. The formula is

$$I_C = \frac{V_{CC}}{R_L}$$

$$= \frac{21 \text{ V}}{560\Omega}$$

$$= 0.0375 \text{ A, or } 37.5 \text{ mA}$$

These two points are connected with a straight line.

Figure 6-10 shows a family of characteristic curves with a load line for the circuit of Fig. 6-9. Development of the load line makes it possible to determine the operating conditions of the amplifier. For linear amplification the operating point should be located near the center of the load line. In our circuit an operating point of 30 μA is used. In the circuit diagram the value of R_B determines I_B. It is calculated by the equation $I_B = V_{CC}/R_B$. The value is

$$I_B = \frac{V_{CC}}{R_B}$$

$$= \frac{21 \text{ V}}{700 \text{ k}\Omega}$$

$$= 0.00003 \text{ A, or } 30\ \mu\text{A}$$

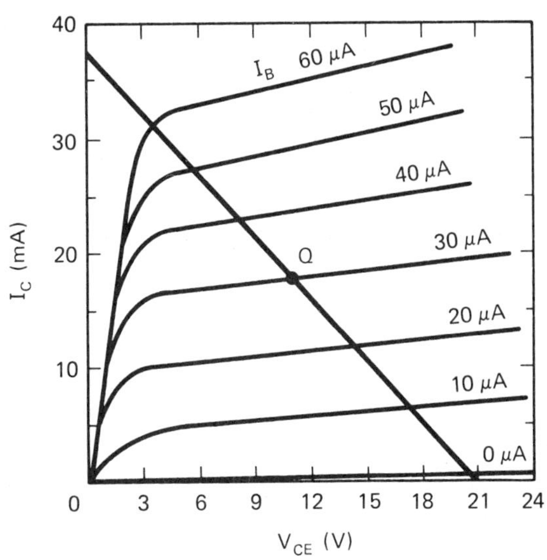

FIGURE 6-10 Family of characteristic curves for Fig. 6-9.

The operating point for this value is located at the intersection of the load line and the 30-μA I_B curve. It is indicated as point Q. Knowing this much about an amplifier shows how it will respond in its steady, or static, state. The Q point shows how the amplifier will respond without an applied signal.

Operation of our amplifier in its static state is displayed by the family of curves in Fig. 6-10. Projecting a line from the Q point to the I_C scale shows the resulting collector current. In this case the I_C is 17.5 mA. The dc beta for the transistor at this point is determined by the formula $\beta = I_C/I_B$. This value is

$$\beta = \frac{I_C}{I_B}$$

$$= \frac{17.5 \text{ mA}}{30 \ \mu\text{A}}$$

$$= \frac{0.0175 \text{ A}}{0.000030 \text{ A}}$$

$$= 583.3$$

The resulting V_{CE} that will occur for the amplifier can also be determined graphically. Projecting a line directly down from the Q point will show the value of V_{CE}. In our circuit V_{CE} is approximately 11 V. This means that 10 V will appear across R_L when the transistor is in its static state.

Dynamic Load-Line Analysis

Dynamic load-line analysis shows how an amplifier will respond to an ac signal. In this case the collector curves and circuit of Fig. 6-11 are used. A load line and Q point for the circuit have been developed on the curves. This establishes the static operation of the amplifier. Note the values of V_{CE} and I_C in the static state.

Assume now that a 0.1-V peak-to-peak ac signal is applied to the input of the amplifier. In this case the signal will cause a 20-μA peak-to-peak change in I_B. During the positive alternation of the input, I_B will change from 30 μA to 40 μA. This is shown as point P on the load line. During the negative alternation, I_B will drop from 30 μA to 20 μA. This is indicated as point N on the load line. In effect, this means that 0.1 V p-p causes I_B to change 20 μA p-p. The I_B signal extends to the right of the load line. Its value is shown as ΔI_B.

To show how a change in I_B influences I_C, lines are projected to the left of the load line. Note the projection of lines P, Q, and N toward the I_C values. The changing value of I_C is indicated as ΔI_C. An increase and decrease in I_B causes a corresponding increase and decrease in I_C. This shows that I_C and I_B are in phase.

The ac beta of the amplifier can be determined by ΔI_C and ΔI_B. First determine the peak-to-peak ΔI_C and ΔI_B values. Then divide ΔI_C by ΔI_B. Determine the ac beta of the amplifier circuit. Using the same procedure, determine the dc beta of the transistor

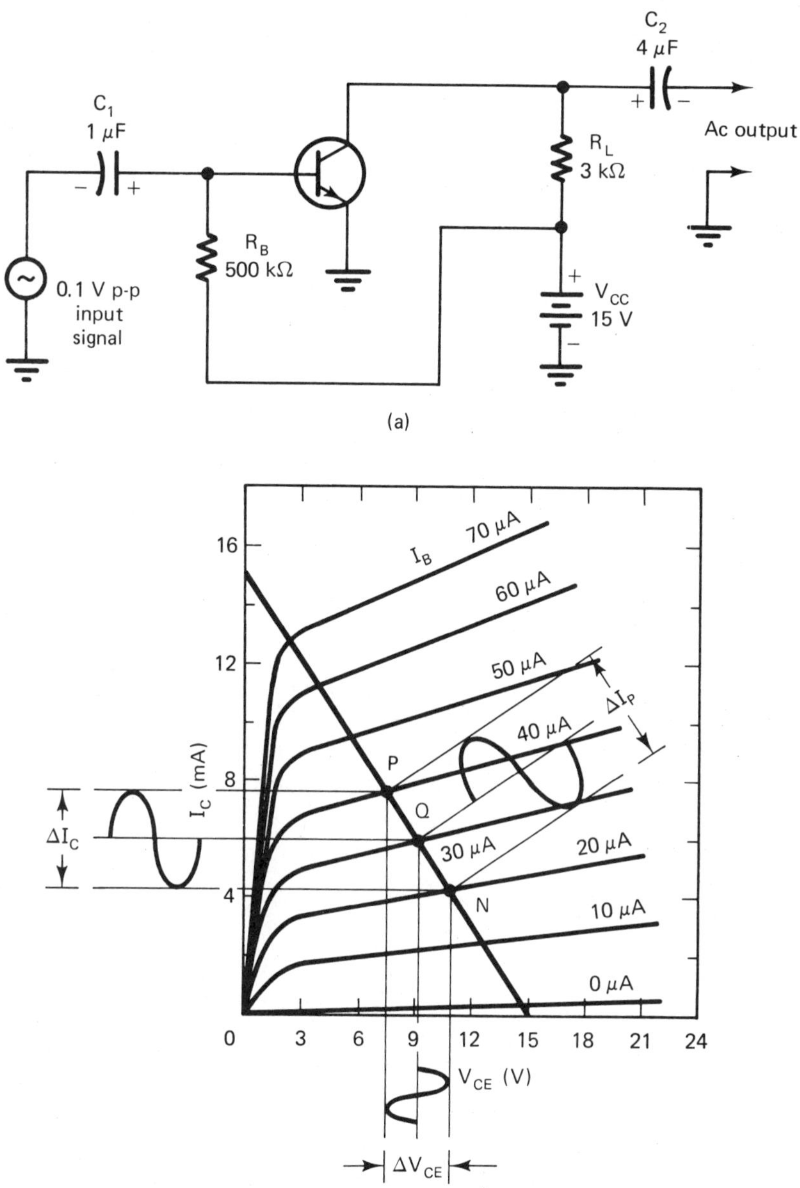

FIGURE 6-11 Dynamic load line: (a) circuit; (b) characteristic curves.

at point Q. How do the ac and dc beta values of this amplifier compare?

Projecting points P, Q, and N downward from the load line shows how V_{CE} changes with I_B. The value of V_{CE} is indicated as ΔV_{CE}. Note that an increase in I_B causes a decrease in the value of V_{CE}. A decrease in I_B causes V_{CE} to increase. This shows that I_B and V_{CE} are 180° out of phase. The difference in V_{CE} at any point appears across R_L.

The ac voltage gain of the amplifier can be determined from the dynamic load line. Remember that 0.1 V p-p input caused a change of 20 μA p-p in the I_B signal. The ac voltage gain can be determined by dividing ΔV_{CE} by ΔV_B. The ΔV_B value is 0.1 V p-p. Using the ΔV_{CE} value from the graph and ΔV_B, determine the ac voltage gain of the amplifier circuit.

Linear and Nonlinear Operation

The V_{CE}, or output, of an amplifier should be a duplicate of the input with some gain. When a sine wave is applied, the output should develop a sine wave. When an amplifier operates in this manner it is considered to be *linear.* For this to be achieved, the amplifier must operate in or near the center or linear area of the collector curves. As a rule, nonlinearity occurs near the saturation and cutoff regions. If the operating point is adjusted near these regions it usually causes the output to be distorted. Normally, this is called *nonlinear distortion.*

Figure 6-12 shows how an amplifier will respond at three different operating points. Part (a) shows linear operation. The input and output are duplicates in this case. Part (b) shows operation in the saturation region. Note that the top of the input wave distorts the negative alternation of the output voltage. Part (c) shows operation near the cutoff region. The lower part of the input wave causes distortion of the positive alternation. As a general rule, nonlinear distortion can be tolerated in some applications. In other applications nonlinear distortion is very obvious.

An input signal that is too large can also produce distortion. Figure 6-13 shows how an amplifier will respond to a large input signal. Note that the input signal swings into the saturation region during the positive alternation and the cutoff region during the negative alternation. This condition causes distortion of both alternations of the output. Normally, this condition is described as *overdriving.* In a radio receiver or stereo amplifier, overdriving causes speech or music to sound bad. This usually occurs when the volume control is turned too high. The audio or sound signal has its peaks clipped. The volume level may be higher, but the quality of reproduction is usually very poor when this occurs. It is interesting to note that the operating point is near the center of the active region. Overdriving can occur even with a properly selected operating point.

Classes of Amplification

Transistor amplifiers are frequently classified according to their bias operating point. This means of classification describes the shape of the output wave. Three general groups of amplifiers are class A, class B, and class C. Figure 6-14 shows a graphical display of these three amplifier classes.

Class A amplifiers generally have linear operation. The bias operating point is set near the center of the active region. With a sine wave applied to the input, the output is a complete sine wave. Figure 6-14(a) shows the input-output waveforms of a class A amplifier. This type of amplifier is used when a true reproduction of the input signal is required.

Figure 6-14(b) shows the input-output waveforms of a class B amplifier. The operating point of this amplifier is adjusted near

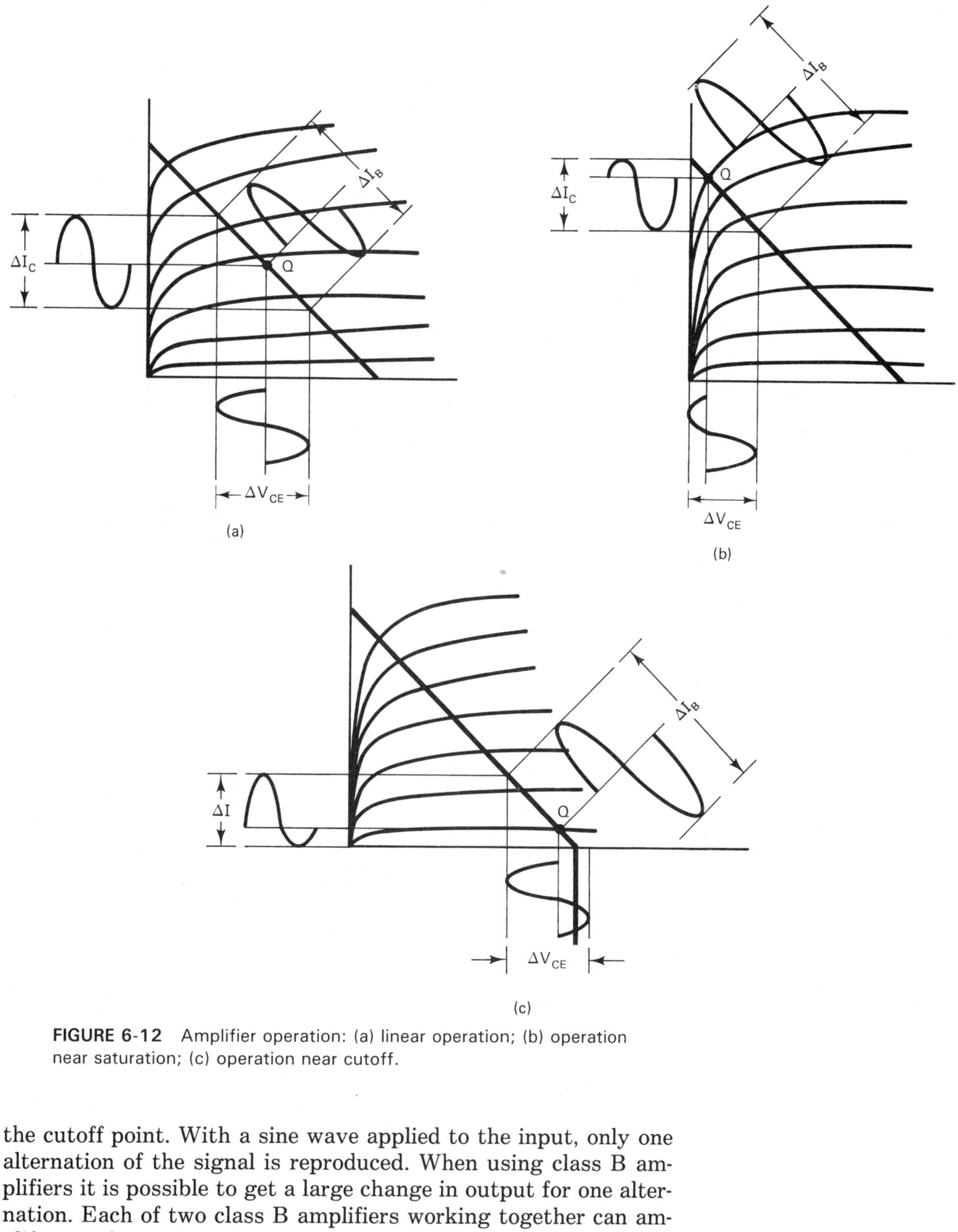

FIGURE 6-12 Amplifier operation: (a) linear operation; (b) operation near saturation; (c) operation near cutoff.

the cutoff point. With a sine wave applied to the input, only one alternation of the signal is reproduced. When using class B amplifiers it is possible to get a large change in output for one alternation. Each of two class B amplifiers working together can amplify one alternation of a sine wave. When the wave is restored, a complete sine wave is developed. Class B amplifiers are commonly used in push-pull audio output circuits. These amplifiers are usually concerned with power amplification.

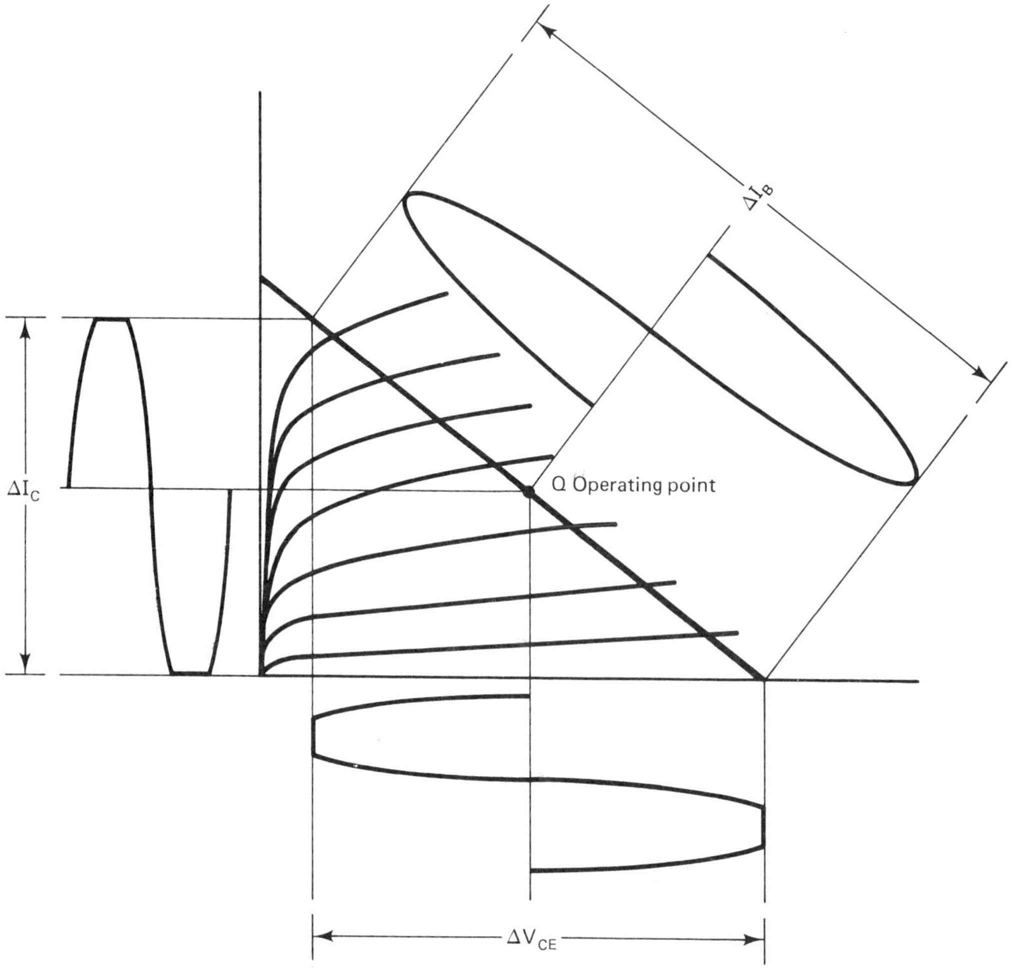

FIGURE 6-13 Overdriven amplifier.

A class C amplifier is shown in Fig. 6-14(c). In this amplifier the bias operating point is below cutoff. With a sine wave applied to the input, the output is less than half of one alternation. The primary application of class C amplifiers today is to radio-frequency circuits. The operational efficiency of this amplifier is quite high. It consumes energy only for a small portion of the applied sine wave.

SELF-EXAMINATION

18. With the load-line method of bipolar transistor analysis, it is possible to predict how a device will respond ______________.
19. The ______________ ______________ rating of a bipolar transistor is the product of collector-emitter voltage and collector current.
20. The ______________ ______________ rating of a bipolar

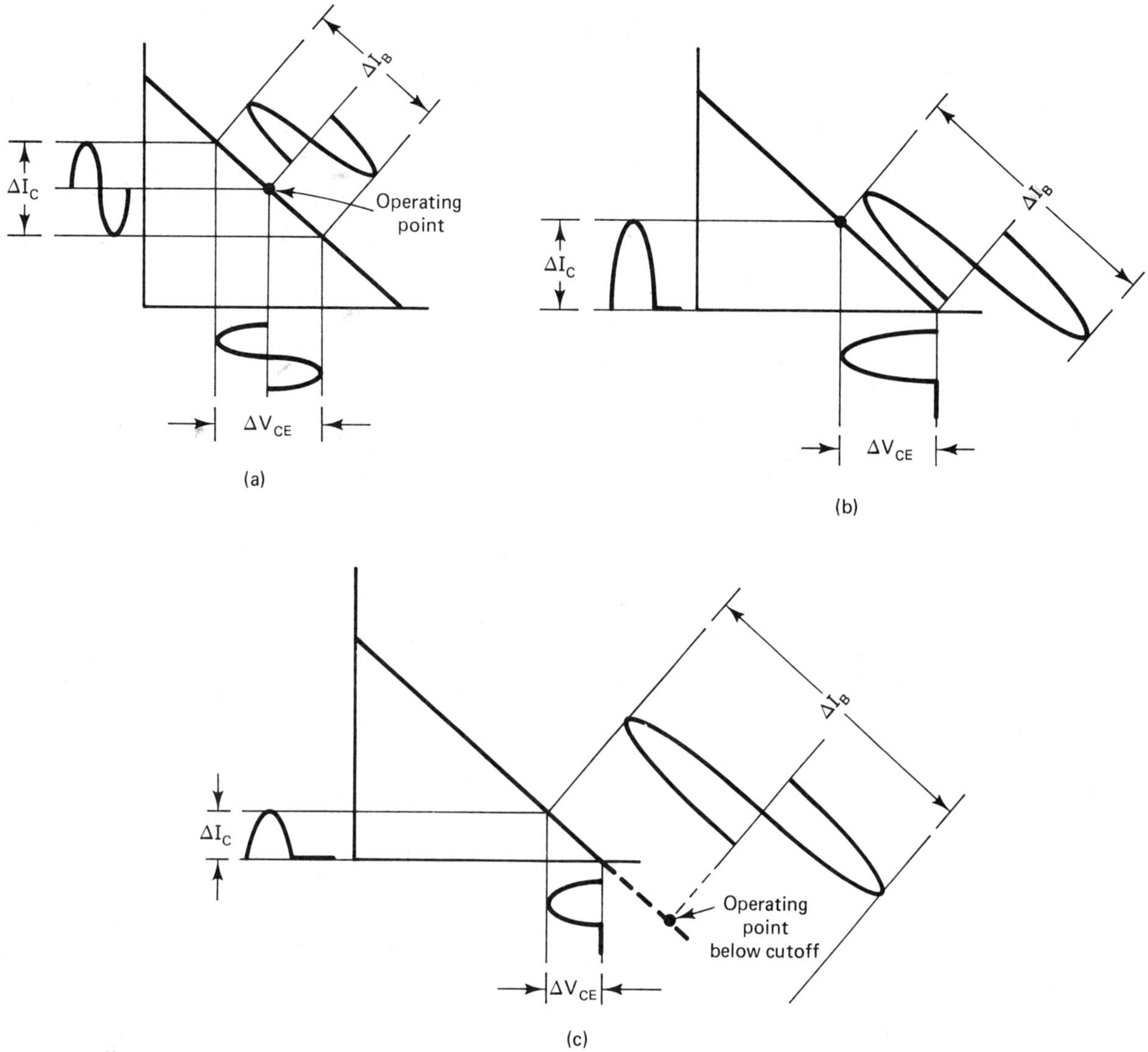

FIGURE 6-14 Classes of amplification: (a) class A; (b) class B; (c) class C.

transistor refers to the maximum heat that can be given off by the base-collector junction.

21. When a load line is set up for a bipolar transistor, the two extreme conditions of operation are ____________ and ____________.
22. When a bipolar transistor is used as a linear amplifier, operation should be near the ____________ of the load line.
23. Nonlinear amplification occurs when operation takes place in or near the ____________ and ____________ regions.
24. ____________ amplification is best achieved when a transistor operates near the center of the load line.

25. If the input signal of a bipolar amplifier is too large, it causes the output signal to be ___________.
26. A class (A, B, C) amplifier has sine-wave input and full, undistorted sine-wave output.
27. The bias operating point of a class (A, B, C) amplifier is near the center of the active region.
28. A class (A, B, C) amplifier has sine-wave input and half-sine wave or one alternation of output.
29. A class (A, B, C) amplifier has its bias operating point adjusted below cutoff.

ANSWERS

18. Graphically	19. Power dissipation
20. Power dissipation	21. Cutoff, saturation
22. Center	23. Saturation, cutoff
24. Linear	25. Distorted
26. A	27. A
28. B	29. C

TRANSISTOR CIRCUIT CONFIGURATIONS

The elements of a transistor can be connected together in one of three different circuit configurations. These are usually described as common-emitter, common-base, and common-collector. Of the three transistor leads, one is connected to the input and one to the output. The third lead is commonly connected to both the input and output. The common lead is generally used as a reference point for the circuit. It is usually connected to the circuit ground or common point. This has brought about the terms *grounded emitter, grounded base,* and *grounded collector.* The terms common and ground mean the same thing. In some circuit configurations the emitter, base, or collector may be connected directly to ground. When this occurs, the lead is at both dc and ac ground potential. When the lead goes to ground through a battery or resistor that is bypassed by a capacitor, it is at an ac ground only.

Common-Emitter Amplifier

The common-emitter amplifier is a very important transistor circuit. A very high percentage of all amplifiers in use today are of the common-emitter type. The input signal of this amplifier is applied to the base, and the output is taken from the collector. The emitter is the common or grounded element.

Figure 6-15 shows a circuit diagram of the common-emitter amplifier. This circuit is very similar to the basic amplifier used in the first part of the unit. In effect, we have used this circuit to

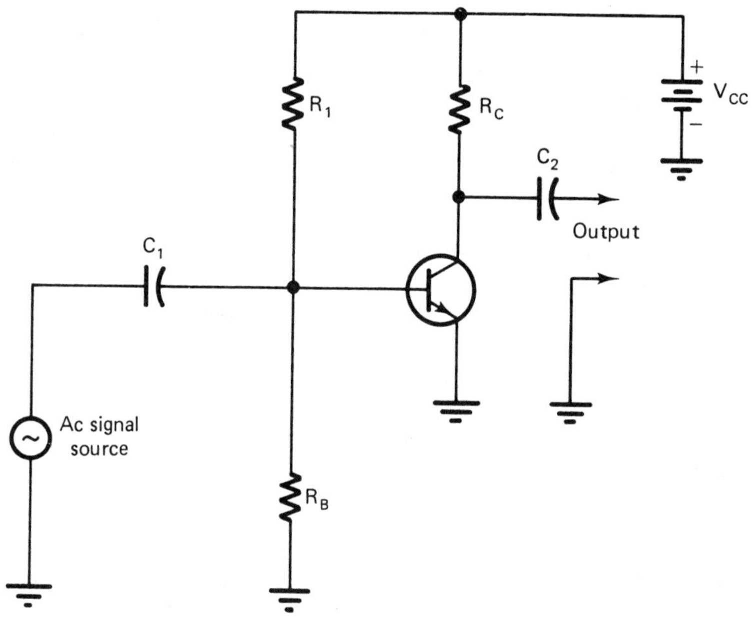

FIGURE 6-15 Common-emitter amplifier.

acquaint you with amplifiers in general. It is presented here to make a comparison with the other circuit configurations.

The signal being amplified by the common-emitter amplifier is applied to the emitter-base junction. This signal is superimposed on the dc bias of the emitter-base. The base current then varies at an ac rate. This action causes a corresponding change in collector current. The output voltage developed across the collector-emitter is inverted 180°. The current gain of the circuit is determined by the beta. Typical beta values are in the range of 50. Voltage gain of the circuit ranges from 250 to 500. Power gain is in the range of 20,000. Input impedance is moderately high, typical values being 100 Ω. Output impedance is moderate, typical values being 2000 Ω. In general, common-emitter amplifiers are used in small-signal applications or as voltage amplifiers.

Common-Base Amplifiers

A common-base amplifier is shown in Fig. 6-16. In this type of amplifier, the emitter-base junction is forward biased and the collector-base junction is reverse biased. In this circuit configuration the emitter is the input. An applied input signal changes the circuit value of I_E. The output signal is developed across R_L by changes in collector current. For each value change in I_E there is a corresponding change in I_C.

In a common-base amplifier the current gain is called alpha. Alpha is determined by the formula I_C/I_E. In a common-base amplifier the value of the gain is always less than 1. Remember that $I_E = I_B + I_C$. The I_C will, therefore, always be slightly less than I_E by the value of I_B. Typical values of alpha are 0.98 to 0.99.

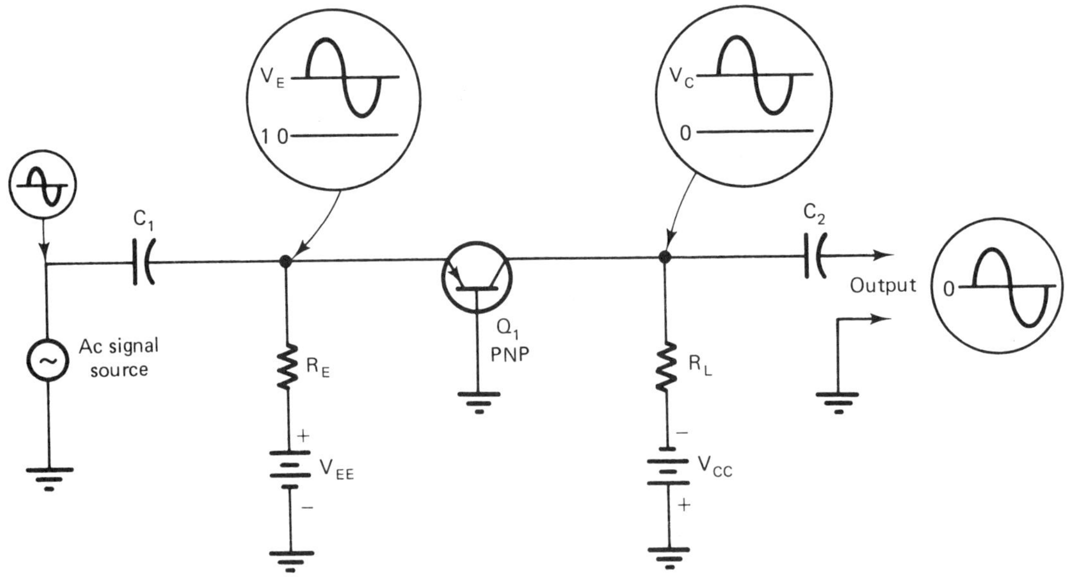

FIGURE 6-16 Common-base amplifier

In Fig. 6-16 V_{EE} forward-biases the emitter-base, whereas V_{CC} reverse-biases the collector-base. Resistor R_E is an emitter-current-limiting resistor, and R_L is the load resistor. Note that the transistor is a PNP type.

When an ac signal is applied to the input, it is added to the dc operating value of V_E. The positive alternation therefore adds to the forward-bias voltage of V_E. This condition causes an increase in I_E. A corresponding increase in I_C also takes place. With more I_C through R_L, there is an increase in its voltage drop. The collector therefore becomes less negative, or swings positive. In effect, the positive alternation of the input produces a positive alternation in the output. The input and output of this amplifier are in phase. See the waveform inserts in the diagram. The negative alternation causes the same reaction, the only difference being a reverse in polarity.

A common-base amplifier has a number of unique characteristics that should be considered when this amplifier is being selected for an application. Its current gain is called alpha (α). The current gain is always less than 1. Typical values are 0.98 to 0.99. Voltage gain is usually very high. Typical values range from 100 to 2500 depending on the value of R_L. Power gain is the same as voltage gain. The input impedance is very low. Values of 10 to 200 Ω are very common. The output impedance of the amplifier is somewhat moderate. Values range from 10 to 40 kΩ. This amplifier does not invert the applied signal.

Common-base amplifiers are used primarily to match a low-impedance input device to a circuit. This type of circuit configuration is used in radio-frequency amplifier applications. As a general rule, the common-base amplifier is not used very commonly today.

Common-Collector Amplifiers

A common-collector amplifier is shown in Fig. 6-17. In this circuit the base serves as the signal input point. The input of this amplifier is primarily the same as the common-emitter circuit. The collector is connected to ground through V_{CC}. Note that the input, output, and collector are all commonly connected. The unique part of this circuit is the output. It is developed across the load resistor (R_L) in the emitter. There is no resistor in the collector circuit.

When an ac signal is applied to the input, it adds to the base current. The positive alternation increases the value of I_B above its static operating point. An increase in I_B causes a corresponding increase in I_E and I_C. With an increase in I_E, there is more voltage drop across R_L. The top side of R_L and the emitter becomes more positive. A positive input alternation therefore causes a positive output alternation across R_L. Essentially, this means that the input and output are in phase. The negative alternation reduces I_B, I_E, and I_C. With less I_E through R_L, the output swings negative. The output is again in phase for this alternation.

The common-collector amplifier is capable of current gain. A small change in I_B causes a large change in I_E. This current gain

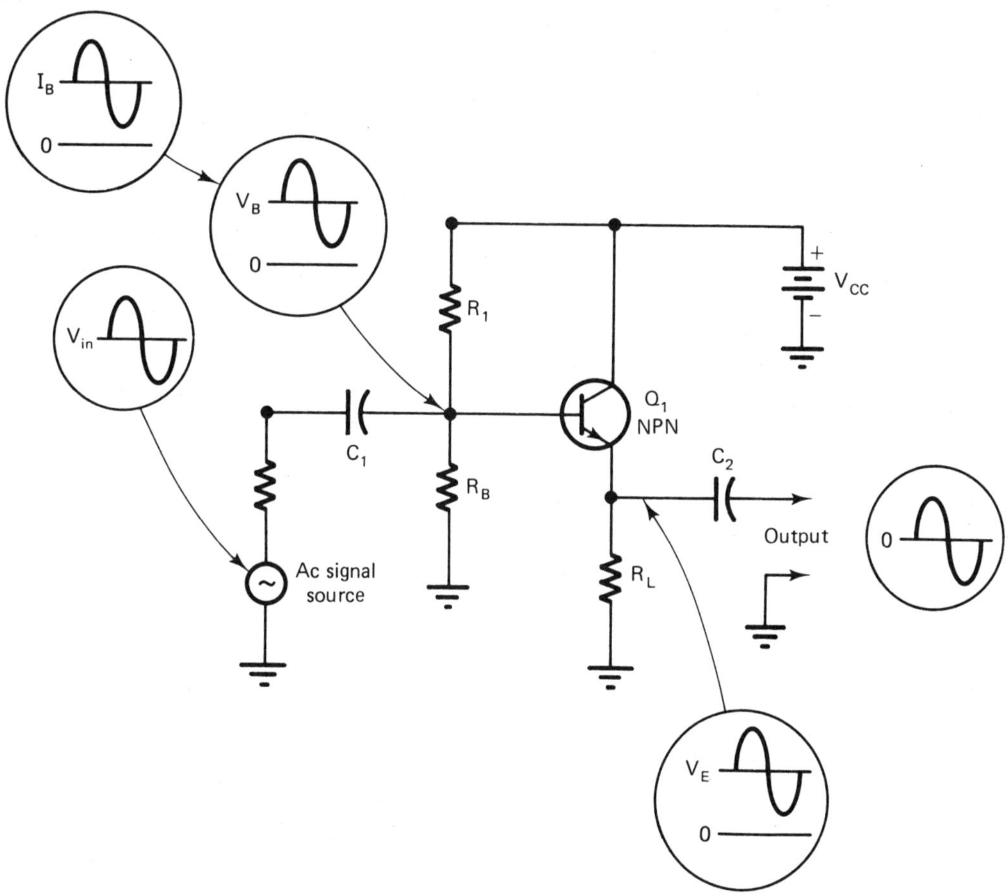

FIGURE 6-17 Common-collector amplifier.

does not have a descriptive term such as alpha or beta. In general, the current gain is assumed to be greater than 1. Values approximate those of the common-emitter amplifier.

Other factors of importance are voltage gain, power gain, input impedance, and output impedance. Voltage gain is less than 1. Power gain is moderately high, with typical values in the range of 50. Input impedance is in the range of 50 kΩ. The output impedance is quite low because of the location of R_L. Typical values are about 50 Ω.

Common-collector amplifiers are used primarily as impedance-matching devices. They can match a high-impedance device to a low-impedance load. Practical applications include preamplifiers operated from a high-impedance microphone or phonograph pickup. Common-collector amplifiers are also used as driver transistors for the last stage of amplification. In this application power transistors generally require large amounts of input current to deliver maximum power to the load device. Since the emitter output follows the base, this type of amplifier is often called an *emitter follower.*

SELF-EXAMINATION

30. The input of a common-emitter amplifier is connected to the ____________.
31. The output of a common-emitter amplifier is connected to the ____________.
32. The input and output of a common-emitter amplifier are ____________ out of phase.
33. The current gain of a common-emitter amplifier is determined by the transistor's ____________.
34. The input signal of a common-base amplifier is applied to the ____________, and the output is removed from the ____________.
35. The input and output of a common-base amplifier are ____________ phase.
36. The current gain of a common-base amplifier is called ____________.
37. The current gain of a common-base amplifier is always (less, more) than 1.
38. The input of a common-collector amplifier is applied to the ____________, and the output is derived from the ____________.
39. The input and output of a common-collector are ____________ phase.
40. The voltage gain of a common-collector amplifier is (less, more) than 1.

ANSWERS

30. Base	31. Collector
32. 180°	33. Beta
34. Emitter, collector	35. In
36. Alpha	37. Less
38. Base, emitter	39. In
40. Less	

SUMMARY

Voltage amplification is a process where the output signal voltage is greater than the input signal voltage. Current amplification deals with circuits that increase the input signal current to a greater output current. Power amplification deals with a combination of voltage and current gain in a transistor amplifier.

In the operation of a bipolar transistor amplifier, the emitter-base junction must be forward biased and the collector-base reverse biased. Specific operating voltages are selected that will permit amplification. If amplification and reproduction are to be achieved the device must operate in the center of its active region.

Signal gain is an important function of a bipolar transistor amplifier. When an ac signal is applied to the input, it causes the dc base current to change at an ac rate. This in turn causes the dc collector current to change at an ac rate. In effect, the amplifier has an ac component riding on the dc. Capacitors placed in the signal path block dc and pass ac. This permits an ac signal to be amplified by the transistor.

The operation of an amplifier can be analyzed graphically through the use of a load-line drawn on a family of collector curves. This line represents two extreme conditions of operation. One extreme occurs at the cutoff point. The other extreme condition of operation occurs when the transistor is in saturation or full conduction. Connecting these two points establishes the load-line for a particular transistor. Operation can be predicted by establishing a Q point and projecting lines to the collector current and collector-emitter voltage values. Selection of the operating, or Q, point shows if amplification will be linear or if distortion will occur.

Bipolar transistor amplifiers are classified according to their bias operating point. This classification is related to the shape of the resulting output waveform. Three general groups of amplifiers are class A, class B, and class C. The application of a specific amplifier determines the classification being used.

The common-emitter circuit has the input connected to the emitter-base junction and the output taken from the emitter-collector. The output voltage developed across the collector is inverted 180°. Typical values of beta are in the range of 50. Voltage gain is 250 to 500. Power gain is in the range of 2000. Input impedance is 100, and output impedance is typically 2 kΩ.

The common-base amplifier has the input connected to the emitter-base junction and the output connected to the base-collector junction. The current gain of the circuit is called alpha. The input and output of a common-base amplifier are in phase. The voltage gain is in a range of 100 to 2500. The power gain is the same as the voltage gain. Input impedance is in the range of 10 to 200 Ω, with the output impedance in the range of 10 to 40 kΩ.

The common-collector amplifier has the input connected to the base-emitter junction and the output removed from the emitter-collector junction. Power gain is in the range of 50. Voltage gain is less than 1. The input impedance is 50 kΩ and the output impedance is in the range of 50 Ω. This amplifier does not invert the signal. Common-collector amplifiers are normally used as impedance-matching devices.

EXPERIMENT 6-1
LOAD-LINE TRANSISTOR ANALYSIS

Selecting a transistor for a specific circuit application usually necessitates some evaluation of its static or dc operating conditions. With specific information of this type one can easily design a circuit to perform a particular function or predict how it will respond in specific situations. A family of collector characteristic curves is used in this procedure. A load line is drawn on the curves and used to predict how the transistor will respond in a circuit.

This procedure is used to demonstrate how the transistor will respond. The load line of a bipolar transistor indicates the two extreme conditions of operation. These are cutoff and full conduction. A power dissipation curve is calculated to show the maximum permissible operating range of the device. An operating point is then selected, and the general operation of the device can be predicted. Static characteristics and dynamic characteristics can be observed graphically. The operation of the transistor can readily be predicted from this procedure.

OBJECTIVES

1. To develop a load-line from a family of bipolar transistor curves.
2. To determine a bias operating point for an amplifier.
3. To predict how a bipolar transistor responds in a circuit graphically.

EQUIPMENT

None needed for this activity.

PROCEDURE

Part A: Static Load-Line Analysis

1. Refer to the family of collector characteristic curves in Fig. 6-1A. Assume that the transistor of Fig. 6-1B was used to develop these curves.
2. As a general rule a power dissipation curve for the transistor is determined first. This then gives some indication of the general location of the load line on the characteristic curve. Here the power dissipation of the

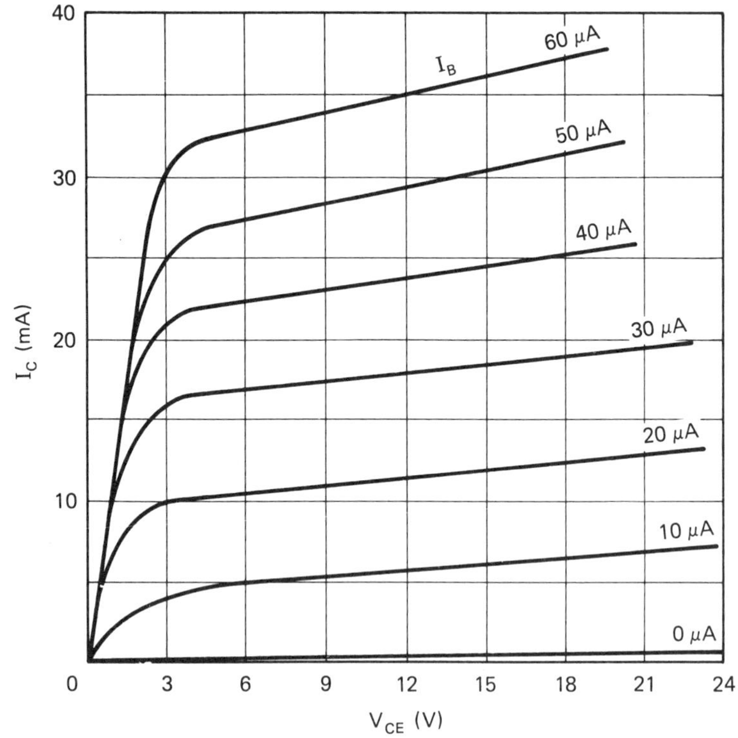

FIGURE 6-1A Collector characteristic curves (static).

transistor is obtained from the manufacturer's data. A power dissipation, or P_D, of 300 mW is used in this case.

3. Calculate the P_D curve for each value of V_{CE} or I_C by the formulas $I_C = P_D/V_{CE}$ or $V_{CE} = P_D/I_C$. If I_C and P_D are used the V_{CE} points are located. Locate these points on the graph and connect them with a curved line.
4. Ideally a load line represents the two extreme conditions of circuit operation. One extreme is full conduction of I_C with nearly zero resistance and zero V_{CE} across the transistor. This represents the X location point or vertical connecting point. The Y location point represents the nonconducting extreme of the circuit. When this occurs, I_C is zero, and V_{CE} equals source voltage V_{CC}. This is the horizontal extreme of the display.

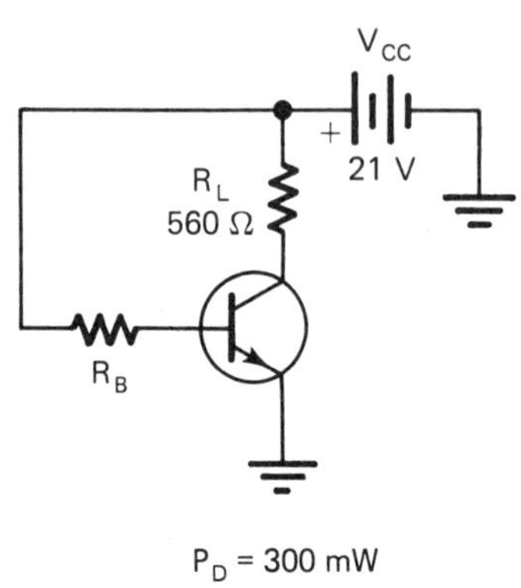

FIGURE 6-1B Transistor circuit for Fig. 6-1A.

5. To locate the Y position one simply uses the value of V_{CC} and zero I_C on the V_{CE} line. Indicate this point with a mark and label it point Y.
6. The X point of the load line is determined by calculation of the maximum I_C through the load resistor (R_L) with the applied V_{CC} value across R_L. If R_L is not known, select an I_C point on the graph at $V_{CE} = 0$ and label it X.
7. Connect points X and Y with a straight edge. Label this line the 560-Ω R_L load line.
8. With the load line developed, it now becomes possible to determine different operating conditions of the amplifier. If the base current (I_B) for this amplifier is 30 μA, then the intersection of $I_B = 30$ μA and the load line is called the quiescent, or Q, point. The Q point represents a condition when the circuit is operating with no outside signal applied. Locate the intersection of the 30-μA base-current line and the load line and label this point Q.
9. If a line is projected horizontally from the point Q to the I_C scale, this indicates the resulting I_C that will occur when an I_B of 30 μA occurs: I_C = ____________ mA.
10. The resulting V_{CE} that will occur when the transistor circuit is in this condition is determined by projecting a line from the Q point down to the V_{CE} scale: V_{CE} = ____________ V.
11. The value of the base resistor needed to produce a 30-μA Q point is determined by the formula $R_B = V_{CC}/I_B$. Calculate the value of R_B of the circuit: R_B = ____________ Ω.

Part B: Dynamic Load-Line Analysis

1. Using the family of I_C curves of Fig. 6-1C and circuit values of Fig. 6-1D, plot a dc load line.
2. Select a Q point at the 30-μA I_B line and label this point Q. The values of I_C and V_{CE} at Q are: V_{CE} = ____________ V; I_C = ____________ mA.
3. Assume now that a 20-μA peak-to-peak change in I_B occurs when an ac signal is applied to the base. During the positive alternation of the signal, I_B will change from 30 uA to 40 μA. Move up the load line to its intersection with the 40-μA I_B line and label this point P. Move down the load line to the 20-μA intersection, and label this point N. Project lines from points P, Q, and N at right angles and to the right of the load line. Draw a representative sine wave between the P and N lines with Q serving as the zero reference line.
4. Project points P, N, and Q horizontally to the I_C scale on the left of the graph. Draw a resulting I_C sine wave between these two points, with the Q point serving as the zero reference line.

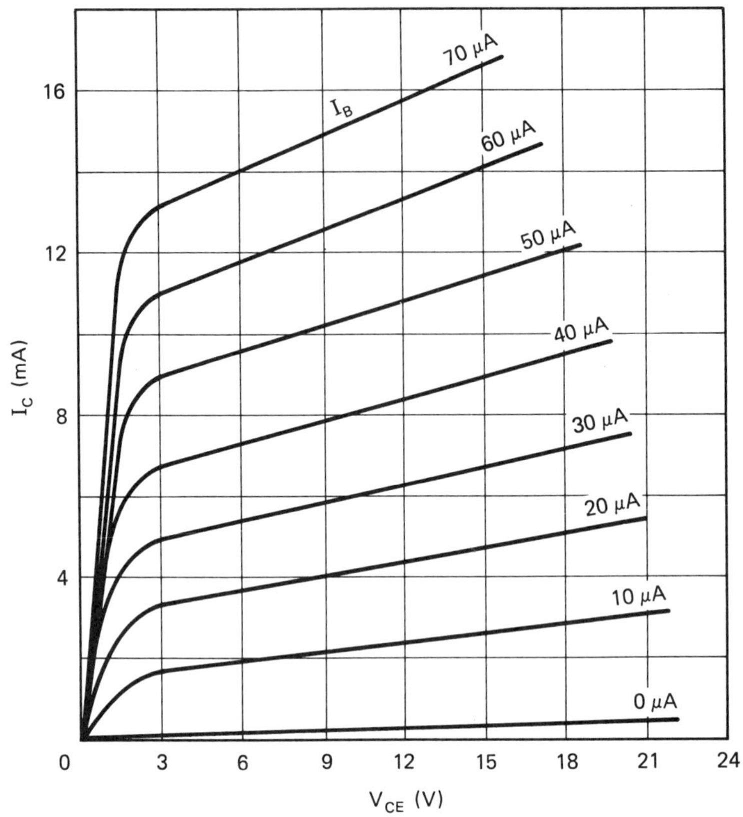

FIGURE 6-1C Collector characteristic curves (dynamic).

5. Project points P, Q, and N downward to the V_{CE} scale. Draw a resulting V_{CE} sine wave between P and N with Q as the zero reference line.
6. Calculate the ac beta (i_C/i_B) for this circuit: Ac beta = ____________.

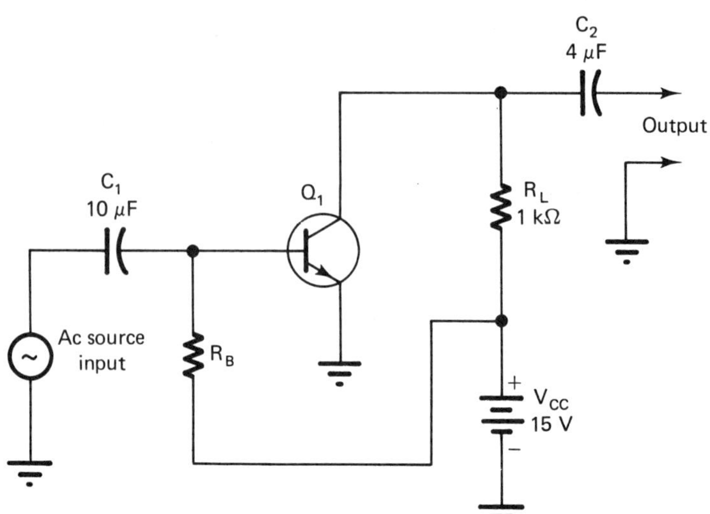

FIGURE 6-1D Transistor circuit for Fig. 6-1C.

7. If the ac signal to the base is caused by 0.1 V p-p, how much voltage gain is achieved by the amplifier? Ac voltage gain = ____________.
8. Calculate the value of R_B needed to establish the Q point of this circuit: R_B = ____________ Ω.

DISCUSSION

In this experiment a family of characteristic curves was used to develop a load-line for a specific circuit application. A power dissipation curve was developed first to establish the transistor's maximum range of operation. The load-line of a circuit must be located to the left of the dissipation curve. The extreme operating points were then located. Full conduction occurs when the transistor is low resistant and has maximum collector current. The V_{CE} is approximately zero and the transistor responds as a closed switch. Cutoff occurs when I_C is zero, and the full value of the source voltage appears across the transistor. This condition is representative of an open switch. The load-line shows how the transistor will respond between these two extreme conditions.

Once the load-line of a transistor has been developed, it can be used to predict circuit operation graphically. Dc or static conditions of operation are obtained by projecting points from a selected operating point. The bias operating point is determined by the value of the base current. This determines the Q point and is used to show how the transistor will respond when voltages are applied but no signal. When an ac signal is applied, the transistor changes from its static state to a dynamic condition of operation. This is displayed graphically by drawing sine waves for selected operating conditions.

ANALYSIS

1. What are the two extreme operating conditions represented by a load line?
2. If a smaller V_{CC} value were used in a circuit, how would it alter the load line?
3. Define the term Q point.
4. What does the power dissipation curve of a transistor show?
5. If the base current of Part B were increased by a stronger signal to 80 μA p-p, how would it alter circuit operation?
6. What is the phase relationship of I_B, I_C, and V_{CE}?

EXPERIMENT 6-2
DC TRANSISTOR AMPLIFIERS

Transistors are often used to amplify dc voltages and signals. In this circuit application the signal path between input and output must not be obstructed by capacitors or inductors. Both dc voltage and current can effectively be increased by a dc amplifier.

A very important type of dc circuit configuration has the emitter of a transistor connected to the ground or common side of both the input and output. When a dc signal is applied between the ground and base of a transistor, it will either increase or decrease the base current. A small change in base current will then cause a corresponding change in collector current. As a result, current passing through the load resistor will produce a corresponding voltage drop. An increase in collector current will therefore cause a reduction in collector-emitter voltage. A decrease in I_C will then cause a corresponding increase in the value of V_{CE}. The input and output voltages of a common-emitter amplifier therefore perform an inverting function.

OBJECTIVES

1. To construct a bipolar transistor circuit that will amplify a dc voltage signal.
2. To measure the input and output dc voltages of a bipolar transistor amplifier.
3. To evaluate the operation of a bipolar transistor amplifier.
4. To determine the current gain and voltage gain of a dc amplifier.

EQUIPMENT

NPN transistor, Motorola MPSA20 or Heath 417-801
10-kΩ, $\frac{1}{4}$-Watt resistor
1-kΩ, $\frac{1}{4}$-Watt resistor
100-kΩ, 1-W potentiometer (linear taper)
Variable dc source
Volt-ohm-milliammeter (VOM), standard or digital type
Oscilloscope

PROCEDURE

1. Connect the common-emitter dc amplifier of Fig. 6-2A.
2. Before applying dc power to the circuit from V_{CC}, adjust the potentiometer to its midrange position. Then turn on the power source and adjust it to 9 V.
3. The base current of this circuit can be measured with a VOM connected in series with resistor R_B or calculated by using the measured voltage across R_B. The voltage measurement and calculation procedure will be outlined for this circuit. Measure the voltage across resistor R_B. Adjust potentiometer R_1 to produce 0.5 V dc. Then calculate the base current ($I_B = V_{RB}/R_B$): I_B = ______.
4. Measure and record the dc emitter-base voltage: V_{BE} = ______ V.
5. Measure and record the dc emitter-collector voltage: V_{CE} = ______ V.
6. Determine the voltage gain of the transistor circuit: $A_v = V_{CE}/V_{BE}$. A_v = ______.
7. Measure and record the voltage across resistor R_L: V_{RL} = ______. Calculate the collector current of the circuit ($I_C = V_{RL}/R_L$): I_C = ______.
8. Determine the current gain or beta of the circuit: $\beta = I_C/I_B$. I_B = ______.
9. If a calibrated dc oscilloscope is available, prepare it to measure dc voltage. Connect it across the emitter-base junction of the transistor. Note the direction of horizontal trace movement when the oscilloscope is connected to the circuit. This should be up or down according to the calibrated value of the vertical amplifier of the oscilloscope. The direction of deflection is ______. The amount of deflection represents ______ V.
10. Measure the collector-emitter voltage of the transistor with the oscilloscope. The amount of deflection is ______ V. The direction of deflection is ______.

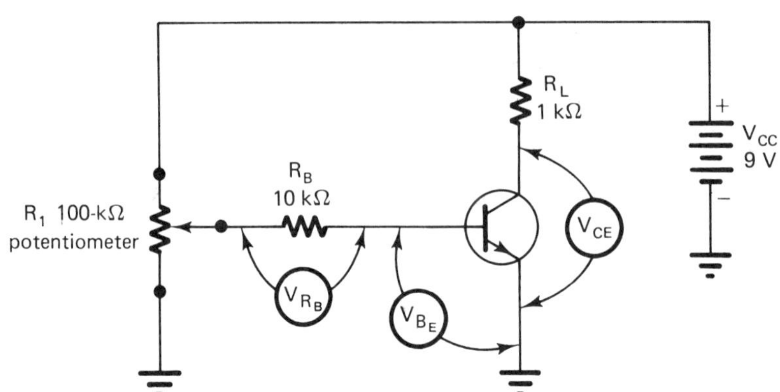

FIGURE 6-2A Dc transistor amplifier.

11. Using the procedure outlined in steps 1 through 10, adjust potentiometer R_1 to produce an R_B voltage of 0.2 V. Measure and record the designated voltages, and calculate the indicated values.

I_B =	____________	V_{BE} =	____________
V_{CE} =	____________	A_v =	____________
V_{RL} =	____________	I_C =	____________
Beta =	____________		

Direction of deflection = ____________. Amount of deflection = ____________.

V_{CE} deflection = ____________. Direction of deflection = ____________.

DISCUSSION

In this experiment you constructed a dc bipolar transistor amplifier in a common-emitter circuit configuration. Emitter-base voltage was measured and served as the input voltage of the amplifier. A change in the value of V_B caused a corresponding change in the value of V_{CE}. The voltage gain or A_v of the amplifier was calculated using output voltage (V_{out}) over input voltage (V_{in}). The current gain or beta of the amplifier was also determined. Measured voltage values across resistors were used to determine base current and collector current.

The circuit of this experiment was a common-emitter dc amplifier. A small change in base-emitter voltage caused a rather substantial change in collector-emitter voltage. This represents the voltage gain or A_v of the circuit. The amplifier also produced a change in the polarity of the input and output voltage values. This was displayed by the direction of oscilloscope display movement or by the deflection polarity of the VOM. Dc transistor amplifiers are widely used in electronic signal control circuits.

ANALYSIS

1. What determines the dc voltage amplification capabilities of a transistor amplifier?
2. Explain why an increase in I_B causes an increase in I_C and a reduction in the value of V_{CE}.
3. What is meant by the term common-emitter amplifier?

EXPERIMENT 6-3
AC TRANSISTOR AMPLIFIERS

Many of the transistor amplifier circuits in operation today are used to increase the strength of alternating signals. This type of amplifier is different from the dc type of transistor amplifier. Dc voltage is used to energize the emitter, base, collector, and other components of the transistor amplifier. When an ac signal is applied, it causes each dc value to be changed at an ac rate. As a result, this amplifier actually has an ac component riding on a specific dc component. The ac voltage gain and beta are usually somewhat different from the dc values.

When an ac signal is applied to the transistor amplifier, it causes the base voltage to fluctuate at an ac rate. The positive half of the applied signal causes an increase in the value of base voltage. This in turn increases the base current and causes a corresponding increase in both emitter current and collector current. As a result, the collector-emitter voltage is reduced because of the increased voltage drop across the load resistor. The negative alternation of an ac signal causes a decrease in base current, which then causes a corresponding decrease in emitter and collector current. As a result, the collector-emitter voltage rises in value due to the reduced collector current through the load resistor. The input and output signals of a common-emitter amplifier are 180 degrees out of phase.

OBJECTIVES

1. To construct a bipolar transistor ac amplifier and test its operation.
2. To measure the operating voltages of a bipolar transistor ac amplifier.
3. To observe the input and output signals of a bipolar transistor ac amplifier with an oscilloscope.

EQUIPMENT

NPN transistor, Motorola MPSA20 or Heath 417-801
100-Ω, $\frac{1}{4}$-W resistor
1-kΩ, $\frac{1}{4}$-W resistor
4.7-kΩ, $\frac{1}{4}$-W resistor
100-kΩ, 1-W potentiometer (linear taper)
Variable dc source

Volt-ohm-milliammeter (VOM), standard or digital
Oscilloscope (optional)
Signal generator or ac signal source
10-μF, 50-V capacitor
0.1-μF, 50-V capacitor

PROCEDURE

1. Construct the ac common-emitter amplifier of Fig. 6-3A.
2. Before connecting the ac signal source to the circuit, turn on the dc power source and adjust it to produce 9 V dc. Connect the VOM across resistor R_3. Adjust potentiometer R_2 to produce a voltage value of 1.4 V. Then connect the ac source to the input and adjust it to produce a signal of 60 Hz with a value of approximately 0.34 V rms, or 1 V p-p. This signal should be measured across R_3 with an ac voltmeter or an oscilloscope. The signal can be derived from a signal generator or a low-voltage transformer. Figure 6-3A shows two possible ac sources.
3. Connect an ac voltmeter or oscilloscope to the designated output of the circuit. Adjust the input signal to produce the greatest output signal. Record the ac volt-

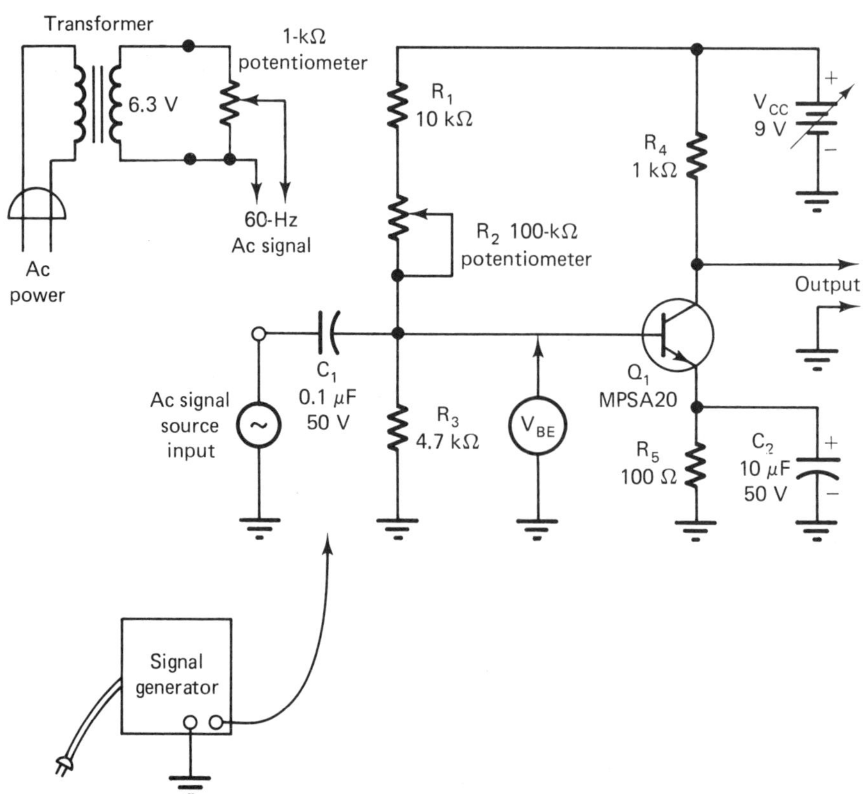

FIGURE 6-3A Common-emitter ac amplifier circuit.

age value of the output signal: ____________ Vac. Measure the input signal voltage across resistor R_3. Record the input voltage value: ac input voltage = ______ V.

3A. (Optional step). If an oscilloscope is available, connect it to the output and adjust the input to produce the greatest output with a minimum of distortion. Make a sketch of the output wave in the provided space of Fig. 6-3B. Record the peak-to-peak voltage value of the output signal. Measure the input signal voltage across resistor R_3 with the oscilloscope. Make a sketch of the input wave in the provided space of Fig. 6-3B. Record the peak-to-peak voltage value of the input signal.

4. Using the input and output voltage values of the previous step, determine the ac voltage gain. A_v = output voltage/input voltage: A_v = ____________.

4A. (Optional step). Connect the oscilloscope across the des-

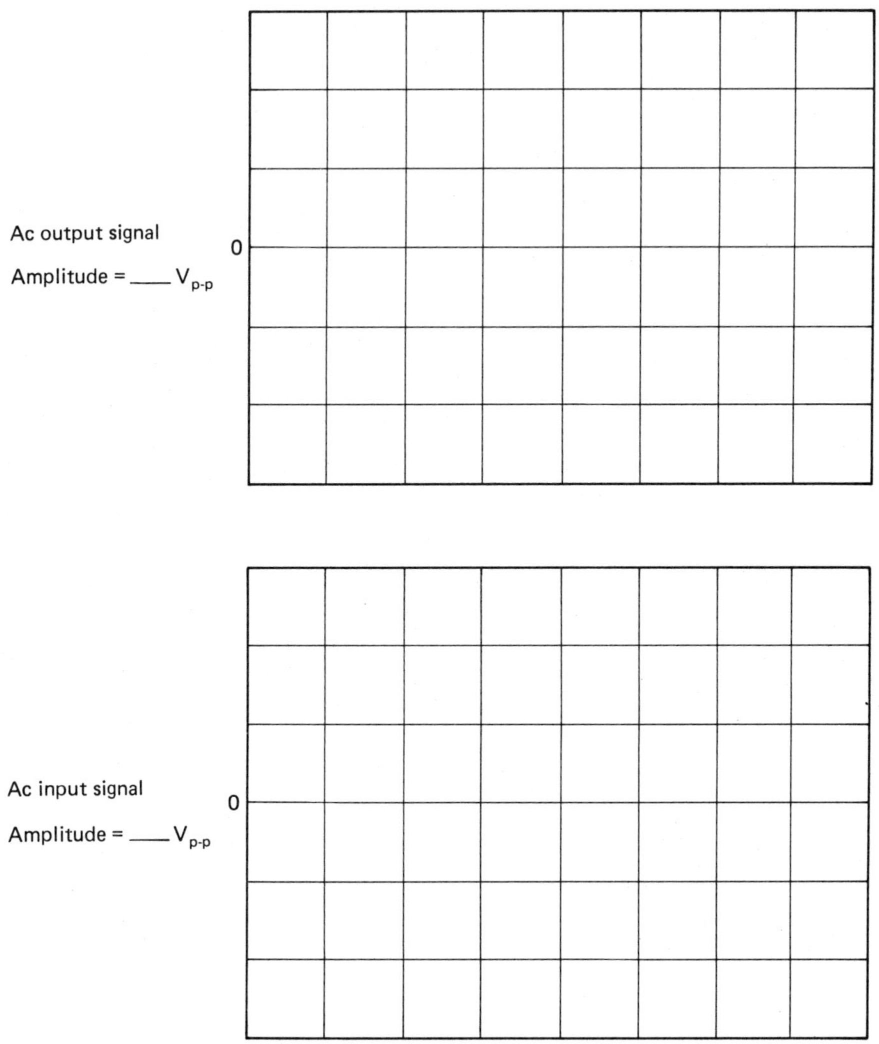

FIGURE 6-3B Ac input and output signals.

ignated output and adjust the input signal until distortion occurs. Describe the signal appearance. __________

____.

4B. (Optional step). Reduce the input signal level until a minimum of distortion occurs. Prepare the oscilloscope for external sync operation. Connect the external sync input of the oscilloscope to the transistor output. Connect the vertical input of the oscilloscope to the designated output. Note the phase of the output signal. Move the vertical input probe of the oscilloscope to the input of the amplifier, which appears across resistor R_3. Note the phase of the input signal. Connect the vertical probe of the oscilloscope across the emitter resistor (R_E). Note the phase of the emitter voltage signal. Describe the phase relationship of the input, output, and emitter signals of the common-emitter amplifier.

5. Disconnect the ac input signal from the circuit. With the VOM measure and record the following dc voltage values:
   V_{R1} = ____________, V_{R2} = ____________,
   V_{R3} = ____________, V_{RE} = ____________,
   V_{RL} = ____________, V_{CE} = ____________.
6. Using the value of V_{RL}, calculate the collector current ($I_C = V_{RL}/R_L$): I_C = ____________.
7. Using the value of V_{RE}, calculate the emitter current ($I_E = V_{RE}/R_E$): I_E = ____________.
8. Base current can be determined by subtracting collector current from emitter current ($I_B = I_E - I_C$): I_B = ____________.
9. Determine the beta of the transistor amplifier: Beta = ____________.

DISCUSSION

In this experiment you constructed a common-emitter amplifier that employs a bias method independent of transistor beta. With dc energy supplied to the circuit and no ac signal, R_1, R_2, and R_3 formed a voltage divider across the dc source. R_1 and R_2 were combined and served as one resistor of the divider. The base voltage V_B appeared across resistor R_3 of the divider. This was determined by the formula $V_B = (R_3/R_1 + R_2 + R_3) \times V_{CC}$. This voltage was adjustable because of the potentiometer used for R_2. In effect 1.4 V should have appeared across R_3 as V_B. The transistor of this experiment needed approximately 0.5 V of emitter-base voltage to be conductive. Emitter voltage V_E was developed across the emitter resistor.

When an ac signal was applied to the input of the amplifier, it caused a change in the emitter-base voltage. This voltage change caused the base current to increase or decrease. A change in base current caused a corresponding change in collector current. In effect, the base current controlled the collector current to produce current gain. The voltage drop across R_L caused a decrease in collector-emitter voltage. This was responsible for causing the emitter-base voltage and collector-emitter voltage to be out of phase with each other. If an oscilloscope was used, you would have been able to see the phase relationship of V_{BE}, V_E, and V_{CE}. The voltage gain of the transistor was also displayed by the oscilloscope.

ANALYSIS

1. Explain how the amplifier of this activity achieves ac amplification with respect to changes in the value of V_{BE}, I_C, I_B, I_E, and V_{CE}.

UNIT 6 EXAMINATION
BIPOLAR TRANSISTOR AMPLIFIERS

Instructions: Circle the response that most correctly completes each statement.

1. When a bipolar transistor is used as an amplifier it is:
   a. Energized by ac
   b. Energized by dc
   c. Energized from the signal being processed
   d. Not energized from an external source
2. When the input of a bipolar transistor shows an output signal that is the same but has an increase in amplitude, it shows:
   a. Reproduction only
   b. Amplification only
   c. Combined amplification and reproduction
   d. No signal change
3. The input voltage of a bipolar transistor amplifier is 0.1 V p-p, and the output is 6 V p-p. The A_v of this amplifier is in the range of:
   a. 0 to 0.02
   b. 0.021 to 1.0
   c. 1.01 to 10
   d. 10.01 to 100
   e. 100.01 or more
4. The base current of a bipolar transistor is 100 μA, and the collector current is 10 mA. In what range does the value of beta fall?
   a. 0.1 to 1.0
   b. 1.01 to 10
   c. 10.01 to 100
   d. 100.1 to 1000
5. Which of the following is not considered to be a beta dependent method of biasing a bipolar transistor?
   a. Fixed biasing
   b. Self-biasing
   c. Emitter biasing
   d. Resistor divider biasing
6. Which of the following biasing methods is considered to be independent of beta?

a. Fixed biasing
b. Self-biasing
c. Resistor divider biasing
d. Self-emitter biasing

7. A bipolar transistor has a power dissipation rating of 300 mW. What is the maximum collector current that can be achieved with a collector-emitter voltage of 15 V?
   a. 1.0 to 15 mA
   b. 15.1 to 30 mA
   c. 30.1 to 45 mA
   d. 45.1 or more
8. The load line of a bipolar transistor represents the extreme operating conditions of:
   a. Full conduction to saturation
   b. Cutoff to the active region
   c. Saturation to the active region
   d. Saturation to cutoff
9. Dynamic operation of a bipolar transistor represents:
   a. Dc operation
   b. Ac operation
   c. Active region operation
   d. Nonlinear operation
10. Class A bipolar transistor amplifiers have the bias operating point set:
    a. Near the center of the active region
    b. At cutoff
    c. At saturation
    d. In the lower third of the active region
11. Class B bipolar transistor amplifiers have the bias operating point set:
    a. Near the center of the active region
    b. At cutoff
    c. At saturation
    d. In the lower third of the active region
12. In a common-emitter amplifier the output signal would be:
    a. Reduced because of reverse biasing
    b. The same phase as the input
    c. Unchanged when compared with the input
    d. A 180° inversion of the input signal
13. In a common-emitter circuit configuration an increase in base current would cause:
    a. An increase in V_{CE}
    b. An increase in I_C

c. A decrease in I_C
d. An increase in V_E

14. A 180° phase inversion between the input and output is accomplished in:
a. All transistor circuit configurations
b. Only the common-base circuit configuration
c. Only the common-collector circuit configuration
d. Only the common-emitter circuit configuration

15. The current gain or alpha of which amplifier is less than 1?
a. Common-base
b. Common-emitter
c. Common-collector
d. All transistor circuit configurations

16. The current gain of a common-emitter amplifier is:
a. I_C/I_E
b. Alpha
c. Always less than 1
d. I_C/I_B

17. The output of a common-collector amplifier is developed:
a. Across the emitter-collector
b. At the collector
c. Across the collector resistor
d. Across the emitter resistor

18. The voltage gain of a common-collector amplifier is:
a. Less than 1
b. Equivalent to the beta
c. Typically 100
d. Always greater than 1

19. In a common-base amplifier the input impedance is:
a. Very low
b. Very high
c. The same as the other two circuit configurations
d. Variable because of the biasing circuit

20. In the three transistor amplifier configurations the input circuit is:
a. The same for all
b. The same for the common-emitter and the common-collector circuits
c. Different for only the common-emitter circuit
d. Different for only the common-collector circuit

UNIT SEVEN

Field-Effect Transistors

UNIT INTRODUCTION

A transistor that has only one P-N junction in its construction is called a *unipolar* device. Unlike the bipolar transistor, a unipolar transistor conducts current through a single piece of semiconductor material. The current carriers of this device have one polarity.

Field-effect transistors (FETs) are unipolar devices. These devices represent a unique part of the semiconductor field. An FET is primarily a three-terminal device. It is capable of amplification, switching, and most of the functions of a bipolar transistor. FETs are used as discrete devices or can be fabricated into integrated circuits.

FETs fall into two general classifications. One type is called a *junction field-effect transistor* (JFET). This type of device was developed first. The second group is known as an *insulated gate field-effect transistor* (IGFET). The construction of this type of device is quite different. The gate is insulated from the other terminals. This type of device is more commonly referred to as a *metal-oxide semiconductor field-effect transistor* (MOSFET). Both classifications of the FET play an important role in solid-state electronics.

UNIT OBJECTIVES

Upon completion of this unit, you will be able to:

1. Describe the physical construction of junction field-effect transistors and metal-oxide field-effect transistors.
2. Understand the fundamental operation of JFETs and MOSFETs.
3. Explain how the elements of a JFET and MOSFET are biased to make the device operational.
4. Define some of the basic terms associated with FETs, such as pinch-off voltage, modes of operation, transconductance, depletion, and enhancement.
5. Inspect the characteristic curves of an FET and explain the relationship of drain current, gate voltage, and source-drain voltage.
6. Test an FET to determine if it is good or defective and identify its leads.
7. Identify the three basic FET circuit configurations and evaluate the characteristic differences.

IMPORTANT TERMS

Before reading Unit 7, review the following terms. These terms provide a basic understanding of terms associated with field-effect transistors.

Channel The controlled conduction path of a field-effect transistor.

Conductance A measure of the ease with which current carriers pass through a device. The unit of conductance is the mho or siemens.

Depletion region The area of a P-N junction where the number of current carriers has diminished.

Discrete Electronic circuits built from separate or finished components.

Drain The output terminal of a field-effect transistor.

Enhancement A conduction function where current carriers are pulled from the substrate into the channel of a MOSFET.

Gate The control element of a field-effect transistor.

IGFET A field-effect transistor whose gate is insulated from the semiconductor by a thin film of insulating material.

Induced channel The channel of an E-MOSFET that is developed by gate voltage.

MOSFET A field-effect transistor containing a metal gate that is placed over a thin layer of insulating material that separates it from the channel.

Parameter A characteristic value or element.

Transconductance A measure of the ease with which current carriers move through a device.

Transfer characteristics A graphic display that shows how a change in input voltage (gate-source voltage) causes a change in output current (drain current).

Source Along with the gate and drain, one of the three regions of a field-effect transistor.

Substrate A piece of underlying N or P semiconductor material upon which a device or circuit is constructed. The construction foundation of a MOSFET.

Unipolar transistor A transistor such as a FET whose action depends on the conduction of majority carriers.

JUNCTION FIELD-EFFECT TRANSISTORS

The junction field-effect transistor (JFET) is a three-element electronic device. Its operation is based on the conduction of current carriers through a single piece of semiconductor material. This piece of material is called the *channel.* An additional piece of semiconductor material is diffused into the channel. This element is called the *gate.* The entire unit is built on a third piece of semiconductor known as the *substrate.* The assembled device is housed in a package. The physical appearance of a JFET in its housing is very similar to that of a bipolar transistor.

N-Channel JFETs

Figure 7-1 shows the crystal structure, element names, and schematic symbol of an N-channel JFET. It is constructed of a thin channel of N-material formed on the P-substrate. The P-material of the gate is then formed on top of the N-channel. Lead wires are attached to each end of the channel and the gate. No connection is made to the substrate of this device.

The schematic symbol of an N-channel JFET is somewhat representative of its construction. The bar part of the symbol refers to the channel. The drain (D) and source (S) are attached to the channel. The gate (G) has an arrow. This shows that it forms a P-N junction. The arrowhead of an N-channel symbol Points iN toward the channel. This indicates that it is a P-N junction. The arrowhead or gate is P-material, and the channel is N-material. This part of the device responds as a junction diode.

The operation of a JFET is somewhat unusual when compared with that of a bipolar transistor. Figure 7-2 shows the source and drain leads of a JFET connected to a dc voltage source. In this case, maximum current will flow through the channel. The I_S and I_D meters will show the same amount of current. The value of V_{DD} and the internal resistance of the channel determine the amount of channel current flow. Typical source-drain resistance values of a JFET are several hundred ohms. Essentially this means that full conduction will take place in the channel even when the gate is open. A JFET is considered to be a normally on device.

Current carriers passing through the channel of a JFET are controlled by the amount of bias voltage applied to the gate. In normal circuit operation the gate is reverse biased with respect to the source. Reverse biasing of the gate-source of a P-N junction increases the size of its depletion region. In effect, this restricts or depletes the number of majority carriers that can pass through the channel. This means that drain current (I_D) is controlled by the value of V_{GS}. If V_{GS} becomes great enough, no I_D will be permitted to flow through the channel. The voltage that causes this condition is called the *cutoff voltage*. I_D can be controlled any-

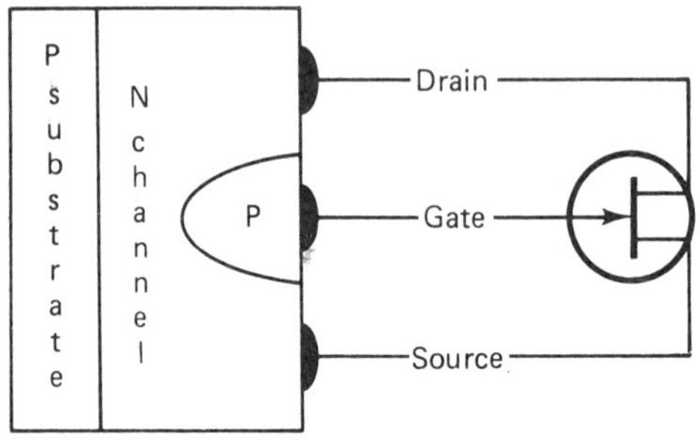

FIGURE 7-1 N-channel junction field-effect transistor.

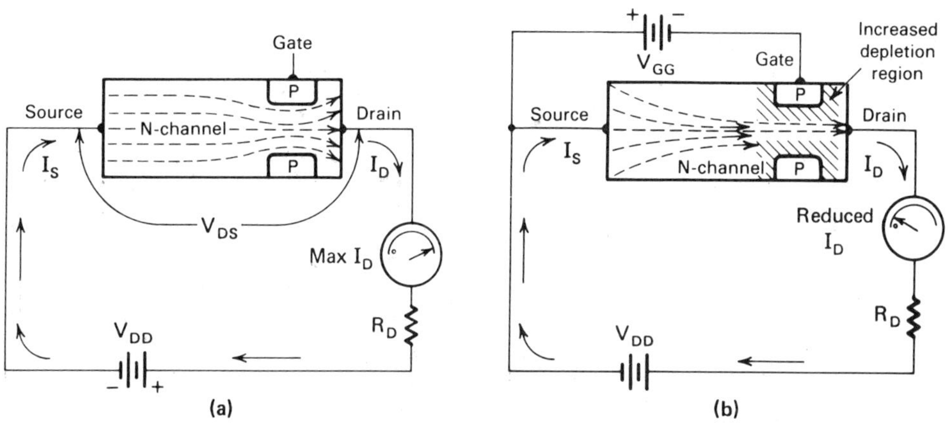

FIGURE 7-2 JFET cross-sectional view.

where between full conduction and cutoff by a small change in gate voltage.

Figure 7-2(b) shows how the JFET will respond when V_{DD} and V_{GG} are connected simultaneously. Note that the gate-source junction is reverse biased by the V_{GG} battery. This voltage causes an increase in the area of the depletion region near the gate. As a result, the number of current carriers in the channel is reduced. An increase in V_{GS} causes a corresponding decrease in I_D. Reducing the value of V_{GS} causes a substantial increase in I_D. Thus a change in the value of V_{GS} can be used to control the internal resistance of the channel. Variations in V_{GS} are therefore used to control channel current, or I_D.

P-Channel JFETs

The crystal structure, element names, and schematic symbol of a P-channel JFET are shown in Fig. 7-3. This device is constructed of a thin channel of P material formed on an N-substrate. The N-material of the gate is then formed on top of the P-channel. Lead wires are attached to each end of the channel and to the gate. Other construction details are the same as those of the N-channel device.

The schematic symbol of a JFET is different only in the gate element. In a P-channel device the arrow is Not Pointing toward the channel. This means that the gate is N-material, and the channel is P-material. In conventional operation the gate is made pos-

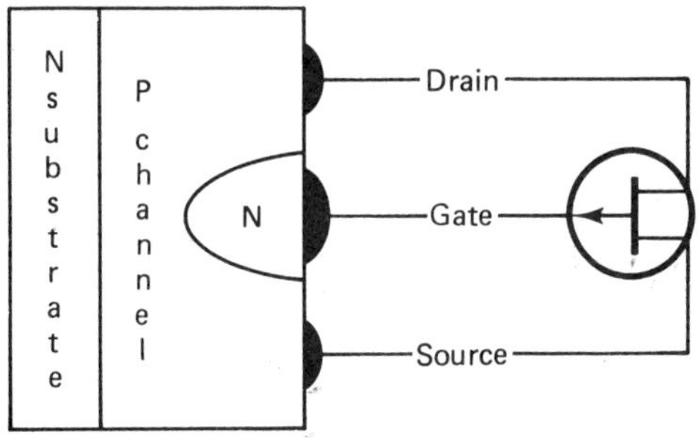

FIGURE 7-3 P-channel junction field-effect transistor.

itive with respect to the source. Varying values of reverse bias gate voltage will change the size of the P-N junction depletion zone. Current flow through the channel can be altered between cutoff and full conduction. P-channel and N-channel JFETs cannot be used in a circuit without a modification in the source voltage polarity.

JFET CHARACTERISTIC CURVES

The JFET is very different from other solid-state components. A small change in gate voltage will, for example, cause a substantial change in drain current. The JFET is therefore classified as a voltage-sensitive device. By comparison, bipolar transistors are classified as current-sensitive devices. A JFET has a rather unusual set of characteristics compared with other solid-state devices.

A family of JFET characteristic curves is shown in Fig. 7-4. The horizontal part of the graph shows the voltage appearing across the source-drain as V_{DS}. The vertical axis shows the drain current (I_D) in milliamperes. Individual curves of the graph show different values of gate voltage (V_{GS}). The cutoff voltage of this

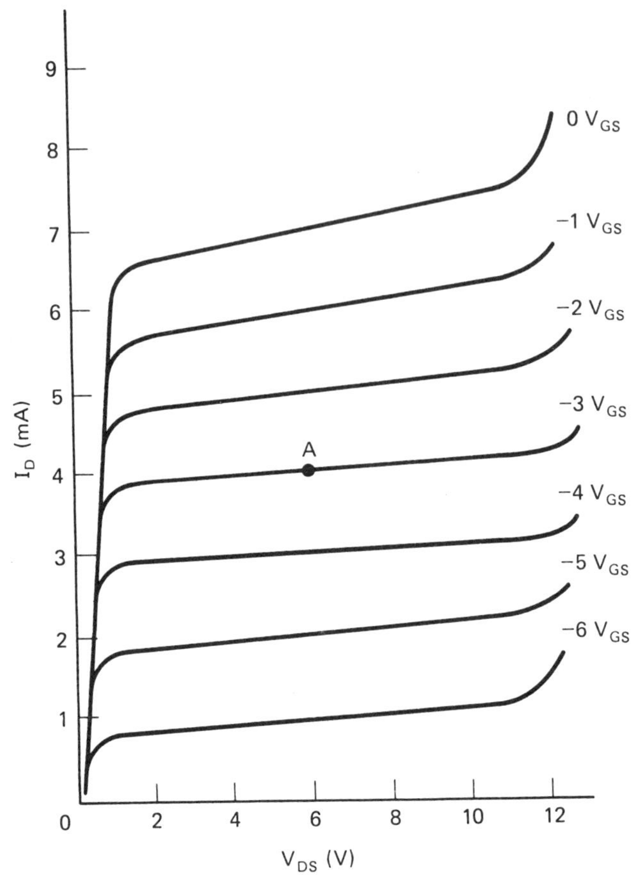

FIGURE 7-4 Family of drain curves.

device is approximately −7 V. The control range of the gate is from 0 to −6 V.

A drain family of characteristic curves tells a great deal about the operation of a JFET. Refer to point A on the $-3V_{GS}$ curve. If the device has −3 V applied to its gate and a V_{DS} of 6 V, approximately 4 mA of I_D flows through the channel. This is determined by projecting a line to the left of the intersection of $-3V_{GS}$ and $6V_{DS}$. Any combination of I_D, V_{GS}, and V_{DS} can be determined through the family of curves.

Developing JFET Characteristic Curves

A special circuit is used to develop the data for a drain family of characteristic curves. Figure 7-5 shows a circuit that is used to find the data points of a characteristic curve for an N-channel JFET. Three meters are used to monitor this data. Gate voltage is monitored by a V_{GS} meter connected between the gate-source leads. Drain current is measured by a milliampere meter connected in series with the drain. V_{DS} is measured with a voltmeter connected across the source and drain. V_{GS} and V_{DS} are adjusted to different values while I_D is monitored. Two variable dc power supplies are used in the test circuit.

The data points of a single curve are developed by first adjusting V_{GS} to 0 V. Normally, V_{DS} is increased in 0.1-V steps up to 1 V. I_D increases very quickly during this operational time. V_{DS} is then increased in 1-V steps while the I_D values are recorded. Corresponding I_D and V_{GS} values are then plotted on a graph as the 0-V curve.

To develop the second curve, V_{DS} must be returned to zero. V_{GS} is then adjusted to a new value. 0.5 V would be a suitable value for most JFETs. V_{DS} is again adjusted through its range while I_D is monitored. Data for the second curve are then recorded on the graph.

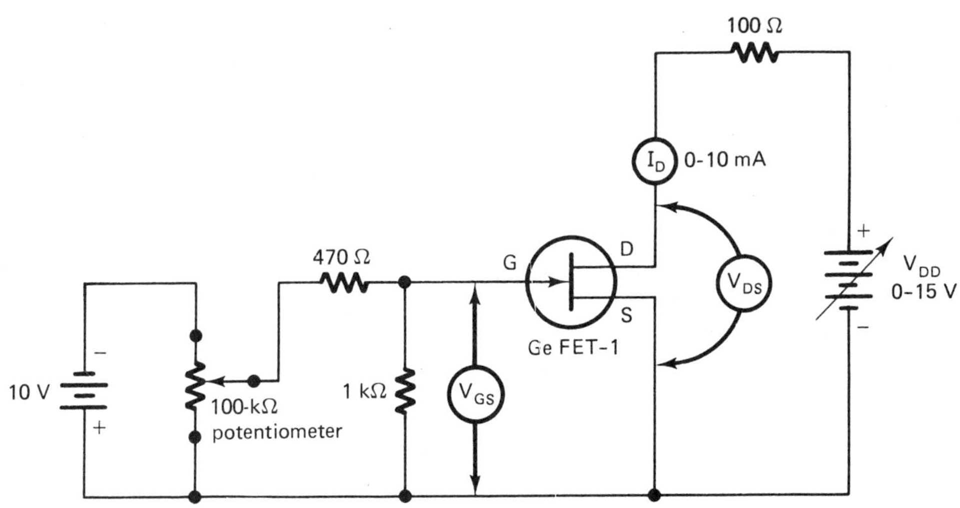

FIGURE 7-5 Drain curve test circuit for N-channel JFET.

To obtain a complete family of drain curves, the process would be repeated for several other V_{GS} values. A typical family of curves may have 8 to 10 different values. The step values of V_{DS}, V_{GS}, and I_D obviously change with different devices. Full conduction is usually determined first. This gives an approximation of representative I_D and V_{DS} values. Normally, I_D levels off to a rather constant value when V_{DS} is increased. V_{DS} can be increased in value to a point where I_D starts a slight increase. Generally, this indicates the beginning of the breakdown region. JFETs are usually destroyed if conduction occurs beyond this area of operation. Maximum V_{DS} values for a specific device are available from the manufacturer. It is a good practice to avoid operation in or near the breakdown region of the device.

Load-Line Analysis

Figure 7-6 shows a small-signal amplifier and a family of curves for the JFET used in the circuit. A load line is plotted using the same procedure outlined for the bipolar transistor. The two intersecting points of the load line show how the JFET responds when it is either fully conductive or cut off. Full conduction occurs when $I_D = V_{DD}/R_L$ and V_{DS} is zero. Cutoff occurs when $V_{DD} = V_{DS}$ and $I_D = 0$ mA.

The operating point of a JFET amplifier is determined by different values of gate-source voltage. For the amplifier of Fig. 7-6, V_{GS} is supplied by the V_{GG} battery. V_{GS} in this case is equal to V_{GG}, or -2.5 V. As a rule, the static operation of the device is established by the value of V_{GS}. This operating state compares

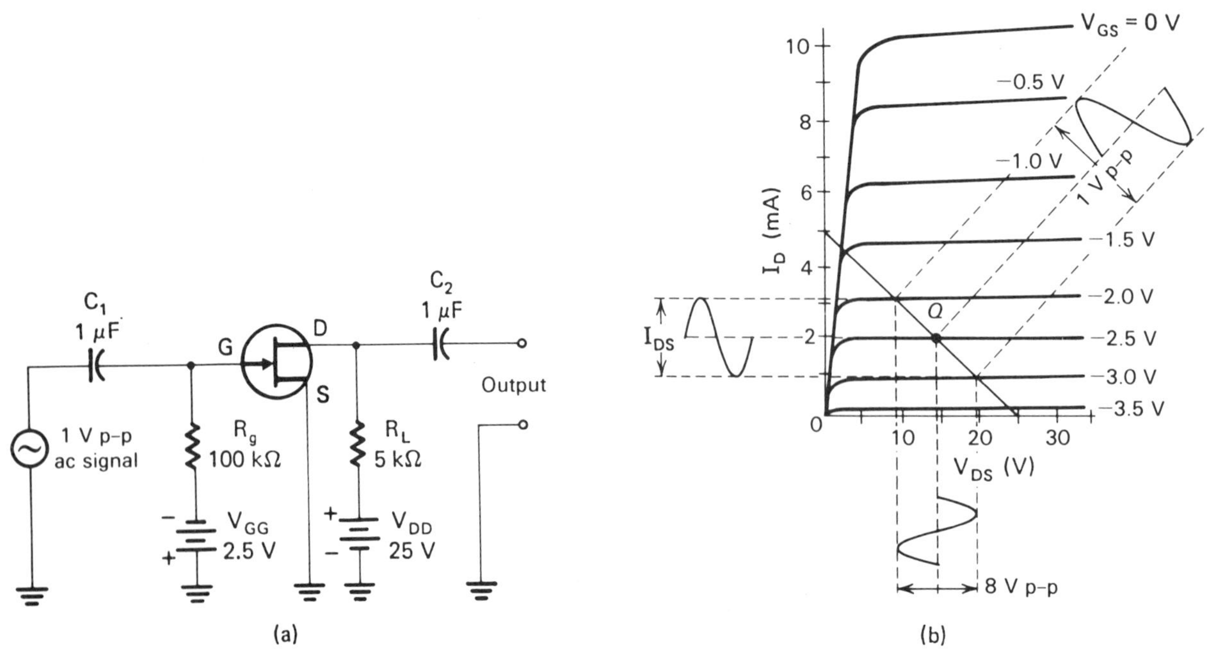

FIGURE 7-6 Family of *I/V* curves and JFET amplifier: (a) N-channel JFET amplifier; (b) drain family of curves.

with that of a bipolar transistor. The term quiescent, or Q, point is also used to denote this condition of operation.

When an ac signal is applied to a JFET amplifier, it causes the current and voltage values to be in a continuous state of change. An amplifier operating in this manner is considered to be in its dynamic state. An ac signal must be applied to the input of the device in order for this type of operation to take place. This signal will cause a change in the value of V_{GS}. The reverse bias voltage of the gate will then change in value at an ac rate. This in turn changes the value of the gain current passing through the channel. A variation in I_D causes a corresponding change in V_{DS}. The output of the amplifier is based on the changing value of V_{DS}. Note the relationship of V_{GS}, I_D, and V_{DS} on the I_D-V_{DS} curves of Fig. 7-6.

Transfer Characteristics

Another way to analyze the operation of a JFET is through the study of its dynamic transfer characteristics. A typical dynamic transfer curve is shown in Fig. 7-7. This curve plots drain current and gate-source voltage for a given value of drain-source voltage. A constant value of $10 V_{DS}$ is used for this particular curve.

A dynamic transfer curve is developed from the data of an I_D-V_{DS} family of curves. Refer to the $10 V_{DS}$ line of the curves in Fig. 7-6. The intersection of each V_{GS} curve and the $10 V_{DS}$ line represents an I_D value. These I_D values are then plotted on a graph for each value of V_{GS}. The intersecting points are then connected to form the transfer characteristic. Essentially, this characteristic shows how a change in V_{GS} causes a change in I_D. Figure 7-7 shows a representative dynamic transfer curve for a JFET.

For the amplifier of Fig. 7-6, the Q point is $-2.5 V_{GS}$. Note the location of this point on the transfer curve. A 1-V p-p change in V_{GS} will cause a change of 2 mA in I_D. This shows a graphic display of the input/output characteristics of a JFET. Nonlinear dis-

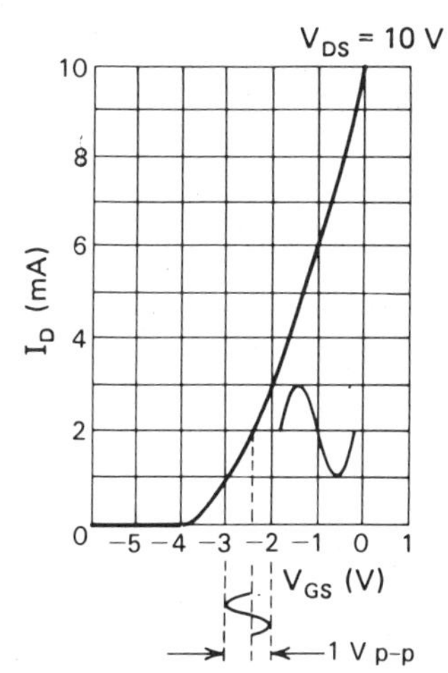

FIGURE 7-7 Dynamic transfer characteristic or input curve.

tortion of the output wave can readily be observed with this type of graph.

A very useful operating parameter of the JFET can be obtained from a dynamic transfer curve. This property is described as the *forward transconductance* (*gm*). Transconductance is a measure of the ease with which current passes through the channel of a JFET. It is defined as the ratio of a small change in drain current to a small change in gate-source voltage for a given value of V_{DS} and is expressed by the formula

$$\text{transconductance} = \frac{\text{drain-current change}}{\text{gate-source voltage change}}$$

or

$$gm = \frac{\Delta I_D}{\Delta V_{GS}}$$

Transconductance is indicated in siemens (S), which is the fundamental unit of conductance. Previously this unit was known as the mho, the reciprocal of an ohm.

SELF-EXAMINATION

1. A transistor that has only one P-N junction is classified as a ____________ device.
2. Current carriers pass through the ____________ of a JFET.
3. The bar or straight line of a JFET symbol represents the ____________.
4. The ____________ and ____________ leads of a JFET are attached to the channel.
5. The arrow of a JFET symbol represents the ________.
6. When the gate of a JFET is open or not biased the channel is ____________ conductive.
7. The gate-source junction of a JFET is ____________ biased in normal operation.
8. An increase in gate-source voltage will cause a(n) (increase, decrease) in drain current.
9. The two intersecting or connecting points of a load-line for a JFET are ________ conduction and ________ conduction.
10. The P-N junction of a JFET represents the ________ and ____________.
11. The three elements or leads of a JFET are the ______, ____________, and ____________.
12. The current flowing through the channel of a JFET is called ____________ current.

13. ____________ is a measure of the ease with which current passes through the channel of a JFET.
14. An N-channel JFET is properly biased when the gate is made ____________ with respect to the source.
15. When the arrow of a JFET symbol points inward it indicates a(n) (N, P) channel.
16. ____________ of a JFET can be obtained from a dynamic transfer curve.
17. (True or False) A JFET must have the gate forward biased in order to achieve control of drain current.
18. (True or False) If the gate of a JFET is not biased, the drain current will be of a maximum value.
19. (True or False) A JFET is considered to be a normally on device.
20. (True or False) Transconductance of a JFET is measured in ohms.

ANSWERS

1. Unipolar	2. Channel
3. Channel	4. Source, drain
5. Gate	6. Fully
7. Reverse	8. Decrease
9. Full, cutoff	10. Gate, channel
11. Source, gate, channel	12. Drain
13. Transconductance	14. Negative
15. Negative	16. Transconductance
17. False	18. True
19. True	20. False

METAL-OXIDE SEMICONDUCTOR FIELD-EFFECT TRANSISTORS

Metal-oxide semiconductor FETs, or MOSFETs, are a unique variation of the basic transistor family. MOSFETs are designed so that the gate is completely insulated from the channel by a thin layer of silicon dioxide. This type of transistor is classified as an insulated gate device (IGFET). As a rule, the term MOSFET is more widely used than IGFET.

Operation of a MOSFET is based on the application of a correct voltage value and polarity to the gate. The gate is simply a coating of metal deposited on a layer of insulation attached to the channel. Voltage applied to the gate causes it to develop an electrostatic charge. No current is permitted to flow between the gate and the channel. The control of current carriers through the channel is based on the polarity and value of the gate voltage.

The channel of a MOSFET has an unusual construction feature. One type of MOSFET responds to the depletion of current carriers that pass through its channel. Depletion-type MOSFETs (D-MOSFETs) have an interconnecting channel built in another piece of semiconductor material known as the substrate. The source and drain are directly connected by the channel. Channel construction of this device is very similar to that of a JFET.

A second type of MOSFET is called an enhancement-type device (E-MOSFET). The channel of this device is produced when it operates. The source and drain are not directly connected through a common channel. Current carriers are pulled from the substrate and form an induced channel. Gate voltage is used to control the size of the induced channel. The channel is present only when the device is energized. E- and D-MOSFETs are both widely used in electronic circuits.

E-MOSFETs

Figure 7-8 shows a cross-sectional view of the crystal structure, element names, and schematic symbols of the P- and N-channel E-MOSFETs. A defined channel connection does not exist between the source and drain. The schematic symbol shows this by having independent source, substrate, and drain leads. The middle electrode, which has an arrowhead, shows the polarity to the substrate. When the arrow Points iN toward the channel, it indicates that the substrate is P and the channel is N. When the arrow does

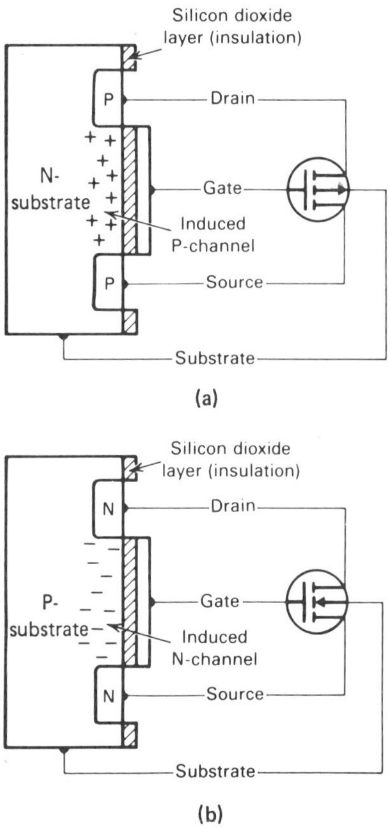

FIGURE 7-8 P- and N-channel E-MOSFETs: (a) P-channel E-MOSFET; (b) N-channel E-MOSFET.

Not Point toward the channel, it shows that the substrate is N and the channel is P. The entire device is built on the substrate. Gate construction is designed to span the entire space between the source and drain. A thin layer of metal is used in the construction of the gate.

An E-MOSFET is classified as a normally off device. This means that no drain current flows through the channel without gate voltage applied. When the gate of an N-channel device is made positive, electrons are pulled from the P-substrate into the source-drain region. This condition causes the channel region to have a high concentration of current carriers. Completion of the channel provides a conduction path for current carriers to flow between the source and drain. The drain current, in effect, is aided or enhanced by the gate voltage. The absence of gate voltage (V_G) does not permit the channel to form. The size of the channel and density of the current carriers are directly related to the value of V_G.

The operation of an E-MOSFET is primarily based on the voltage value and polarity of its elements. Figure 7-9 shows the correct voltage polarity for N- and P-channel E-MOSFETs. The P-channel device has the source and drain energized by V_{DD}. The source is positive, and the drain is negative. The gate is made negative with respect to the source by battery V_{GG}. A negative gate will pull holes from the N-substrate material, forming the induced channel. Current flow through the channel is achieved by hole movement. The P-N junction formed by the P-source and drain and the N-substrate must be reverse biased. The substrate is usually connected to the same polarity as the source.

A family of characteristic curves for an E-MOSFET is shown in Fig. 7-10. This set of curves is for an N-channel device. A set of curves for a P-channel device would be very similar. The primary difference is the polarity of the gate voltage. Note that an increase in V_{GS} causes a corresponding increase in drain current. An individual curve is represented for each value of V_{GS}. Increasing the source-drain voltage causes an increase in I_D. The leveling off of I_D indicates the *pinch-off region* of the device. The pinch-off region and the ohmic region of this device are similar to those of the JFET. A family of I_D-V_{DS} curves of this type could be used to develop a load-line for a representative amplifier. The procedure to be followed is primarily the same as that for a JFET.

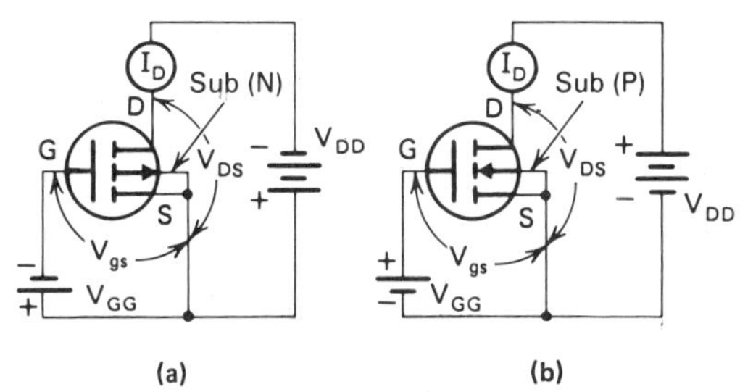

FIGURE 7-9 E-MOSFET voltage polarities: (a) P-channel; (b) N-channel.

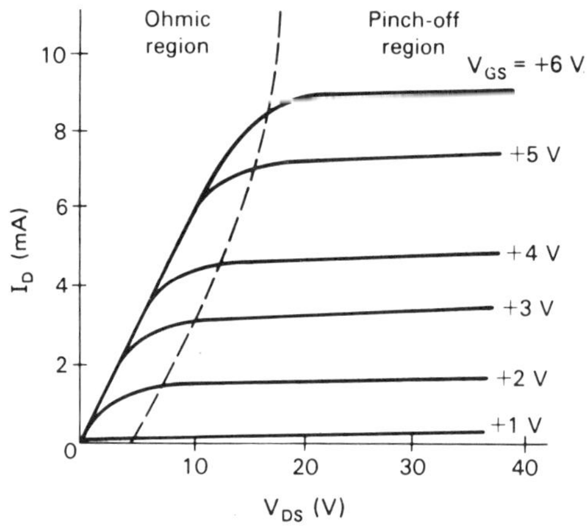

FIGURE 7-10 I_D/V_{DS} curves for an N-channel E-MOSFET.

Depletion-Type MOSFETs

Figure 7-11 shows a cross-sectional view of the crystal structure, element names, and symbols of P- and N-channel D-MOSFETs. In the N-channel device shown, a channel of N-material is formed on a P-substrate. The source and drain are connected by this channel. A layer of silicon dioxide (SiO_2) insulates the gate from the channel. The gate is simply a strip of metal that has been deposited on the SiO_2 layer. The entire assembly is built on a P-substrate. The schematic symbol of the device shows the source, substrate, and drain commonly connected together. This indicates that the source and drain are connected by a piece of channel material that is formed on a piece of substrate material. The substrate of the symbol is represented by an arrow. When the arrow Points iN toward

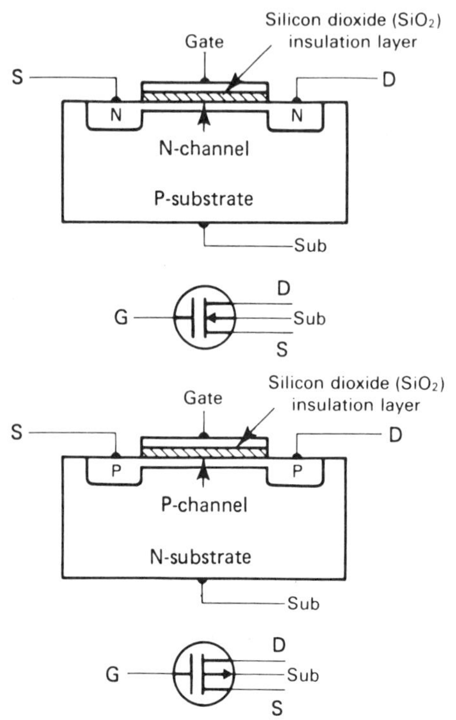

FIGURE 7-11 N- and P-channel D-MOSFETS.

the channel, it indicates that the substrate is P and the channel is N. The current carriers of an N-channel device are electrons.

P-channel device construction is very similar to that of the N-channel device. The crystal material of the channel and substrate are of a different polarity. In a P-channel device the current carriers are holes. The schematic symbol shows that the source, drain, and substrate are commonly connected. The substrate arrow does Not Point toward the channel. This indicates that the substrate is N- and the channel is P-material.

The operation of a D-MOSFET is quite unusual when compared with other active devices. With voltage applied to the source and drain only, drain current flows through the channel. In this condition of operation, the gate does not need voltage to produce conduction. A D-MOSFET is therefore classified as a normally on device. Channel current (I_D) is, however, controlled by different values of gate voltage. When the channel and gate voltage are of the same polarity, there is a reduction or depletion of current carriers. This condition of operation increases the depletion region of the channel. Depleting the number of current carriers in the channel causes a corresponding reduction in I_D. The depletion mode of operation represents only one form of control that can be achieved.

If the gate and channel of a D-MOSFET have a different polarity, the number of channel current carriers increases. This means that the device is capable of both depletion and enhancement modes of operation. Making the gate positive in an N-channel device increases the number of current carriers. In effect, this pulls electrons into the channel from the substrate. A negative gate voltage depletes the number of current carriers. This action increases the depletion region of the channel, which reduces I_D. A zero V_{GS} divides the depletion and enhancement modes of operation. In a strict sense this device could be called an enhancement-depletion MOSFET.

A family of characteristic curves for an N-channel D-MOSFET is shown in Fig. 7-12. A family of P-channel curves would be the same except for the gate polarity. The top V_{GS} lines

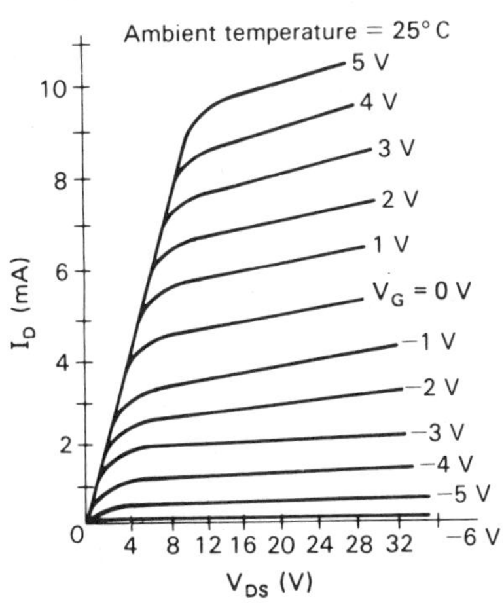

FIGURE 7-12 Characteristic curves for an N-channel D-MOSFET.

would be negative values. Note in this display that $0V_{GS}$ is near the center of the curves. This divides the enhancement and depletion modes of operation. For ac amplification an ideal operating point would be $0V_{GS}$. For the N-channel display, the positive alternation of an input signal causes an increase in I_D. The negative alternation produces a decrease in I_D.

Today, P-channel D-MOSFETs are not as readily available as N-channel devices, because P-channel devices are somewhat more difficult to fabricate than N-channel devices and are, in general, more costly to produce. Most applications tend to favor the N-channel D-MOSFET. N-channel devices have better current mobility. This is a desirable feature in high-frequency ac signal amplification, where these devices are widely used.

V-MOS Field-Effect Transistors

A new generation of the field-effect transistor is finding its way into solid-state active device technology. This type of device is largely used to replace the bipolar power transistor. Power FETs are now being used in power supplies and solid-state switching applications. Collectively these devices are called V-MOSFETs or V-FETs. The *V* designation refers to vertical-groove MOS technology. This device has a V-shaped groove etched into the substrate. Construction of this type requires less area than a horizontally assembled device. The geometry of a V-MOS device also permits greater heat dissipation and high-density channel areas. V-MOSFETs have fast switching speeds and lower channel resistance.

Figure 7-13 shows a cross-sectional view of the crystal structure, element names, and schematic symbol of an N-channel V-MOSFET. Notice that a V-groove is etched in the surface of the structure. From the top, the V-cut penetrates through N^+, P, and N^- layers and stops near the N^+ substrate. The two N^+ layers are heavily doped, and the N^- layer is lightly doped. A thin layer of silicon dioxide covers both the horizontal surface and the V-

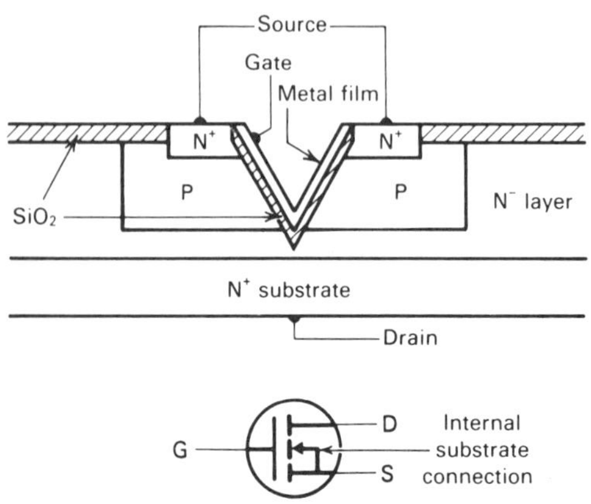

FIGURE 7-13 Cross-sectional view of an N-channel V-MOSFET.

groove. The source leads on each side of the groove are connected internally. The bottom layer of N^+ material serves as a combined substrate and drain. Current carriers move between the source and drain vertically.

A V-MOSFET responds as an enhancement MOSFET. No current carriers exist in the source and drain regions until the gate is energized. An N-channel device, such as the one of Fig. 7-13, does not conduct until the gate is made positive with respect to the source. When this occurs, an N channel is induced between the two N^+ areas near the groove. Current carriers can then flow through the vertical channel from source to drain. When the gate of an N-channel device is made negative, no channel exists, and the current carriers cease to flow.

P-channel V-MOSFETs are also available today for use in electronic circuit applications. The current carriers of a P-channel device are holes. In general, holes are less mobile than electrons. These devices do not respond as well to high-frequency ac. The primary characteristics and theory of operation are, however, very similar to those of an N-channel device. Differences exist only in voltage polarity and current-carrier flow. As a rule, a wider variety of N-channel devices are manufactured today; industrial applications tend to favor the N-channel devices over the P-channel units. Only N-channel devices will be discussed here.

Characteristics of V-MOSFETs

The characteristics of a V-MOSFET are very similar to those of an enhancement MOSFET. Figure 7-14 shows the I_D-V_{DS} curves and the transfer characteristics of an N-channel device. This particular device is a power V-FET. The I_D-V_{DS} characteristic shows that V_{GS} must be positive with respect to the source if there is to be an increase in I_D. The value of V_{GS} also controls I_D for an increase in V_{DS}. Note that V-MOSFETs have ohmic and pinch-off regions very similar to those of an E-MOSFET.

The transfer characteristic of a V-MOSFET plots V_{GS} and I_D for a single V_{DS} value. A change in V_{GS} is considered to be a variation in the input voltage. I_D is used to denote an output condi-

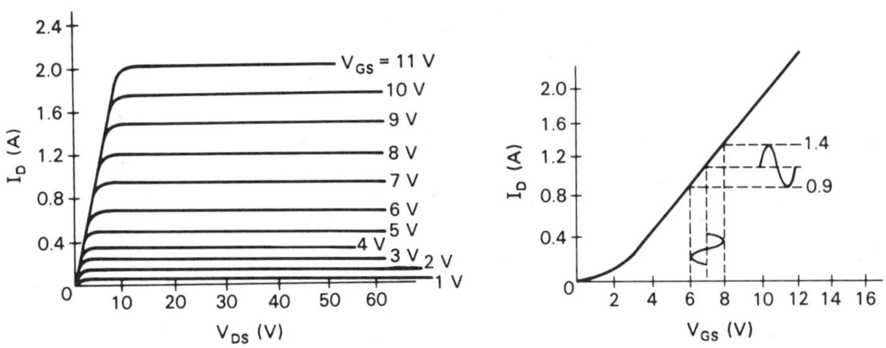

FIGURE 7-14 E-type V-MOSFET characteristics.

tion. A transfer curve shows how the output responds to a change in input voltage. This curve is nearly linear over its entire length. The forward transconductance (*gm*) can be obtained from the data of this curve.

To determine the forward transconductance of a V-MOSFET, refer to Fig. 7-14. In this case the transfer curve shows an input voltage change of 2 V p-p with an operating point of 7 V_{GS}. Locate the 7 V_{GS} operating point. A 2-V p-p input signal will cause the operating point to change from 6 to 8 V. Trace the V_{GS} values upward to the curve. The resulting I_D values are 0.9 A to 1.4 A. This represents an I_D of 0.5 A. The *gm* is determined to be

$$\text{transconductance} = \frac{\text{change in drain current}}{\text{change in gate-source voltage}}$$

or

$$gm = \frac{\Delta I_D}{\Delta V_{GS}}$$

$$= \frac{0.9 \text{ to } 1.4 \text{ A}}{6 \text{ to } 8 \text{ V}}$$

$$= 0.5 \text{ A}/2 \text{ V}$$

$$= 0.25 \text{ S, or } 250 \text{ mS}$$

Note that the *gm* of a V-MOSFET is quite large. Compared with other FETs, this is a very significant condition of operation. The voltage gain (A_v) of an FET amplifier is based on $gm \times R_L$. With high values of gm, it is possible to achieve a great deal more voltage gain with a V-MOSFET than with other types of FETs. This feature, its power-handling capability, and high switching speed make the V-MOSFET an extremely attractive amplitude-control device for electronic circuits.

SELF-EXAMINATION

21. MOSFETs are designed so that the ____________ is electrically insulated from the channel.
22. The ____________ of a MOSFET is a strip of metal instead of a piece of semiconductor material.
23. Voltage applied to the ____________ of a MOSFET causes it to develop an electrostatic charge.
24. In MOSFET operation no current is permitted to flow between the gate and the ____________.
25. A MOSFET is built on a piece of semiconductor material called the ____________.
26. The source and drain of an enhancement MOSFET are not directly connected through a ____________.
27. The source and drain of an enhancement MOSFET are

connected by an ____________ channel when the device is in operation.

28. A depletion MOSFET has an interconnecting channel built in another piece of semiconductor material known as the ____________.
29. The schematic symbol of a(n) (enhancement, depletion) MOSFET has independent lines representing the source, substrate, and drain.
30. The majority current carriers of the drain of an N-channel depletion type of MOSFET are ____________.
31. The schematic symbol of a(n) (enhancement, depletion) MOSFET has the source, substrate, and gate commonly connected by a line.
32. The majority current carriers passing through the drain of a P-channel enhancement MOSFET are ____________.
33. When (enhancement, depletion) MOSFETs have voltage applied to the source and drain only, drain current will flow through the channel.
34. When the ____________ of an E-MOSFET is energized, it pulls current carriers from the substrate to form an induced channel.
35. (Enhancement, Depletion) MOSFETs can operate in either the depletion or enhancement mode of operation.
36. (Enhancement, Depletion) MOSFETs have a 0 V_{GS} value near the centers of their characteristic curve.
37. In an N-channel enhancement type of MOSFET, the polarity of the gate must (match, be opposite) the polarity of the induced channel to produce conduction.
38. V-MOSFETs involve ____________ ____________ MOS technology.
39. A change of drain current of 1 A caused by a gate-source voltage change of 2 V will produce a transconductance of ____________ S.
40. V-MOSFETs respond as (enhancement/depletion) MOSFETs.

ANSWERS

21. Gate	22. Gate
23. Gate	24. Channel
25. Substrate	26. Channel
27. Induced	28. Substrate
29. Enhancement	30. Electrons
31. Depletion	32. Holes
33. Enhancement	34. Gate
35. Depletion	36. Depletion
37. Be opposite	38. Vertical-groove
39. 0.5	40. Enhancement

FIELD-EFFECT TRANSISTOR AMPLIFIERS

The primary function of an amplifier is to reproduce the applied signal and provide some level of amplification. Unipolar transistors can be used to achieve this function. JFETs and MOSFETs are examples of unipolar amplifying devices. Amplification depends on the device being used and its circuit components. The composite circuit is normally energized by dc voltage. An ac signal is then applied to the input of the amplifier. After passing through the device, an amplified version of the signal appears in the output. Operating conditions of the device and the circuit combine to determine the level of amplification to be achieved.

For a FET to respond as an amplifier, it must be energized with dc voltage. Specific voltage values and polarities must be applied to each type of device. Forward and reverse biasing are not as meaningful in FET circuits. Remember that current conduction is through a channel. The value of voltage and its polarity determine the amount of control being achieved.

An illustration of the supply voltages and polarities needed for different types of FETs is shown in Fig. 7-15. JFETs must have the gate-source junction reverse biased to achieve control. The gate voltage of an E-MOSFET is opposite that of the source. D-MOSFETs have some flexibility in the polarity of the gate voltage. Some of these devices have a zero value of V_{GS} and swing positive or negative during operation.

Basic FET Amplifier Operation

Let us examine the circuitry of a basic FET amplifier. In this example an N-channel JFET is used. A P-channel device could also have been selected. The primary difference in operation is the polarity of the source voltage and the type of current carrier in the channel.

Figure 7-16 shows an N-channel JFET amplifier connected in the common-source configuration. The source of this device is common to both the input signal and the output. This circuit is very similar to the common-emitter amplifier. V_{DD} serves as the dc energy source for the source and drain.

V_{GS} reverse biases the gate with respect to the source. The value of V_{GS} establishes the static operating point of the circuit. R_g is normally extremely high. No gate current flows through capacitor C. Selecting a large value of R_g keeps the input impedance of the amplifier high. If R_g is 1 MΩ, the signal source sees an impedance of 1 MΩ.

Let us now consider the operation of the JFET amplifier of Fig. 7-16 in its static state. The family of drain curves of Fig. 7-17 will be used in this explanation. A load line for the circuit must be established. Remember that two extreme conditions of operation are needed for the load line. In a JFET these extremes are full conduction and cutoff. At the cutoff point, no current

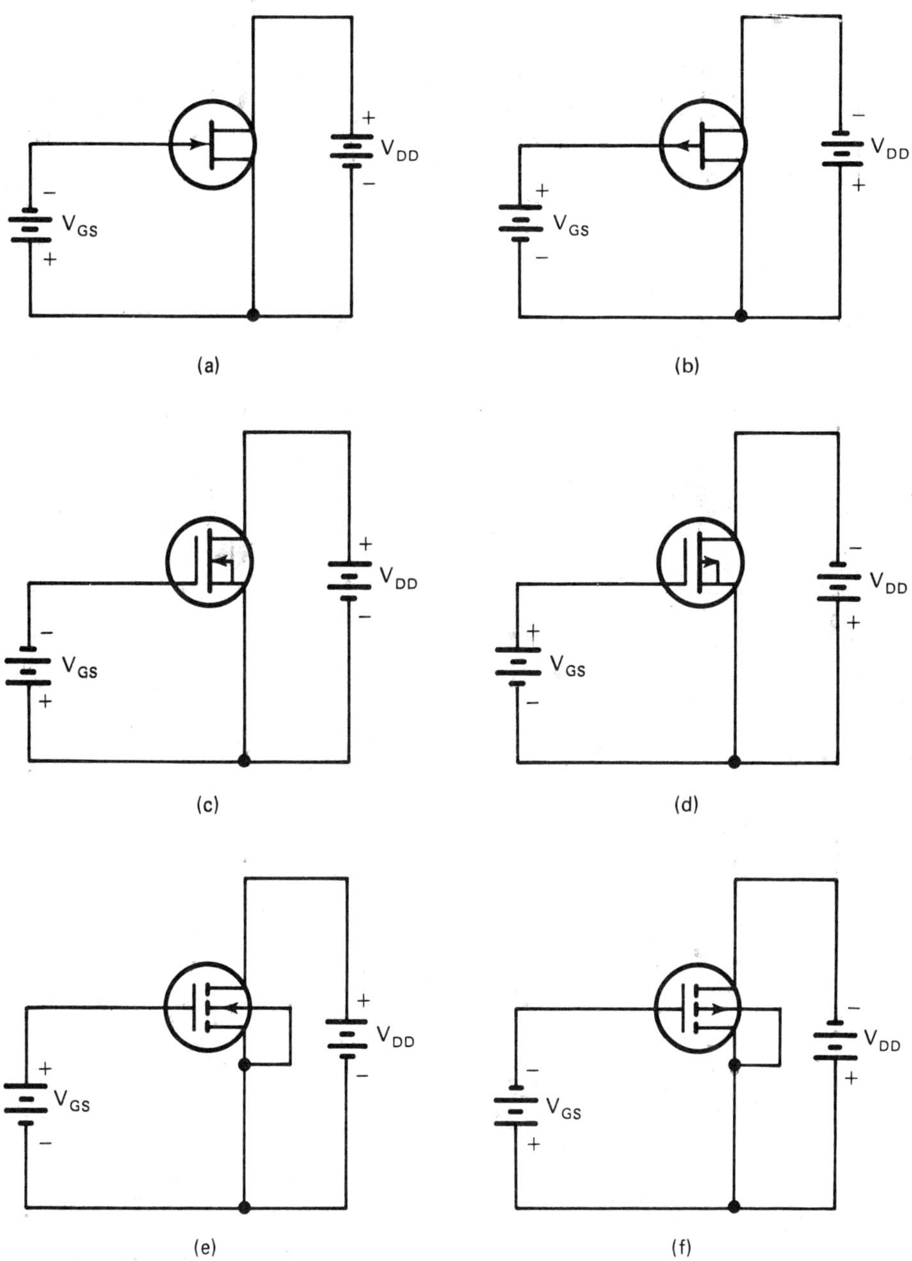

FIGURE 7-15 FET operational voltages: (a) N-channel JFET; (b) P-channel JFET; (c) N-channel D-MOSFET; (d) P-channel D-MOSFET; (e) N-channel E-MOSFET; (f) P-channel E-MOSFET.

passes through the channel. The V_{DD} voltage will appear across V_{DS}. Full conduction occurs when a maximum value of I_D flows through R_L. This value is determined by the formula

$$\text{drain current} = \frac{\text{source voltage}}{\text{load resistance}}$$

or

$$I_D = \frac{V_{DD}}{R_L}$$

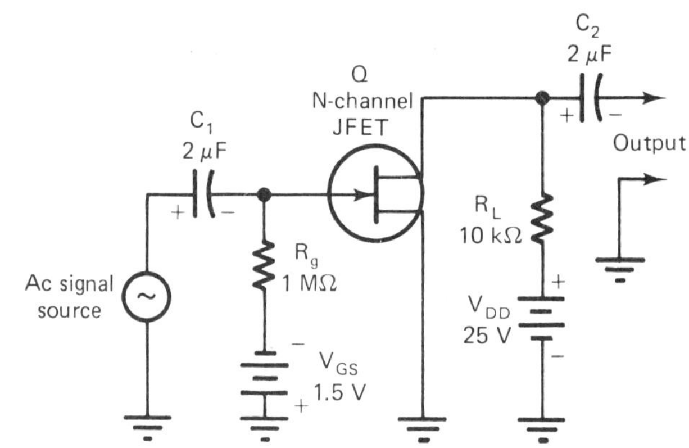

FIGURE 7-16 Common-source N-channel JFET amplifier.

For the amplifier circuit, the maximum value of I_D is

$$I_D = \frac{25 \text{ V}}{10 \text{ k}\Omega}$$

$$= \frac{25 \text{ V}}{10 \times 10^3 \Omega}$$

$$= 0.0025 \text{ A, or } 2.5 \text{ mA}$$

With this conduction through R_L, the V_{DS} will be zero. The two locating points of the load line are V_{DS} = 25 V with 0 mA of I_D, and V_{DS} = 0 V at 2.5 mA I_D. Check these two locations on the load line. A straight line connects the points.

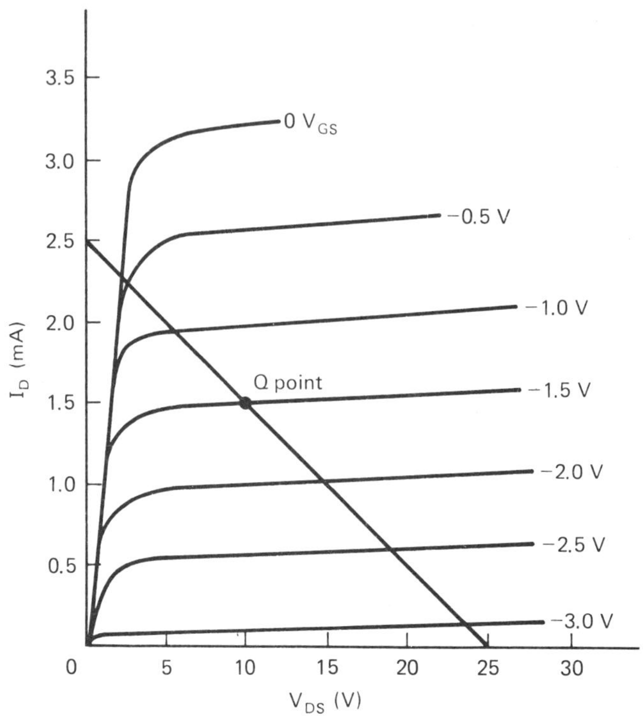

FIGURE 7-17 Family of drain curves.

With the load line developed, it is possible to see how the JFET will respond in its static state. Static operation occurs without a signal applied. For linear operation the amplifier should respond near the center of the active region. In our basic circuit, V_{GS} is -1.5 V. Point Q on the load line denotes the operating point. With an appropriate input signal, the amplifier should respond in the linear region.

To see how the JFET responds in its static state, lines are projected from the Q point. Projecting a line from the Q point to the I_D scale shows the resulting drain current. In this case, the I_D is 1.5 mA. Projecting a line from the Q point down shows the drain-source voltage. V_{DS} is approximately 10 V. This means that 15 V appears across R_L at this point of operation. The dc voltage gain can be determined with this value. Voltage amplification (A_v) equals the drain-source voltage (V_{DS}) divided by the gate-source voltage. For our circuit the gain is

$$A_v = \frac{V_{DS}}{V_{GS}} = \frac{10\text{ V}}{1.5\text{ V}} = 6.667$$

Dynamic JFET Amplifier Analysis

Dynamic JFET amplifier analysis shows how the device will respond when an ac signal is applied. For this explanation we will use the drain curves and circuit of Fig. 7-18. A load-line and Q point for the circuit have been developed on the curves. This shows how the JFET responds in its static state. Check the load-line location points and Q point for accuracy.

The circuit diagram shows a 1-V p-p ac signal applied to the input of the JFET amplifier. With this signal the V_{GS} operating point will change in value from -2.0 V to -3.0 V. For the positive alternation, V_{GS} will swing from -2.5 V to -2.0 V. This change is shown as point P on the load line. During the negative alternation, V_{GS} will drop from -2.5 V to -3 V. This value is shown as point N on the load line. In effect, this means that a 1-V p-p input signal will cause V_{GS} to change from -2.0 to -3.0 V. This is considered to be the ΔV_{GS} value.

To show how ΔV_{GS} alters I_D, lines are projected to the left of points P, Q, and N. Note the projection of these points toward the I_D values in Fig. 7-18. The changing value of I_D is shown as ΔI_D. An increase or decrease in V_{GS} causes a corresponding change in I_D. This shows that V_{DS} and I_D are in phase.

Projecting lines P, Q, and N downward from the load line shows how V_{DS} changes with V_{GS}. The value of V_{DS} is shown as ΔV_{DS}. Note that an increase in V_{GS} causes a decrease in V_{DS}. Note that an increase in V_{GS} causes a decrease in V_{DS}. This tells us that

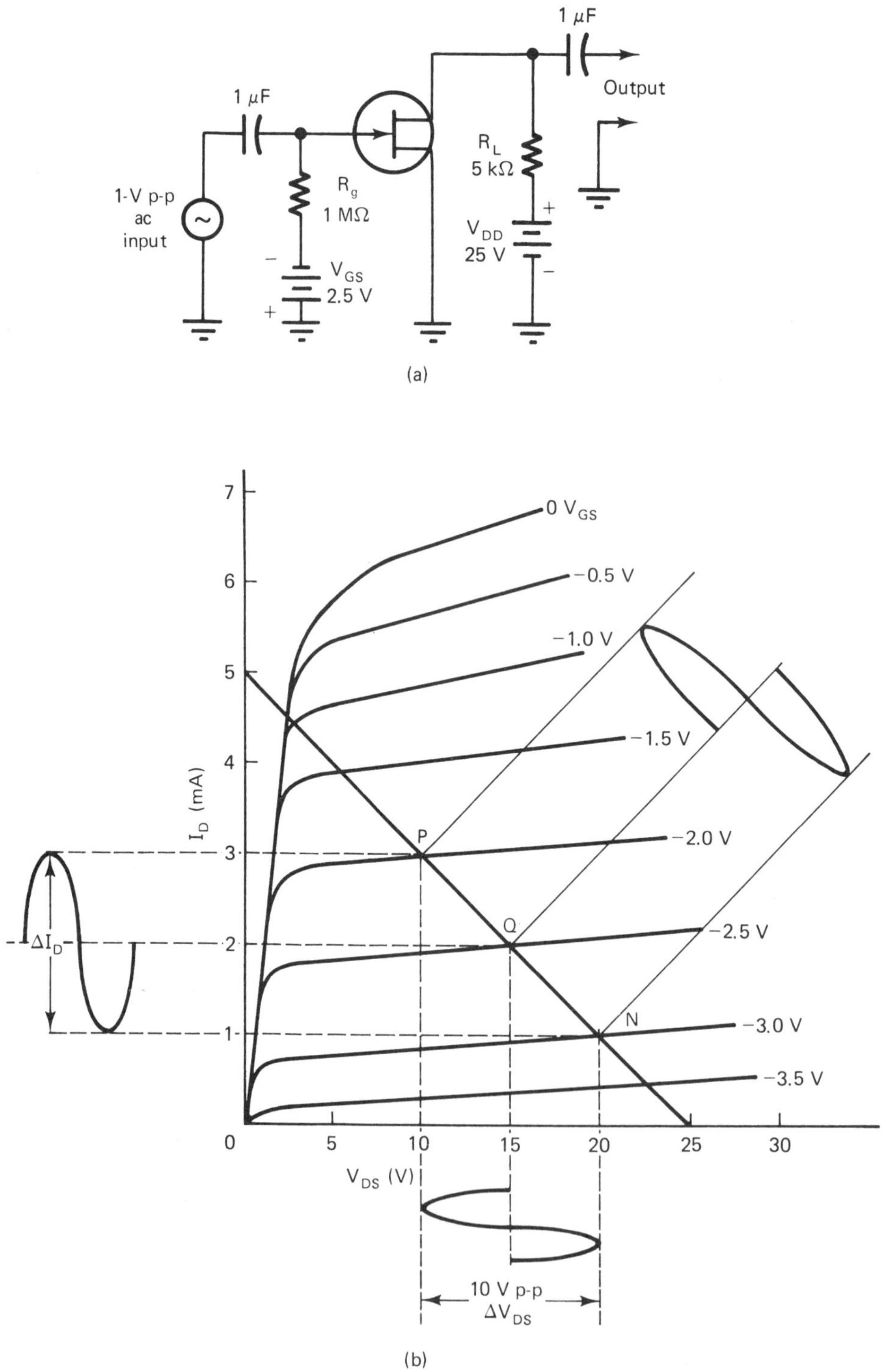

FIGURE 7-18 JFET amplifier; (a) N-channel amplifier; (b) drain family of curves.

V_{GS} and V_{DS} are 180° out of phase. The value difference in V_{DS} and V_{DD} appears across the load resistor as V_{RL}.

The voltage gain of a JFET amplifier can readily be observed by load-line data. A ΔV_{GS} of 1 V p-p causes a ΔV_{DS} of approximately 18 v p-p. This shows an obvious gain of 18 from the formula

$$A_v = \frac{\Delta V_{DS}}{\Delta V_{GS}}$$

$$= \frac{18 \text{ V p-p}}{1 \text{ V p-p}}$$

$$= 18$$

MOSFET Circuit Operation

D-type and E-type MOSFET operation is very similar to that of the JFET. There is an obvious difference in the required operational voltage values and polarity. A load-line analysis of either amplifier shows the same basic principle of operation. The gate is highly resistant or infinite for all devices. Current flow in the channel is controlled by the voltage value and polarity of the gate signal. The only difference of any concern is with the D-MOSFET. It is frequently biased at the 0 V_{GS} point. An ac input signal would cause V_{GS} to vary above and below the zero value. It is important to remember that all FETs are voltage-sensitive devices. They produce only voltage amplification; no current flows in the gate circuit.

FET Biasing Methods

For FET operation, biasing refers to the necessary dc voltage needed by the gate with respect to the source lead, that is, the development of a suitable V_{GS} value. By selection of a proper bias voltage, it becomes possible to force a device to operate at a chosen Q point. The V_{DD} voltage applied to the source-drain is generally not considered to be bias voltage. The biasing method selected for a specific circuit depends on the type of device involved. E-MOSFETs and D-MOSFETs require different biasing procedures.

Fixed Biasing. The simplest way to bias an FET is with fixed biasing. This method of biasing can be used equally well for E- or D-MOSFETs or for JFETs. In fixed biasing the correct voltage value and polarity is supplied by a small battery or cell. An electronic power supply could also be used to perform the same operation. Figure 7-19 shows fixed biasing applied to three different N-channel FET types. If P-channel devices were used, the polarity of V_{CC} and V_{DD} would be reversed. V_{CC} is used as the fixed bias source for the three FET types.

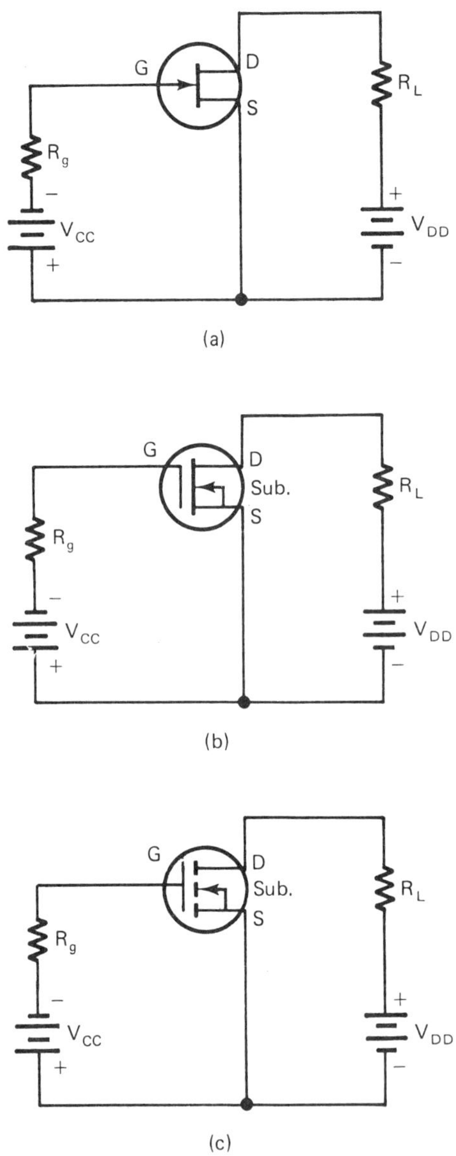

FIGURE 7-19 Methods of fixed biasing: (a) N-channel JFET; (b) N-channel D-type MOSFET; (c) N-channel E-type MOSFET.

If an enhancement-depletion type of MOSFET is to be fixed biased, the circuit of Fig. 7-20 would be used. In this type of device the operating point is usually $V_{GS} = 0$ V. No voltage source is needed to establish the operating point. R_g is normally of a high value. Typically, values of 1 to 10 MΩ are used. R_g must be in-

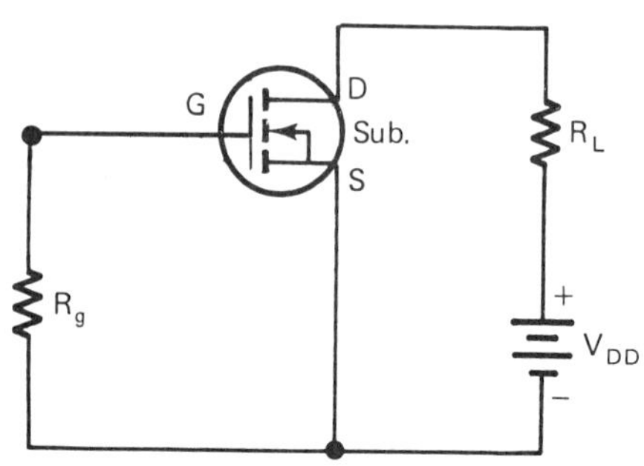

FIGURE 7-20 Fixed bias of an N-channel D-MOSFET.

cluded in the circuit; it serves as a return for the gate to the common connection point.

Voltage-Divider Biasing. The voltage-divider method of biasing FETs is rather easy to achieve. It is applicable only to E-MOSFETs. With these devices V_{GS} and V_{DD} are of the same polarity. As a result of this condition, a single voltage source can be used as a supply. The V_{GS} voltage is only a fraction of the value of V_{DD}. The divider resistors are used to develop the necessary voltage values. Figure 7-21 shows the divider method of biasing for N- and P-channel E-MOSFETs.

The values of R_1 and R_g must be properly chosen to establish a desired operating point. By the voltage-divider equation we find that the bias voltage is

$$\text{gate-source voltage} = R_g \text{ voltage}$$

$$= \text{source voltage}\left(\frac{R_g}{R_1 + R_g}\right)$$

or

$$V_{GS} = V_{Rg} = V_{DD}\left(\frac{R_g}{R_1 + R_g}\right)$$

Self-Biasing. Self-biasing is commonly called source biasing. Self-biasing is rather easy to recognize. The source current of an

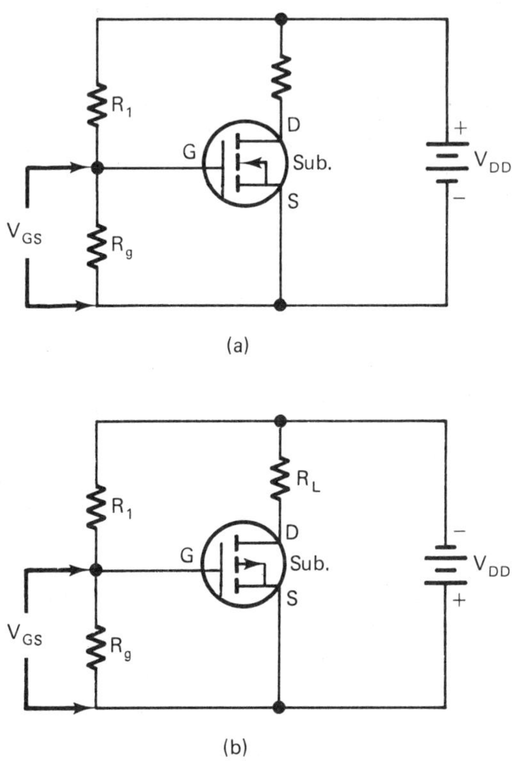

FIGURE 7-21 Voltage-divider method of biasing for E-MOSFETs: (a) N-channel; (b) P-channel.

FET is used to develop the bias voltage. The voltage developed is achieved by a resistor (R_s) connected in series with the source. Current flow through the source-drain causes a voltage drop across R_s. The polarity of the developed voltage is based on the direction of current flow through R_s. Figure 7-22 shows biasing for N- and P-channel JFETs. D-MOSFETs can also be biased by the same method.

For the N-channel circuit, current flow through R_s causes the source to be slightly positive with respect to the gate. With R_g connected to the lower side of R_s, this makes R_g negative and the source positive. This value of R_s and drain current (I_D) determines the bias operating point of the circuit. Note the polarity difference in the P-channel circuit.

FET CIRCUIT CONFIGURATIONS

FET amplifiers, like their bipolar counterparts, can be connected in three different circuit configurations. These are described as common-source, common-gate, and common-drain circuits. One lead is connected to the input, and a second lead is connected to the output. The third lead is commonly connected to both the input and output. This lead is used as a circuit reference point and is often connected to the circuit ground—hence the terms *grounded source, grounded gate,* and *grounded drain.* The terms common and ground refer to the same type of connection.

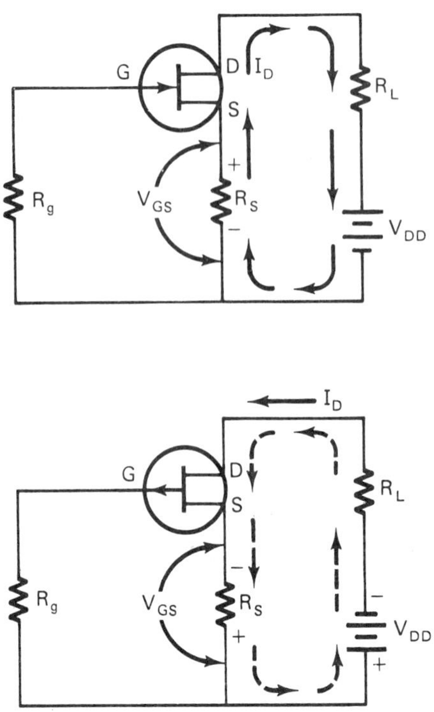

FIGURE 7-22 Self-biasing of JFETs: (a) N-channel; (b) P-channel.

Common-Source Amplifiers

The common-source amplifier is the most widely used FET circuit configuration. This circuit is similar in many respects to the common-emitter bipolar amplifier. The input signal is applied to the gate-source, and the output signal is taken from the drain-source. The source lead is common to both input and output.

A practical common-source amplifier is shown in Fig. 7-23. This circuit is essentially the same as the one used in our basic JFET amplifier. It is shown here so that a comparison can be made with the other circuit configurations. The device can be a JFET, a D-MOSFET, or an E-MOSFET. Circuit characteristics are very similar for all three devices.

The signal being processed by the common-source amplifier is applied to the gate-source. Self-biasing of the circuit is achieved by the source resistor R_2. This voltage establishes the static operating point. The incoming signal voltage is superimposed on the gate voltage. This causes the gate voltage to vary at an ac rate, which causes a corresponding change in drain current. The output voltage developed across the source and drain is inverted 180°. Voltage gain A_v is V_{DS}/V_{GS}. Typical A_v values are 5 to 10. The input impedance is extremely high for nearly any signal source. One to several megohms is common. The output impedance (Z_{out}) is moderately high. Typical values are in the range of 2 to 10 kΩ. Z_{out} is dependent primarily on the value of R_L.

A common-source amplifier has a very good ratio between input and output impedance. Circuits of this type are extremely valuable as impedance-matching devices. Common-source amplifiers are used exclusively as voltage amplifiers. They respond well in radio-frequency signal applications.

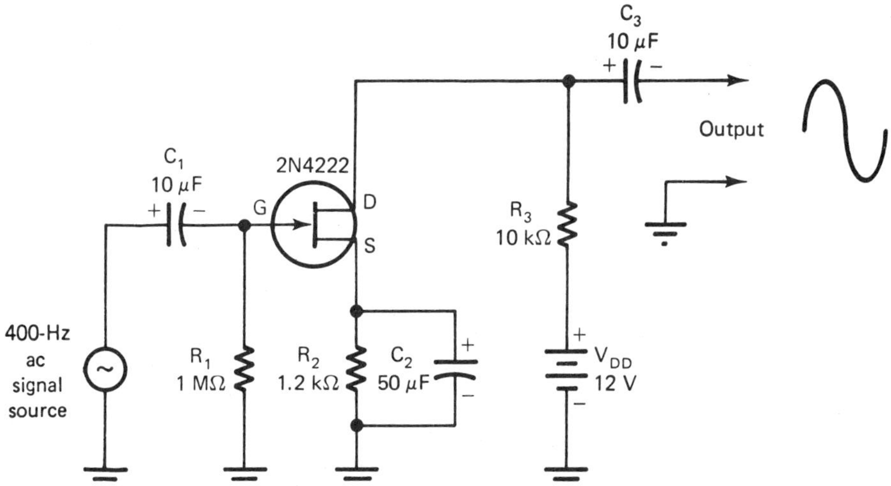

FIGURE 7-23 Common-source JFET amplifier.

Common-Gate Amplifiers

The common-gate FET amplifier is similar in many respects to the common-base bipolar transistor circuit. The input signal is applied to the source-gate, and the output appears across the drain-gate. The common-gate amplifier is capable of a rather significant amount of voltage gain. Current gain is not an important circuit consideration. The gate does not ordinarily have any current flow. JFETs, E-MOSFETs, and D-MOSFETs may all be used as common-gate amplifiers.

A practical common-gate amplifier is shown in Fig. 7-24. This circuit employs an N-channel JFET. The operating voltages are the same as those of the common-source circuit. Self-biasing is achieved by the source resistor R_1. This voltage is used to establish the static operating point. An input signal is applied to R_1 through capacitor C_1. A variation in signal voltage causes a corresponding change in source voltage. Making the source more positive has the same effect on I_D as making the gate more negative. The positive alternation of the input signal will make the source more positive. This, in turn, reduces drain current. With less I_D there is a smaller voltage drop across the load resistor R_2. The drain or output voltage therefore swings positive. The negative alternation of the input reduces the source voltage by an equal amount. This is the same as making the gate less negative. As a result, I_D is increased. The voltage drop across R_2 similarly increases. This, in turn, causes the drain or output to be less positive or negative going. The input signal is therefore in phase with the output signal.

The common-gate amplifier has a number of rather unusual characteristics. Its voltage gain is somewhat less than that of the common-source amplifier. Representative values are 2 to 5. The common-gate circuit has very low input impedance (Z_{in}). The output impedance (Z_{out}) is rather moderate. Typical Z_{in} values are 200

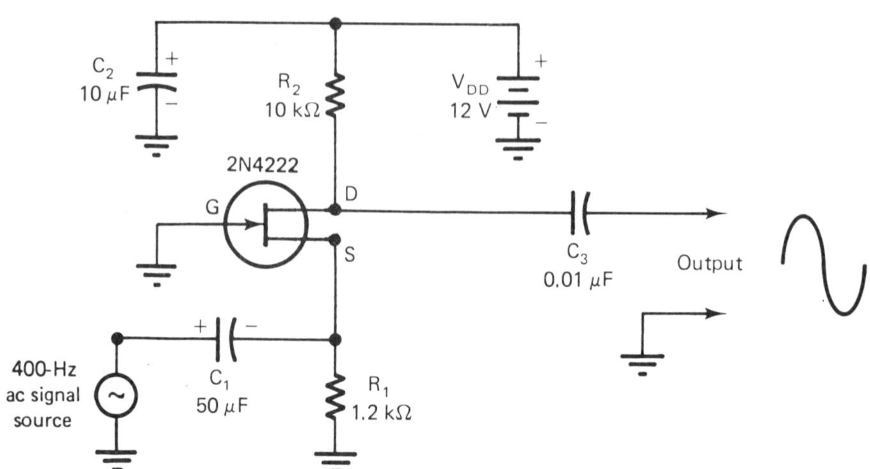

FIGURE 7-24 Common-gate JFET amplifier.

to 1500 Ω, with Z_{out} being 5 to 15 kΩ. This type of circuit configuration is often used to amplify radio-frequency signals. Amplification levels are very stable for radio frequencies, without feedback between the input and output.

Common-Drain Amplifiers

A common-drain amplifier has the input signal applied to the gate and the output signal removed from the source. The drain is commonly connected to one side of the input and output. Common-drain amplifiers are also called *source followers.* This circuit has similar characteristics to those of the common-collector bipolar amplifier.

Figure 7-25 shows a practical common-drain amplifier using an N-channel JFET. The input of this amplifier is primarily the same as that of the common-source amplifier. The input impedance is therefore very high. Z_{in} is determined largely by the value of R_1. The operating point of our amplifier is determined by the value of R_2. Essentially, this circuit has the same operating point as our other circuit configurations. Resistor R_3 has been switched from the drain to the source in this circuit. Resistors R_2 and R_3 are combined to form the load resistance. The output impedance is based on this value.

When an ac signal is applied to the input of the amplifier, it changes the gate voltage. The dc operating point is established by the value of source resistor R_2. The positive alternation of an input signal makes the gate less negative. This causes an N-channel device to be more conductive. With more current through R_3 and R_2, the source swings positive. The negative alternation of the input then makes the gate more negative. This action causes the channel to be less conductive. A smaller current therefore causes the source to swing negative. In effect, this means that the input and output voltage values of the amplifier are in step with each other. These signals are in phase in a common-drain amplifier.

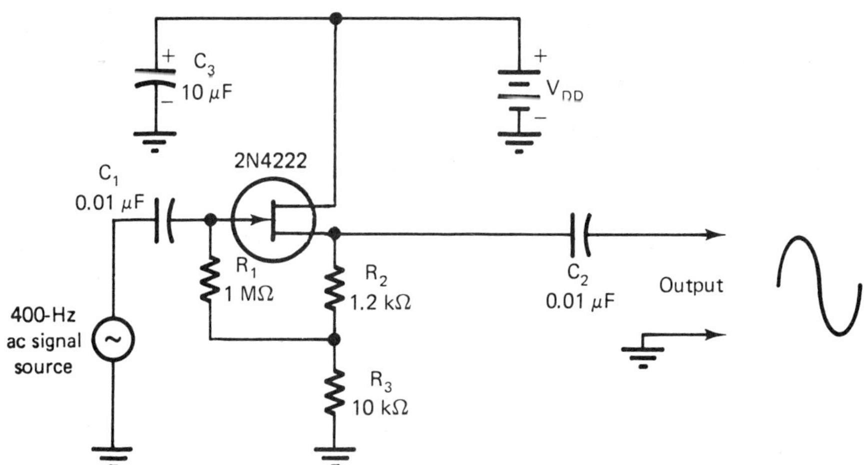

FIGURE 7-25 Common-drain JFET amplifier.

Common-drain amplifiers are primarily used today as impedance-matching devices. They are used to match a high-impedance device to a low-impedance load. This type of circuit is capable of handling a high-input signal level without causing distortion. The input impedance places a minimum load on the signal source. Common-drain amplifiers are used frequently today to match high-impedance devices such as microphones and phonograph pickups to the input of an audio amplifier.

SELF-EXAMINATION

41. JFET and MOSFETs are examples of ____________ transistors.
42. The two extreme conditions of JFET amplifier operation represented by a load line are ____________ and ____________.

For the next four items, refer to Fig. 7-18.

43. Full conduction of the transistor of Fig. 7-18 occurs at ____________ of I_D and ____________ V_{DS}.
44. Cutoff occurs at ____________ I_D and ____________ V_{DS}.
45. The Q point or static operating condition occurs at ____________ V_{GS}.
46. The ac voltage gain (A_v) of the amplifier in Fig. 7-18 is ____________.
47. JFET amplifiers are considered to be ____________ sensitive devices.
48. No ____________ flows into the gate of a normally operating MOSFET.
49. In fixed biasing of a MOSFET, voltage is supplied to the ____________ by a battery or dc power supply.
50. The voltage-divider method of biasing applies only to (enhancement, depletion) MOSFETs.
51. Source biasing has a ____________ connected in series with the source of a JFET to develop bias voltage.
52. In a common-(source, gate, drain) amplifier the input is applied to the gate-source, and the output is developed across the source-drain.
53. The input and output signals of a common-source amplifier are inverted ____________.
54. The input signal of a common-(source, gate, drain) amplifier is applied to the source-gate, and the output appears across the drain-gate.
55. The input signal of a common-(source, gate, drain) amplifier is applied to the gate, and the output signal is removed from the source.

ANSWERS

41. Unipolar	42. Cutoff or full conduction
43. 5 mA, 0 V	44. 0 mA, 25 V
45. 2.5 V	46. 18
47. Voltage	48. Current
49. Gate	50. Enhancement
51. Resistor	52. Source
53. 180°	54. Gate
55. Drain	

MOSFET-HANDLING PROCEDURES

MOSFETs must receive special care in handling when they are used in an electronic circuit. The silicon dioxide layer that insulates the gate from the source, drain, and substrate is very thin. A static charge or voltage less than 100 V between the gate and channel will cause permanent damage to the insulation material. This generally causes the oxide layer to rupture. As a result, gate leakage current occurs between the gate and channel. This generally causes permanent damage to the device.

The voltage supplied to an operating MOSFET generally comes from a dc power source or supply. The voltage used in most circuit applications must not exceed the breakdown rating of the device. In general, this can be avoided by checking the manufacturer's maximum ratings so that the device is not damaged by excessively high operating voltage values. The correct source voltage values must not be exceeded. The device should not be connected to an energized or operating circuit. As a rule, these precautions apply to nearly all solid-state devices.

Static-charge damage to a MOSFET presents a number of problems that do not occur in other solid-state devices. Friction caused by rubbing the device in a plastic bag or on a piece of fabric during shipping can be damaging. Static buildup on the body of a person handling the MOSFET can also cause it to be damaged. Soldering a device into a circuit with an ungrounded soldering device can cause some form of damage. All of this requires some understanding of the special handling procedure that must be followed. As a rule, static damage is the most troubling problem in the use of MOSFETs.

Most manufacturers ship MOSFETs with their leads shorted together by means of a shorting ring or wire or pressed into a conducting foam material. The shorting method should not be removed from the device until it is installed in its respective circuit. This precaution is extremely important. It is extremely important, however, that the shorting device be completely removed from the device in order to avoid damage to the circuit. The following precautions should be observed when handling MOSFETs:

1. Turn off the power source of the circuit before inserting a MOSFET into it. Voltage transients developed by the circuit will cause the device to be permanently damaged.
2. Use special grounded-tip soldering devices for circuit connections. Do not use soldering guns. The tip of a soldering device can be grounded with a clip lead to reduce this problem.
3. Neutralize electrostatic body voltage by using a grounding wristband, or connect a clip lead to your metal watchband.
4. Let the shorting ring or conductive foam remain on a MOSFET until all circuit connections are complete and the device is installed in the circuit. Do not forget to remove the shorting material from the device.

Most new MOSFETs have zener diode protection built into the device to avoid electrostatic problems. Figure 7-26 shows how these diodes are connected into the MOSFET internally. Protection is achieved when the diodes become conductive. This occurs when the voltage exceeds 15 V. The zener diodes are connected in reverse order so that conduction will occur in either polarity of applied voltage. This type of built-in protection reduces most of the in-circuit and out-of-circuit transient problems associated with MOSFET installation. However, it is still recommended that the general precautions for handling MOSFETs be followed in order to avoid unnecessary damage.

FET TESTING

FETs, like bipolar transistors, periodically require testing. The condition of the device can be tested with a curve-tracer oscillo-

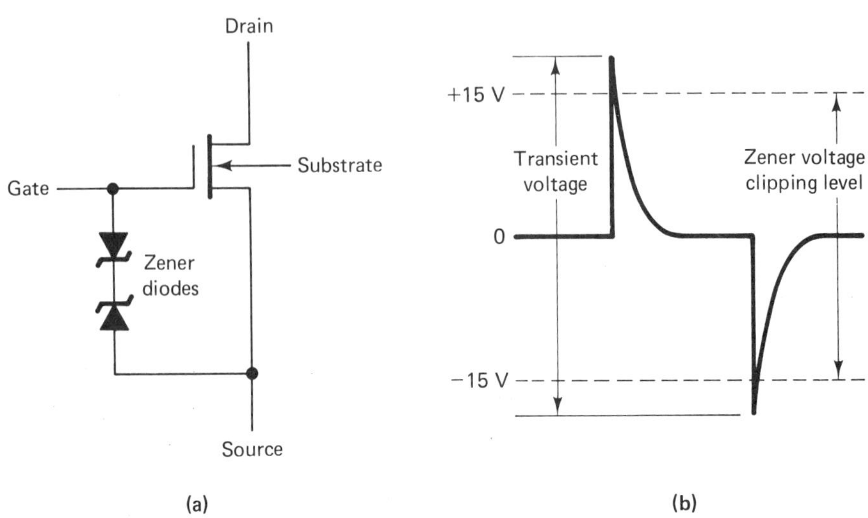

FIGURE 7-26 Zener diode gate protection of (a) MOSFET; (b) transient voltage.

scope, sound-producing instruments, or in-circuit/out-of-circuit testers. This equipment is normally used to evaluate a device and to identify leads. Test procedures are rather easy and are a reliable indicator of device condition. If equipment of this type is available, become familiar with its operation by reviewing the operational manual.

The ohmmeter of a VOM or DVM can also be used to evaluate field-effect transistors. This type of instrument is more readily available than specialized testers. Lead identification, shorted devices, open conditions, and gain can be evaluated. Before using the ohmmeter it is important that the polarity of its voltage source be known. In the following procedure, we describe an ohmmeter with straight polarity. The black or common lead is negative, and the probe or red lead is positive. If the meter polarity is unknown, it may be tested by connecting a voltmeter across its leads.

JFET testing with an ohmmeter is relatively easy. Remember that a JFET is a single bar of silicon with one PN junction. To identify the leads, first select any two of the three leads. Connect the ohmmeter to these leads. Note the resistance. If it is 100 to 1000 Ω, the two leads are the source and drain. Reverse the ohmmeter and connect it to the same leads. If the two leads are actually the source and drain, the same resistance value will be indicated. The third lead is the gate. If the two selected leads show a different resistance in each direction, one must be the gate. Select the third lead and one of the previous leads. Repeat the procedure. A good JFET must show the same resistance in each direction between two leads. This represents the source-drain channel connections.

The gate of a JFET responds as a diode. It will be low resistant in one direction and high resistant in the reverse direction. If the device shows low resistance when the positive probe is connected to the gate and negative to the source or drain, the JFET is an N-channel device. A P-channel device will show low resistance when the gate is made negative and the source or drain positive. The resistance ratio in the forward and reverse direction must be at least 1:1000 for a good device. Figure 7-27 indicates some representative resistance values for good JFETs.

D-MOSFET testing is very similar to the JFET. The gate of the D-MOSFET is insulated from the channel. It will show infinite resistance between each lead regardless of the polarity of the ohmmeter. Figure 7-28 shows some representative values for good D-MOSFETs. Three lead devices have the substrate and source internally connected. Four lead devices have an independent substrate lead.

A four-lead device can be identified as an N-channel or a P-channel device. A P-channel device will show low resistance when the positive lead is connected to the substrate and the negative lead is connected to either the source or drain. An N-channel device has low resistance when the negative lead is connected to the substrate and the positive lead to either the source or drain.

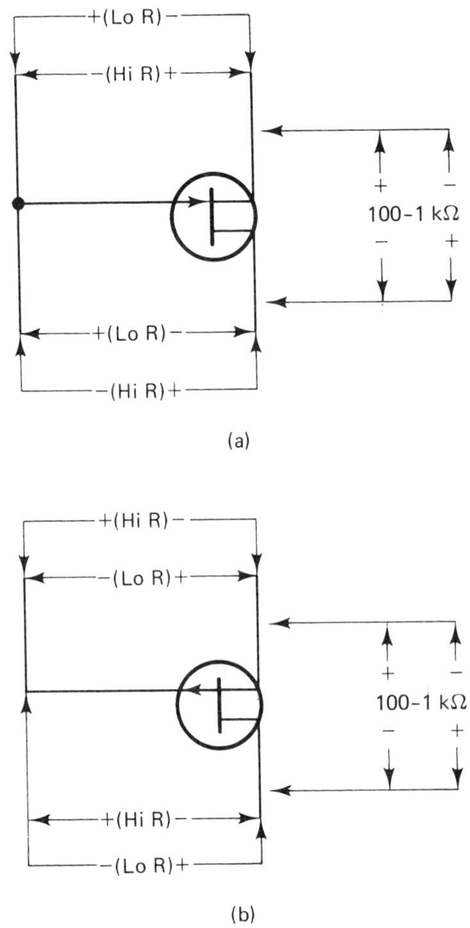

FIGURE 7-27 Resistance values of a JFET: (a) N-channel; (b) P-channel.

E-MOSFET testing with an ohmmeter is not very meaningful. The structure of this device does not interact effectively between different leads. The gate, source, and drain of a good device will show infinite resistance between these leads. A four-lead device will respond as a diode between the substrate and the source or drain. Polarity of the ohmmeter can be used to identify the substrate and the channel. A three-lead device responds as a diode between the source and drain. Figure 7-29 shows some representative ohmmeter resistances for good E-MOSFETs.

SELF-EXAMINATION

56. MOSFETs are susceptible to electrostatic damage because of ____________ breakdown.
57. New MOSFETs have ____________ diodes built in to the structure to protect it from electrostatic damage.
58. A (good, bad) JFET must show the same resistance in each direction between two leads.
59. The gate and channel of a good JFET respond as a ____________.

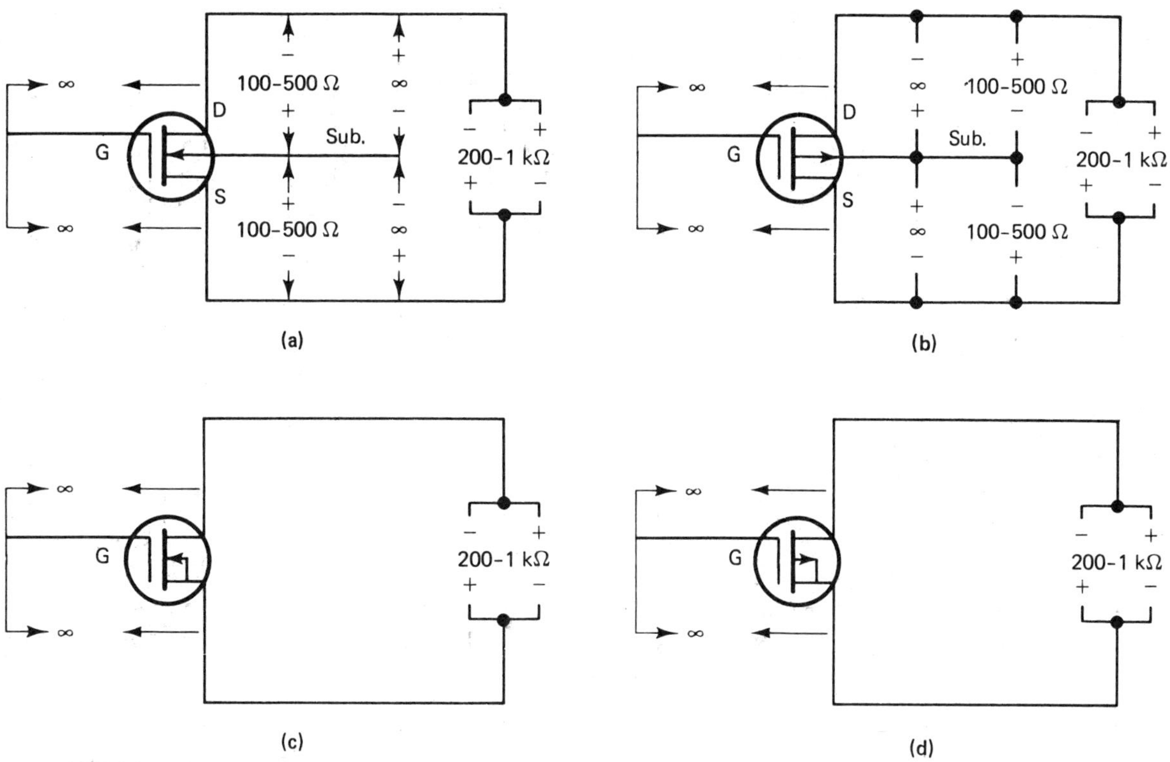

FIGURE 7-28 D-MOSFET resistances: (a) N-channel; (b) P-channel; (c) N-channel; (d) P-channel.

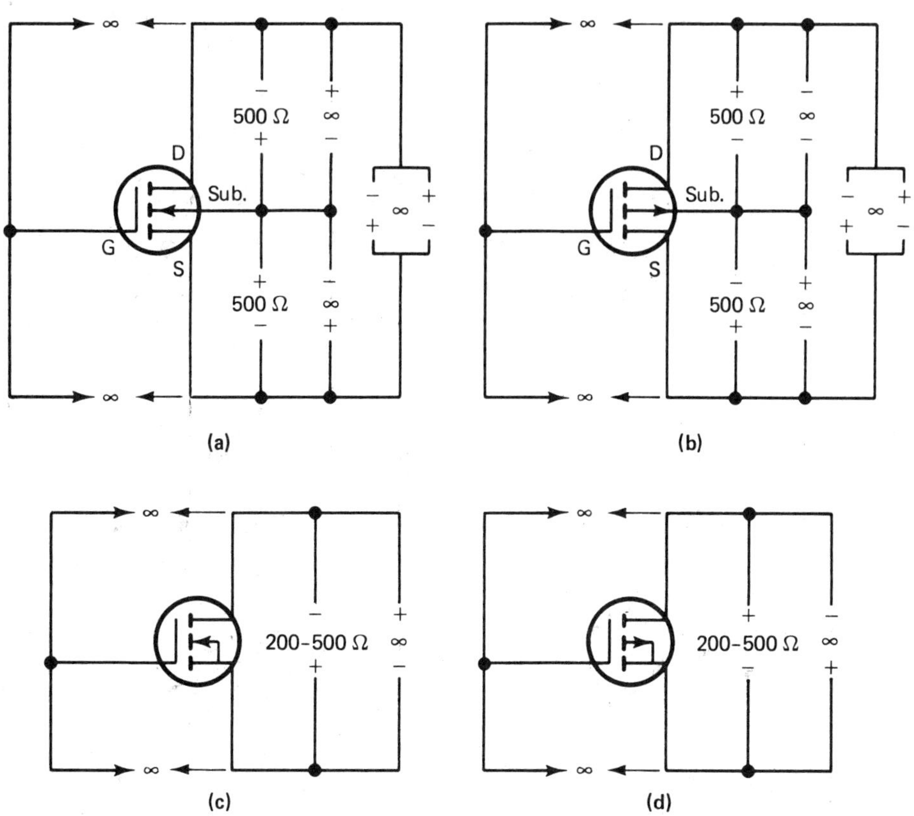

FIGURE 7-29 Resistances of E-MOSFETs: (a) N-channel; (b) P-channel; (c) N-channel; (d) P-channel.

60. The source and drain of a (good, bad) D-MOSFET will show low resistance in either direction of ohmmeter polarity.

ANSWERS

56. Insulation	57. Zener
58. Good	59. Diode
60. Good	

SUMMARY

Field-effect transistors are unipolar devices, which are classified as junction field-effect transistors and metal-oxide semiconductor field-effect transistors. These devices are more commonly called JFETs and MOSFETs, respectively.

JFETs are three-element electronic devices. The elements are source, drain, and gate. The channel is a single piece of semiconductor material built on a substrate. The gate is a piece of semiconductor material diffused into the channel. Operation is achieved by applying energy to the channel through the source and drain connections. A JFET will conduct drain current without gate voltage applied. These devices are considered to have normally on conduction. When the gate is reverse biased, it increases the depletion region of the channel and reduces drain current. Drain current is controlled by the value of reverse-biased gate voltage. JFETs are manufactured as N- or P-channel devices.

Characteristic curves are used to analyze the operation of field-effect transistors. For an FET this type of curve shows drain current and drain-source voltage for different values of gate-source voltage. A load line can be developed that will show how the FET operates in a representative circuit. The extreme conditions of operation are full conduction and cutoff. When an operating point is established, it permits the load line to show how the transistor will perform in a circuit. A dynamic transfer curve can be developed from I_D-V_{DS} curves to show input characteristics. Transconductance can be determined from this type of curve. Transconductance is a measure of the ease with which current carriers pass through the channel of an FET. The siemens (S) is the unit of transconductance.

MOSFETs are a unique variation of the basic field-effect transistor. MOSFETs have the gate insulated from the channel by a thin layer of silicon dioxide. The gate is a strip of metal and not a semiconductor material. Operation is achieved by energizing the channel. The gate must be energized in order to cause current carriers to pass through the channel in an enhancement or E-MOSFET. Operation of this device relies on an induced channel.

When the gate is energized, it causes current carriers to move into the channel from the substrate. An E-MOSFET is considered to be a normally off device. Gate voltage must be applied in order for this device to be conductive.

Depletion-MOSFETs, or D-MOSFETs, have a different structure compared with the E-MOSFET devices. D-MOSFETs have a channel formed on the substrate. The gate is a piece of metal insulated from the channel by a thin layer of silicon dioxide. Operation is based on energizing the channel and applying voltage to the gate. An E-MOSFET is conductive without gate voltage applied. This type of device is considered to have normally on conduction. A D-MOSFET is somewhat unusual with its gate biasing voltage. Normally, the center line of a family of characteristic curves is 0 V. For an N-channel device a positive voltage change will cause an increase in drain current. A negative gate-source voltage will cause a decrease in drain current. A D-MOSFET can be operated as a D- or E-MOSFET according to the polarity of the gate voltage.

Vertical-groove technology is used in construction of V-MOSFETs. A V-groove is etched in the surface of the substrate of this device. The channel and gate are then deposited in the groove. Construction of this type permits greater heat dissipation and high-density channel areas. V-MOSFETs respond as enhancement MOSFETs. The transconductance of this device is quite large compared with that of other FETs. This makes it possible to achieve a great deal more voltage gain. V-MOSFETs are extremely popular amplitude-control devices.

The primary function of an amplifier is to reproduce the applied signal and provide some level of amplification. JFETs, MOSFETs, and V-MOSFETs can all be used for this purpose. A JFET must have energy supplied to the channel by an outside source and the gate reverse biased. Control of drain current is achieved by changing the value of the reverse bias voltage applied to the gate. MOSFETs operate in primarily the same manner. E-MOSFETs have an induced channel. The conduction of drain current is based on the application of gate voltage. D-MOSFETs have a thin channel built in the substrate that connects the source and drain. Operation of a D-MOSFET is very similar to that of a JFET. The gate does not need to be energized to produce conduction through the channel. A D-MOSFET can essentially be used as E- or D-MOSFET. V-MOSFETs respond the same as D-MOSFETs.

FETs, like their bipolar counterparts, can be connected in three different circuit configurations. These are described as common-source, common-gate, and common-drain. The common-source amplifier is the most widely used. The signal being processed is applied to the gate-source and the output developed across the source-drain. The output signal is inverted 180°. Voltage gain or A_v is in the range of 5, with an input impedance of several megohms and an output impedance of 2 to 10 kΩ. The common-gate amplifier has the input applied to the source-gate and the output developed across the drain-gate. The voltage gain is less than that of the common-source amplifier, and the input impedance is very low, with a moderately high output impedance. Common-drain amplifiers are primarily used as impedance-matching circuits. They can match high-impedance devices to a low-impedance output. The voltage gain is less than 1, and the output is developed across the source resistor. Common-drain amplifiers are also called source followers.

MOSFETs require special care in handling. They are susceptible to damage due to breakdown of the gate insulating material. These devices are shipped with a shorting strip or placed in a conducting foam to prevent electrostatic damage when handling. When using these devices do not attach them to energized circuits, solder them into circuits with ungrounded soldering equipment, or handle them without discharging body static.

EXPERIMENT 7-1
FIELD-EFFECT TRANSISTOR TESTING

The internal circuitry of an ohmmeter has a voltage source connected to its leads when it is used to measure resistance. This voltage can be used to bias a field-effect transistor. The status of the terminals, the condition of the device as good or bad, and to some extent the leads can be identified with an ohmmeter. An ohmmeter is a quick and easy method of field-effect transistor evaluation. This method of evaluation is rather commonly used in electronic troubleshooting procedures.

OBJECTIVES

1. To test the status of a field-effect transistor with an ohmmeter.
2. To identify the polarity of the construction material of a field-effect transistor.
3. To identify the leads of a field-effect transistor.

EQUIPMENT

Volt-ohm-milliammeter (VOM), standard or digital type

N-channel JFET, Motorola 2N5459, Heath 417-291, or equivalent

N-channel MOSFET, Motorola 2N3796, Heath 417-935, or equivalent

100-kΩ, $\frac{1}{4}$-W resistor

PROCEDURE

1. Determine the lead polarity of the ohmmeter by placing it in the Ohms position and measuring the probe voltage with a voltmeter. Measure the voltage on the R × 1000, 2 kΩ, or equivalent range. Record the voltage of the ohmmeter: ____________ V. If the common lead is negative and the probe lead is positive, the meter has forward polarity. An opposite polarity indicates reverse ohmmeter polarity. The meter used for this test has (forward, reverse) polarity.
2. Select the 2N5459 JFET. Make a sketch of the bottom view of the transistor lead layout in the right-hand margin.

3. First determine the source (S) and drain (D) leads of the transistor by connecting the ohmmeter between two randomly selected leads. If the two selected leads indicate a resistance, reverse the ohmmeter probes. Two leads that show the same resistance in either direction of polarity are the source and drain. It may be necessary to try several different two-lead combinations in order to determine the source and drain of the JFET. When the leads have been identified, indicate your findings on the JFET sketch of step 2. The measured source-drain resistance is ______________ Ω.
4. The gate-channel of a JFET responds as a diode. This can be observed by connecting the ohmmeter between two leads and then reversing the ohmmeter leads. The gate-source or gate-drain will respond as a diode by showing low forward resistance and high or infinite reverse resistance.
5. If the gate shows low resistance when the positive probe is connected to the gate and the negative to either the source or drain, the JFET is an N-channel device. A P-channel device will produce low resistance when the gate is negative and the source or drain is positive. The resistance ratio between forward and reverse connections should be 1:100 for a device to be good.
6. An approximate test of the JFET's ability to amplify is displayed by connecting the ohmmeter to the source and drain. The polarity of the ohmmeter is not significant. Touch the gate of the JFET with your finger. It should cause a decrease in resistance.
7. A more significant resistance change occurs when a 100-kΩ resistor is connected between the gate and either the source or drain. Very little or no change in resistance during this test indicates an open device or improper lead identification.
8. Using the same procedure outlined in steps 2 through 6, test the 2N3796 MOSFET. This device has a fourth lead that is connected to the substrate. The substrate and the source or drain leads will respond as a diode. The gate of a MOSFET will show an infinite resistance between all leads in either direction of ohmmeter polarity.

DISCUSSION

This experiment showed how to evaluate the status of a unipolar transistor with an ohmmeter. In the first part of the experiment the ohmmeter was used to identify a JFET. This procedure used the polarity of the ohmmeter to bias the device. The leads can be identified by the response produced by the materials. The source and drain will show the same resistance in either polarity direction of the ohmmeter. The gate-source and gate-drain both respond as a diode. They will show low resistance in the forward direction and high or infinite resistance in the reverse direction. The polarity of the ohmmeter was used to identify the device as P- or N-channel.

The ability of a JFET to amplify a signal can be evaluated to some extent with an ohmmeter. This was demonstrated by connecting the ohmmeter to the source and drain, then touching the gate with a finger to one side of the meter. A resistor connecting the meter to the gate better indicates the device's amplification ability. A good device will show a rather significant decrease in meter resistance.

MOSFETs were tested in the same way. A MOSFET has its gate open to all leads regardless of the meter polarity and lead connection combination. The substrate of a MOSFET responds as a diode between the source and the drain. The polarity of the ohmmeter was used to determine the polarity of the channel and substrate. All other tests responded in the same way as those of the JFET.

ANALYSIS

1. Explain how an ohmmeter can be used to evaluate the status of a JFET.
2. How would an ohmmeter test distinguish between an N-channel and a P-channel JFET?
3. What are some of the differences that an ohmmeter will show between a JFET and a MOSFET?

EXPERIMENT 7-2
JUNCTION FET CHARACTERISTICS

In this experiment you will make a family of I_D-V_{DS} curves for a JFET. The family of curves is then used to show the operating characteristics of the JFET in a circuit. A load line is plotted for the circuit and its operation is predicted and tested for evaluation. Voltage gain and transconductance for the transistor is then determined by calculation. These values are used to predict how the transistor will respond in an operating circuit. Through this procedure you will be able to see how a JFET operates in a representative circuit and graphically predict how it will respond.

OBJECTIVES

1. To demonstrate the operating characteristics of a JFET by measuring the source-drain voltage and calculating the drain current.
2. To plot a partial family of I_D-V_{DS} curves for a JFET.
3. Determine a load line for a JFET and plot it on a family of I_D-V_{DS} curves.

EQUIPMENT

N-channel JFET, Motorola 2N5459, Heath 417-291, or an equivalent
100-Ω, $\frac{1}{4}$-W resistor
1-kΩ, $\frac{1}{4}$-W resistor
4.7-kΩ, $\frac{1}{4}$-W resistor
1-kΩ, 1-W potentiometer, (linear taper)
100-kΩ, 1-W potentiometer, (linear taper)
Dual-section variable dc power supply ±15 V, or two 9-V batteries
Volt-ohm-milliammeter (VOM), standard or digital
Circuit construction board

PROCEDURE

Part A: Characteristic Curve Data Collection

1. Construct the JFET characteristic curve test circuit of Fig. 7-2A. The leads, or pin-out, of the JFET are shown in the diagram.

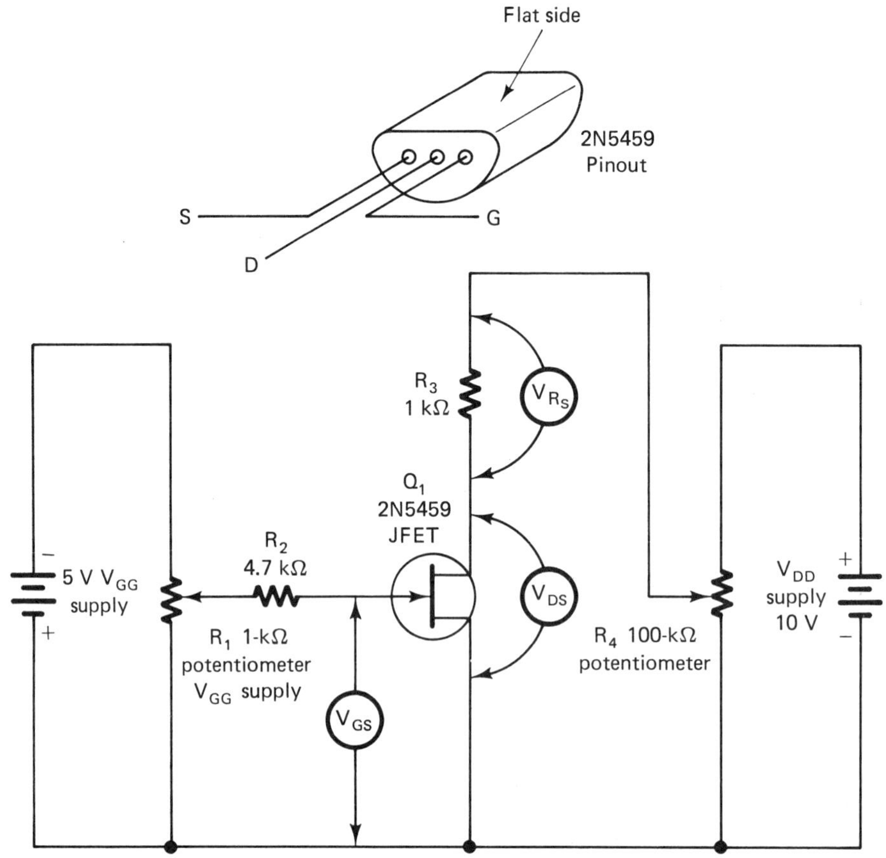

FIGURE 7-2A JFET characteristic test circuit.

2. Adjust the two power supplies to the 0 V position; then turn on the power supply. Adjust the V_{DD} supply to produce 10 V. This is measured across the two outside leads of potentiometer R_4. Adjust the V_{GG} supply to produce 5 V. This is measured across the two outside leads of potentiometer R_1. Potentiometers R_1 and R_4 are used to change the value of the gate-source voltage and the source-drain voltage of the transistor. Voltage measured across the drain resistor R_3 is used to determine the drain current by calculation. $I_D = V_{R3}/R_3$. This calculation will be made for each V_{DS} voltage value and recorded in the data chart as I_D. V_{GS} is measured across the gate and source of the transistor. This voltage is adjusted to a specified value and then the V_{DS} voltage is adjusted to the values indicated in the chart of Fig. 7-2B.
3. Connect the VOM across the gate-source of the transistor and adjust R_1 to produce 0 V_{GS}. Connect the VOM across the source-drain and adjust potentiometer R_4 to produce 0 V_{DS}. Measure the voltage across resistor R_3. Calculate the value of drain current and record it in the chart of Fig. 7-2B.
4. Increase V_{DS} to each of the indicated voltage values of

V_{DS}	$V_{GS} = 0$ V	$V_{GS} = -0.5$ V	$V_{GS} = -1$ V	$V_{GS} = -2$ V	$V_{GS} = -3$ V
0 V	0 mA	0 mA	0 mA	0 mA	0 mA
1 V					
2 V					
3 V					
4 V					
5 V					
6 V					
7 V					
8 V					
9 V					

FIGURE 7-2B I_D-V_{DS} chart.

the chart and measure the voltage across R_3. Calculate the drain current for each value of V_{R3} and record this value in the chart as I_D. Return the V_{DS} voltage to 0 V.

5. Increase the value of the gate voltage V_{GS} to -0.5 V. Repeat the procedure outlined in steps 3 and 4. Record the calculated drain-current values for each V_{GS} value.
6. Adjust V_{GS} to -1.0 V and repeat the procedure for each of the V_{DS} values. Record the calculated drain-current values in the chart.
7. Complete the chart by repeating the same procedure for V_{GS} values of -2.0 and -3.0. Record the calculated drain-current values for each V_{DS} value.
8. Using the data collected in the chart of Fig. 7-2B, plot an I_D-V_{DS} family of curves for the JFET in Fig. 7-2C.

Part B: Load-Line Analysis

1. Plot a load line on the I_D-V_{DS} family of curves of Fig. 7-2C for the circuit in Fig. 7-2D.
2. Resistor R_D serves as the load and V_{DD} as the source. The two conditions of operation are conduction and nonconduction.
3. Select a Q point based on the value of V_{GS}. Mark the Q point on the load line.
4. Project lines from the Q point to the I_D scale and to the V_{DS} scale of the graph:

I_D = ____________ mA

V_{DS} = ____________ V

5. With the indicated ac signal, the total change in V_{DS} is ____________ V to ____________ V.
6. The output V_{DS} resulting from this input is ____________

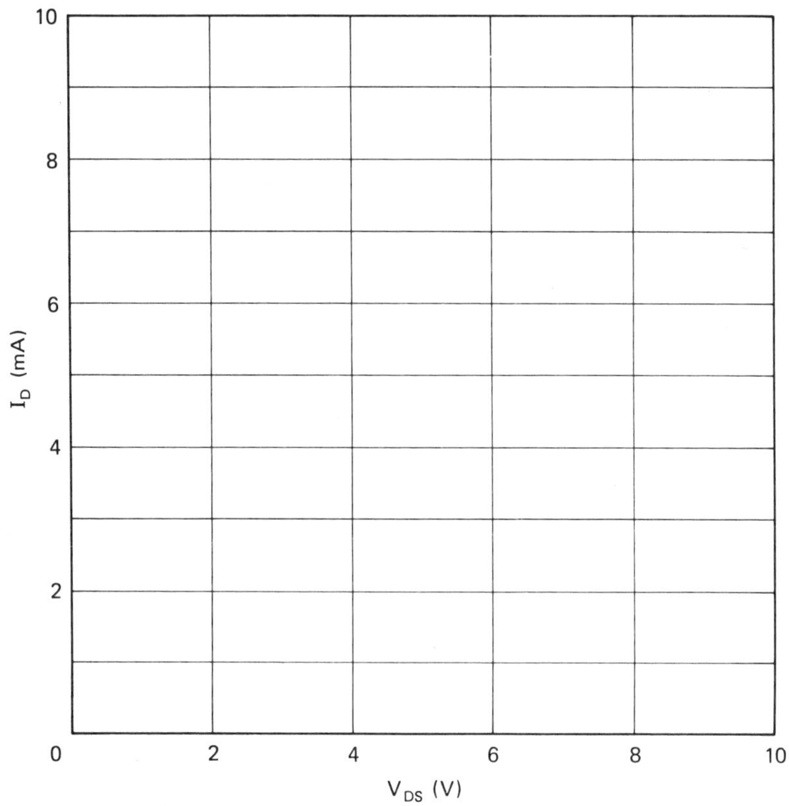

FIGURE 7-2C I_D versus V_{DS} graph.

V p-p. Note the polarity of V_{DS} compared with V_{GS}. What does this indicate? ______

7. The resulting I_D that occurs for this input is ______ mA, peak to peak. What is the polarity of I_D with respect to V_{GS}? ______
8. Calculate the ac voltage gain: A_v = ______.
9. Transconductance (*gm*) is a measure of the ease with which current carriers pass through the channel of a

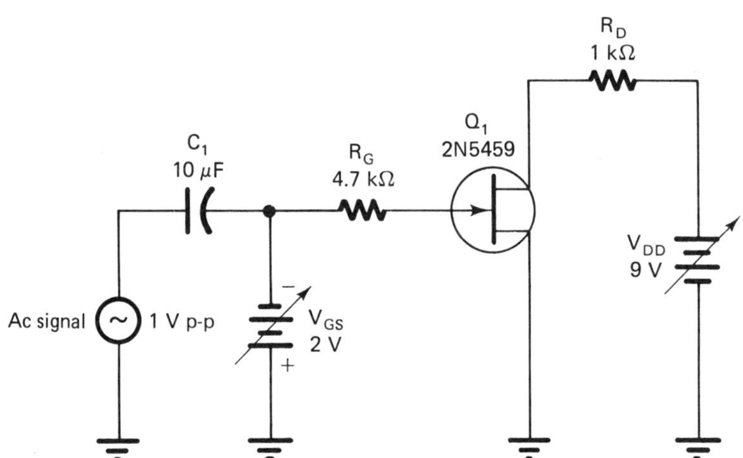

FIGURE 7-2D Load-line circuit.

JFET. It is used to describe the effectiveness of a JFET as an amplifier. The higher the transconductance, the greater its amplifying capability. $gm = I_D/V_{GS}$. Determine the transconductance of the JFET: gm = ______ ______ S.

DISCUSSION

In the first part of this experiment, you were asked to determine the I_D-V_{DS} characteristics of a JFET. In this procedure V_{GS} was set to a specific value, and the V_{DS} and V_{R3} was measured. The drain current was then calculated for specific values of V_{DS}. This procedure was repeated for different values of V_{DS}. The measured V_{DS} values and calculated I_D values were then used to plot a family of I_D-V_{DS} curves for the JFET. This family of curves shows how the JFET will respond in a circuit.

In the second part of this experiment the family of I_D-V_{DS} curves were used to plot a load line for a transistor amplifier. The load line shows graphically how the transistor responds. An operating point was established, and I_D and V_{DS} values were projected from the plotted points to show representative voltage and current values. This graphic presentation also permitted you to see the phase relationship of V_{GS}, I_D, and V_{DS}. Transconductance was then calculated. The higher the value of gm, the more effective the JFET is as an amplifier.

ANALYSIS

1. How would an increase in the resistance of the drain resistor change the operation of a JFET amplifier?
2. Explain how a JFET achieves amplification.
3. How does the operation of a JFET differ from that of a bipolar transistor?

EXPERIMENT 7-3
MOSFET CHARACTERISTICS

In this experiment you will develop a partial I_D-V_{DS} family of characteristic curves for a MOSFET. The curves are derived by setting the gate voltage to a specific value and adjusting the source-drain voltage to specific values. The drain-resistor voltage is then measured for each V_{DS} value and used to calculate the drain current. The V_{DS} and I_D values are then used to plot an I_D-V_{DS} curve for the MOSFET. Several other V_{GS} values are used to plot representative curves. The polarity of the gate voltage is then reversed to show how the MOSFET will respond with a different polarity of gate voltage. This curve is also plotted.

The developed family of I_D-V_{DS} curves of the MOSFET are then used to show the operational characteristics of the device in a representative circuit. A load line is plotted for the circuit. An operating point is established according to the circuit's data, and characteristic values of drain current and source-drain voltage are determined from the curves. These values are used to calculate transconductance and voltage gain. This procedure shows how the amplifier circuit responds graphically.

OBJECTIVES

1. To demonstrate the operating characteristics of a MOSFET by measuring some representative source-drain voltages and calculating the drain current.
2. To plot a partial family of I_D-V_{DS} characteristic curves for a MOSFET.
3. Determine a load line for a representative circuit and plot it on the partial characteristic curves.
4. Evaluate the operation of a MOSFET circuit with a family of I_D-V_{DS} curves and determine voltage gain, transconductance, and amplifier phase relationships.

EQUIPMENT

Volt-ohm-milliammeter (VOM), standard or digital

N-channel MOSFET, Motorola 2N3796, Heath 417-935, or equivalent

100-Ω, $\frac{1}{4}$-W resistor

1-kΩ, $\frac{1}{4}$-W resistor

4.7-kΩ, $\frac{1}{4}$-W resistor

1-kΩ, 1-W potentiometer (linear taper)
100-kΩ, 1-W potentiometer (linear taper)
Dual-section variable power supply, ±15 V, or two 9-V batteries
Circuit construction board

PROCEDURE

Part A: Characteristic Curve Data Collection

1. Construct the MOSFET characteristic curve test circuit of Fig. 7-3A. The leads or pin-out of the MOSFET are shown in the diagram.
2. With circuit construction complete, remove the MOSFET shorting ring or conducting foam.
3. Adjust the two power supply voltage controls to the 0-V position and turn on the power supply. Adjust the V_{DD} supply to produce 10 V. This is measured across the two outside leads of potentiometer R_4. Adjust the V_{GG} supply to produce 5 V. This is measured across the two outside leads of potentiometer R_1. Potentiometers R_1 and R_4 are used to change the value of the gate-source voltage and the source-drain voltage of the MOSFET.
4. Voltage measured across the drain resistor R_3 is used to determine the drain current by calculation. $I_D = V_{R3}/R_3$.

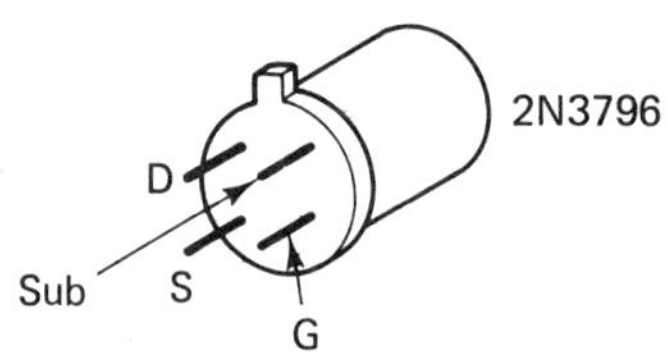

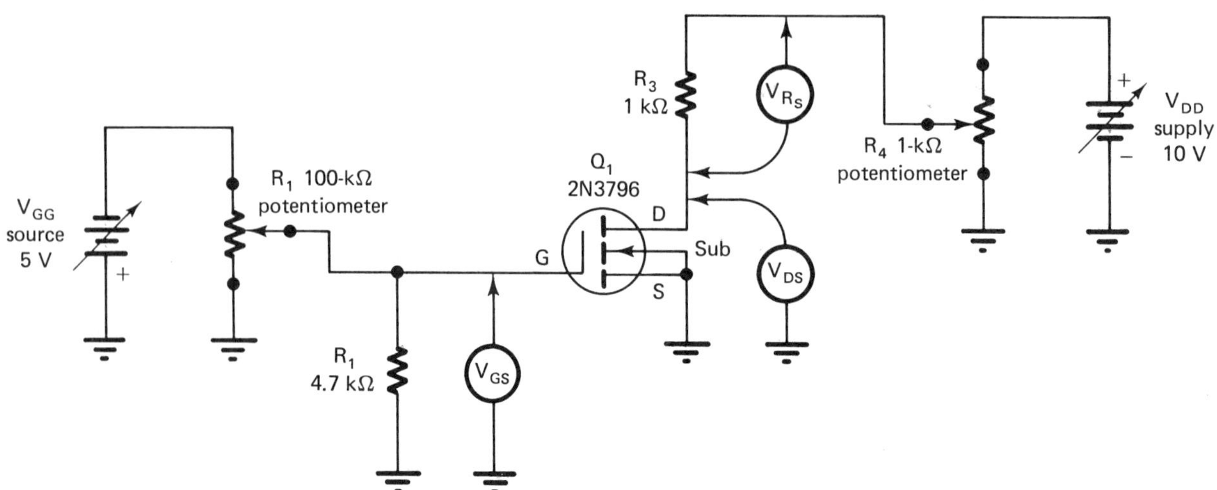

FIGURE 7-3A MOSFET I_D-V_{DS} test circuit.

This calculation will be made for each V_{DS} voltage value and recorded in the data column of the chart as I_D.

5. V_{GS} is measured across the gate-source resistor R_2. This voltage is adjusted to a specified value and then V_{DS} is adjusted to the values indicated in the chart of Fig. 7-3B.
6. Connect the VOM across the gate-source of the MOSFET and adjust R_1 to produce 0 V_{GS}. Connect the VOM across the source-drain and adjust potentiometer R_4 to produce 0 V_{DS}. Measure and record the voltage across resistor R_3. Calculate the drain current for this value. Adjust potentiometer R_4 to produce a V_{DS} of 1 V. Measure and record the voltage across resistor R_3 for this V_{DS} setting. Calculate the drain current and record its value in the chart. Repeat this procedure for V_{DS} values of 2, 4, 6, 8, and 10 V.
7. Return the value of V_{DS} to 0 V.
8. Increase the value of V_{GS} to -1 V. Repeat the procedure outlined in step 6. Record your data values in the -1-V_{GS} column of the chart.
9. Return the two potentiometers to 0 V. Disconnect the negative lead of the power supply connected to potentiometer R_1. Connect this potentiometer lead to the positive side of the V_{DD} power supply. Adjust the V_{DD} power supply to 10 V.
10. Adjust potentiometer R_1 to produce $+1$ V_{GS}. Repeat the procedure outlined in step 6 for different values of V_{DS} for 0 to 10 V. Record your data in the $+1$-V_{GS} column of the chart of Fig. 7-3B.
11. Using the data collected in the chart of Fig. 7-3B, plot a partial set of I_D-V_{DS} curves for the MOSFET in Fig. 7-3C.

	$V_{GS} = 0$ V		$V_{GS} = -1$ V		$V_{GS} = +1$ V	
V_{DS}	Measured V_{R_3}	Calculated I_D	Measured V_{R_3}	Calculated I_D	Measured V_{R_3}	Calculated I_D
0 V						
1 V						
2 V						
4 V						
6 V						
8 V						
10 V						

FIGURE 7-3B Chart of I_D-V_{DS} values.

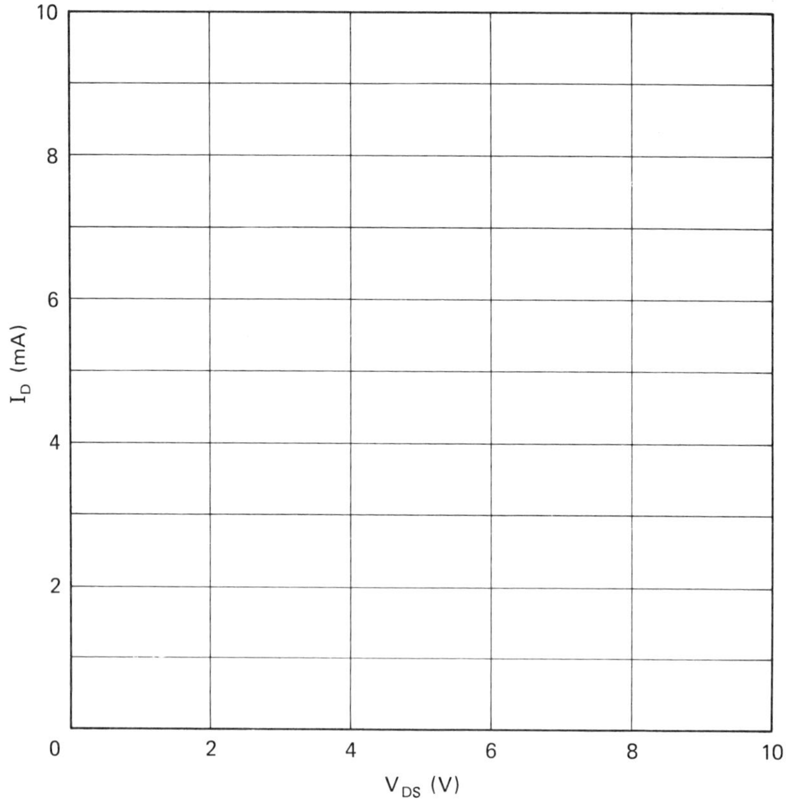

FIGURE 7-3C I_D versus V_{DS} graph.

Part B: Load-Line Analysis

1. Plot a load line on the partial set of I_D-V_{DS} curves of Fig. 7-3C for the circuit of Fig. 7-3D.
2. Resistor R_D serves as the load, and V_{DD} serves as the source. The two conditions of operation are full conduction and nonconduction.
3. The operating point of this circuit is zero, based on the value of V_{GS}. Mark the Q point on the load line.

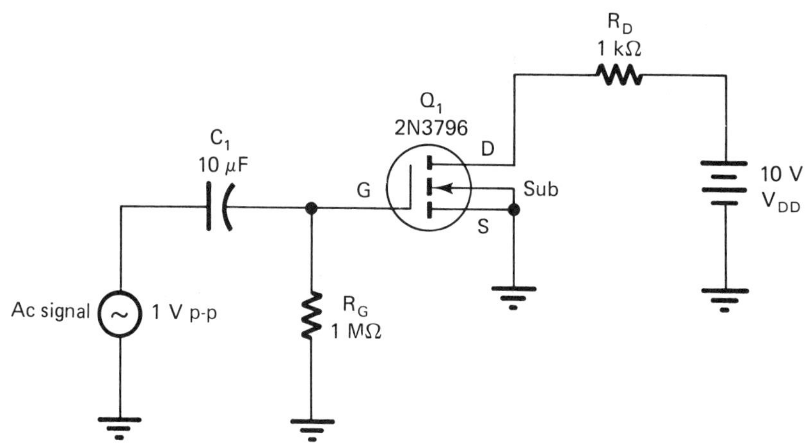

FIGURE 7-3D MOSFET load-line circuit.

4. Project lines from the Q point to the I_D scale and to the V_{DS} scale of the graph: I_D = ____________ mA; V_{DS} = ____________ V.
5. With the indicated ac signal, the total change in V_{GS} is ____________ peak to peak. The resulting output from this input is ____________ V p-p. Note the polarity of V_{DS} compared with V_{GS}. What does this indicate? ________.
6. The resulting I_D that occurs for this input is ________ mA p-p. How does the polarity of I_D compare with V_{DS}? ____________________________________.
7. Using the value of the indicated input and the output voltage, calculate the voltage gain (A_v) for the circuit. A_v = ____________.
8. Transconductance is a measure of the effectiveness of this device to amplify a signal. Transconductance (gm) = I_D-V_{GS}. Determine the transconductance of the amplifier circuit using your data: gm = ____________ S.

DISCUSSION

In the first part of this experiment, you were asked to determine three different V_{GS} values for a partial set of I_D-V_{DS} curves for a MOSFET. In this procedure V_{GS} was set to a specific value, and the V_{DS} and V_{RL} values were measured. The drain current was then calculated for specific values of V_{DS}. This procedure was repeated for different values of V_{GS}. The measured values of V_{DS} and calculated values of I_D were then used to plot a partial family of curves for the MOSFET. This family of curves shows how the MOSFET will respond in a circuit.

In the second part of this experiment the partial family of I_D-V_{DS} curves were used to plot a load line for a MOSFET amplifier. The load line graphically shows how the transistor responds. An operating point is established, and I_D and V_{DS} values are projected from the plotted points to show representative voltage and current values. This graphic presentation also permits you to see the phase relationship of V_{DS}, I_D, and V_{GS}. Transconductance and amplification were then calculated from the derived values. This experiment permits you to see how a MOSFET will respond graphically as an amplifier.

ANALYSIS

1. Explain how an N-channel E-MOSFET achieves amplification.
2. How would an increase in the resistance of the drain resistor or load change the operation of a MOSFET?
3. What are the operational differences between an E-MOSFET and a JFET amplifier?

UNIT 7 EXAMINATION
FIELD-EFFECT TRANSISTORS

Instructions: Circle the response that most correctly completes each statement.

1. For proper operation, the drain of an N-channel MOSFET must be:
   a. Positive with respect to the source
   b. Forward biased with respect to the gate
   c. At the same potential as the source
   d. Reverse biased with respect to the gate
2. For the greatest conduction of an N-channel JFET to take place, the gate must:
   a. Be slightly forward biased
   b. Be reverse biased
   c. Not be biased
   d. Be heavily reverse biased
   e. Be heavily forward biased
3. Conduction of channel current in an N-channel JFET is based on:
   a. The depletion of current carriers
   b. The enhancement of current carriers
   c. Forward biasing of the gate-source junction
   d. The combined depletion-enhancement of current carriers
4. In a JFET the resistance measured between source and drain with an ohmmeter should be:
   a. High in both directions
   b. Low in both directions
   c. Either high or low according to polarity
   d. Infinite in both directions
5. A 180° phase inversion between the input and output signal occurs when using:
   a. The common-drain circuit configuration.
   b. The common-source circuit configuration.
   c. The common-gate circuit configuration.
   d. All JFET circuit configurations.
6. Voltage amplification of a MOSFET is based on:
   a. V_{DS}/V_{GS}
   b. V_{DD}/V_{GG}

c. V_{GS}/V_{DS}
d. V_{DS}/V_{DG}

7. For ac signal amplification a depletion-type MOSFET can be biased:
   a. At 0 V
   b. With positive voltage
   c. With negative voltage
   d. With all the above alternatives
   e. None of the above alternatives
8. If a particular JFET has a 0.5-V change in V_{GS} and a corresponding 2.5-mA change in I_D, the *gm* value is in the range of:
   a. 0.001 to 10 μS
   b. 11 to 100 μS
   c. 101 to 1000 μS
   d. 1001 to 10,000 μS
9. A JFET amplifier has a 10-V_{DD} source and 5 V appearing across a 1-kΩ drain resistor. The resulting I_D that occurs falls in the range of:
   a. 0.1 to 1 mA
   b. 1.1 to 10 mA
   c. 10.1 to 100 mA
   d. 101 to 1000 mA
   e. 1001 mA or more
10. The drain current of a properly biased N-channel JFET will be the greatest value when the gate is:
    a. Made slightly positive with respect to the source
    b. Made highly negative with respect to the source
    c. Made slightly negative with respect to the source
    d. Shorted to the source to reduce V_{GS} to zero
    e. Shorted to the drain to increase the depletion region
11. The gate of an N-channel E-MOSFET is made of:
    a. A P-type semiconductor material
    b. An N-type of semiconductor material
    c. Deposited metal
    d. Silicon dioxide
    e. The substrate material
12. An N-channel E-MOSFET is built on:
    a. A layer of silicon dioxide
    b. An intrinsic crystal
    c. A substrate
    d. A layer of the source-drain material
13. A field-effect transistor that has a 0-V_{GS} line near the center of its I_D-V_{DS} curves is:

a. An N-channel JFET
b. A P-channel JFET
c. An E-MOSFET
d. A D-MOSFET
e. A V-MOSFET

14. The major difference between an E-MOSFET and a D-MOSFET is:
a. Substrate formation
b. Source-drain construction
c. The gate material
d. Channel formation
e. V-MOS technology

15. A V-MOSFET responds the same as:
a. A P-channel E-MOSFET
b. An N-channel JFET
c. A D-MOSFET
d. An N-channel E-MOSFET
e. A combined E-MOSFET and a D-MOSFET

UNIT EIGHT

Thyristors

UNIT INTRODUCTION

The term *thyristor* refers to a rather general classification of solid-state devices that are used as electronic switches and power-control devices. Thyristors can be two-, three-, or four-terminal devices. In this unit we are concerned with the conductivity capabilities of these devices. Devices placed in this classification operate through a type of regenerative feedback. When conduction is initiated, the device will latch or hold in its on state. Momentarily removing or reducing the energy level of the source will cause non-conduction or switching to the off state. In general, this type of device is classified as a voltage-controlled switch.

Thyristors should not be confused with bipolar junction transistors or field-effect transistors. Transistors in general are capable of performing switching operations. As a rule, these devices are not as efficient as thyristors and do not have the power-handling capability. Thyristors are used as power-control devices, whereas transistors are primarily used in amplifying applications.

Thyristors are used widely in electronic circuits. One of the first to be developed was the silicon-controlled rectifier or SCR. The SCR is a solid-state counterpart of the older thyratron gaseous tube. The term thyristor was derived from the words *thyra*tron and trans*istor.* An SCR is classified as a reverse-blocking triode thyristor. The term *triode* means three electrodes. Reverse blocking refers to conduction in only one direction. An SCR is, therefore, a three-electrode device with unidirectional power-control capabilities. Unijunction transistors (UJTs), programmable unijunction transistors (PUTs), and gate turn-off switches (GTO) are other devices of this classification.

Bidirectional thyristors are another classification of solid-state power control devices. In this group of devices we have the triac and the diac. A *triac* is a *tri*ode *ac* power-control device. The *diac* is a *di*ode for *ac.* These devices provide some rather unusual power-control capabilities that cannot be duplicated by other devices. Bidirectional thyristors provide some very efficient power-control capabilities.

UNIT OBJECTIVES

Upon completion of this unit, you will be able to:

1. Describe the physical construction of silicon-controlled rectifiers, unijunction transistors, programmable unijunction transistors, triacs, and diacs.
2. Explain the fundamental operation of an SCR, UJT, PUT, triac, diac, GTO, and SCS.
3. Explain how the elements of different electronic power-control devices are energized or biased to make the device operational.
4. Define some of the basic terms associated with electronic power-control devices such as gate current, holding current, latching, intrinsic standoff ratio, forward breakover voltage, reverse blocking voltage, and triggering.
5. Inspect the characteristics of various power-control devices and be able to interpret their meaning.
6. Test different electronic power-control devices, tell something about the status of a particular device, and be able to identify its leads.
7. Identify electronic power-control devices in a specific circuit application.

IMPORTANT TERMS

Before reading Unit 8, review the following terms. These terms provide a basic understanding of terms associated with thyristors.

Alternation Half of an ac sine wave. There is a positive alternation and a negative alternation for each ac cycle.

Bidirectional triggering The triggering or conduction that can be achieved in either direction of an applied ac wave. Diacs and triacs are of this classification.

Breakover voltage The voltage across the anode-cathode of an SCR or terminal 1 and terminal 2 of a triac that causes it to turn on.

Circuit switching An on-off control function. Conduction occurs during the on time with no conduction during the off time.

Commutation The changing from on to off or from one state to the opposite state.

Conduction time The time when a solid-state device is turned on or in its conductive state.

Contact bounce When a mechanical switch is turned on, the contacts are forced together. This causes the contacts to open and close several times before making firm contact.

Delay time The time that an alternation is held in an off state before it turns on.

Diac A solid-state diode device that will trigger in either direction when its breakover voltage is exceeded. A bidirectional triggering device.

Forward breakover voltage The voltage at which an SCR or triac goes into conduction in quadrant one of its *I-V* characteristics.

Gate The control electrode of a triac, SCR, or FET.

Gate current (I_G) The resulting current that flows in the gate when it is forward biased.

Holding current (I_H) A current level that must be achieved when an SCR or triac latches or holds in the conductive state.

Interbase resistance The resistance between base 1 and base 2 of a unijunction transistor.

Internal resistance The resistance between the anode and cathode of an SCR or between terminals 1 and 2 of a triac. Also called the dynamic resistance.

Latch A holding state in which an SCR or triac turns on and stays in conduction when the gate current is removed.

Main 1 or 2 The connection points on a triac where the load current passes. Also called terminal 1 and terminal 2.

Negative resistance A UJT characteristic where an increase in emitter current causes a decrease in emitter voltage.

Off-state resistance The resistance of an SCR or triac when it is not conducting.

Peak voltage point (P_v) A characteristic point of the unijunction transistor at which the emitter voltage rises to a peak value.

Phase A position or part of an ac wave that is compared with another wave. A wave can be in phase or out of phase with respect to another.

Phase shifter A circuit that changes the phase of an ac waveform.

Quadrant One-fourth or a quarter part of a circle or a graph.

Reverse breakdown The conduction that occurs in quadrant III of the *I-V* characteristics of a triac.

Switching device An electronic device that can be turned on and off very rapidly. This type of device is capable of achieving control by switching.

Terminal (T_1 or T_2) The connecting points where load current passes through a triac. These two connection points are also called main 1 or 2.

Triggering The process of causing an NPNP device to switch states from off to on.

Turn-on time The time of an ac waveform when an alternation occurs.

Valley voltage (V_v) A characteristic voltage point of the UJT where the voltage value drops to its lowest level.

SILICON-CONTROLLED RECTIFIERS

A *silicon-controlled rectifier* (SCR) is probably the most popular electronic power-control device today. The SCR is used primarily as a switching device. Power control is achieved by switching the SCR on and off during one alternation of the ac source voltage. For 60-Hz ac the SCR is switched on and off 60 times per second. Control of electrical power is achieved by altering or delaying the turn-on time of an alternation.

An SCR, as the name implies, is a solid-state rectifier device. It conducts current in only one direction. It is similar in size to a comparable silicon power diode. SCRs are usually small, are rather inexpensive, waste very little power, and require practically no maintenance. The SCR is available today in a full range of types and sizes to meet nearly any power-control application, in current ratings from less than 1 A to over 1400 A. Voltage values range from 15 to 2600 V.

SCR Construction

An SCR is a solid-state device made of four alternating layers of P- and N-type silicon. Three P-N junctions are formed by the structure. Each SCR has three leads or terminals. The anode and cathode terminals are similar to those of a regular silicon diode. The third lead is called the gate. This lead determines when the device switches from its off to on state. An SCR will usually not go into conduction by simply forward biasing the anode-cathode. The gate must be forward biased at the same time. When these conditions occur, the SCR becomes conductive. The internal resistance of a conductive SCR is less than 1Ω. Its reverse or off-state resistance is generally in excess of 1 MΩ.

A schematic symbol and the crystal structure of an SCR are shown in Fig. 8-1. Note that the device has a PNPN structure from anode to cathode. Three distinct P-N junctions are formed. When the anode is made positive and the cathode negative, junctions 1 and 3 (J_1 and J_3) are forward biased, and J_2 is reverse biased. Reversing the polarity of the source alters this condition: J_1 and J_3 become reverse biased, and J_2 becomes forward biased, which will not permit conduction. Conduction will occur only when the anode, cathode, and gate are all forward biased simultaneously.

Some representative SCRs are shown in Fig. 8-2. Only a few of the more popular packages are shown here. As a general rule, the anode is connected to the largest electrode if there is a difference in their physical size. The gate is usually smaller than the other electrodes. Only a small gate current is needed to achieve control. In some packages the SCR symbol is used for lead identification.

SCR Operation

Operation of an SCR is not easily explained by using the four-layer PNPN structure of Fig. 8-1. A rather simplified method has been developed that describes the crystal structure as equivalent to two interconnected transistors. Figure 8-3 shows this imaginary division of the crystal. Notice that the top three cyrstals form an NPN device. Two parts of each transistor are interconnected.

Figure 8-4 shows how the two-transistor equivalent circuit and the PNPN structure of an SCR will respond when voltage is applied. Note that the anode is made positive and the cathode negative. This condition forward biases the emitter of each transistor and reverse biases the collector. Without base voltage applied to the NPN transistor, the equivalent circuit is nonconductive. No

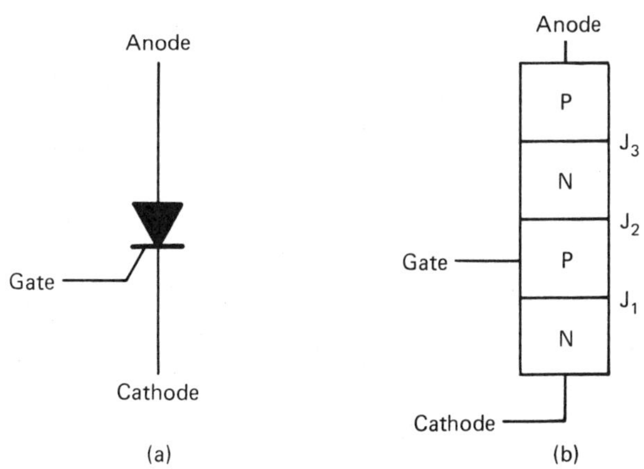

FIGURE 8-1 SCR crystal: (a) symbol; (b) structure.

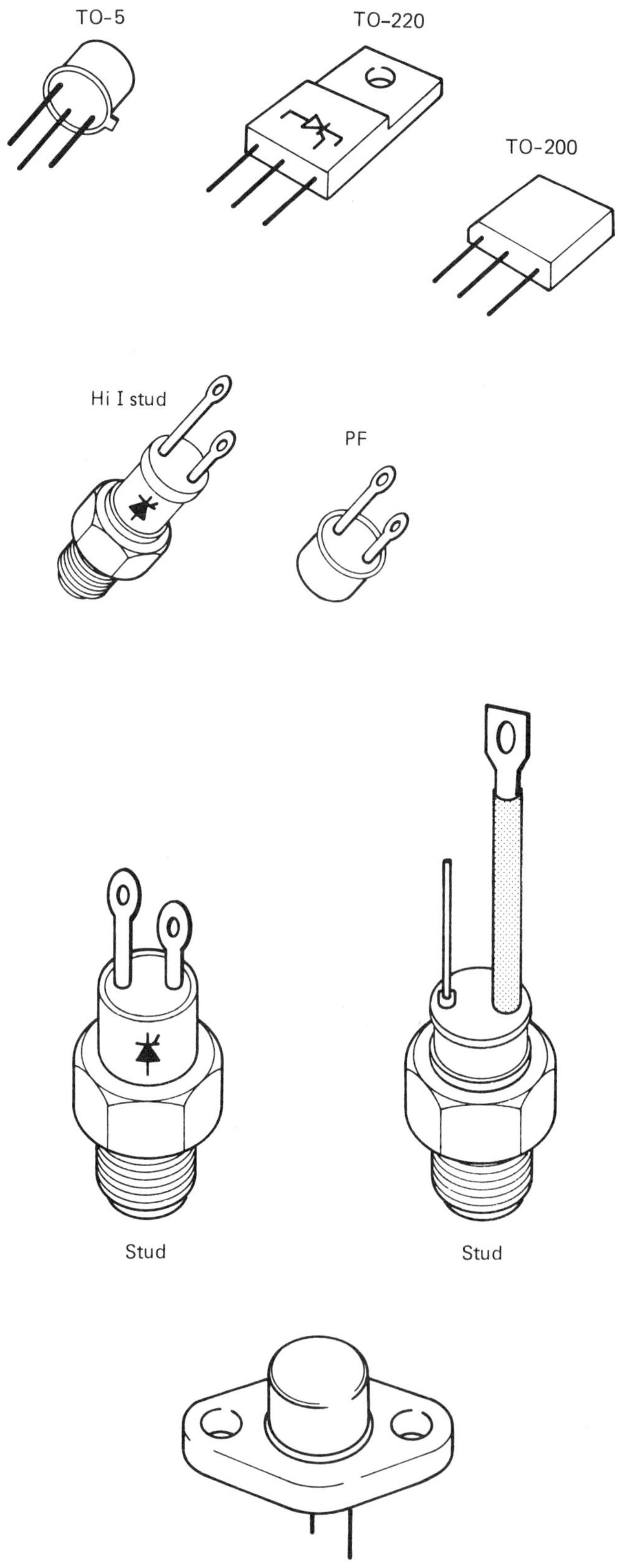

FIGURE 8-2 Representative SCR packages.

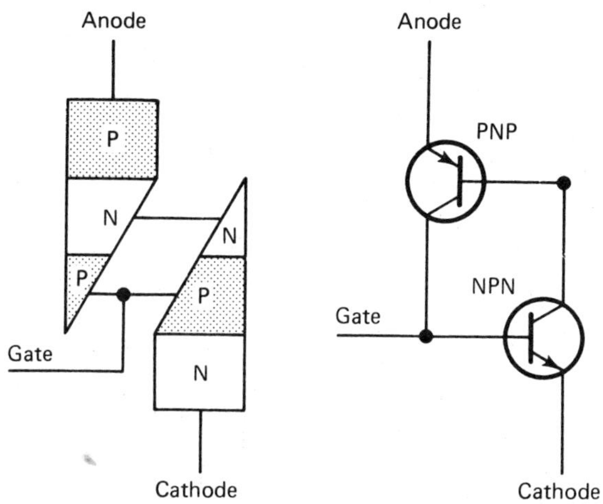

FIGURE 8-3 Equivalent SCR.

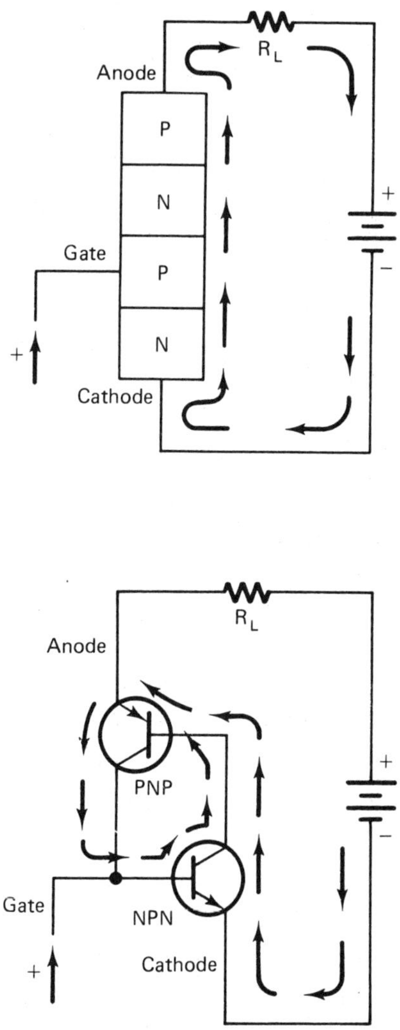

FIGURE 8-4 SCR response.

current will flow through the load resistor. Reversing the polarity of the voltage source will not alter the conduction of the two circuits. The emitter of each transistor will be reverse biased by this condition. This means that no conduction will take place when the anode and cathode are reverse biased. With only anode-cathode voltage applied, no conduction will occur in either direction.

Assume now that the anode and cathode of an SCR are forward biased, as indicated in Fig. 8-4. If the gate is momentarily made positive with respect to the cathode, the emitter-base junction of the NPN transistor becomes forward biased. This action immediately causes collector current to flow in the NPN transistor. As a result, base current also flows into the PNP transistor. This in turn causes a corresponding collector current to flow. PNP collector current now causes base current to flow into the NPN transistor. The two transistors therefore latch or hold when the gate is disconnected. In effect, when conduction occurs, the device latches in the on state. Current will continue to flow through the SCR as long as the anode-cathode voltage is of the correct polarity.

To turn off a conductive SCR it is necessary to momentarily remove or reduce the anode-cathode voltage. This action will turn off the two transistors. The device will then remain in this state until the anode, cathode, and gate are all forward biased again. With ac applied to an SCR, it will automatically turn off during one alternation of the input. Control is achieved by altering the turn-on time during the conductive or on alternation.

SCR Current-Voltage Characteristics

The current-voltage characteristics of an SCR tell a great deal about its operation. The *I-V* characteristics of Fig. 8-5 show that an SCR has two conduction states. Quadrant I shows conduction in the forward direction. This shows how conduction occurs when the forward breakover voltage (V_{BO}) is exceeded. Note that the curve returns to approximately zero after the V_{BO} has been exceeded. When conduction occurs, the internal resistance of the SCR drops to an extremely small value. This value is similar to that of a forward-biased silicon diode. The conduction current (I_{AK}) must be limited by an external resistor. This current, however, must be great enough to maintain conduction when it starts. The holding current or I_H level must be exceeded in order for this to take place. Note that the I_H level is just above the knee of the I_{AK} curve after it returns to the center.

Quadrant III of the *I-V* characteristic curve shows the reverse breakdown condition of operation. This characteristic of an SCR is very similar to that of a silicon diode. Conduction occurs when the peak reverse voltage (P_{RV}) value is reached. Normally, an SCR will be permanently damaged if the P_{RV} is exceeded. Today, SCRs have P_{RV} ratings of 25 to 2000 V.

For an SCR to be used as a power-control device, the forward V_{BO} must be altered. Changes in gate current will cause a decrease

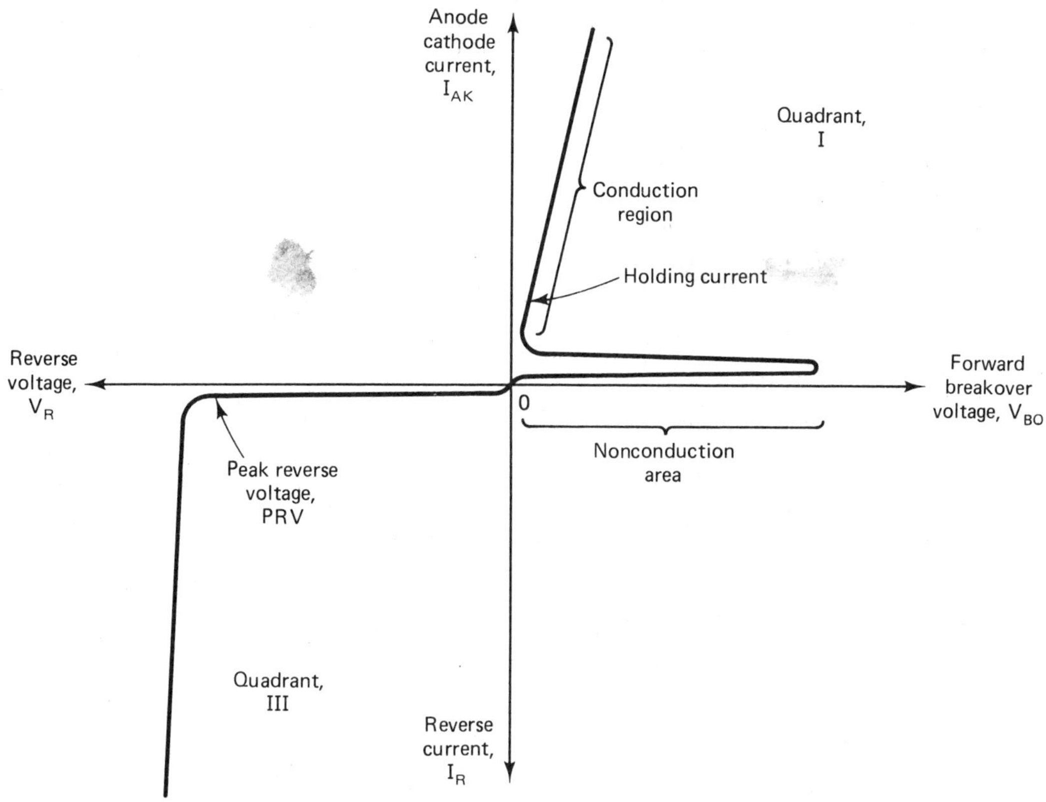

FIGURE 8-5 *I-V* characteristics of an SCR.

in the V_{BO}. This occurs when the gate is forward biased. An increase in I_G will cause a large reduction in the forward V_{BO}. An enlargement of quadrant I of the *I-V* characteristic is shown in Fig. 8-6. This shows how different values of I_G change the V_{BO}. With zero I_G it takes a V_{BO} of 400 V to produce conduction. An increase in I_G reduces this quite significantly. With 7 mA of I_G the SCR conducts as a forward-biased silicon diode. Lesser values of I_G will cause an increase in the V_{BO} needed to produce conduction.

The gate-current characteristic of an SCR shows a very important electrical operating condition. For any value of I_G there is a specific V_{BO} that must be reached before conduction can occur. This means that an SCR can be turned on when a proper combination of I_G and V_{BO} is achieved. This characteristic is used to control conduction when the SCR is used as a power-control device.

DC Power Control with SCRs

When an SCR is used as a power-control device, it responds primarily as a switch. When the applied source voltage is below the forward breakover voltage, control is achieved by increasing the gate current. Gate current is usually made large enough to ensure that the SCR will turn on at the proper time. Gate current is generally applied for only a short period of time. In many applications this may be in the form of a short-duration pulse. Continuous I_G

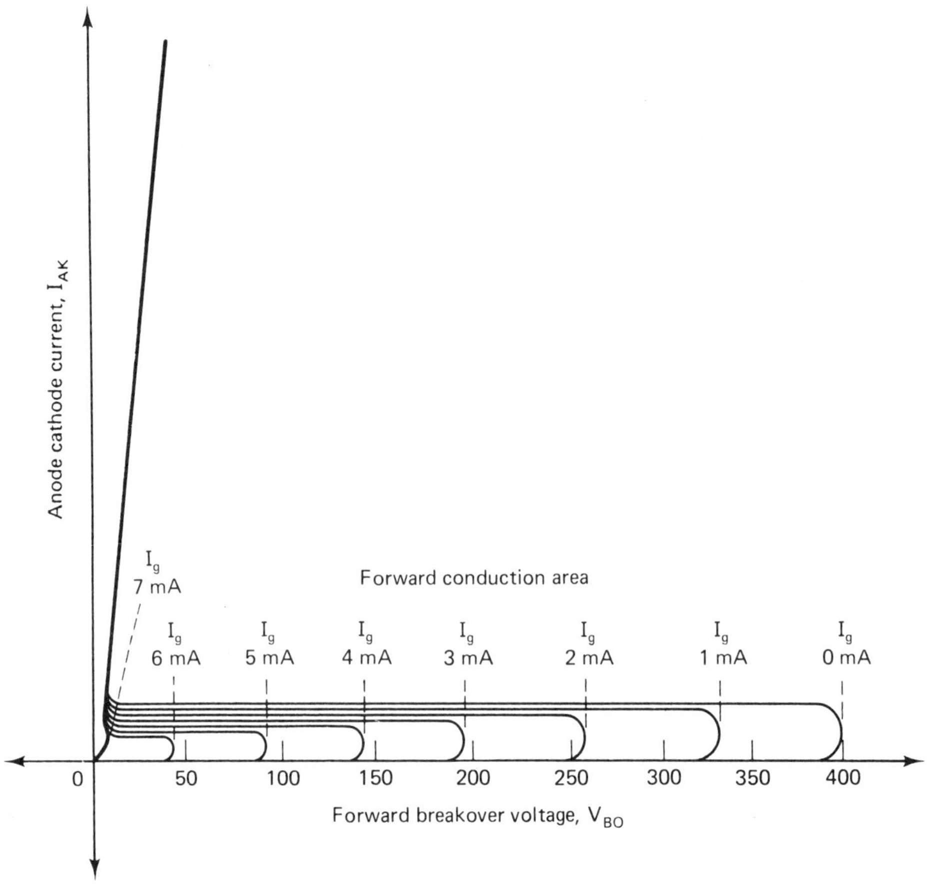

FIGURE 8-6 Gate-current characteristics of an SCR.

is not needed to trigger an SCR into conduction. After conduction occurs, the SCR will not turn off until the I_{AK} drops to zero.

Figure 8-7 shows an SCR used as a dc power-control switch. In this type of circuit a rather high load current is controlled by a small gate current. Note that the electrical power source (V_S) is

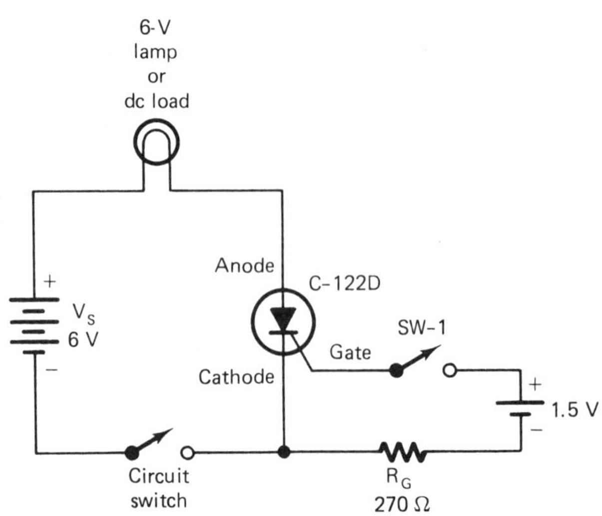

FIGURE 8-7 Dc power-control switch.

controlled by the SCR. This is achieved by making the anode positive and the cathode negative.

When the circuit switch is initially turned on, the load is not energized. In this situation the V_{BO} is greater than the V_S voltage. Power control is achieved by turning on SW-1. This forward biases the gate. If a suitable value of I_G occurs, it will lower the V_{BO} and turn on the SCR. The I_G can be removed and the SCR will remain in conduction. To turn the circuit off, momentarily open the circuit switch. With the circuit switch on again, the SCR will remain in the off state. It will go into conduction again by closing SW-1.

Dc power-control applications of the SCR require two switches to achieve control. This application of the SCR is not very practical. The circuit switch would need to be capable of handling the load current. The gate switch could be rated at an extremely small value. If several switches were needed to control the load from different locations, this circuit would be more practical. More practical dc power circuits can be achieved by adding a number of additional components. Figure 8-8 shows a dc power-control circuit with one SCR being controlled by a second SCR. SCR-1 controls the dc load current. SCR-2 controls the conduction of SCR-1. In this circuit, switching of a high-current load is achieved with two small low-current switches. SCR-1 can be rated to handle the load current. SCR-2 can have a rather small current-handling capacity. Control of this type can probably be achieved for less than a circuit employing a large electrical contactor switch.

Operation of the dc control circuit of Fig. 8-8 is based on the conduction of SCR-1 and SCR-2. To turn on the load, the on pushbutton is momentarily closed. This forward biases the gate of SCR-1. The V_{BO} is reduced, and SCR-1 goes into its conduction state. This action also causes C_1 to charge to the indicated polarity. The load remains energized as long as power is supplied to the circuit.

Turn-off of SCR-1 is achieved by pushing the stop button. This momentarily applies I_G to SCR-2 and causes it to be conductive. This in turn causes C_1 to be connected across SCR-1. The charge on C_1 is momentarily applied to the anode-cathode of SCR-1. This reduces the V_{AK} of SCR-1 and causes it to turn off. The circuit will

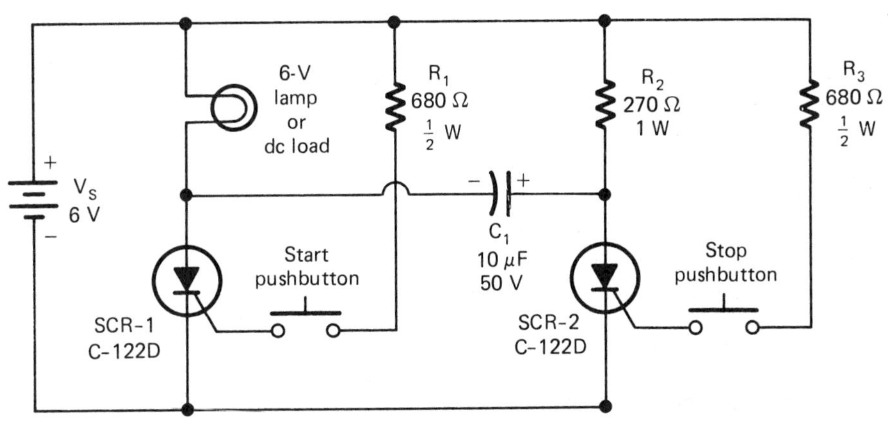

FIGURE 8-8 Dc power-control circuit.

remain in the off state until it is energized by the start button. An SCR power circuit of this type can be controlled with two small pushbuttons. As a rule, control of this type is more reliable and less expensive than a dc electrical contactor circuit.

AC Power Control with SCRs

Ac power-control applications of an SCR are very common. As a rule, control is very easy to achieve. The SCR automatically turns off during one alternation of the ac input. This eliminates the turn-off problem with the dc circuit. The load of an ac circuit will see current only for one alternation of the ac input cycle. In effect, an SCR power-control circuit has half-wave output. The conduction time of an alternation can be varied with an SCR circuit. We can have variable output through this method of control.

A simple SCR power-control switch is shown in Fig. 8-9. Connected in this manner, conduction of ac will only occur when the anode is positive and the cathode is negative. Conduction will not occur until SW-1 is closed. When this takes place, there is gate current. The value of I_G lowers the V_{BO} to the point where the SCR becomes conductive. R_g of the gate circuit limits the peak value of I_G. Diode D_1 prevents reverse voltage from being applied between the gate and cathode of the SCR. With SW-1 closed, the gate will be forward biased for only one alternation. This is the same alternation that forward biases the anode-cathode. With a suitable value of I_G and correct anode-cathode voltage (V_{AK}), the SCR will become conductive.

The ac power-control switch of Fig. 8-9 is designed primarily to take the place of a mechanical switch. With a circuit of this type, it is possible to control a rather large amount of electrical power with a rather small switch. Control of this type is very reliable. The switch does not have contacts that spark and arc when changes in load current occur. Control of this type, however, is only an on-off function.

SCRs are used widely to control the amount of electrical power supplied to a load device. Circuits of this type respond very

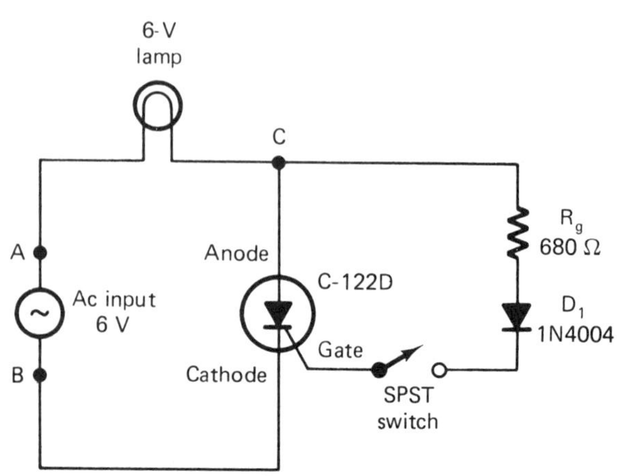

FIGURE 8-9 SCR ac power-control switch.

well to 60-Hz ac. This type of circuit is called a half-wave SCR phase shifter. Figure 8-10 shows a simplified version of the variable power control circuit.

Operation of the variable power control circuit is based on the charge and discharge of capacitor C_1. Assume now that the positive alternation occurs. Point A will be positive, and point B will be negative. This will forward bias the SCR. It will not turn on immediately, however, because there is no gate current. Capacitor C_1 begins to charge through R_1. In a specific time, the gate voltage builds up to where the gate current is great enough to turn on the SCR. Diode D_1 is forward biased by the capacitor voltage. I_G then flows through D_1. Once the SCR goes into conduction it continues for the remainder of the alternation. Current flows through the load. In effect, turn-on of the SCR is delayed by the value of RC.

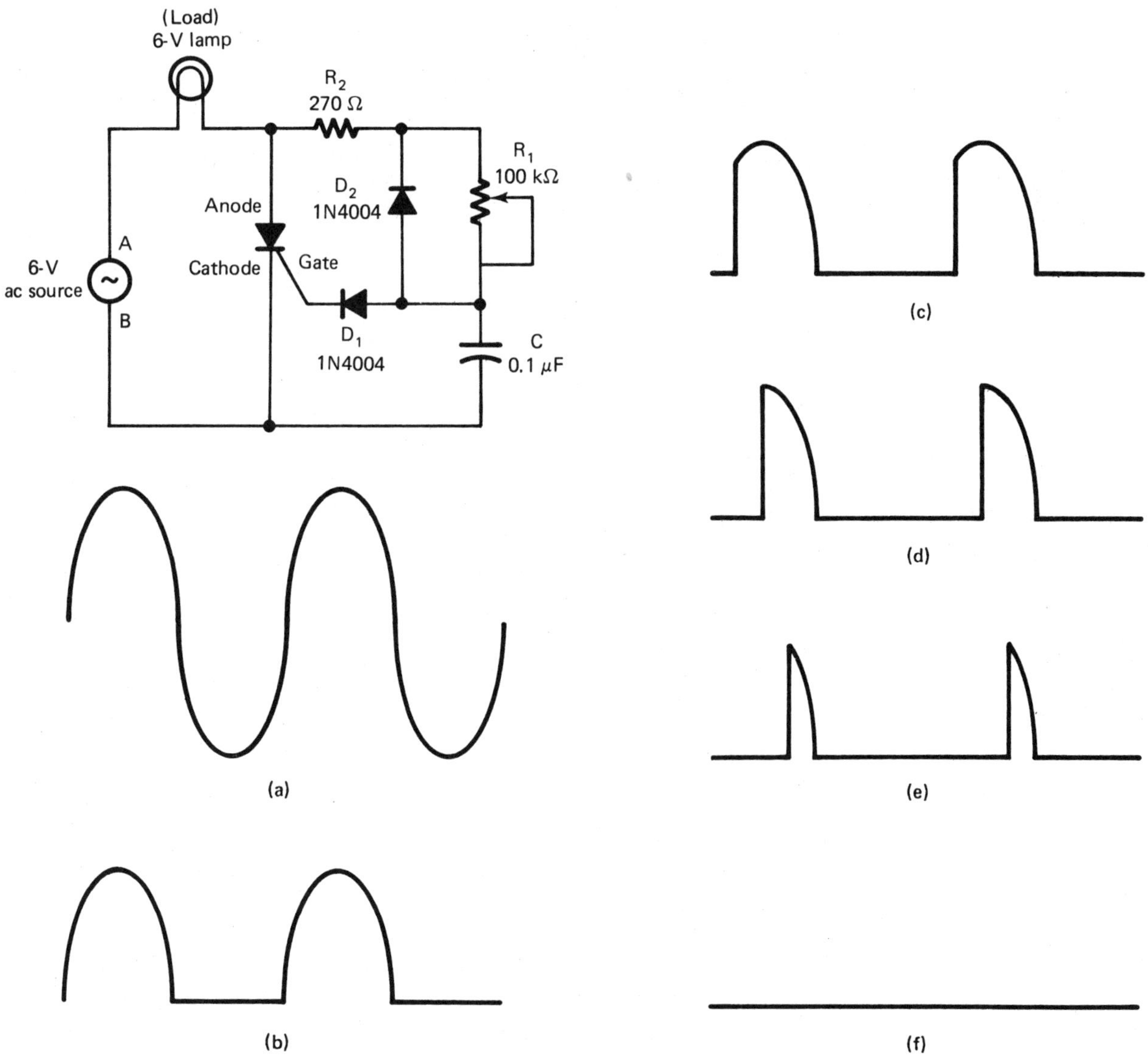

FIGURE 8-10 Phase-control SCR circuit and waveforms: (a) applied ac; (b) 0° delay; (c) 45° delay; (d) 90° delay; (e) 135° delay; (f) 180° delay.

When the negative alternation occurs, point A becomes negative, and point B goes positive. This reverse biases the SCR. No current flows through the load for this alternation. Diode D_2 is, however, forward biased by this voltage. Capacitor C_1 then discharges very quickly through D_2. This bypasses the resistance of R_1. The capacitor is reset or in a ready state for the next alternation. The process repeats itself for each succeeding cycle.

A change in the resistance of R_1 will alter the conduction time of the SCR power control circuit. When R_1 is increased, C_1 cannot charge as quickly during the positive alternation. It therefore takes more time for C_1 to build up its charge voltage. This means that the turn-on of I_G will be delayed for a longer time. As a result, conduction of the SCR will be delayed at the beginning of the alternation. See the waveform in Fig. 8-10(c). The resulting load current for this condition will be somewhat less than that of the waveform in (b). Variable control of the load is achieved by altering the conduction time of the SCR.

A further increase in the resistance of R_1 will cause more delay in the turn-on of the SCR during the positive alternation. Waveforms in (d), (e), and (f) of Fig. 8-10 show the result of this change. The waveform in (d) has a 90° delay of SCR conduction time. The waveform in (e) shows a 135° delay in the conduction time. If the value of R_1 is great enough, the SCR will delay its turn-on for the full alternation. The waveform in (f) shows a 180° turn-on delay. Note that the load does not receive any current for this condition. Control of this circuit is from full conduction, as in waveform (b), to 180° delay, as in (f).

Power control with an SCR is very efficient. If no delay of the conduction time occurs, the load develops full power. In this case, full power occurs only for one alternation. If conduction is delayed 90°, the load develops only half of its potential power. If the conduction delay is 135°, the load develops only 25% of its potential power. It is important to note that nearly all the power controlled by the circuit is applied to the load device. Very little of it is consumed by the SCR. In variable power control applications it is extremely important that the load be supplied to its full value of power. With an SCR control circuit, no power is consumed by the other components. By delaying the turn-on time of conduction, power can be controlled with the highest level of efficiency.

SELF-EXAMINATION

1. An SCR has three leads, which are known as the ________, ____________, and ____________.
2. An SCR will not go into conduction by simply ______ biasing the anode-cathode.
3. An SCR has ____________ junctions or semiconductor layers.

4. The internal resistance of an SCR is (high, low) after triggering.
5. The internal resistance of an SCR is (high, low) before it is triggered.
6. In order to turn off a conductive SCR, the ________ ________ voltage must be momentarily removed or reduced.
7. When ac is applied to the anode-cathode of an SCR, it will automatically turn off during one ________.
8. Conduction of an SCR occurs during quadrant (I, II, III, IV) of the *I-V* characteristics.
9. Conduction of an SCR occurs when the ________ ________ voltage is exceeded.
10. Reverse breakdown of an SCR occurs in quadrant (I, II, III, IV) of the *I-V* characteristics.
11. An ________ in gate current will cause a decrease in the forward breakover voltage of an SCR.
12. The gate-cathode of an SCR must be ________ biased to produce gate current.
13. A suitable value of ________ current lowers the forward breakover voltage of an SCR to produce conduction.
14. The ________ loses control over the anode-cathode after an SCR has been triggered into conduction.
15. When the ________ current of an SCR is reached or exceeded it will latch or continue conduction.

ANSWERS

1. Anode, cathode, and gate (in any order)	2. Forward
3. Three	4. Low
5. High	6. Forward breakover
7. Alternation	8. I
9. Forward breakover	10. III
11. Increase	12. Forward
13. Gate	14. Gate
15. Holding	

BIDIRECTIONAL TRIODE THYRISTORS (TRIACS)

Alternating current power can be efficiently controlled with a special device that switches conduction on and off during each alternation. Control of this type is accomplished with a special semiconductor device known as a triac. A triac is classified as a bidirectional triode thyristor. A triode is a three-terminal device. Thyristor refers to the older gaseous tube thyratron that this device has replaced. Bidirectional refers to the conductivity characteristics of the device; it conducts during each alternation of the ac input.

The term triac is more commonly used today than bidirectional triode thyristor. A triac is better described as a triode ac switch. Conduction is triggered by a gate signal. A triac is the equivalent of two reverse-parallel-connected SCRs with one common gate. Conduction can be achieved in either direction with an appropriate gate current. Ac can be controlled efficiently and accurately with this device.

Functionally a triac is classified as a gate-controlled ac switch. Conduction is achieved by selecting an appropriate crystal combination. Selection depends on the polarity of the source. During one alternation, conduction is through a PNPN combination. Conduction for the next alternation is by an NPNP combination. The crystal selection process is achieved automatically. P- and N-materials are jointly connected to each terminal. Selection is made according to the polarity of a specific alternation.

Figure 8-11 shows the crystal structure, junction diagram, and schematic symbol of a triac. Notice that the crystal structure of this device forms an N1P1N2P2 structure between T_1 and T_2. The gate is commonly connected to the P1N4 junction. From T_2 to T_1, the crystal structure is N3P2N2P1. The gate of this structure utilizes N_4 of the P1N4 junction. This shows that T_1, T_2, and G are in either an NPNP or a PNPN combination depending on the voltage of T_1 and T_2. Selection of the crystal combination is based on the polarity of the applied voltage. When T_1 is negative, N_1 is forward biased, and P_1 is reverse biased. Current carriers will flow easily through the forward-biased material and not into the reverse-biased material. This means that the NPNP or PNPN

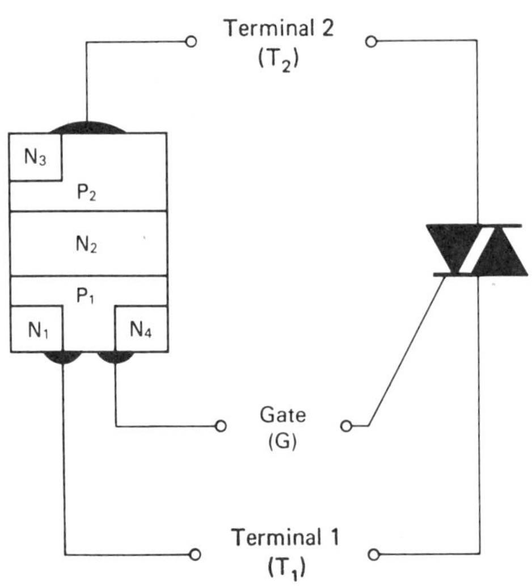

FIGURE 8-11 Junction diagram and schematic symbol of a triac.

combination and the appropriate gate voltage for initiating conduction are selected by source voltage polarity. For ac the polarity changes with each alternation. Thus one structure combination responds for one alternation and the other combination for the second alternation.

I-V Characteristics

When ac is applied to a triac, T_2 is positive and T_1 is negative for one alternation. For the next alternation, T_2 is negative and T_1 is positive. The current-voltage characteristic of a triac must therefore show how it will respond during both alternations of the ac source. Figure 8-12 shows a typical I-V characteristic of a triac. Operation in quadrant I occurs when T_2 is positive and T_1 is negative. This represents the forward conduction mode of operation. Quadrant III denotes reverse conduction. T_2 is made negative and T_1 positive for operation in quadrant III. The breakover voltage in quadrant I or III is usually quite large when the gate current is zero. Ordinarily, a triac is not designed for conduction when the gate is zero.

When gate current is applied to a triac, it lowers the V_{BO}. This action permits conduction to be achieved at a lower voltage. In quadrant I, either a positive or negative gate voltage produces gate current. Positive gate voltage tends to be more sensitive to triggering than negative values. When a suitable value of I_G occurs, the triac is triggered into conduction. I_T flows through T_1, the P_1 material, and the N_2-P_2 region, which is forward biased by T_2. This current continues for the remainder of the alternation, when it moves into the holding current region of the I-V characteristic.

During the negative alternation of the source, T_1 is positive and T_2 is negative. Quadrant III displays this condition of operation. With no gate current, the reverse V_{BO} is quite large. No conduction occurs unless an extremely high value of V_{BO} is

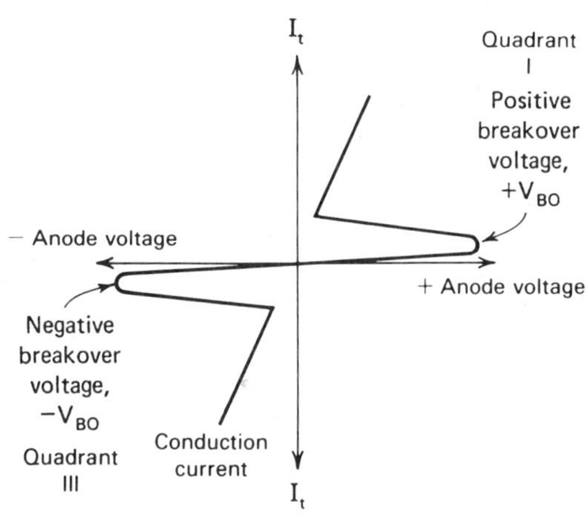

FIGURE 8-12 *I-V* characteristics of a triac.

reached. This voltage value, like that of quadrant I, is usually avoided in normal circuit operation. When I_G occurs, it lowers the reverse V_{BO} to a value that can cause conduction. In some triacs either a positive or negative gate voltage may be used to bias the gate. Ordinarily, a positive V_G is avoided in most applications. A negative V_G is therefore used to bias the gate and produce I_G. Conduction is triggered and I_G flows from T_2, through N_3, P_2, N_2, P_1, and T_1. The N_2-P_1 junction then sees a significant increase in current when I_G flows in the gate. This current added to that of the N_3-P_2 junction lowers the V_{BO} and causes full conduction to occur. The resulting *I-V* characteristic is thus a mirror image of conduction in quadrant I.

Due to the dual polarity of the gate, T_1 and T_2 of a triac have four possible trigger polarity combinations. The table of Fig. 8-13 lists these combinations as trigger modes and representative gate current (I_G) triggering values. Quadrant I operation has T_1 negative and T_2 positive. Note that the positive gate voltage produces an I_G that is smaller than that of the negative voltage I_G. This indicates that the triac is more sensitive to the positive voltage value in quadrant I. Quadrant III operation occurs best when T_1 is positive and T_2 is negative. In this quadrant the triac is more sensitive to a negative voltage polarity. Ordinarily, the positive V_G mode of operation is avoided in most circuit applications. Some manufacturers recommend that their triacs not be used in this triggering mode. A common practice used in selecting the most sensitive mode of operation is to match the polarity of the gate with the polarity of terminal 2.

Some manufacturers have special-function triacs that are designed for operation in the +III operational mode. This device generally has a special number designation or code to denote its unusual sensitivity. General Electric Company lists a standard triac as an SC141B. A device designed for +III operation is numbered SC141B13. The specialized device is usually more expensive than a standard unit.

TRIAC OPERATION

In normal operation the circuit of a triac must be designed so that it goes into conduction only by action of the gate. The applied source voltage must therefore be of a value that is somewhat less than that normally needed to overcome the operating V_{BO}. This

Quadrant	Element polarities			Required I_G
I +	T_2 +	T_1 −	G +	5-10 mA
I −	T_2 +	T_1 −	G −	10-20 mA
III −	T_2 −	T_1 +	G −	7-15 mA
III +	T_2 −	T_1 +	G +	> 40 mA

FIGURE 8-13 Trigger modes of a triac.

characteristic therefore becomes a critical factor when selecting a triac for a specific control function. Transient pulses or uncontrolled line voltage spikes must also be limited in order to avoid uncontrolled triggering of a triac. Special bidirectional triggering devices known as diacs are generally used in triac circuits to limit voltage spikes to amplitude levels that will not damage or trigger the gate junction. These devices will be discussed in the next part of this unit.

DC Control

Dc control of a triac is achieved the same way that it was for an SCR. In general, dc gate control of a triac is not very practical. Conduction can be delayed for only the first 90° of one alternation. After this point of the alternation, conduction will either be on or off for the last 90° depending on the gate current. The resulting output of the two alternations is quite different. Quadrant I and III do not have the same sensitivity to gate current. In quadrant III the gate current must be greater to cause triggering. This, in effect, causes a slightly different triggering level for each alternation. In most devices a positive gate voltage in quadrant III will not permit any conduction during the negative alternation. Figure 8-14 shows a dc circuit of this type and some of its representative waveforms.

As a general rule, there is no reason to use dc control to change the operation of a triac. The variable dc source would need to be supplied from a battery or by a rectified power supply and still produce half-wave dc output. Control of this type is only about 25% effective. Ac control of a triac, by comparison, is 100% effective and can be supplied from the same source.

AC Control

With ac control of a triac, it is possible to achieve a 180° delay of the conduction time. For this to be achieved, two distinct ac signals must be applied to the triac. The ac source represents one of these signals. It is applied to terminals 1 and 2 and must not be shifted in phase or altered. The second ac signal is used to control the conduction time of a triac by altering $+V_{BO}$ and $-V_{BO}$. This signal is controlled by a phase shifter and can be changed from 0° to 90° or 0° to 180°, depending on its design. This type of control produces a change in the phase relationship that exists between the two signals. The gate control signal of a triac must be made to lag behind the source voltage in order to be used effectively as a control factor.

Figure 8-15 shows two distinct phase control circuits used to alter the conduction time of a triac. The single-leg circuit of Fig. 8-15(a) achieves approximately 90° of conduction time delay. The bridge phase-shifter circuit of Fig. 8-15(b) permits 100% control from full conduction to 180° of delay, or nonconduction. A diac is

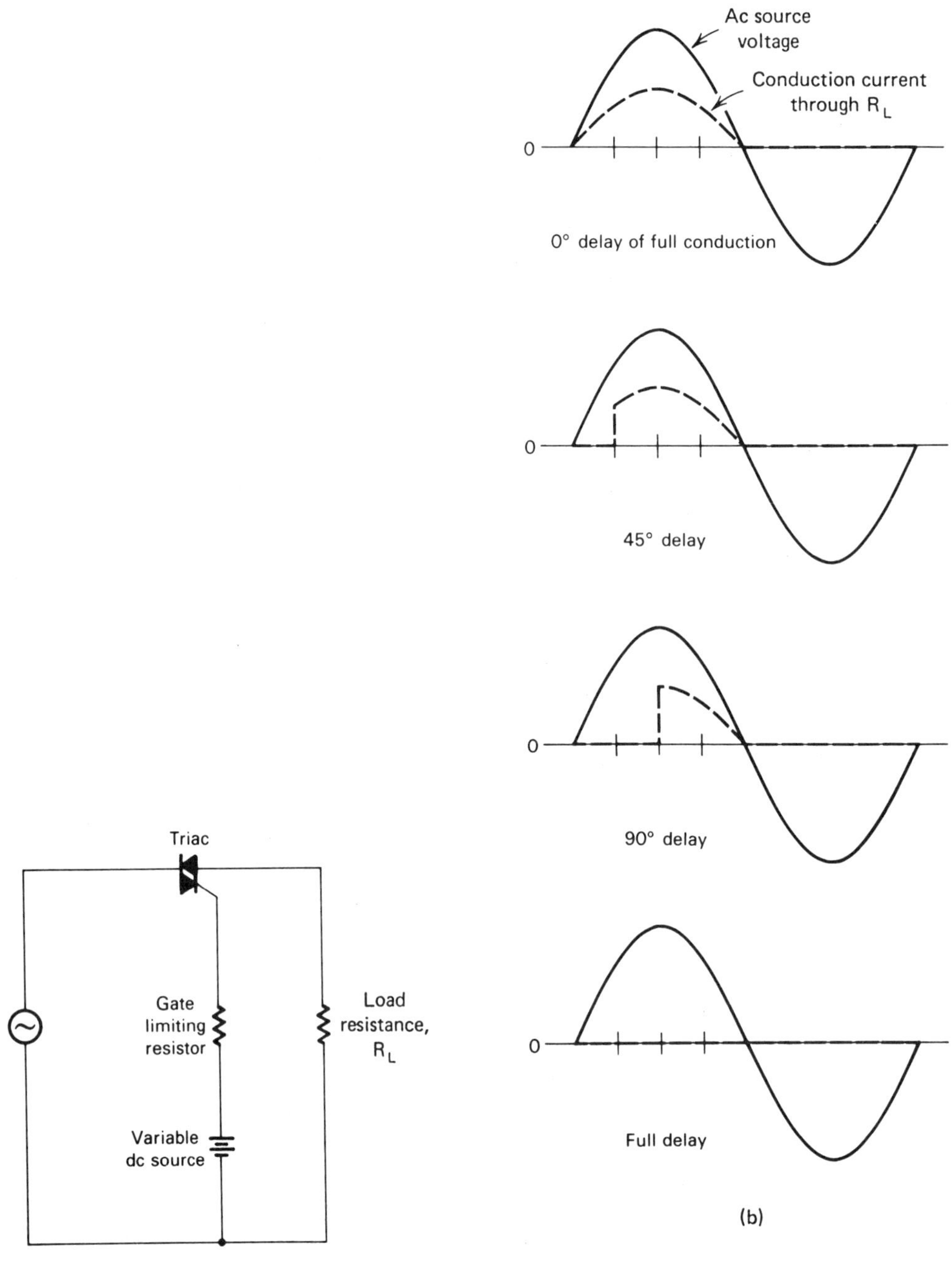

FIGURE 8-14 Dc control circuit for a triac: (a) base circuit; (b) source voltage and conduction current waveforms.

used in the gate of both circuits. This device permits the gate to be triggered into conduction in either direction when the applied voltage reaches a specific value.

Figure 8-16 shows six different conditions of operation of an ac bridge-controlled triac. In the 0° delay condition of operation, a full sine wave is developed as an output signal. The gate voltage and terminal voltage are in phase, which causes immediate conduction, without delay. The power output *P*, which is determined

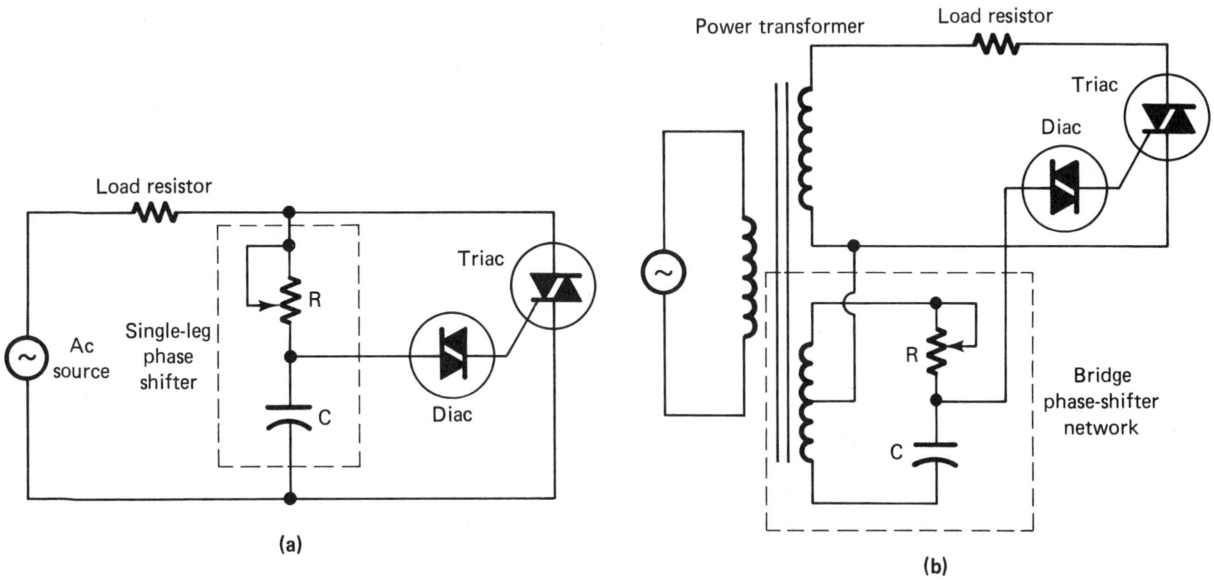

FIGURE 8-15 Ac phase control of a triac: (a) with single-leg phase shifter; (b) with bridge phase shifter.

by multiplying the terminal voltage V_T times the terminal current I_T, is at its maximum output level during this condition of operation.

If the gate voltage applied to a triac is shifted 45° behind V_T, the output current will be delayed. At the 45° point, the gate signal reaches the zero voltage level and causes conduction. Conduction current in this case is reduced somewhat because I_T occurs for less than the full sine wave.

Delay of conduction time is achieved by the same process for the remaining four conditions of operation. In the 180° delay condition, however, no output is achieved, because I_T is zero. The triac is nonconductive for this condition of operation. By shifting the phase of the gate voltage with respect to terminals 1 and 2, it is possible for a triac to achieve full control of the ac power output. Control of this type is easy to achieve, efficient, and can be done effectively with a minimum of costly components.

Triac Power Control

Triacs are widely used to control ac power. In this application the triac is designed to respond as a switch. Through this switching action, it is possible to control the ac source for a portion of each alternation. If a triac is in conduction for both alternations of the source voltage, 100% of source power is delivered to the load device. Conduction for half of each alternation permits 50% control of the power. When conduction is for one-fourth of each alternation, the load receives only 25% of its normal power. It is possible through this device to control from 0 to 100% of the electrical power being applied to a load device. Electrical power control with

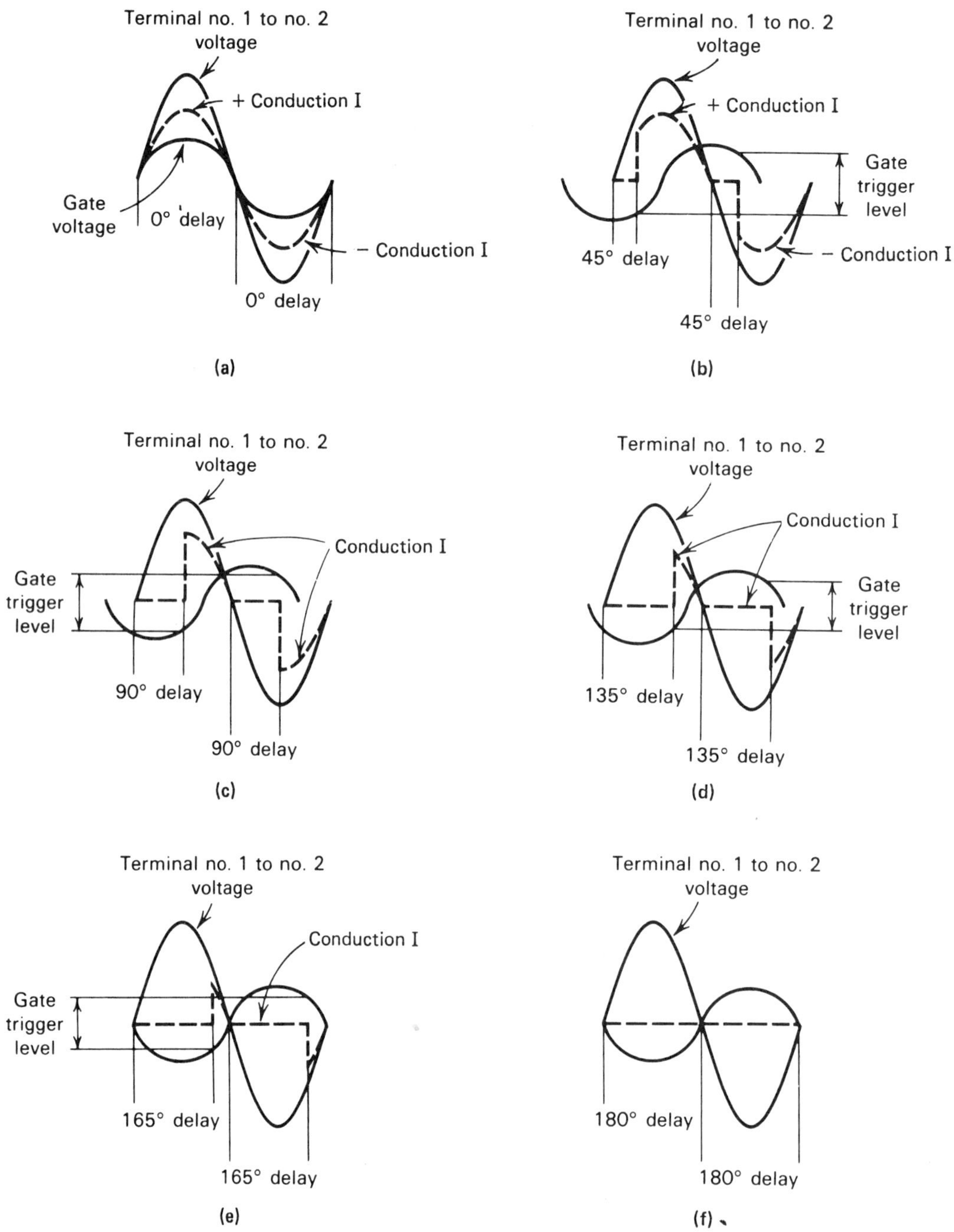

FIGURE 8-16 Ac control conditions of a triac: (a) 0° delay; (b) 45° delay; (c) 90° delay; (d) 135° delay; (e) 165° delay; (f) 180° delay.

this device is very efficient. Very little power is actually consumed by a triac when it is controlling.

Static Switching

The use of a triac as a static switch in ac power control is primarily an on-off operation. Control of this type has a number of advantages over mechanical load-switching techniques. A large load current can, for example, be controlled with a very small gate current switch. This type of switching has no contact bounce problems. Contact arcing and switch deterioration are reduced to a minimum. Static switching is usually very easy to achieve. Only a small number of components are needed to achieve control. In general, static triac switching has no moving parts, is very efficient, and can be achieved at a nominal cost. It is a very important alternative to power control switching.

Two very elementary triac static switch applications are shown in Fig. 8-17. Circuit A shows a load being controlled by a triac that is energized into conduction by a small SPST toggle switch. This switching operation could be accomplished with a wide variety of small switching devices. When the switch is closed, ac is applied to the gate. Resistor R_1 limits the gate current to a reasonable operating value. With ac applied to the gate, current flows during both alternations. During the positive alternation, T_2 and G are positive, and T_1 is negative. Gate current and T_2-T_1 voltage occur at the same time. The triac is triggered into conduction without a delay. Conduction continues for the remainder of the alternation. At the zero crossover point, the triac is turned off. The

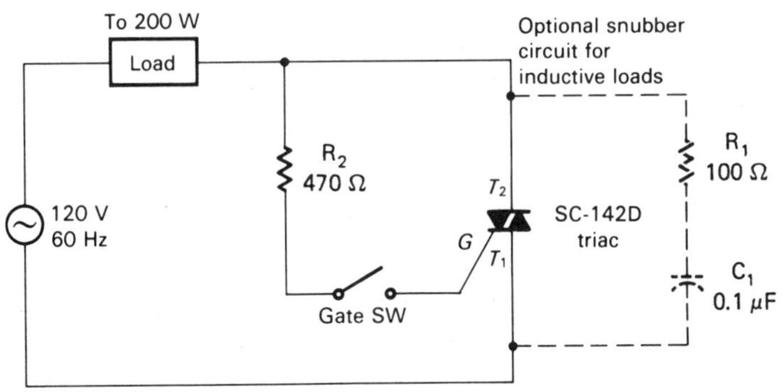

(a)

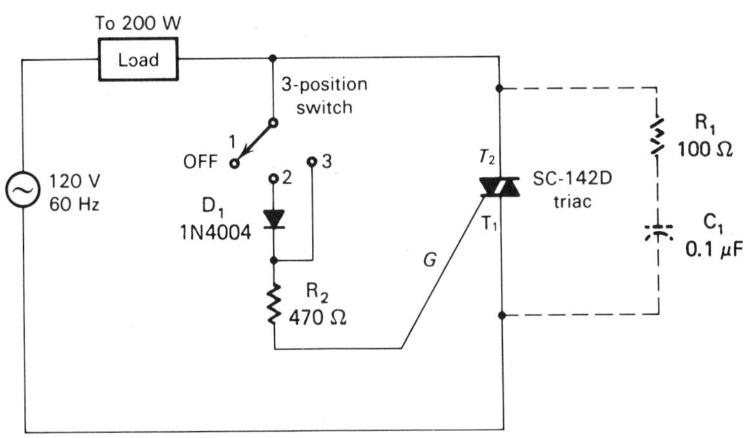

(b)

FIGURE 8-17 Triac static switch circuits: (a) static switch; (b) three-position static switch.

negative alternation then causes T_2 and G to be negative and T_1 to be positive. Conduction is again initiated without a delay. Conduction continues for the remainder of the negative alternation. At the zero crossover point or end of the negative alternation, the triac is again turned off. The process is then repeated for each succeeding alternation. Control is continuous as long as the gate switch is on and T_1-T_2 is energized. Turn-off is achieved by opening the SPST switch.

Figure 8-17(b) is a very simple three-position static switch. In position 1, the gate is open and the power to the load is off. In position 2, gate current flows only during the positive alternation. The triac is triggered into conduction only during the positive alternation. The load is energized for only one alternation or has a half-wave output. Only 50% of the source energy is supplied to the load in this switch position. In position 3, gate current flows for both alternations of the ac source. The load is energized for both alternations and receives full power from the source. The load of this circuit could be one or more incandescent lamps, a heating element, or a universal motor.

SELF-EXAMINATION

16. A triac is the equivalent of two reverse-parallel-connected __________ with a common gate.
17. In a triac crystal selection is achieved automatically according to the __________ of the applied source voltage.
18. According to Fig. 8-11 the crystal structure from T_1 to T_2 is __________.
19. According to Fig. 8-11 the crystal structure from T_2 to T_1 is __________.
20. The breakover voltage of a triac in quadrant I and III is at its highest value when the gate current is ______.
21. Gate current __________ the breakover voltage of a triac in either direction of operation.
22. According to the trigger mode chart of Fig. 8-13, triac operation is more sensitive when T_2 is +, T_1 is −, and G is (+, −).
23. According to the trigger mode chart of Fig. 8-13, triac conduction in quadrant III is best when T_2 is −, T_1 is +, and G is (+, −).
24. By shifting the phase of the __________ voltage with respect to T_1 and T_2, it is possible to achieve full control of ac power output.
25. When a triac is not conducting, the voltage across T_1 and T_2 is equal to the __________ voltage.

ANSWERS

16. SCRs	17. Polarity
18. N1P1N2P2	19. N3P2N2P1
20. Zero	21. Lowers
22. +	23. −
24. Gate	25. Source or supply

TRIAC TRIGGERING

The gate of a triac responds as a nonlinear low-impedance junction similar to that of a forward-biased diode. Triggering of the gate must therefore be achieved by some type of low-impedance source. The gate junction usually requires a rather sizable current to produce triggering. This means that the G-T_1 junction is current sensitive rather than voltage sensitive. Most triacs, for example, require a hundred or more milliamperes of I_G for a few microseconds to be triggered into conduction. Several things can be done to develop this amount of I_G. A capacitor is usually discharged into the gate to initiate the triggering process. As a rule, this necessitates some type of low-power on-off switching device to produce a suitable value of I_G. The device must respond equally well to both alternations of the ac input. Special ac diodes, known as diacs, have been developed for triac triggering.

Diacs

A diac is a bidirectional diode that can be triggered into conduction with voltage. The crystal structure of this device is basically the same as an NPN transistor with no base connection. N_1 and N_2 are identical in nearly all respects so that the two junctions will have matched operating characteristics in either direction. A diac goes into conduction at precisely the same positive and negative voltage level. A device of this type is triggered into conduction by reaching a specific voltage value. For voltage values less than the trigger level, the device is nonconductive.

Figure 8-18 shows the crystal structure and schematic symbol of a diac. Notice that the two leads are labeled terminals 1 and 2 instead of the conventional anode-cathode designations. When T_1 is negative and T_2 is positive, the N_1-P junction is forward biased and N_2-P is reverse biased. When the breakover voltage of N_2-P is reached, the entire structure is triggered into conduction. This is the equivalent of reaching the zener voltage of a zener diode. Reversing the polarity of the source voltage causes T_1 to be positive and T_2 negative. The N_2-P junction is now forward biased, with the N_1-P junction being reverse biased. When the breakover

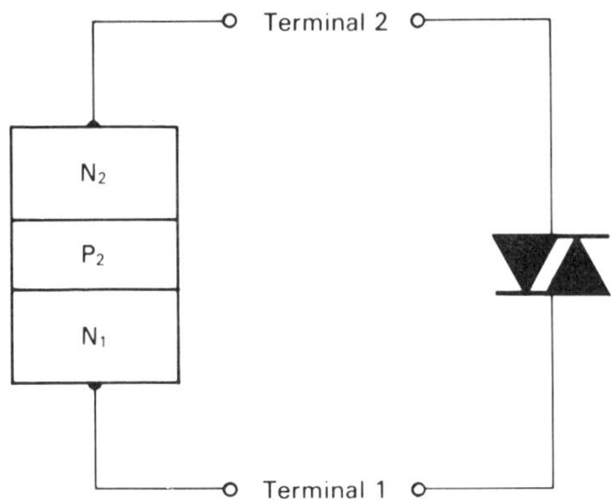

FIGURE 8-18 Crystal structure and schematic symbol of a diac.

voltage in this direction is reached, the structure again becomes conductive. Ordinarily, the breakover voltage is symmetrical in either direction.

The current-voltage characteristic of a diac is shown in Fig. 8-19. Notice that the breakover voltage and resulting current are the same in quadrants I and III. When T_1 is made positive and T_2 negative, this device does not conduct until the forward breakover voltage V_{BO} is reached. For a General Electric type ST-2 diac, triggering occurs at 28 V ± 10%. Reversing the polarity of T_1 and T_2 causes an identical triggering action as shown in quadrant III.

After the breakover voltage of a diac has been reached, current conduction occurs very quickly. A resistor placed in series with the device limits the conduction current to a safe operating level in either direction. The voltage across a diac also decreases quite readily as it increases in conduction. This conduction is known as the negative-resistance characteristic of the diac. The pronounced curve that takes place after V_{BO} is reached shows this characteristic in both quadrants.

Diac-Triac Power Control

An elementary full-wave ac power control circuit employing a diac and a triac is shown in Figure 8-20. The diac is used in an *RC*

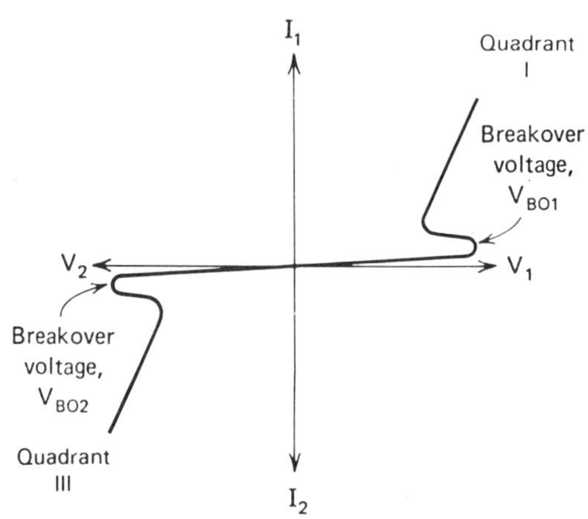

FIGURE 8-19 *I-V* characteristics of a diac.

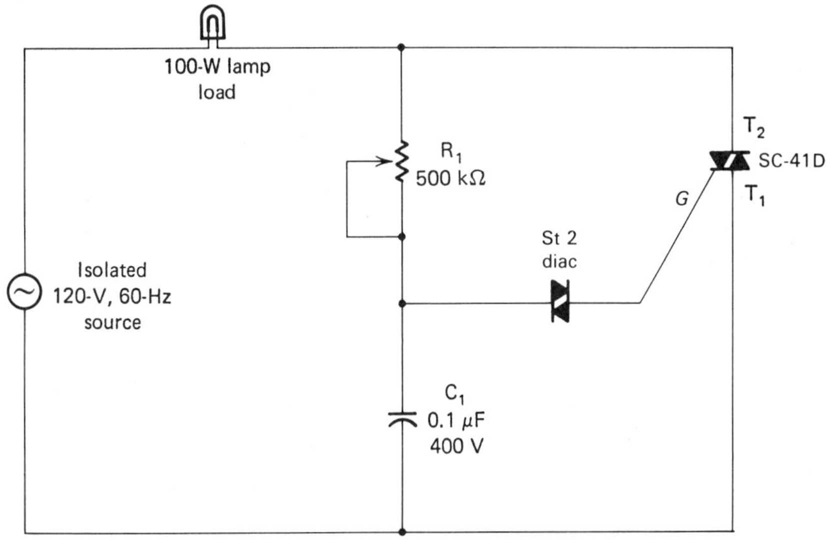

FIGURE 8-20 Elementary full-wave triac power control circuit.

circuit to trigger the triac into conduction. Operation is based on the polarity of the applied ac source. The circuit is connected so that T_2 and the gate of the triac are of the same polarity. The gate is connected to capacitor C_1 through the diac. Gate current is developed when the capacitor voltage builds up to a value that will trigger the breakover voltage of the diac. Discharging C_1 causes gate current. The polarity of the charge on C_1 is determined by the alternation of the source.

Assume now that the positive alternation of the source occurs first. This will make point A positive and B negative. Capacitor C_1 begins to charge to the voltage source through R_1. At the same time, T_2 is made positive and T_1 negative. When the charge voltage of the capacitor exceeds the breakover voltage on the diac, C_1 discharges. The component values of C_1, R_1, and the ac source determine the amount of gate current and the duration of the discharge pulse. Typical values of I_G are 50 mA or more for approximately 30 μs. The polarity of the charge on the top plate of C_1 is positive for this alternation. The discharge of C_1 causes current to flow from the bottom plate of C_1 to T_1, through the gate, the diac, and to the top plate of C_1. This discharge current, or I_G, triggers the triac into conduction. Current then flows into the load during this alternation. Delay of the conduction time is controlled by the resistance value of R_1. A low-resistance setting causes zero or no delay, whereas a high-resistance setting may cause delay for some value up to the end of the alternation.

For the negative alternation of the source, A is negative and B is positive. Capacitor C_1 again charges to its source voltage through R_1. In this case the charge polarity is reversed because of the alternation change in the source. At the same moment, T_2 is made negative and T_1 positive. When capacitor voltage (V_C) exceeds the breakover voltage of the diac, C_1 discharges. Discharge current flows from the top plate of C_1, through the diac, the gate, T1, and returns to the lower plate of C_1. This I_G triggers the triac into conduction during the negative alternation. Current flows through the lamp and into the triac from the source. Delay of conduction time is again based on the value setting of R_1. Control of this alternation is nearly the same as that of the positive alternation.

SELF-EXAMINATION

26. The gate-T_1 junction of a triac responds as a ________ -biased diode.
27. A ____________ is a bidirectional diode that can be triggered into conduction with voltage.
28. The two leads of a diac are labeled ____________ and ____________.
29. The breakover voltage of a diac is the same in quadrants ____________ and ____________ of the *I-V* characteristics.
30. In Fig. 8-20 the diac is connected so that the ________ and T_2 are of the same polarity.
31. In Fig. 8-20 the gate voltage triggering time is determined by the values of ____________ and ____________.
32. In Fig. 8-20 when the capacitor voltage (V_C) exceeds the V_{BO} of the diac, it triggers the gate of the triac to produce ____________.

ANSWERS

26. Forward
27. Diac
28. T_1 and T_2
29. Same
30. Gate
31. Resistor (R_1) and capacitor (C_1)
32. Conduction or current

PULSE TRIGGERING

A positive pulse of energy can be used to trigger an SCR or triac into conduction instead of a continuous form of ac or dc. Triggering by this method takes advantage of the latching characteristic of the device. When conduction is initiated, it will be made continuous for the remainder of the alternation. The duration of the triggering pulse may be extremely short. This method of triggering in general consumes very little energy from the source. The operational efficiency of a pulse-triggered power control circuit is usually much higher than that of other gate-triggering methods.

To trigger an SCR into conduction, the magnitude of the pulse must be several times greater than the normal I_G triggering level. A high-amplitude voltage pulse discharges into the gate and causes an I_G. For pulses above 20 μs in duration, the triggering level is approximately the same as that of a dc source. Below 20 μs, the magnitude of the pulse should be at least equal to the dc gate-triggering level. Pulse widths of less than 1 μs duration will not trigger a device into conduction.

The waveforms of a pulse-triggered SCR are shown in Fig. 8-21. The anode-cathode voltage (V_{AK}) is ac. The gate current triggering pulse is labeled I_G. The triggering level of I_G is for dc control values. Note that the amplitude of the pulse exceeds the triggering level by a substantial amount. The delay of conduction is based on the position of the pulse with respect to the positive alternation. Load current (I_{AK}) flows during the remainder of the alternation after the gate pulse initiates conduction. By varying the time or position of the pulse, conduction can be altered from 0 to 180°.

Unijunction Transistor Pulse Triggering

Pulse triggering of an SCR can be accomplished with a unijunction transistor, or UJT. This device is commonly described as a voltage-controlled diode. When used to trigger an SCR into conduction, UJTs can be designed to respond to low-level signals from other inputs. The output of photocells, thermocouples, and other transducers can be used to energize the input of a UJT, which in turn will trigger the conduction of an SCR. Through this type of circuitry, it is possible to control a large SCR current with an extremely small input signal.

A unijunction transistor is a three-terminal, single-junction solid-state device that has unidirectional conductivity. As illustrated in Fig. 8-22, a small bar of N-type silicon is mounted on a ceramic base. Leads attached to the silicon base are called base 1

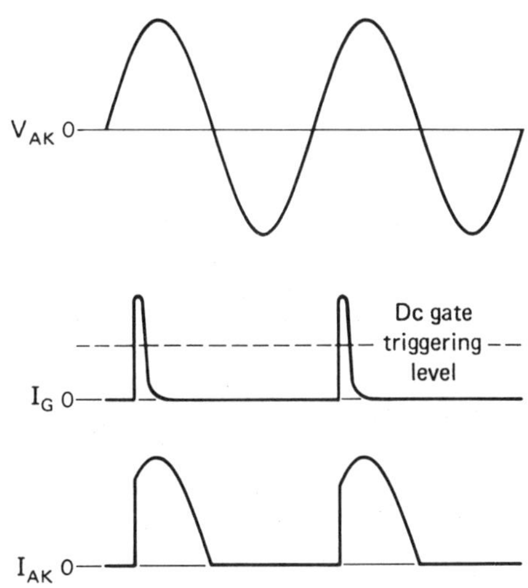

FIGURE 8-21 Waveforms of a pulse-triggered SCR.

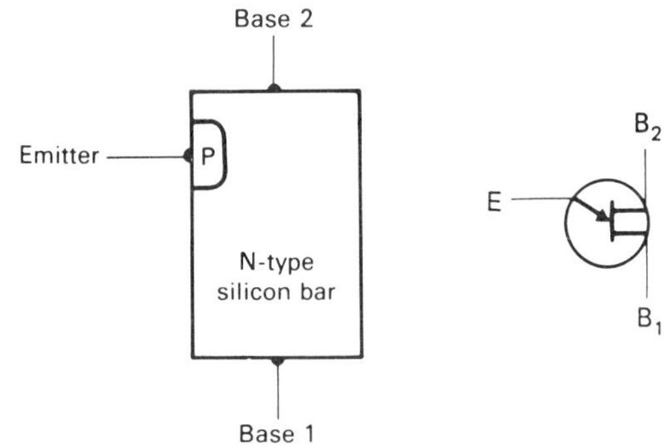

FIGURE 8-22 Unijunction transistor crystal structure and schematic symbol.

(B_1) and base 2 (B_2). The emitter (E) is formed by fusing an aluminum wire to the opposite side of the silicon bar. The emitter is oriented so that it is closer to B_2 than B_1. A P-N junction is formed by the emitter and the silicon bar. The arrow of the UJT symbol Points iN, which indicates that the emitter is P and the silicon bar is N material. The arrow of the symbol is slanted to distinguish the UJT from the N-channel JFET symbol.

An equivalent circuit of the UJT is shown in Fig. 8-23. When the E-B_1 junction is reverse biased, R_{B1} and R_{B2} are of practically the same resistance value. Forward biasing of the E-B_1 junction, however, causes the resistance of the junction to change to a rather small value. The E-B_2 junction resistance is not changed by the action of the E-B_1 junction. As a result, E-B_1 responds as a voltage-controlled variable resistor, whereas E-B_2 remains at a fixed value. The interbase resistance (R_{BB}) is in the range of 4 to 10 kΩ. The interbase voltage (V_{BB}) appears across B_1 and B_2. Emitter voltage (V_E) causes an emitter current (I_E) when E-B_1 is forward biased. No I_E flows when E-B_1 is reverse biased.

The operation of a UJT is quite different from other solid-state devices. When the emitter is reverse biased, B_1-B_2 responds as a simple voltage divider. The resistance of R_{BB} is at its maximum value. Any I_E that flows at this time is due to the leakage current of E-B_1. When the emitter is forward biased, R_{BB} drops to a very small value. A change in E-B_1 voltage therefore causes a significant change in R_{BB}. The current flow between B_1 and B_2 increases, and V_{BB} decreases in value.

A characteristic curve for a typical UJT is shown in Fig. 8-24. The vertical part of this characteristic indicates different val-

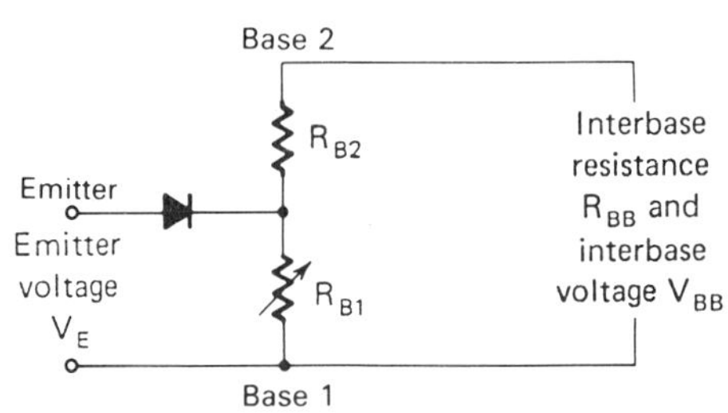

FIGURE 8-23 Equivalent of a UJT.

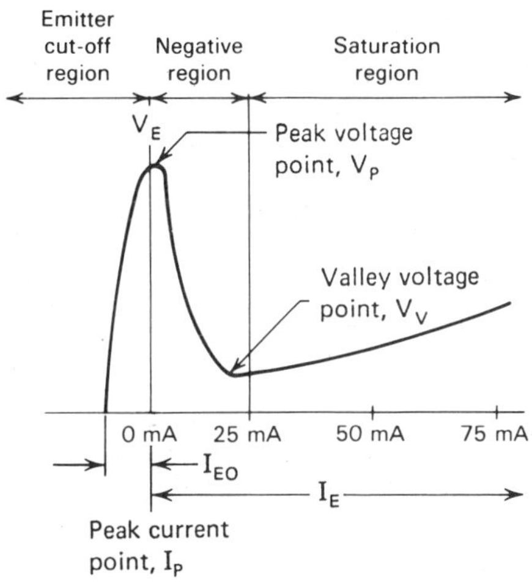

FIGURE 8-24 Characteristic curve of a UJT.

ues of emitter voltage. An increase in the value of V_E causes a sharp vertical rise in the curve. Emitter current is represented by the horizontal part of the graph. Note that the first application of V_E is to the left of a vertical line. This indicates the reverse bias condition of operation. Any current flow that occurs here is due to the leakage of the E-B_1 diode junction. After a very nominal leakage current (I_{EO}), there is a significant increase in I_E, shown by the curve extending to the right. Forward biasing of E-B_1 is shown where the vertical line is crossed. The peak emitter voltage point (V_P) shows where the UJT is triggered into conduction. The peak-point current (I_P) is the minimum I_E needed for triggering. A further increase in I_E causes V_E to drop to the valley voltage (V_V) point. The area between V_P and V_V is called the negative resistance region. An increase in I_E causes a decrease in V_E in this part of the graph. A device that has a negative resistance region is capable of regeneration or oscillation. This part of the characteristic curve is an important operating area for the UJT in pulse trigger generation applications.

The triggering of a UJT is dependent on changes in the conductivity of the E-B_1 junction. When the E-B_1 junction is forward biased, electrons leave the silicon bar and move into the P material of the emitter. This action causes holes to appear in the silicon bar at the B_1 region. Electrons pulled from the negative side of the source attached to B_1 fill the injected holes. In effect, this causes an increase in the conduction of the E-B_1 junction. This region then becomes low resistant. A sudden change in R_{B1} causes a corresponding change in B_1-B_2 current. This change in current can be used to trigger the gate of an SCR.

Normal operation of the UJT has B1 grounded or connected to the negative side of the source voltage and B2 connected to the positive side. This voltage causes an interbase current flow between B_1-B_2. Internally, V_{BB} divides across the interbase resistance of R_{B1} and R_{B2}. For example, if R_{B1} is 4 kΩ and R_{B2} is 6 kΩ, the voltage across R_{B1} is 0.4 of V_{BB}, and R_{B2} is 0.6 of V_{BB}. It is not possible to measure these two voltage values because the junction point of R_{B1} and R_{B2} does not exist. The fractional value of voltage that appears across R_{B1} is called the *intrinsic standoff ratio* (η). Typical values are in the range from 0.45 to 0.85 of V_{BB}. This value determines the emitter triggering voltage for a constant V_{BB}. If V_E is less than ηV_{BB}, the emitter junction is reverse biased and only leakage current (I_{EO}) flows. If V_E exceeds ηV_{BB} by a value slightly greater than the voltage drop across the E-B_1 diode, triggering occurs. R_{B1} is then reduced to a few ohms, and the interbase current I_{BB} increases.

Figure 8-25 shows a UJT relaxation oscillator that is frequently used to trigger an SCR. In this circuit, two parallel paths are formed by the components. Resistors R_1, R_2, and C form the charge path, whereas R_3 and the UJT form the discharge path. When the switch is turned on, approximately 4 V appears across R_3 and 6 V across the UJT and R_4. This voltage reverse biases the E-B_1 junction by 6 V.

Resistor R_1 and capacitor C_1 receive energy from the source at the same time that the base path does. Capacitor C_1 begins to develop voltage across it at a rate based on the RC values of R_1 and C_1. When the charge voltage at C_1 reaches 6 V, it overcomes the reverse biasing of the E-B_1 junction. When this occurs, the junction becomes very low resistant, and C_1 discharges very quickly. The E-B_1 junction immediately becomes reverse biased because of the reduced emitter voltage. The capacitor then recharges and the process is repeated. A sawtooth waveform appears across the capacitor, as indicated in Fig. 8-25. The discharge voltage across R_3 produces a spiked pulse that shows the discharge time of C_1. The rise time of the sawtooth wave is adjusted to some extent by different values of R_1. When the charging action of the capacitor is maintained in the first time constant, very accurate pulse generation can be achieved by this circuit.

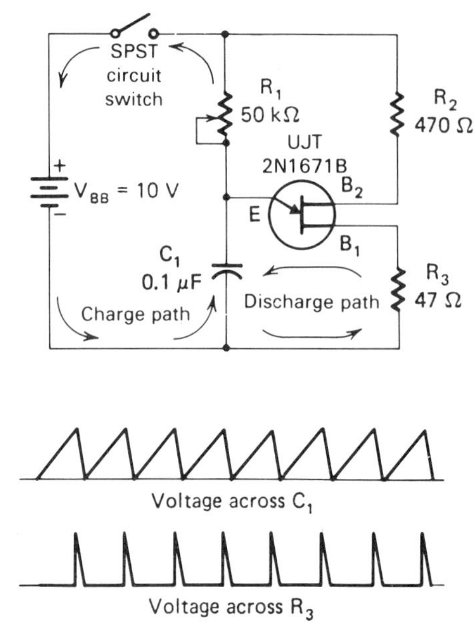

FIGURE 8-25 UJT sawtooth oscillator.

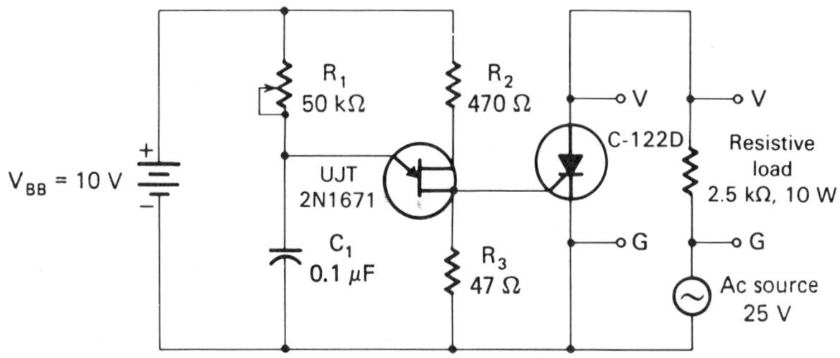

FIGURE 8-26 UJT pulse-trigger control circuit.

A UJT pulse trigger control circuit for an SCR is shown in Fig. 8-26. Adjustment of the variable resistor (R_1) determines the pulse triggering rate of the SCR. A value change in R_1 alters the RC of the UJT trigger circuit. When the charge voltage of C_1 reaches the V_P value, it discharges through the E-B_1 junction of the UJT and R_3. The top is positive and the bottom is negative. This forward biases the gate-cathode of the SCR and causes it to be triggered into conduction. The load will receive current only during the positive alternation of the applied ac. A pulse trigger circuit of this type can be used to effectively delay the conduction time of the positive alternation from 0° to 180°.

SELF-EXAMINATION

33. A unijunction transistor is described as a voltage-controlled ____________.
34. A UJT is a ____________ terminal device.
35. UJTs are commonly used to ____________ the conduction of an SCR.
36. Two leads connected to the common N-type silicon bar of a UJT are called ____________ and ____________.
37. The lead attached to the P-type element of a UJT is called the ____________.
38. When the E-B_1 junction is open or reverse biased, the resistance of B_1 and B_2 are practically the ____________.
39. When the E-B_1 junction of a UJT is forward biased by an appropriate voltage value, the internal base resistance ____________ in value.
40. The area between V_P and V_V of a UJT's characteristic curve is called the ____________ ____________ region.
41. The fractional value of source voltage that appears across R_{B1} of a UJT is called the ____________ ____________ ratio or eta.
42. In the UJT circuit of Fig. 8-26, when the voltage across capacitor C_1 builds up to a value that exceeds the in-

trinsic standoff ratio, it causes ____________ of the E-B_1 junction.

ANSWERS

33. Diode	34. Three
35. Trigger	36. Base 1 and base 2
37. Emitter	38. Same
39. Drops	40. Negative resistance
41. Intrinsic standoff	42. Triggering

PROGRAMMABLE UNIJUNCTION TRANSISTOR TRIGGERING

A programmable unijunction transistor (PUT) is frequently used to generate trigger pulses for SCR control applications. A PUT is actually a thyristor that responds as a UJT that has a variable trigger voltage. This voltage can be adjusted to a desired value by changing two external voltage-divider resistors. The trigger voltage level can therefore be set or programmed to respond to a specific value.

Figure 8-27 shows the crystal structure, schematic symbol, and element names of a PUT. The crystal structure and schematic symbol of the PUT are very similar to the SCR. The gate junction is the primary difference. In a PUT, the gate (*G*) is connected to the N material nearest the anode. A P-N junction is formed by the anode-gate. Conduction of the device is controlled by the bias voltage of A-G.

The polarity of the bias voltage of a PUT is referenced with respect to the cathode. The cathode is usually connected to the ground or the negative side of the power source. The gate is then made positive relative to the cathode. This is the gate voltage (V_G). The anode voltage (V_A) is also made positive with respect to the cathode. Conduction of the PUT is based on the difference in positive voltage between the gate and the anode. When the gate is

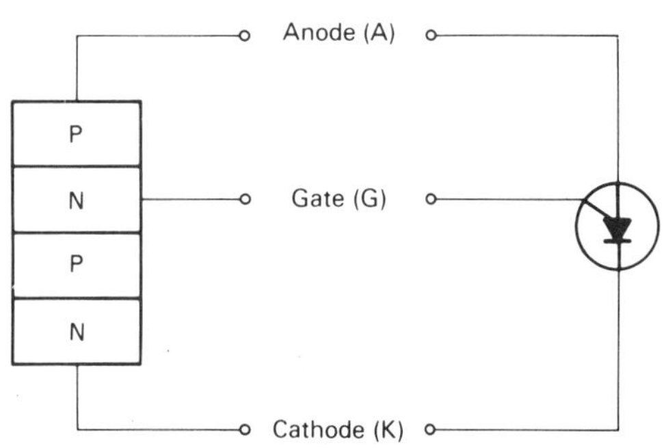

FIGURE 8-27 Programmable unijunction transistor.

more positive than the anode, the *A-G* junction is reverse biased. This condition causes the device to be nonconductive or in its off state. The anode-cathode has infinite resistance, and the device responds as an open switch. When the anode becomes more positive than the gate by 0.5 V, it forward biases the *A-G* junction. This condition causes gate current to flow, and the device is triggered into conduction. In the on state the anode-cathode resistance drops to a very low value. The device then responds as a switch in the on state.

Refer to the PUT circuit and characteristic curve in Fig. 8-28. Gate voltage (V_G) is developed by resistors R_1 and R_2 connected across V_{GG}. The anode-cathode voltage (V_{AK}) is supplied by V_s. When V_{AK} is made more positive than V_G, the PUT is triggered into conduction. The characteristic curve shows this by the changing value of V_{AK}. Triggering occurs when V_{AK} reaches the peak voltage point (V_P). The voltage then drops to the valley voltage point (V_V). Conduction current (I_{AK}) is indicated by expansion of the curve to the right. The PUT has a negative resistance region between V_P and V_V. The I_{AK}-V_{AK} characteristic of a PUT is very similar to that of a UJT.

The anode of a PUT responds as the emitter of a UJT. Resistors R_1 and R_2 are similar to the interbase resistance of the UJT. The intrinsic standoff ratio (η) of a PUT is determined by the value of the gate voltage with respect to the anode voltage. V_G can be altered externally by changing the values of R_1 and R_2. This per-

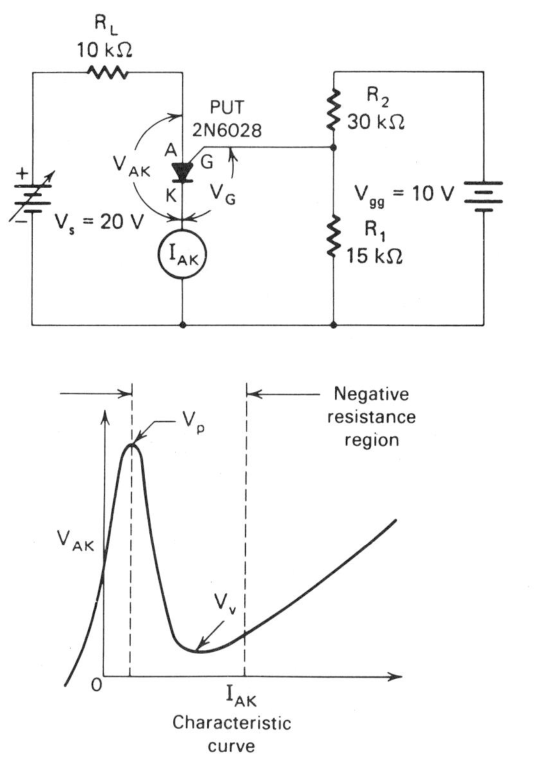

FIGURE 8-28 PUT circuit and characteristic curve.

mits the PUT to be programmed or adjusted to different trigger voltage values.

Figure 8-29 shows a PUT relaxation oscillator that can be used to trigger an SCR. In this circuit two parallel paths are formed by the components and the device. Resistor R_3 and capacitor C_1 form the charge path for the capacitor and the anode voltage. The PUT and resistor R_4 form a discharge path for C_1. A trigger pulse is developed across R_4 when C_1 discharges through the anode-cathode to the PUT.

When the circuit switch of Fig. 8-29 is turned on, it energizes the PUT from the power source V_s. Resistors R_1 and R_2 form a voltage-divider network across the V_{GG} source. In this case, R_1 is 15 kΩ, and R_2 is 30 kΩ. The intrinsic standoff ratio (η) is determined as follows:

$$\eta = \frac{R_1}{R_1 + R_2}$$

$$= \frac{15\ \text{k}\Omega}{15\ \text{k}\Omega} + 30\ \text{k}\Omega$$

$$= 0.333$$

The gate voltage (V_G) of the PUT is then determined by $\eta \times V_s$. V_G for this circuit is

$$V_G = \eta \times V_s$$

$$= 0.333 \times 20\ \text{V}$$

$$= 6.66\ \text{V}$$

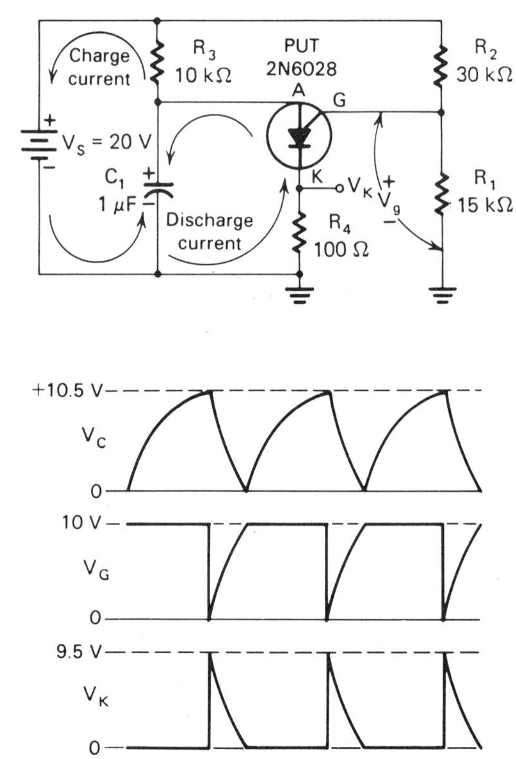

FIGURE 8-29 PUT relaxation oscillator.

Initially, V_G makes the gate more positive than the anode by 6.66 V. The PUT is in its nonconductive state or off.

After a short period of time, capacitor C_1 begins to charge to the source voltage through R_1. When the charge potential exceeds 6.66 V, it makes the anode more positive than the gate. This condition triggers the PUT into conduction. C_1 then discharges through the low resistance of the PUT and R_4. Discharge takes place very quickly due to the low resistance. The resulting current produces a positive-going voltage pulse across R_4. Triggering the PUT into conduction also causes a drop in the value of V_G. This is due to the low resistance of the forward-biased AG junction. The waveforms of Fig. 8-29 show how the relaxation oscillator responds for three operational cycles.

SCR Triggering with a PUT

A programmable unijunction transistor is frequently used to control the conduction of an SCR. The PUT is primarily responsible for generating trigger pulses. These pulses are developed by a relaxation oscillator. The pulses are then applied to the gate of an SCR to initiate the conduction process. Only one pulse is needed during a selected alternation to initiate conduction. A circuit of this type has synchronization of the trigger pulse and the alternation that forward biases the SCR. This is accomplished by energizing the SCR and PUT from the same ac source. Conduction of the SCR occurs only when the anode-cathode is forward biased. Control by synchronized pulse triggering is very precise and has a wide range of adjustment capabilities.

A PUT pulse-triggered SCR power control circuit is shown in Fig. 8-30. The outer part of the circuit is controlled by the SCR. Notice the location of the load device, SCR, and ac power source. The center part of the circuit is used to develop trigger pulses that control the conduction of the SCR. The PUT, in this case, is used as a relaxation oscillator. Operation of the oscillator is essentially the same as Fig. 8-29.

The PUT is energized by a dc source. A 15-V zener diode is used in this case as a rectifier-regulator. It is energized by the same ac source that supplies the SCR load. A series dropping resistor connects the ac to the zener diode. During the positive alternation, the zener diode clips the source voltage to a peak value of 15 V. The negative alternation causes forward conduction of the diode, and the voltage drops to approximately zero. The dc supply is essentially a pulsating dc of 15 V. See the V_z dc supply and the ac source waveforms. Notice the phase relationship of the ac source and the PUT supply voltage. When the positive alternation and the dc supply pulse occur at the same time, this relationship permits the trigger pulse to have synchronized control of the SCR. The time of the trigger pulse can be changed, however, by altering the resistance of R_1.

The waveforms of Fig. 8-30 show how the circuit responds for

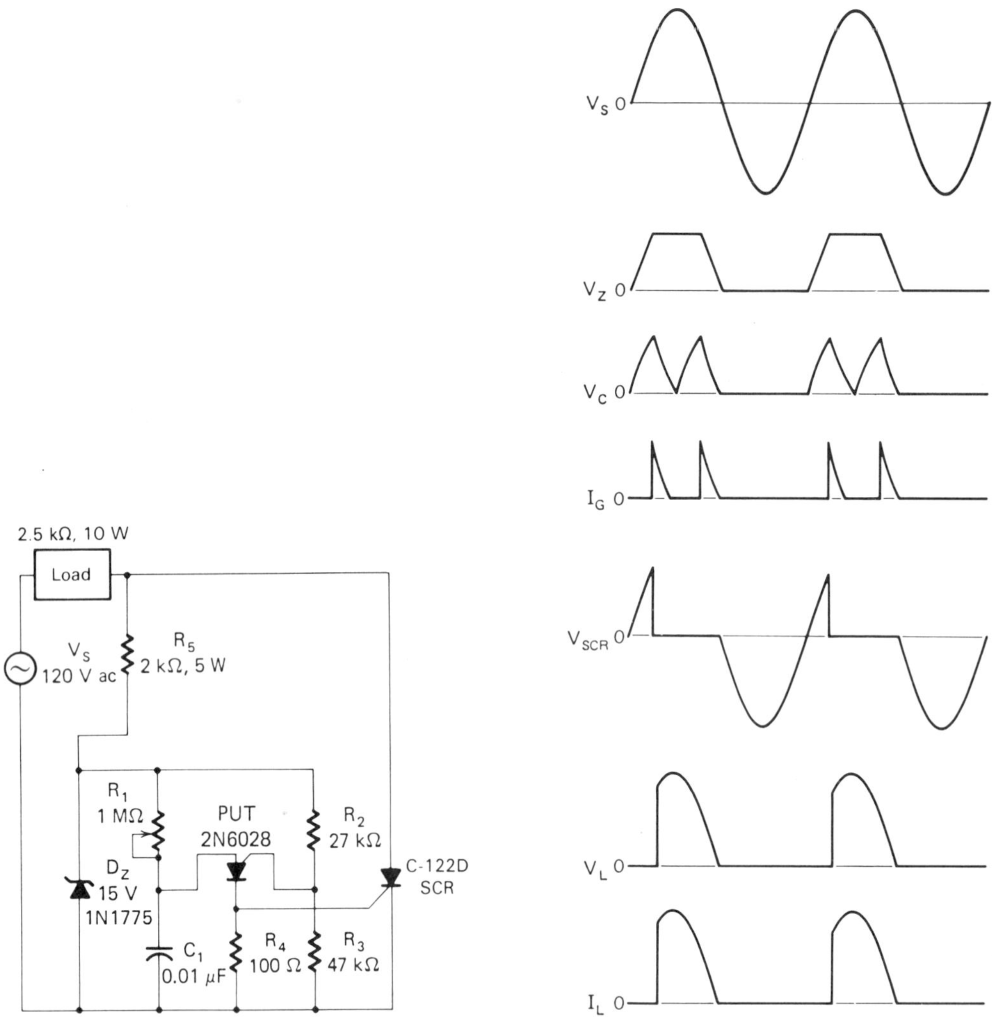

FIGURE 8-30 PUT pulse-triggered SCR power control.

two complete operational cycles. Note that more than one triggering pulse may be generated by the PUT for a given dc source pulse. The number of pulses produced is based on the time constant of R_1 and C_1. A low-resistance setting causes several pulses to occur. Higher resistance values may permit only one pulse to be generated. The SCR is triggered into conduction by only the first pulse. With a suitable value of holding current, conduction continues for the remainder of the alternation. The SCR is nonconductive during the negative alternation. Refer to the SCR voltage, load current, and load voltage waveforms. V_{AK} of the SCR and V_L are primarily the reverse of each other. Conduction causes the amplitude of V_{AK} to drop to a straight line. Conduction of the load current and voltage is shown as a rise in amplitude. Delay of conduction time is altered by changing the value of R_1.

SELF-EXAMINATION

43. A ____________ ____________ ____________ is a thyristor that responds as a UJT.

44. The crystal structure of a PUT is similar to that of an ___________ ___________ ___________.
45. In a PUT the ___________ is connected to a piece of N-material next to the anode.
46. Conduction of a PUT is based on the difference in positive voltage between the ___________ and ___________.
47. When the anode becomes more ___________ than the gate by 0.5 V, it forward biases the anode-gate junction.
48. When a PUT is triggered into conduction, its internal resistance is very ___________.
49. Before a PUT is triggered into conduction, its internal resistance is very ___________.
50. The ___________ ___________ ___________ (eta) of a PUT is determined by the value of the gate voltage with respect to the anode voltage.
51. The intrinsic standoff ratio of a PUT can be ___________ by altering the resistance of a divider network.
52. A PUT is commonly used as a ___________ source for an SCR.

ANSWERS

43. Programmable unijunction transistor	44. Silicon-controlled rectifier
45. Gate	46. Gate and anode
47. Positive	48. Low
49. High	50. Intrinsic standoff ratio
51. Programmed or changed	52. Trigger

GATE TURN-OFF THYRISTORS

Gate turn-off thyristors are unique power control devices that can be triggered into conduction or turned off by a gate signal. Characteristically, this type of thyristor responds as an SCR. It has unidirectional current conductivity and responds only in quadrant I of the *I-V* characteristic. Construction is of the PNPN type of configuration. This thyristor can be triggered into conduction when the gate is forward biased. Gate current lowers the breakover voltage (V_{BO}), and latching is maintained as long as the initial holding current level is exceeded. The unique feature of this type of device is its gate turn-off capability. The gate can be used to extinguish conduction and turn off the device when it is reverse biased.

The gate turn-off principle of operation is found in two distinct types of thyristors. The silicon-controlled switch (SCS) is a low-power, low-current control device. The gate turn-off switch (GTO) is used to control larger amounts of power. These devices are primarily used in computer logic circuits, oscillators, and as trigger devices for other thyristors.

Silicon-Controlled Switch

The silicon-controlled switch (SCS), like the silicon-controlled rectifier, is a four-layer PNPN thyristor. External connection is made to all four layers of the SCS. It has an anode, cathode, anode-gate, and a cathode-gate. The crystal structure, schematic symbol, and two-transistor equivalent of an SCS are shown in Fig. 8-31.

The current-voltage characteristic of an SCS is primarily the same as that of an SCR. Conductivity occurs in quadrant I. This is achieved by making the anode positive and the cathode negative. Quadrant III represents the reverse-biased condition of operation. Ordinarily, the SCS is not designed to operate in this condition. The peak reverse voltage (PRV) rating of the device is represented by quadrant III.

Operation of the SCS is quite different from the SCR. An SCS can be triggered into conduction or turned off by an appropriate signal applied to either gate. A positive voltage or pulse applied to the cathode-gate (G_C) with respect to the cathode causes gate current to flow. This I_{GC} lowers the forward breakover voltage and permits conduction. A negative voltage with respect to the anode forward biases the anode-gate (G_A). This reaction causes a corre-

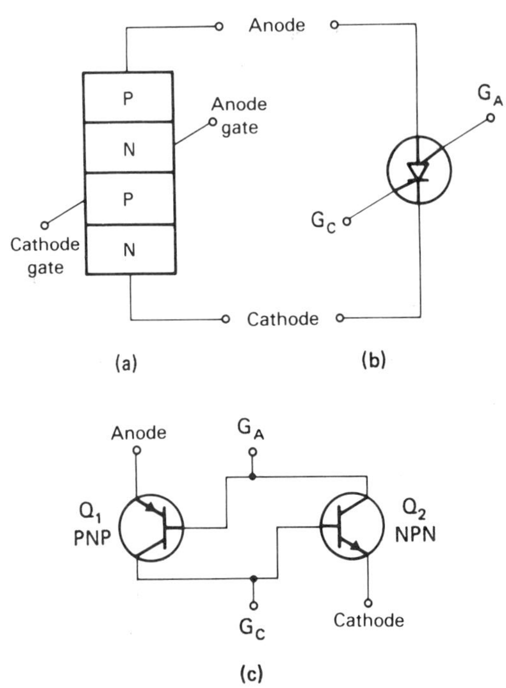

FIGURE 8-31 Silicon-controlled switch: (a) crystal structure; (b) schematic symbol; (c) two-transistor equivalent.

sponding anode-cathode current to flow. I_{GA} then lowers the V_{BO} and causes triggering.

The polarity of the gate-triggering voltage is more meaningful if we study the two-transistor equivalent of the SCS in Fig. 8-31(c). Note that the cathode-gate is connected to the base of the NPN transistor Q_2. It should be obvious that the base of Q_2 must be positive in order to cause forward biasing. When this occurs, some of the output of Q_2 is supplied to the base of Q_1, thus causing regeneration. The entire assembly then becomes conductive. A negative voltage or trigger pulse applied to G_A will also cause forward biasing of the PNP transistor Q_1. Some of this output is then returned to the base of Q_2, which causes regeneration and the assembly to be conductive. In effect, a negative voltage forward biases G_A, and a positive voltage forward biases G_C. In most SCSs the cathode-gate responds to a lower triggering current than the anode-gate.

An SCS can be turned off or brought out of conduction in three different ways. The most obvious method is to reduce the anode current below the holding current (I_H) level. This response is the same as that of an SCR. Second, a negative pulse may be applied to the cathode-gate. This would reverse bias the NPN transistor (Q_2) of the equivalent circuit. The third method of turn off is to apply a positive-going pulse to the anode-gate. This condition reverse biases the base of the PNP transistor (Q_1). In most SCSs, it takes more gate current to bring the device into conduction. The off triggering current must overcome a larger anode-cathode current in order to stop conduction. Ordinarily, this is not a real problem in most SCS applications. Maximum I_{AK} values are usually rather small. Typical I_{AK} values are 100 to 300 mA, with power dissipation ratings of 100 to 500 mW.

Gate Turn-Off Switches

The gate turn-off switch (GTO) is a thyristor that is very similar in construction to the SCR. Its operation is different in that conduction can be triggered on or turned off by a single gate. A positive gate pulse causes the device to be conductive. Conduction continues if the holding current is maintained. A negative gate signal turns off a conductive GTO.

The crystal structure and schematic symbol of a GTO are shown in Fig. 8-32. The crystal structure is primarily the same as an SCR. The gate of a GTO is, however, normally made thicker than that of an SCR. Reverse biasing of the gate increases the size of the depletion region to a point that it will extinguish conduction. The schematic symbol of a GTO differs from that of an SCR by employing a small line across the gate lead. This denotes the on-off switching function of the gate. Operation other than this is primarily the same as for an SCR. It has unidirectional conductivity and will latch when made conductive.

The most obvious advantage of a GTO over an SCR is its gate

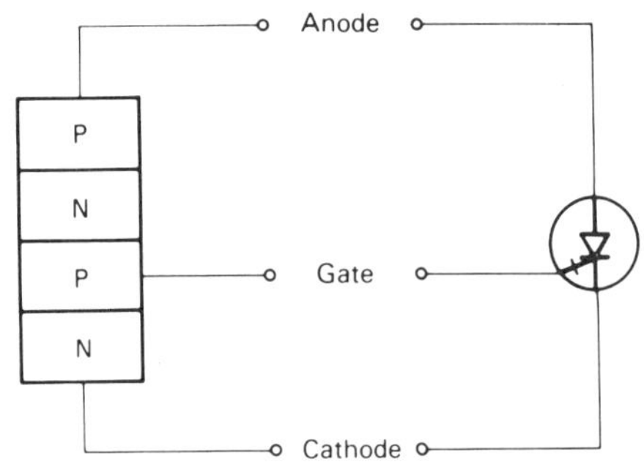

FIGURE 8-32 Gate turn-off thyristor.

turn-off capability. A consequence of this function is an increase in the amount of I_G needed to produce triggering. For an SCR and GTO with similar I_{AK} ratings, an SCR can be triggered with 1 mA, while the I_G of the GTO must be 25 mA. The turn-off current of a GTO is significantly greater than the triggering current. GTOs are available that can control up to 3 A with a power dissipation of 20 W.

The switching time of a GTO is another important characteristic. The trigger time of a GTO is primarily the same as for an SCR. Typical trigger-time values are 1 μs. The turn-off time of a GTO is, however, approximately the same as its trigger time. An SCR must be turned off by reducing I_{AK} below the holding current level. Typical turn-off values may take from 5 to 30 μs. The rapid on-off characteristic of a GTO permits it to be used in high-speed switching applications.

Figure 8-33 shows a GTO used as a sawtooth generator. When the source voltage is applied, the GTO turns on immediately. The gate is made positive with respect to the cathode by the value of

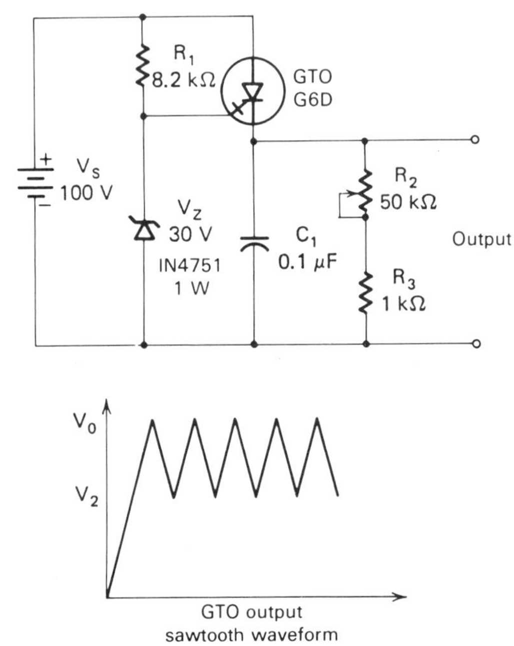

FIGURE 8-33 GTO sawtooth generator.

the zener voltage (V_z). Conduction of the GTO permits capacitor C_1 to charge to the source voltage. When the voltage across C_1 reaches V_z, it reverse biases the gate. This action increases the depletion region of the cathode-gate junction. The holding current (I_H) drops in value to turn off the GTO. C_1 then discharges through the combination of R_2 and R_3. The discharge time is determined by $(R_2 + R_3)C_1$. Proper selection of the R and C values will result in a sawtooth waveform. When the value of V_{C1} drops below V_z, the GTO will turn on and the process will be repeated.

SELF-EXAMINATION

53. A gate turn-off thyristor is similar in construction to a ____________ ____________ ____________.
54. A silicon-controlled switch has ____________ gates.
55. A (+ or −) voltage pulse applied to the cathode-gate will trigger an SCS into conduction.
56. A (+ or −) voltage pulse applied to the anode-gate will trigger an SCS into conduction.
57. An SCS can be turned off in ____________ ways.
58. In a GTO a (+, −) voltage pulse will cause triggering and a (+, −) voltage pulse will cause turn-off.
59. A GTO generally takes more ____________ current to produce triggering than an equivalent SCR.

ANSWERS

53. Silicon-controlled rectifier	54. Two
55. +	56. −
57. Three	58. +, −
59. Gate	

SUMMARY

A thyristor is a general classification for solid-state devices that are used for power control applications. The term thyristor is a contraction of the words (thyr)atron trans(istor). Thyristors can be two-, three-, or four-element devices. One group of devices in this classification is the reverse-blocking triode thyristor. Silicon-controlled rectifiers, unijunction transistors, programmable unijunction transistors, gate turn-off switches, and silicon-controlled switches are included in this group of devices. Bidirectional thyristors are another group of power control devices. Triacs and diacs are included in this classification.

An SCR is a solid-state device made of four alternate layers of P- and N-type silicon. Three junctions are formed by the structure. The leads are the anode, cathode, and gate. The gate is used to trigger conduction current between the anode and cathode. An SCR has unidirectional conductivity. When the anode is positive, the cathode negative, and the gate positive, an SCR will become conductive. When conduction reaches the holding current level, the SCR will latch and hold its conduction. This device responds as a switch. To stop conduction, the anode-cathode current must be lowered below the holding current level or the voltage must be momentarily removed. When the gate of an SCR is forward biased, the resulting gate current causes a reduction in the forward breakover voltage. This is called triggering.

The *I-V* characteristics of an SCR tell a great deal about its operation as a power control device. Quadrant I shows operation in the forward conduction direction. This shows how conduction occurs when the forward breakover voltage is exceeded. Quadrant III shows the reverse breakdown condition of operation. This characteristic is similar to that of a conventional silicon diode. When the PRV is exceeded, an SCR is permanently damaged.

When an SCR is used as a power control device it primarily responds as a switch. When the applied source voltage is below the forward breakover voltage, control is achieved by increasing the gate current. Dc power control requires two switches to achieve control. One turns on the source voltage, and the second switch controls the gate. Dc power control by an SCR generally necessitates some type of commutation circuit to turn the device off. Ac power control has automatic commutation or turn-off because of the reversal of each alternation. The conduction time of an alternation can be changed to make a circuit have variable ac power control. This is achieved with a phase shifter, an *RC* circuit, or a pulse-control device.

A triac is functionally classified as a gate-controlled ac switch. It responds as two reverse-parallel-connected SCRs with one common gate. Conduction is achieved by selecting an appropriate crystal combination. Selection depends on the polarity of the source. Each terminal connection of the triac is jointly connected to an N-P material combination. The voltage polarity of the source alternately changes and selects the correct combination to produce conduction. The leads of a triac are called terminal 1, terminal 2, and the gate. The dual polarity of each connection causes the triac to have four possible trigger combinations. The most sensitive modes of operation occur when the gate polarity matches the polarity of terminal 2.

The *I-V* characteristics of a triac show a great deal about its operation. Conduction in quadrants I and III is identical. This means that a triac has bidirectional conductivity. With ac control of the gate it is possible to achieve 180° delay of the conduction time for each alternation. This means that the output of a triac power control circuit is ac. Conduction time of the ac waveform can be varied to achieve 100% control of a power source.

A diac is a bidirectional diode used to trigger a triac. It is the equivalent of an NPN transistor with no base connection. Conduction is the same in each direction. The *I-V* characteristic of a diac shows that conduction occurs when the breakover voltage is exceeded. This is the same in quadrants I and III.

Pulse triggering of an SCR can be accomplished with a unijunction transistor. A UJT is commonly described as a voltage-controlled diode. It responds as a three-terminal, single-junction solid-state device. A small bar of N-type silicon has base 1 and base 2 attached to the ends of the bar. An emitter is formed by fusing an aluminum wire to the approximate center of the bar. Operation is based on the bias voltage applied to the emitter-base 1 junction. When E-B1 is forward biased it lowers the internal B1-B2 resistance. The bias voltage needed to achieve conduction is determined by the intrinsic standoff ratio. UJTs are normally used in pulse-generator circuits that are attached to an SCR. The pulse rate of the circuit is an *RC* function.

A programmable unijunction transistor is commonly used to generate trigger pulses. A PUT is actually a thyristor that responds as a UJT. The intrinsic standoff ratio voltage of this device can be altered or programmed externally by changing the resistance ratio of a voltage-divider network. The crystal structure of a PUT has its gate connected to the N material nearest to the anode. A PN junction is formed by the anode-gate. Conduction of the device is controlled by the bias voltage of A-G.

Gate turn-off thyristors are unique power control devices that can be triggered into conduction or turned off by a gate signal. A GTO thyristor is very similar in construction to the SCR. A positive gate pulse causes the device to be conductive. A negative gate pulse signal turns off a conductive GTO.

The silicon-controlled switch (SCS), is a four-layer PNPN thyristor. An external lead connection is made to all four layers of the device. It has an anode, cathode, anode-gate, and a cathode-gate. The *I-V* characteristics are the same as those of an SCR. Conductivity occurs in quadrant I. This occurs when the anode is positive and the cathode is negative. Triggering is achieved by making the anode-gate negative or the cathode-gate positive. An SCS can be turned off in three ways. The most obvious method of turn-off is to bring the anode current below the holding current level. Second, a negative pulse may be applied to the cathode-gate. The third method is to apply a positive-going pulse to the anode-gate. An SCS is generally not as sensitive to triggering as an SCR.

EXPERIMENT 8-1
SILICON-CONTROLLED RECTIFIER TROUBLESHOOTING

The internal circuitry of an ohmmeter has a voltage source connected to its leads when it is used to measure resistance. This voltage can be used to bias the junctions of an SCR. The good-bad status of the device and its leads can be identified through this process. The ohmmeter is a quick and easy method of evaluating an SCR. This method of evaluation is used widely in electronic troubleshooting procedures.

As a rule, a standard VOM or a hand-deflection meter works very well for this method of evaluation. A digital VOM does not work as well for some of the measurements. Digital instruments generally have a very small current flow when measuring resistance. This may restrict the test procedure by not permitting the meter to trigger an SCR and have it latch or hold in its conduction range. All other test procedures are essentially the same.

OBJECTIVES

1. To test the status of a silicon-controlled rectifier with an ohmmeter.
2. To identify the leads of a silicon-controlled rectifier.

EQUIPMENT

Volt-ohm-milliammeter (VOM), standard or digital type

Silicon-controlled rectifier, General Electric C106F, Heath 57-50, or equivalent

Circuit construction board

PROCEDURE

1. Prepare the VOM to measure resistance in the low resistance range. Use R × 1 for standard VOMs and 2 kΩ or equivalent for a digital instrument.
2. Select any two SCR leads at random and test the forward and reverse resistance ratio. A two-lead combination that responds as a diode (low resistance in one direction and infinite resistance in the other direction) is the cathode-gate. A two-lead combination that shows infinite or high resistance in both directions is either the anode-cathode

or the gate-anode. Find the two-lead combination that responds as a diode. The third or alternate lead of this combination is the anode. For the two-lead diode combination, connect the meter to produce a low resistant reading. This will show the cathode connected to the negative and the gate to the positive lead. This assumes that the ohmmeter has straight polarity.

3. Make a sketch of the SCR in the right margin indicating the lead assignments you have selected.
4. To verify the correctness of the lead assignments, connect the ohmmeter as shown in Fig. 8-1A. This should initially make an infinite resistance indication on the VOM.
5. Connect a wire between the anode and the gate. If the lead assignment is correct, the meter will change to a low resistance value. If the wire is disconnected, the meter should continue to show low resistance. This shows the latching or holding function of the SCR. If a digital meter is used, the SCR may not latch or hold in its conduction state. A standard or hand-deflection VOM will have enough current in the ohmmeter range to latch and hold the SCR into conduction. To alter the latching function, momentarily disconnect one side of the ohmmeter, then return the ohmmeter leads to the anode-cathode. This should show an infinite resistance with the gate lead disconnected. Connect the gate lead again to the anode to see the latching function. Try this test procedure several times to be certain of how it is performed.
6. This test procedure applies only to good SCRs. If a two-lead combination cannot be found that responds as a diode, the gate-cathode junction is open. If any two-lead combiantion shows low resistance in either direction of polarity, the junction or combined junctions are leaky or shorted. SCRs can become open or shorted in normal usage. As a rule, the SCR is rather free of trouble and responds as a very reliable device.

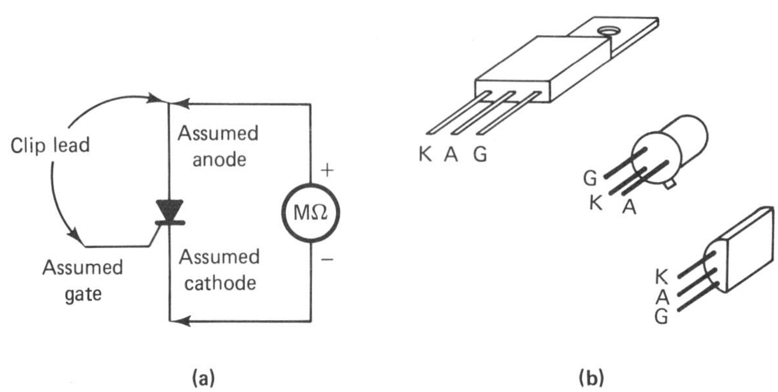

FIGURE 8-1A Ohmmeter connections for SCR testing: (a) ohmmeter connection; (b) lead pin-outs.

DISCUSSION

In this experiment you were asked to test an SCR with an ohmmeter to determine if it is good, shorted, or leaky. The leads of the SCR were also identified, and the device was tested to see if it would latch. This was done to verify the lead identification and to evaluate the status of the device. A standard VOM or hand-deflection instrument works better for this evaluation procedure than a digital instrument. Standard VOMs have a greater current flow when measuring resistance than do digital instruments. The holding current level of an SCR must be reached before it will latch and hold in its conductive state. A digital instrument can be used for all tests other than the latching function.

ANALYSIS

1. Why does an analog or standard VOM work best in evaluating the status of an SCR?
2. Why does the gate-cathode of an SCR show low resistance in one direction and high resistance in the other direction of ohmmeter polarity?
3. Explain why the anode-cathode of a good SCR shows infinite resistance in either direction of ohmmeter polarity.

EXPERIMENT 8-2
SCR CONTROL APPLICATIONS

In this experiment you will construct a test circuit that will permit you to see how an SCR responds when it is conductive or in its nonconductive state. When an SCR is conductive it responds as a closed switch. The anode-cathode resistance is extremely low, and the voltage across the device is very low. Typical voltage values are 1 V. The series resistor connected to the SCR serves as a load device for the circuit. When an SCR is triggered into conduction, a rather large amount of current will flow through the series resistor. This will produce a rather large value of voltage across the resistor. This voltage is a good indicator of the status of the SCR. Resistor voltage indicates that the SCR is conductive. Very small or practically no anode-cathode voltage also indicates conduction of the SCR.

When an SCR is nonconductive, it responds as an open switch. In this operational state, the internal resistance is infinite or an extremely high value. Nearly all the source voltage will appear across the SCR. No voltage will appear across the series resistor. Current calculated by using the series resistor voltage and its resistance value will be zero. A nonconductive SCR is easily recognized by this evaluation procedure.

OBJECTIVES

1. To demonstrate the operation of an SCR as a power control device.
2. To determine the operational characteristics of an SCR when it is conductive and nonconductive.
3. To measure the anode-cathode voltage and series resistor voltage when an SCR is conductive and nonconductive and be able to define the status of the device.

EQUIPMENT

Volt-ohm-milliammeter (VOM), standard or digital

Silicon-controlled rectifier, Motorola C106F, Heath 57-50, or equivalent

Silicon diode, Texas Instruments 1N4001, Heath 56-28, or equivalent

Germanium diode, General Electric 1N4149, Heath 56-89, or equivalent

Variable dc power supply, ± 15V
Ac power source, 12.6 V from a transformer
100-Ω, $\frac{1}{4}$-W resistor
470-Ω, $\frac{1}{4}$-W resistor
4.7-kΩ, $\frac{1}{4}$-W resistor
100-kΩ, 1-W potentiometer, (linear taper)
12-V lamp, ALCO #2182, GTE Sylvania 12ES, Heath 412-601, or equivalent
Capacitor, 0.1-μF 50 Vdc
Circuit construction board

PROCEDURE

Part A: SCR Circuit Control

1. Construct the circuit of Fig. 8-2A. The leads of the SCR were identified in the previous activity. Refer to the lead assignment drawing of step 3 of Experiment 8-1.
2. Before connecting the dc source to the circuit, turn on the power supply and adjust it to produce 10 V. Use the VOM to measure this voltage.
3. The position of the gate resistor (R_g) in the circuit determines the status of the SCR. In the first part of this experiment the gate resistor is connected to cause triggering of the SCR. In this regard, one end of R_g is connected to the positive side of the source to produce gate current. The lamp should be on, indicating that the SCR is triggered and in its conductive state. If your circuit

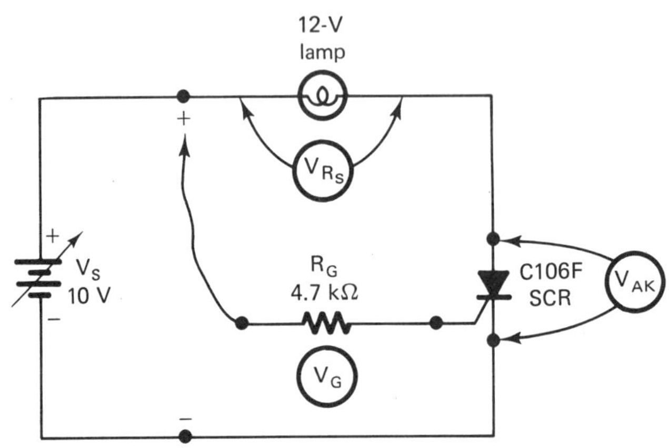

	Measured values			Calculated values		
Step no.	V_{AK}	V_{RS}	V_{RG}	I_{AK}	I_G	SCR status
11						
14						

FIGURE 8-2A SCR circuit operation.

is properly connected, the lamp will be on. Does the circuit respond in this manner? ____________ If it does not respond, check the circuit wiring or lead assignments of the SCR.

4. If the circuit is functioning properly, disconnect the end of R_g from the positive side of the source. If the holding current is great enough, the lamp should continue to be illuminated. This shows that when triggering has occurred, the SCR will continue to be conductive even with the loss of gate current. Does your circuit respond in this manner? ____________
5. With one end of R_g still disconnected, momentarily turn off the dc power source. Then turn it on again. How does this alter the status of the lamp? ____________
6. Momentarily connect the loose end of R_g to the positive side of the source to retrigger the lamp. How does the lamp respond? ____________
7. Turn off the lamp again. This time disconnect one end of the lamp from the circuit momentarily, then return the loose end of the lamp to the circuit. Does this action cause the lamp to turn off? ____________
8. Try triggering the SCR and turning it off several times to be certain of the procedure.
9. Modify the circuit by removing the lamp and replacing it with a 470-Ω resistor. All else remains the same. Voltage will be measured across the SCR, the series resistor (R_s), and the gate resistor (R_g). These measurements will be used to determine the status of the SCR's conduction. Record the measured values in the chart of Fig. 8-2A. Calculate the current values from the measured values.
10. Trigger the SCR into conduction by connecting R_g to the positive side of the source.
11. Measure V_{AK}, V_{RS}, and V_G with the VOM. Record the measured values as data for step 11 in the chart of Fig. 8-2A.
12. Calculate the current values of the chart using the measured voltage values that appear across the appropriate resistors. I_{AK} is V_{RS}/R_s. Gate current I_G is V_G/R_g. Determine these values and record them in the chart.
13. Prepare the SCR to be in its off state with power applied and the gate resistor open.
14. Measure and record V_{AK}, V_{RS}, and V_G. Calculate the current values.
15. Evaluate the data in the chart. Note in particular the status of the SCR in its on and off states and how the measured voltage values and the calculated current values appear for each state.
16. Disconnect the SCR circuit.

DISCUSSION

In this experiment you had an opportunity to see how an SCR is triggered into conduction and turned off. The first part of the experiment permitted you to see what occurs when an SCR is in control of a lamp. When the gate is forward biased by connection to the positive side of the source, it causes gate current. This lowers the forward breakover voltage and causes the SCR to be triggered into conduction. Once triggering has occurred, the gate is no longer needed to cause conduction.

To turn off a conducting SCR the conduction current must drop below the holding current level. This can be achieved by momentarily disconnecting the source voltage. The power can also be momentarily turned off from the source. If the gate is disconnected, the SCR will not be triggered into conduction when the power is restored to the circuit.

In the second part of the experiment you were asked to measure dc voltage values across several components to see how the circuit responds when it is conductive and nonconductive. A conductive SCR will show very little voltage across the anode-cathode. The series resistor or load device will show a rather substantial amount of the source voltage across it. Current passing through the resistor will cause this voltage to be developed. The current of the circuit was calculated by using measured voltage values and resistor values.

Nonconductive SCR operation is a reverse of the conductive function. Voltages measured across the SCR and the series resistor showed the full source voltage across the anode-cathode and no voltage across the series resistor. Nonconduction of the SCR means that its internal resistance is extremely high. All the source voltage will appear across the anode-cathode. No voltage across the series resistor signifies the loss of current through a device. This is an important concept to keep in mind when evaluating the operation of an SCR.

ANALYSIS

1. Explain why the anode-cathode voltage of an SCR is low when it is conductive and high when nonconductive.
2. How is the SCR triggered in the circuit of this activity?
3. Explain how an SCR circuit is turned off after it has been triggered into conduction.

Part B: AC Power Control with an SCR

1. Connect the circuit of Fig. 8-2B. The source of power for this circuit is 12.6 V ac. The circuit is the same as Fig. 8-10. Its operation was explained in the text.
2. With power applied to the circuit the lamp should be illuminated. If it is not, try adjusting potentiometer R_2

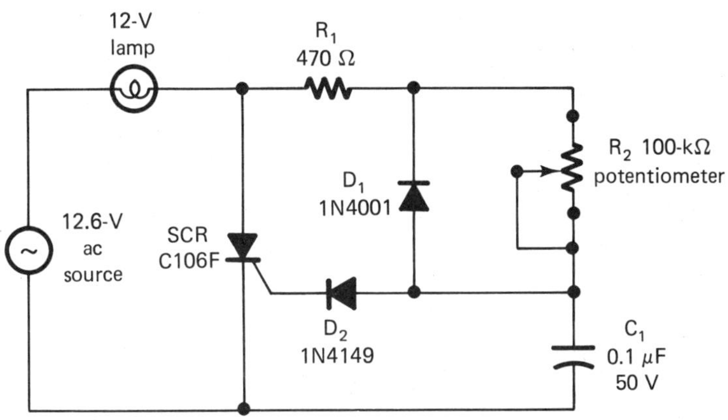

FIGURE 8-2B AC power control with an SCR.

through its range. This adjustment should cause the intensity of the lamp to change according to its setting. If the circuit does not function, recheck the circuit wiring for correctness. Check the SCR for proper lead connections. With the VOM measure the ac source voltage to see if it is of the correct value. The circuit can also be tested for resistance and continuity with the ohmmeter of the VOM. Be certain that the power source is disconnected from the circuit before testing circuit resistance.

3. If the circuit has changeable light intensity, adjust potentiometer R_2 to produce full brightness. Disconnect the ac power from the circuit and measure the resistance of the potentiometer: ____________.
4. Apply ac power to the circuit and adjust potentiometer R_2 to produce no illumination of the lamp. Disconnect the ac power and measure the resistance of the potentiometer: ____________.
5. Apply power to the circuit and adjust the potentiometer through its range. Describe the intensity of the lamp with respect to potentiometer resistance. ______________

   ____
6. Turn off the ac source of power and disconnect the circuit.

DISCUSSION

In Part B of this experiment you constructed an ac power control circuit using an SCR. The energy source was ac and control was provided by the SCR. Since an SCR has unidirectional conductivity, the resulting output of this circuit is dc. Functionally, the output developed across the lamp is a half-wave version of the ac input.

Control of the SCR is achieved by a potentiometer. The resistance of the potentiometer causes the intensity of the lamp to vary. Operation is based on the charge and discharge of capacitor

C_1. When the resistance of the potentiometer is low, C_1 charges very quickly. Essentially there is very little delay in the turn on of the SCR. Lamp intensity is at its brightest level. An increase in resistance causes C_1 to take a longer time to develop its charge voltage. This delays the turn on of the lamp. The lamp is dim for this condition.

Ac power control is determined by the conduction time of one alternation. If the turn on of an alternation is delayed, it will occur for a shorter length of time. The resulting power output is reduced. If the alternation occurs without delay, full power will be developed by the load. The output will be at its greatest value when this occurs. Since an SCR has unidirectional conductivity, the best power output that can be developed by the load device is one alternation, or 50% of the sine wave.

ANALYSIS

1. Why is the output of an ac power control circuit only half of the ac sine wave?
2. How does potentiometer resistance of the ac power control circuit vary the intensity of the lamp?
3. What is the function of the two diodes in the ac power control circuit?

EXPERIMENT 8-3
TESTING A TRIAC WITH AN OHMMETER

The internal circuitry of an ohmmeter has a voltage source connected to its leads when it is used to measure resistance. This voltage source can be used to bias a triac into conduction. The good-bad status of a triac can be tested, and the leads can be identified. An ohmmeter is a quick and easy method of evaluating the status of a triac.

A hand-deflection or standard ohmmeter works best for this method of evaluation. This type of instrument generally has a larger value of current in its ohmmeter. The current value is somewhat critical when triggering a triac into conduction and getting it to latch. Digital VOMs, as a rule, have a very nominal current flow when the meter is used as an ohmmeter. This current value is generally not large enough to cause a triac to latch. All other tests being performed will respond equally with either type of meter.

The test procedure of a triac is primarily the same as that of an equivalent SCR. The triac, however, can be triggered in four different modes or conditions. Each lead has a dual-crystal polarity that is voltage selective. The polarity of the ohmmeter is used to select the appropriate combination of the test. Identification of the respective leads is somewhat more complex than the procedure used in evaluation of the SCR.

OBJECTIVES

1. To test the status of a triac with an ohmmeter.
2. To identify the leads of a triac with an ohmmeter.

EQUIPMENT

Volt-ohm-milliammeter (VOM), standard or digital type
Triac, RCA 2N5754 or equivalent
Circuit construction board

PROCEDURE

1. If a standard or hand-deflection VOM is being used, prepare it to measure ohms in the R × 1 range. If a DVM is used, prepare it to measure resistance in the 2-kΩ or equivalent range. The R × 1 is preferred because it usu-

ally provides the necessary holding current to latch this device into conduction. We will assume that the instrument being used has straight polarity with the negative lead connected to the black or common lead and positive connected to the red or ohms lead.

2. Select any two of the three leads at random and determine the forward-reverse resistance ratio. Terminal 1 to terminal 2 should indicate an infinite resistance or extremely high resistance in both directions. The gate and terminal 1 will show a rather low resistance in either direction. With the two leads that show low resistance in either direction identified, the third lead is terminal 2.
3. Make a sketch of the triac leads in the right-hand margin. Identify T_2 in the sketch. Leads T_1 and the gate have not been identified at this time.
4. The next test procedure is based on a right or wrong selection opportunity of determining the correct lead assignment for T_1 and G. Connect the positive side of the ohmmeter to T_2; then select one of the two remaining leads and connect the negative side of the ohmmeter to it. We will assume in this case that the negative ohmmeter lead is connected to T_1. Now attach a connecting wire to the third lead, which we have assumed to be the gate. Connect the other end of this wire to T_2. If the assumed T_1 and G are correct, the ohmmeter will show a low resistance, which indicates triggering. Momentarily disconnecting the gate from T_2 should cause the triac to stay in its conduction state. If this procedure does not produce a change in resistance, the assumed T_1 and G are reversed. Switch the negative ohmmeter lead from the first assumed T_1 lead to the assumed gate lead. Place the connecting wire to the newly assumed gate lead and try the triggering procedure again. If triggering is produced, the correct lead assignment has been made. Label these two leads as the gate (G) and terminal 1 (T_1). If the triac cannot be triggered by this procedure it may be faulty. Only a good triac will respond to this test.
5. With the lead assignments made, once again try the triggering procedure. Connect the negative lead to T_1 and the positive lead to T_2. Then attach a connecting wire between the gate and T_2. This should cause a resistance change in the VOM and trigger the triac into conduction.
6. If the lead assignments are correct, the triac should also trigger in the reverse polarity direction. In this case, connect the negative side of the VOM to T_2 and the positive side to T_1. Attach a connecting wire to the gate and momentarily connect it to T_2. This action should also trigger the triac into conduction. If the triac is functioning prop-

erly, it can be triggered in either direction of polarity. Test the triac to see if it has this capability.

7. A digital meter or DVM will work for all test procedures except the latching function. The standard VOM will latch the triac only when it is in the R × 1 range of operation.

DISCUSSION

In this experiment you were asked to test the status of a triac with an ohmmeter and to identify its leads. The procedure followed is primarily the same as that for the SCR. A triac, however, has dual-polarity crystals at each lead. The polarity of the material is selected by the polarity of the energizing source. For this procedure the ohmmeter source determines the test polarity. We assumed that the ohmmeter had straight polarity.

Lead identification of a triac is somewhat more complicated than the procedure used for the SCR. The dual polarity of each lead is responsible for the added complexity. A number of test characteristics were observed through this activity. Between T_1 and T_2 of the triac the ohmmeter will indicate infinite or extremely high resistance for either polarity. Between the gate and T_1 an ohmmeter will show the same low resistance for either polarity. A good triac will also show infinite or extremely high resistance between the gate and T_2 for either polarity. Triggering is best achieved by connecting the ohmmeter to T_1 and T_2 and momentarily connecting the gate to T_2. This applies to either ohmmeter polarity. Latching can be achieved only if the ohmmeter is capable of supplying the current needed to reach the holding current level. A digital ohmmeter is usually not capable of triggering and causing the device to latch.

ANALYSIS

1. What are some of the common things to look for in testing an SCR to see if it is faulty?
2. Why does the gate and T_1 of a triac show low resistance in either direction of ohmmeter polarity?
3. Why does T_1 and T_2 of a good triac show infinite resistance in either direction of ohmmeter polarity?

EXPERIMENT 8-4
TRIAC POWER CONTROL

A triac is used widely as an ac power control device. Ac power can be controlled by switching action that causes a circuit to simply be turned on or off. Control can also be variable and changed to a range of different values. Variable control involves a potentiometer that changes the value of gate current needed to trigger the triac into conduction. This method of control actually changes the conduction time of the ac waveform. Control is achieved by altering the amplitude of the applied gate voltage so that it causes a corresponding change in gate current. Variable control of this type is easy to achieve and rather inexpensive to accomplish.

OBJECTIVES

1. To build an ac power control circuit that employs a triac.
2. Evaluate the operation of a triac power control circuit.

EQUIPMENT

Silicon-controlled rectifier, RCA 2N5754, or equivalent

Silicon diode, Texas Instruments 1N4001, Heath 56-28, or equivalent

Ac power source, 12.6 V from a transformer

12-V lamp, ALCO #2182, GTE Sylvania 12ES, Heath 412-601, or equivalent

470-Ω, $\frac{1}{4}$-W resistor

100-kΩ, 1-W potentiometer, (linear taper)

Circuit construction board

PROCEDURE

Part A: Triac Power Switching

1. Construct the triac power switching circuit of Fig. 8-4A. The loose end of the gate resistor R_1 has three possible connection positions. Position A will cause full brightness of the lamp. Position B will cause half-brightness. Position C will cause the circuit to be off by eliminating the gate current. A connecting wire attached to the end of resistor R_1 will serve as the contact position changer for a three-position switch.

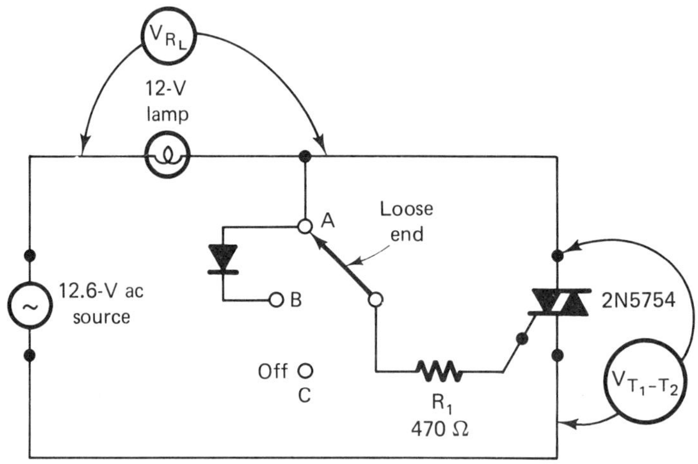

Position	V_{R_L}	$V_{T_1-T_2}$
A		
B		
C		

FIGURE 8-4A Triac power switching circuit.

2. Apply 12.6 V ac to the circuit from a transformer. Momentarily connect the loose end of R_1 to point *A*, then point *B*, and finally to point *C*. If the circuit is working properly there should be a change in the brightness level of the lamp from full, to low level, to off. If the circuit does not respond in this manner, recheck the wiring and the lead connections to the triac. An ohmmeter can be used to test the continuity between different circuit points. Be certain that the ac power source has been disconnected before testing circuit resistance with an ohmmeter.
3. Connect the loose end of R_1 to position A. This should cause the lamp to have full brightness.
4. With the VOM measure the ac voltage across the lamp. Likewise measure the ac voltage across the triac. Record these values in the chart of Fig. 8-4A.
5. Connect the loose end of R_1 to position B. Measure the ac voltage across the lamp and across the triac. Record these values in the chart of Fig. 8-4A.
6. Connect the loose end of R_1 to position C. Measure and record the ac voltage across the lamp and across the triac.
7. Turn off the ac source and disconnect the circuit.

DISCUSSION

In this experiment you constructed a triac power control circuit. Ac power is controlled by triggering the triac into conduction by a switching operation. Control is of a fixed value and determined by the ac voltage applied to the gate. When the loose end of resistor R_1 is connected to position A, it places the ac source voltage to the gate. In this case, ac will cause triggering of the gate at the same time that ac is applied to T1 and T2. Full brightness of the lamp will be achieved by this type of control. The gate voltage across $G\text{-}T_2$ is 1.8 V. The measured ac voltage across the lamp is approximately 11 V. Across the triac it should be 1 to 2 V.

Switching the loose end of resistor R_1 to position B causes the lamp to again be triggered into conduction. In this case the intensity of the lamp is reduced by one-half of the original value. The diode permits the gate to be biased during only one alternation. As a result of this condition, the triac triggers on only one alternation. This causes a 50% reduction in conduction time. The lamp is illuminated at half of its full brightness level. The measured ac voltage across the lamp is approximately 8 V. This is less than the value measured when R_1 was in position A.

Switching the loose end of resistor R_1 to position C causes the lamp to turn off. In this case, the gate current is zero, and the applied ac is not great enough to overcome the breakover voltage of the triac. The triac is in a nonconductive state. Ac voltage measured across the lamp is nearly zero, and across the triac it is equal to the source voltage. This is normal for a nonconductive triac.

ANALYSIS

1. Why is the triac fully conductive when the loose end of resistor R_1 is connected to point *A* of Fig. 8-4A?
2. Explain why connecting the loose end of resistor R_1 to the diode causes the lamp to be at only half of its maximum brightness level.

Part B: Variable Triac Power Control

1. Construct the variable ac power control circuit of Fig. 8-4B.
2. When ac power is applied to the circuit initially, the lamp may not be illuminated. If this occurs, adjust the potentiometer through its range. Adjust R_1 to produce the highest level of brightness. If the circuit does not respond, recheck the circuit wiring and the lead connections to the triac. An ohmmeter can also be used to test the path of the circuit to various components. This must be done with the power turned off.
3. Turn on the ac power source to the circuit and adjust the potentiometer to produce the greatest level of brightness. With the VOM, measure and record the ac voltage across the lamp, the triac, and G-T2. Record these values in the chart of Fig. 8-4B.
4. Adjust the potentiometer to produce half of the maximum brightness level of the lamp. With the VOM mea-

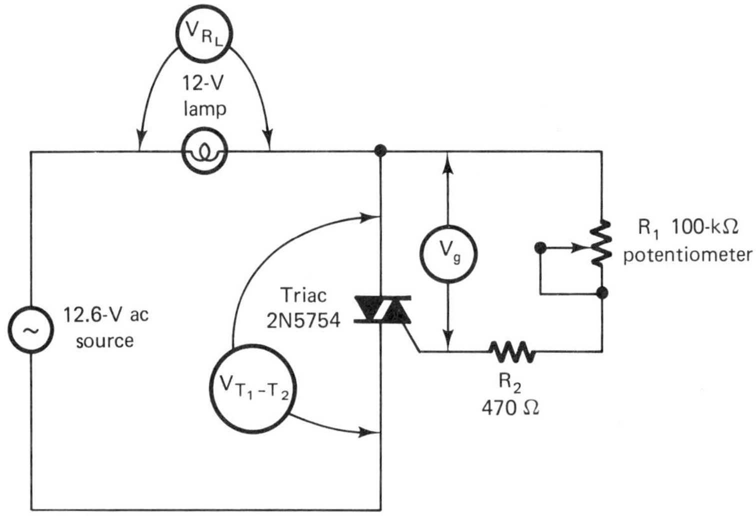

Voltage chart

Lamp status	V_{R_L}	$V_{T_1-T_2}$	V_G
Maximum brightness			
Half-brightness			
Off			

FIGURE 8-4B Triac power control circuit.

sure and record the ac voltage across the lamp, the triac, and G-T_2. Record these values in the chart of Fig. 8-4B.

5. Adjust the potentiometer to a point where the lamp is off. With the VOM measure and record the voltage across the lamp, the triac, and G-T_2. Record these values in the chart of Fig. 8-4B.
6. Turn off the ac source and disconnect the circuit.

DISCUSSION

In this experiment you constructed a variable power control circuit with a triac. The potentiometer of this circuit serves as the variable element for control to the triac by altering the amplitude of the ac applied to the G-T_2 part of the triac. This ac voltage determines the level of conduction achieved by the triac. When the voltage level is at its highest value, it causes the lamp to be at its maximum brightness level. A reduction in G-T_2 voltage reduces the conduction level of the triac and the brightness of the lamp. When the lamp is off, the G-T_2 voltage is at its lowest value, and the triac is nonconductive. Low G-T_2 voltage causes the breakover voltage of the triac to increase, which reduces the possibility of conduction. Amplitude control of a triac is simple, inexpensive to achieve, and fairly effective. Only 50% control of each alternation can be achieved with amplitude control. An oscilloscope connected across the lamp of this circuit would permit a person to see this level of power control.

ANALYSIS

1. How is variable control of a triac achieved in Fig. 8-4B?
2. How does the voltage across a fully conductive triac respond?
3. How does the voltage across a nonconductive triac respond?

EXPERIMENT 8-5
UNIJUNCTION TRANSISTOR TESTING AND TROUBLESHOOTING

The internal construction of an ohmmeter has a voltage source connected to its leads when measuring resistance. This voltage can be used to evaluate the good-bad status of a UJT and to identify its leads. The voltage is used to bias the elements of the UJT. The ohmmeter provides a quick and easy method of evaluating a UJT.

The construction of a UJT has base 1 and base 2 connected to a common bar of doped silicon. The emitter is an aluminum wire fused slightly above the center of the silicon bar. This construction causes the E-B_2 junction to have lower resistance than the E-B_1 junction. The B_1-B_2 resistance is the same in either direction of ohmmeter polarity. The E-B_1 junction of the UJT responds as a diode. The value of the forward-biased resistance of this junction is very important. The E-B_2 junction also responds as a diode. The forward bias resistance of this junction has a lower value than that of E-B_1. This permits B_1 and B_2 to be identified by a resistance measurement.

OBJECTIVES

1. To test the status of a unijunction transistor with an ohmmeter.
2. To identify the leads of a unijunction transistor.

EQUIPMENT

Volt-ohm-milliammeter (VOM), standard or digital type
Unijunction transistor, Motorola 2N2646, Heath 417-183, or equivalent
Circuit construction board

PROCEDURE

1. Prepare the VOM to measure resistance in the 1-kΩ range. Use R × 1 kΩ for a standard VOM and 20 kΩ or equivalent for a digital meter.
2. Select any two leads at random and test the forward and reverse resistance ratio. If the two leads show the same resistance in each direction, they are B_1 and B_2. Normally

this resistance is in the range of 5 kΩ to 10 kΩ. If the two selected leads have a low forward resistance and a high or infinite reverse resistance, they are the E-B_1 or E-B_2. Find two leads that show the same high resistance in either direction of ohmmeter polarity. These leads are B_1 and B_2. The third or remaining lead is the emitter.

3. Make a sketch of the UJT in the right margin. Indicate the position of the emitter in the sketch.
4. Connect the positive side of the ohmmeter to the assumed emitter lead of the UJT. Connect the negative lead of the ohmmeter to one of the remaining two leads. Note the resistance of this lead combination. Switch the negative lead of the ohmmeter to the remaining lead. Note the resistance of this lead combination. The E-B lead combination that has the lowest forward resistance is E-B_2. The highest resistance combination appears across E-B_1. This should identify the location of B_1 and B_2 of the UJT.
5. Indicate the location of B_1 and B_2 in the lead diagram sketch of the UJT.
6. The resistance evaluation procedure of the UJT will work only when the device is good. A faulty UJT will have an open or a shorted E-B_1, E-B_2, or B_1-B_2 junction. A shorted UJT will have a very low resistance across the junction in either direction of ohmmeter polarity. An open UJT will have infinite resistance in either direction across one of its junctions.
7. Perform this test procedure again to identify the leads and to evaluate the device. If an additional UJT is available, test it and identify the leads.

DISCUSSION

In this experiment you were asked to test a UJT with an ohmmeter to identify its leads and to determine if it is good or bad. The B_1-B_2 leads show a few thousand ohms of resistance in either direction of ohmmeter polarity. The emitter-base junction responds as a diode. E-B_1 has a higher forward resistance than E-B_2. This permits B_1 and B_2 to be identified by a difference in resistance.

The UJT can also be tested with an ohmmeter to see if it is faulty. A shorted junction will show low resistance in either direction of ohmmeter polarity. An open junction will show infinite resistance in either direction of ohmmeter polarity. If either the shorted or open condition exists across one of the junctions, it indicates that the UJT is faulty.

ANALYSIS

1. Why does the $E\text{-}B_2$ junction of a UJT show lower resistance than the $E\text{-}B_1$ junction?
2. Why is the resistance of B_1 and B_2 the same in either direction of ohmmeter polarity?
3. What is meant by a shorted junction?

EXPERIMENT 8-6
UNIJUNCTION TRANSISTOR CHARACTERISTICS

In the first part of this experiment you will determine the peak voltage and valley voltage values of an operating UJT. These voltage values are used to calculate the intrinsic standoff ratio of a UJT. These voltage values are determined by connecting a voltmeter across the E-B_1 junction and gradually increasing the applied voltage. An increase in voltage to the emitter first rises to a peak value. When triggering occurs, the emitter voltage drops very quickly to the valley voltage. As a rule, several observations of the voltage change are needed to accurately determine the peak voltage and the valley voltage.

In the second part of this experiment, you will connect the UJT in a pulse-generator circuit. The UJT has three possible outputs available. Across the B_1 resistor the output is a positive-going pulse. The V_{B2} output is negative going. The emitter voltage is a sawtooth waveform. The polarity of the output and the shape of the desired waveform are determined by the device being triggered by the UJT. For SCR control, the positive-going B_1 pulse is commonly used.

The pulse rate of a UJT generator is dependent on the values of R_1 and C_1 in the emitter circuit. Frequency is determined by the expression $f = 1/R_1C_1$. This value is only an approximation of the frequency. The intrinsic standoff ratio alters this value to some extent. As a rule, the emitter voltage of the UJT is not used as a pulse source because it is loaded down by the circuit attached to it, which generally causes a change in pulse frequency. The outputs at B1 and B2 do not effectively influence the frequency of the UJT.

OBJECTIVES

1. To determine the valley voltage, peak voltage, and intrinsic standoff ratio of a UJT.
2. To construct a circuit with the UJT that shows how it can be used as a pulse generator.
3. To demonstrate the reaction of current and voltage when a UJT is triggered into conduction and when it is nonconductive.

EQUIPMENT

Volt-ohm-milliammeter (VOM), standard or digital
Unijunction transistor, Motorola 2N2646, Heath 417-183, or equivalent
Variable dc power supply, +10 V
47-Ω, $\frac{1}{4}$-W resistor
100-Ω, $\frac{1}{4}$-W resistor
220-Ω, $\frac{1}{4}$-W resistor
1-kΩ, $\frac{1}{4}$-W resistor
10-kΩ, $\frac{1}{4}$-W resistor
330-kΩ, $\frac{1}{4}$-W resistor
100-kΩ, 1-W potentiometer, (linear taper)
Capacitor, 10 μF, 50 V dc
Circuit construction board

PROCEDURE

Part A: Characteristic UJT Values

1. Construct the circuit of Fig. 8-6A. The leads of the UJT were identified in the previous activity. Refer to the lead assignment for the UJT.
2. Turn on the power source and adjust it to produce 10 V as measured across the two outside leads of the potentiometer. This will serve as the V_{BB} voltage source and for V_E.
3. Connect the VOM across the emitter (E) and ground as indicated by voltmeter V_E. Adjust the potentiometer to produce a V_E of 0 V. Potentiometer R_1 controls the value of V_E. Resistor R_2 serves as an emitter-current-limiting resistor. Resistor R_3 in series with B_2 is used to prevent excessive base current. Resistor R_4 in series with B_1 is

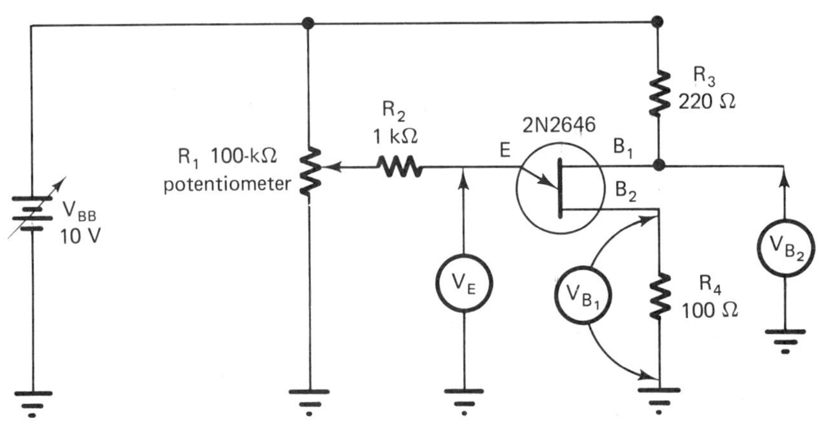

FIGURE 8-6A UJT characteristic test circuit.

also used to prevent excessive current. Two resistors are used in this circuit to develop two outputs. The B_1 output is positive going, and B_2 is negative going.

4. With the VOM connected as V_E, gradually change R_1 while increasing the value of V_E. A definite peak voltage value (V_P) will occur first. Continue to change R_1 while noticing that V_E drops to a definite valley voltage (V_V) value. V_E will then increase slightly if R_1 is continually increased. The V_P and V_V values are important characteristics in the operation of a UJT.
5. Using the procedure just outlined, determine the V_P and V_V points of the UJT. Record these values: V_P = ______; V_V = ______. Since the change from V_P to V_V takes place so rapidly, you may need to try this procedure several times to come up with a very accurate value.
6. The forward voltage (V_F) across the P-N junction of E-B_1 is generally considered to be 0.7 V. This value is used to determine the intrinsic standoff ratio along with V_{BB} and V_P. Eta (η) = $V_P - V_F/V_{BB}$ is the expression for the intrinsic standoff ratio. Using the values for V_F, V_P, and V_{BB}, determine the intrinsic standoff ratio of the UJT: η = ______.
7. It is rather interesting to note that the intrinsic standoff ratio of a UJT is essentially the same for nearly all source voltage values. To show this characteristic, alter the source voltage of the power supply to produce 8 Vdc. Using the same procedure, determine the V_P and V_V values for this new voltage value: V_P = ______; V_V = ______.
8. Using the value of V_{BB} as 8 V and V_F as 0.7 V, determine the value of the intrinsic standoff ratio again: η = ______.
9. Return the source voltage to 10 V and turn off the power source. Do not disconnect the circuit.

DISCUSSION

In Part A of this experiment you had an opportunity to see how the peak voltage and valley voltage of a UJT changed values. You should have noted that the peak voltage was approximately 8 V and that the valley voltage was equal to 2 V. Using a V_F of 0.7 V and a V_{BB} of 10 V, the intrinsic standoff ratio should have been somewhere between 0.6 to 0.8. Slightly lower or higher values can be attributed to inaccuracies in the measurement of V_P and V_V. The manufacturer of the UJT recommended for this activity indicates the eta to be 0.56 to 0.75. Your calculated value of eta should fall in this range.

You were then asked to change the V_{BB} value to 8 V and again measure V_P and V_V. You should have found a V_P of 6.4 and a V_V of 1.6 V. The eta should have been calculated to be approximately the same value as determined previously. This was done to show that the eta of a UJT remains the same for all source voltage values in a reasonable operating range.

ANALYSIS

1. What does the intrinsic standoff ratio of a UJT represent?
2. What does the voltage change area between V_P and V_V of a UJT represent?

Part B: UJT Operational Characteristics

1. The circuit of Part A will be used for this experiment. Turn on the power source and measure the value of the source voltage. It should be 10 V. If necessary, adjust the power source to produce 10 V.
2. The VOM will be used to measure voltages at different places in the circuit. When the meter is connected in these places, it will measure emitter voltage (V_E), base 1 voltage (V_{B1}), and base 2 voltage (V_{B2}). These voltages will be used to determine different current values through calculations. These are measured before triggering and after triggering.
3. Adjust potentiometer R_1 to produce an emitter voltage of 0 V. Measure and record the values of V_{B1} and V_{B2}. Record these values in the chart of Fig. 8-6B. Calculate the values of I_{B1}, I_{B2}, and I_E.
4. Adjust potentiometer R_1 to a voltage value that causes the UJT to be triggered. V_E will show a peak voltage (V_P) value then drop to a low voltage value (V_V) when the UJT is triggered into conduction. Measure and record the values of V_{B1} and V_{B2} after the UJT has been triggered into conduction. Calculate the values of I_{B1}, I_{B2}, and I_E and record these values in the chart.
5. You may want to try these measurements several times to be certain that some degree of accuracy is achieved.
6. Turn off the power and disconnect the circuit.

UJT Condition	V_E	V_{B_1}	V_{B_2}	$I_{B_1} = \frac{V_{B_1}}{R_4}$	$I_{B_2} = \frac{V_{B_1} - V_{B_2}}{R_3}$
Before triggering					
After triggering					

FIGURE 8-6B UJT characteristic operation chart.

DISCUSSION

In this experiment you had an opportunity to see how a UJT responds in its triggered and nonconductive states. Since B_1 and B_2 of a UJT are connected to a common piece of doped silicon, there will be some current flow through this part of the device when it is not triggered. A typical value for V_{B1} is 0.1 V and for V_{B2}, 9.9 V. The emitter current for this condition of operation is zero.

When the UJT is triggered into conduction, V_{B1} changes to 0.5 V, and V_{B2} is 9.5 V. The emitter current for this condition of operation is approximately 2 mA. The UJT is uniquely different from an SCR when it is triggered. A UJT, for example, does not latch in its conductive state. Current flow through the B_1 resistor is significantly greater than that passing through the B_2 resistor. In a sense, the interbase resistance (R_{BB}) of a UJT decreases to half of its value when the device is triggered.

ANALYSIS

1. What is the relationship of I_E and V_E of a UJT when it is not triggered into conduction?
2. What is the relationship of I_E and V_E of a UJT when it has been triggered into conduction?
3. What is the relationship between V_{B1} and V_{B2} when the UJT has been triggered and before it is triggered?

Part C: UJT Pulse Generation

1. Construct the UJT pulse generator of Fig. 8-6C. The source voltage for the circuit is 10 Vdc. The circuit is very similar to those constructed in the first part of this activity. A point of concern in this circuit is the connection of the electrolytic capacitor (C_1) in the emitter. It is essential that the polarity of C_1 be connected as indicated.

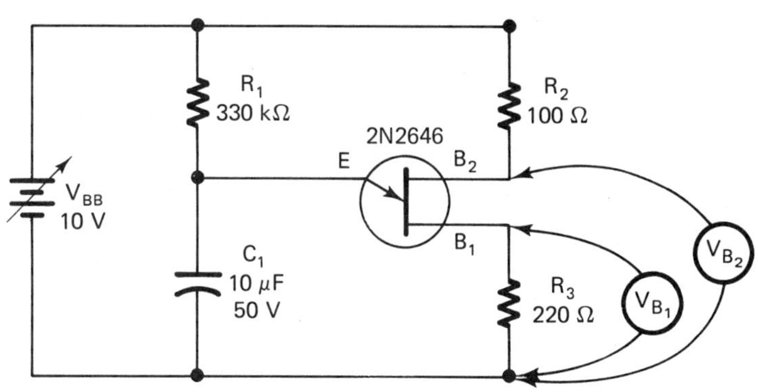

Maximum V_C = ________ V_{B_1} = ________ Polarity (+ or −)

Minimum V_C = ________ V_{B_2} = ________ Polarity (+ or −)

FIGURE 8-6C UJT pulse generator circuit.

2. The pulse generator of this experiment is very similar to the one described in Fig. 8-25. The circuit of this activity has a fixed value of pulse frequency. R_1 and C_1 determine the pulse rate of the generator. In this case the pulse rate is very long. This permits you to observe the voltage changes of the circuit with a voltmeter.
3. Turn on the power source and adjust it to produce 10 V dc. This should be measured with the VOM.
4. Measure and record the voltage across capacitor C_1 with the VOM. This voltage value will rise to a peak value then drop to a very small value during one operating cycle. The resulting waveform produced is in the shape of a sawtooth. Note the shape of the waveform in Fig. 8-25 of the text. Record the maximum and minimum voltage values in the appropriate space of Fig. 8-6C.
5. Measure and record the V_{B1} pulse voltage across resistor R_3. This voltage will occur only in very short duration pulses. Observe the generated pulse for two or three cycles of operation. Record the measured value of the pulse and note its polarity in Fig. 8-6C.
6. Measure the V_{B2} pulse voltage as indicated. Record the voltage value of the pulse and its polarity in Fig. 8-6C. How does this compare with the pulse of step 5? ______ ______________.
7. Turn off the power source. Remove the 330-kΩ resistor and replace it with a 100-kΩ potentiometer. One side of the potentiometer goes to the positive side of the source. The center lead goes to the emitter, and the lower side goes to the capacitor. Connect a jumper wire between the bottom lead and the center lead. Figure 8-25 of the text shows the potentiometer connected in this manner. Adjust the potentiometer to its midrange position.
8. Turn on the power source and adjust it to produce 10 Vdc. Connect the VOM across C_1 and observe a cycle of operation. Is there any difference in the pulse rate displayed in comparison with that of step 4? ______________
Adjust the potentiometer to its full clockwise position and observe the pulse rate. Then adjust the potentiometer to its maximum counterclockwise position. How does the resistance of the potentiometer alter the operational cycle of the generator? ______________________.
9. Turn off the power source and disconnect the circuit.

DISCUSSION

In this part of the experiment you were asked to construct a pulse-generator circuit using a UJT. You were then asked to measure V_C, V_{B1}, and V_{B2}.

The capacitor voltage changed from a minimum value to a

maximum value on a gradual basis. The resulting voltage change produced a sawtooth waveform. The frequency of the waveform was very slow, so that it could be observed on the VOM. Essentially, the V_C voltage will rise in value until it reaches the intrinsic standoff ratio voltage that causes triggering of the emitter. When triggering occurs, C_1 discharges through the E-B_1 junction and drops to a minimum value very quickly. This produces the fall time of the sawtooth waveform.

The measured V_{B1} and V_{B2} voltage values show when conduction occurs in the pulse-generator circuit. The resulting pulse has a very short duration. It occurs only during the time when I_E triggers the UJT into conduction. The V_{B1} pulse has a positive polarity, and the V_{B2} pulse has a negative polarity. These two output pulses are used to trigger external circuits. The two output pulses are completely independent of the sawtooth waveform voltage developed across C_1.

Resistor R_1 of the sawtooth voltage circuit was then replaced with a potentiometer, which was used to alter the pulse rate of the output. When the potentiometer is turned fully clockwise, the pulse rate occurs very quickly. When it is adjusted in the counterclockwise direction, the pulse rate occurs at a slow rate. This addition to the circuit causes the pulse-generator output to be variable.

If an oscilloscope is available, the voltage measurements of this activity could be observed on a screen instead of being measured with a VOM. Steps 4 through 8 could be performed with the oscilloscope. The shape of the sawtooth waveform would be very obvious. The shape and polarity of the output could be readily observed by the same procedure. Waveforms are more meaningful when observed on an oscilloscope.

ANALYSIS

1. At what value does the capacitor voltage cause triggering of the UJT?
2. How does a change in the resistance of R_1 alter the pulse rate of the UJT generator?
3. What is the polarity of outputs V_{B1} and V_{B2}?

UNIT 8 EXAMINATION
THYRISTORS

Instructions: Circle the response that most correctly completes each statement.

1. Thyristors are best described as:
   a. Power supply rectifiers
   b. Amplifying devices
   c. Pulse-generating devices
   d. Voltage-controlled switches
2. How many P-N junctions are there between the anode and cathode of an SCR?
   a. One
   b. Two
   c. Three
   d. Four
3. An SCR energized by dc is triggered into conduction. The conduction can be turned off by:
   a. Momentarily disconnecting the gate
   b. Reducing the positive gate voltage
   c. Applying negative gate voltage
   d. Lowering the holding current below a minimum level
4. An SCR energized by ac is triggered into conduction. This conduction will occur for:
   a. The entire ac cycle of operation
   b. Only the positive alternation
   c. Only the negative alternation
   d. Part of each alternation
5. An ohmmeter connected across two leads of an SCR shows low resistance in one direction and high resistance in the opposite direction. This represents:
   a. The anode-cathode leads
   b. The anode-gate leads
   c. The gate-cathode leads
   d. A faulty response for an SCR
6. The best method of triggering an SCR into conduction is accomplished by:
   a. Increasing the forward breakover voltage
   b. Increasing the peak reverse voltage

c. Increasing the gate current
d. Decreasing the gate current

7. Increasing the gate current of an SCR causes:
a. A reduction in the forward breakover voltage
b. An increase in the peak reverse voltage
c. A lowering of the holding current level
d. The forward breakover voltage to increase in value

8. A device that is triggered into conduction with either a positive or negative gate current is:
a. A triac
b. A diac
c. An SCR
d. A UJT
e. A PUT

9. Before an SCR is triggered into conduction, the V_{AK}:
a. Equals the source voltage
b. Is less than the source voltage
c. Is zero
d. Is the source voltage less the voltage drop across the series load resistor

10. The gate of an SCR loses control after being triggered into conduction because it:
a. Loses the applied voltage
b. Is reverse biased
c. Has a large amount of current flow
d. Has an I_{AK} that is great enough to cause continuous conduction

11. When an SCR goes into conduction, the current needed to cause continuous conduction is called the:
a. Trigger current
b. Breakover current
c. Reverse breakdown current
d. Forward current
e. Holding current

12. An ohmmeter connected across T_1 and T_2 of a good triac will:
a. Show high resistance in either direction of polarity
b. Show low resistance in either direction of polarity
c. Show low resistance in one direction and high resistance in the opposite direction
d. Not tell anything about the condition of an SCR

13. A triode bidirectional conduction device is:
a. An SCR
b. A triac

c. A UJT
d. A PUT
e. A silicon-controlled switch

14. An ohmmeter connected across two leads of a triac shows the same low resistance in both directions of polarity. This represents:
a. T_1 and T_2
b. T_2 and G
c. G and T_1
d. G and T_2
e. A faulty response for the triac

15. A device that is triggered into conduction by either a positive or negative source voltage change is:
a. An SCR
b. A triac
c. A diac
d. A UJT
e. A PUT

16. The operation of a conductive triac is best described by saying that the T_1-T_2 voltage is:
a. High, and the load current is low
b. High, and the load current is high
c. Low, and the load current is low
d. Low, and the load current is high
e. Zero with maximum load current

17. If a silicon diode is used in the gate circuit of a triac power control circuit instead of a diac, the:
a. Circuit will not work
b. Diode will be damaged
c. Output will not be effectively changed
d. Gate will be controlled by dc gate current
e. Output will be ac with reduced delay capabilities

18. An ohmmeter connected across T_1 and G of a good triac will:
a. Read high resistant in either direction of polarity
b. Read low resistant in either direction of polarity
c. Read low resistant in one direction and high resistant in the opposite direction
d. Not tell anything about the condition of the junctions

19. The most sensitive triggering polarities of a standard triac occur when:
a. G and T_2 are the same polarity
b. G and T_2 are of opposite polarity

c. G and T_1 are of the same polarity
d. G and T_1 are of the opposite polarity

20. If the intrinsic standoff ratio of a UJT is 0.45, and the applied V_{BB} is 15 V, the amount of V_E needed to produce triggering is in the range of:
a. 1 to 3 V
b. 3.1 to 6 V
c. 6.1 to 8 V
d. 8.1 to 10 V
e. 10.1 or above

21. An ohmmeter connected across two terminals of a UJT shows a resistance of 5 kΩ in either direction of polarity. This represents:
a. B_1 and B_2
b. B_1 and E
c. E and B_2
d. A faulty condition for a UJT

22. Triggering or conduction of a UJT occurs when the emitter is:
a. Positive with respect to B_1
b. Negative with respect to B_1
c. Zero with respect to V_{BB}
d. Positive with respect to B_2
e. Negative with respect to B_2

23. In operation, a nontriggered UJT responds as:
a. An open circuit
b. An infinite resistance
c. A low resistant diode
d. A 4- to 10-kΩ resistance

24. The negative resistance region of a UJT refers to a characteristic that occurs:
a. Before it is triggered
b. Between the peak and valley points after triggering has occurred
c. Beyond the valley point after triggering
d. In the saturation region

25. A programmable unijunction transistor is a three-junction thyristor that is triggered into conduction when the gate is:
a. Less positive than the anode
b. More positive than the anode
c. More positive than the cathode
d. More negative than the cathode

UNIT NINE

Integrated Circuits

UNIT INTRODUCTION

Electronic circuits have become more and more sophisticated with the development of solid-state electronic technology. The microminiaturization of components and circuits has revolutionized the electronics industry in the last few years. Development of the printed circuit board, packaged-circuit modules, and the invention of the transistor have all paved the way for the microcircuits that we use today.

Space-age technology and military demands for small electronic devices have required that electronics circuitry become smaller and smaller. A computer the size of a book can now outperform and do more work faster than earlier computers, which required the space of a large room. Because of developments now being made in microelectronics, applications are practically limitless.

The expression *integrated circuit,* or *IC,* is a generic term used to describe a group of small electronic components that are constructed and permanently interconnected on or in a piece of semiconductor material. The Institute of Electrical and Electronic Engineers defines an IC as "a combination of interconnected circuit elements inseparably associated on or within a continuous substrate." The semiconductor material serving as the foundation on which integrated circuits are made is called a *substrate.* ICs are often described as microcircuits or microelectronic devices.

UNIT OBJECTIVES

Upon completion of this unit, you will be able to:

1. Describe the physical construction techniques used to develop an integrated circuit.
2. Discuss the advantages and disadvantages of integrated circuits.
3. Explain the construction differences between various families of integrated circuits.
4. Describe the differences between linear and digital ICs.
5. Recognize the basic types of IC packages in use today.
6. Look at IC component data and recognize how they are used to achieve a specific application.

IMPORTANT TERMS

Before reading Unit 9, review the following terms. These terms provide a basic understanding of terms associated with integrated circuits.

Active devices Electronic components that change or alter an applied signal. Transistors and diodes are typical solid-state active devices.

Chip Part of a wafer of silicon upon which the components of an IC are fabricated.

Complementary symmetry A transistor circuit configuration that has similar characteristics but opposite construction, such as one N-channel and one P-channel MOSFET connected together.

Diffusion A process by which the atoms of one material are absorbed or move into another material when subjected to a controlled atmosphere.

Dual in-line package An IC housing that has leads formed in a row on each side of the package. The abbreviation *DIP* is commonly used.

Evaporation process A fabrication process in which materials are vaporized and deposited on the surface of another material. Epitaxial growth is an evaporation process.

Gate A circuit that performs special logic operations such as AND, OR, NOT, NAND, and NOR.

Hybrid IC An integrated circuit that combines thin- or thick-film components with monolithic-constructed devices on an insulated substrate.

Logic A decision-making capability of an electronic digital circuit.

Monolithic IC An IC construction procedure in which all circuit elements are formed and interconnected on or within a single piece of silicon.

Passive devices Resistors, capacitors, and inductors that can effectively pass or transfer electrical energy from one point to another in an electric circuit.

Thick film A monolithic assembly in which the passive circuit parts and interconnections are formed on an insulating material substrate by screen printing. The active components are added in chip form.

Thin film A monolithic assembly in which the passive circuit elements and interconnections are formed by evaporation onto a substrate. The active devices are added as discrete components.

VLSI chips Very large scale integration; refers to chips that have a component density of more than a thousand components.

Yield A percentage ratio of the number of good devices produced to the maximum number of possible devices that can be produced in a production run.

IMPORTANCE OF INTEGRATED CIRCUITS

Integrated circuits (ICs) are actually microminiature circuits. Each may contain resistors, conductors, semiconductors, capacitors, and—in some cases—inductors in a single package. The components are classified as either active or passive. Active components are semiconductor devices such as transistors and diodes. Passive components are resistors, capacitors, and inductors. These components perform the same job that they do in a conventional circuit, but they do not have the same general appearance when housed in an IC. The National Semiconductor LM741C integrated circuit of Fig. 9-1, for example, has 20 transistors, 11 resistors, and a capacitor in its package. All of this is housed on a chip that is approximately $\frac{1}{16}$ in^2.

Integrated circuits have been on the market for several years. The IC was originally designed for the space industry and for military circuits. At first the cost of an IC was very high. It was common for an IC chip to sell for hundreds of dollars. Many ICs now sell for less than \$1, with many of the more sophisticated devices costing in the range of \$15. Microprocessor ICs often sell for less than \$50. Radio and television receivers use ICs in their construction. Complete AM-FM-stereo receivers are now built on a single IC chip. Only a few external components are needed to make the circuit functional. A number of consumer products are now utilizing ICs in their construction. Many manufacturers are making ICs for consumer products. Production cost of a particular chip is primarily determined by component density and construction processes. High-volume production of devices usually reduces the unit cost of a device. Specialized, low-production items generally cost a great deal more than the high-production items. The production cost of ICs has dropped rather significantly in recent years. New developments in photography and fabrication techniques have helped to reduce production costs.

Advantages of ICs

There are a number of advantages that integrated circuits have over standard electronic parts wired in a conventional circuit. These include:

1. *Size.* ICs are becoming so small that hundreds can be manufactured on a piece of material the size of a shirt button. The industry is predicting much smaller and more complex chips in the future. Chips with a component density of 1 million are available today.
2. *Cost.* The cost of an integrated circuit is based largely on its size and component density. By packing more components into less chip area, the cost per component is reduced. Additional savings can be realized when using ICs

Operational Amplifiers

LM741/LM741C operational amplifier

general description

The LM741 and LM741C are general purpose operational amplifiers which feature improved performance over industry standards like the LM709. They are direct, plug-in replacements for the 709C, LM201, MC1439 and 748 in most applications.

The offset voltage and offset current are guaranteed over the entire common mode range. The amplifiers also offer many features which make their application nearly foolproof: overload protection on the input and output, no latch-up when the common mode range is exceeded, as well as freedom from oscillations.

The LM741C is identical to the LM741 except that the LM741C has its performance guaranteed over a 0°C to 70°C temperature range, instead of −55°C to 125°C.

schematic and connection diagrams

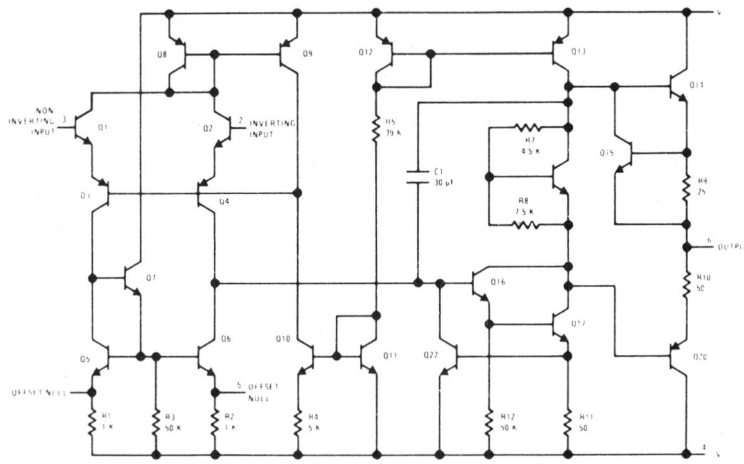

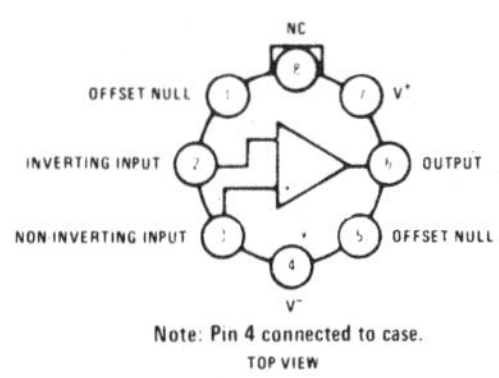

Order Number LM741H or LM741CH
See Package 11

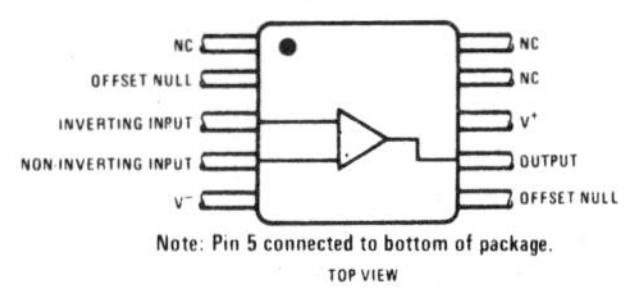

Order Number LM741F
See Package 3

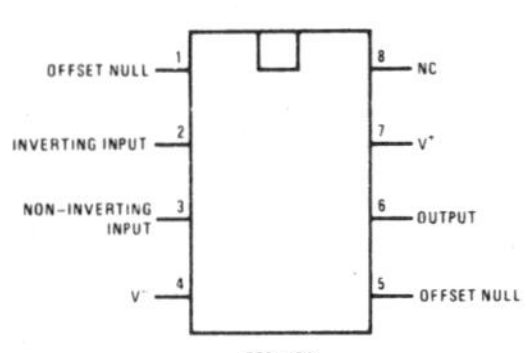

Order Number LM741CN
See Package 20

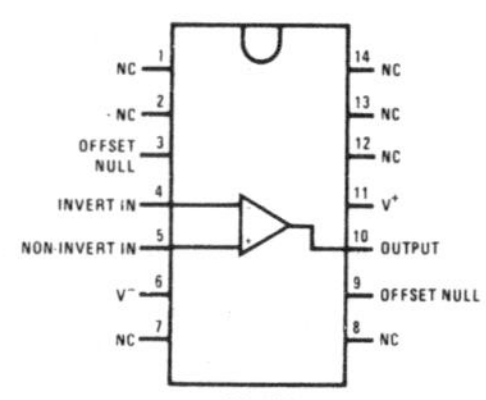

Order Number LM741CD
See Package 1
Order Number LM741CN-14
See Package 22

FIGURE 9-1 Data sheet of an LM741C. [Courtesy of National Semiconductor Corp.]

because fewer parts are needed to order, inventory, and assemble into an operating system.

3. *Reliability.* Perhaps the most important advantage of an IC over discrete component circuits is reliability. Increased reliability is due to a number of factors. The most significant of these is the reduction of interconnections between components. In earlier discrete-component circuits, up to 50% of circuit failures were related to interconnections between components. IC interconnections are achieved on the chip and formed at the same time during the manufacturing process. If the chip has interconnection problems they will be detected before it is placed in a completed package.
4. *Power consumed.* The small size of an IC makes it well suited for low-power operation. The closeness of components in the structure reduces stray electrical pickup, allowing very small signal operation. Low-power operation means lower internal temperature rise, which improves reliability.
5. *Ease of maintenance.* Since the IC is actually a circuit containing many components, servicing can be achieved by plugging in new chips to replace faulty ones. The faulty chip is then discarded, since its repair is not practical.

Integrated Circuit Limitations

In spite of the numerous advantages of ICs, they have some very definite limitations, such as low power, low voltage, and limited component selection. These limitations can be altered to some extent by more costly manufacturing techniques and circuit design changes. IC limitations are often considered to be some of the trade-offs used to accomplish unique advantages.

The power limitation of an IC is primarily due to its physical size. The current-handling capability of a device determines its heat production. Heat concentrated in a small component may produce temperatures great enough to destroy the device. The size advantage of an IC is actually a trade-off for its current-handling capacity. A large number of the ICs being manufactured are used only to process information rather than control power. The data output of an IC is often used to control a signal applied to a discrete transistor that is used to control high values of power.

The voltage used by ICs must be kept very low because of the weak insulation between circuit elements. Typical voltage ratings are in the range of 5 to 30 V. If this voltage is exceeded, it generally causes insulation breakdown in some part of the chip. IC voltage values limit applications to signal processing or data manipulation. Although new developments in IC manufacturing have reduced this limitation to some extent, ICs are still considered to be low-voltage devices.

The component limitation of an IC is primarily due to the use of silicon in its construction. Silicon is ideal for the fabrication of diodes and transistors, but it does not work very satisfactorily for other components. Resistors, for example, cannot be fabricated to very precise values with silicon. Resistance value tolerances are very high when silicon is used in their production. As a rule, higher resistance values occupy more space than small values. A 40-kΩ resistor, for example, takes the same amount of space as that needed for five transistors. Capacitance poses an even more significant space problem in IC design than a resistor. Capacitance of any size occupies a rather significant amount of space. A 20-pF capacitor will take the same space needed to fabricate 10 or more transistors. In general, most ICs do not employ capacitors except for those of an extremely small value. Similarly, inductors and transformers are almost impossible to construct on an IC chip. Component limitations are one of the serious drawbacks of IC utilization. Unique circuit designs have solved these limitations to some extent.

Another limitation of the IC is based on the repair of its components. When one specific component becomes defective, it cannot be replaced. This means that the entire IC must be replaced when only one component becomes faulty. The large number of good components that remain on the faulty chip cannot be recovered and used for other applications. The faulty chip must be discarded and replaced with a good one in order to make the circuit functional again. This disadvantage is not as bad as it may first appear. First, the process of locating a faulty IC is much easier than locating a faulty component in a complex circuit. As a rule, the additional cost of a new IC is offset by reduced labor costs in the servicing procedure. This simplifies the servicing procedure and reduces the down time of equipment. It also reduces the equipment and spare-part inventory.

When all factors are considered, the advantages of an IC outweigh the disadvantages. In general, circuits that have a great deal of redundancy or duplication are best suited for IC applications. Powerful digital computers are a classic example of this trend. Complex circuits that contain a large number of discrete components are also well suited for IC design considerations. The dollar value of ICs sold today far exceeds the value of discrete circuit components. The IC is truly the device of the future.

SELF-EXAMINATION

1. ICs are built on a ____________.
2. ICs are actually ____________ circuits built on a tiny chip.
3. Size, cost, reliability, power consumption, and ease of maintenance are some of the unique ____________ of an IC.

4. When ICs are constructed in large quantities, the production ______ is less than that of an equivalent discrete component circuit.
5. ICs generally require low ______ and ______ to operate.
6. When a single component of an IC becomes faulty, the entire chip must be ______.
7. The electrical ______ consumed by an IC is generally quite low.
8. Most ICs are designed to process ______ instead of controlling power.
9. The operational voltages on an IC are in the range of ______ to ______ volts.
10. The internal circuitry of an IC relies very heavily on the use of ______ and ______.

ANSWERS

1. Substrate	2. Small
3. Advantages	4. Cost
5. Voltage, current	6. Discarded
7. Power	8. Information, data, or signals
9. 5 to 30	10. Diodes, transistors

IC CONSTRUCTION

The technology used in the construction of ICs is not particularly new to the solid-state field. We discussed such things as substrates, epitaxial growth, and diffusion in units dealing with diodes and transistors. IC technology is an extension of the manufacturing techniques used in bipolar transistors and FETs. In fact, these two devices are the construction basis of many ICs today. This technology is continually improving, which permits extremely large numbers of components to be fabricated on a single chip. The technology is very basic, but the manufacturing techniques used to achieve a specific chip may be held in strict secrecy. As a result, we will only look at some of the basics of IC fabrication.

The making of an IC usually starts with a piece of circular silicon called a *wafer.* The physical size of the wafer is either 1, 2, 4, or 6 in. in diameter with a thickness of 0.006 in. The wafer serves as the substrate of the IC. Only a small portion of the wafer is needed for one chip. Many ICs are formed simultaneously on one wafer. The ICs are independent chips but of the same type. After individual ICs are formed, the wafer is cut into small squares or chips. Figure 9-2 shows some examples of this process. Each chip

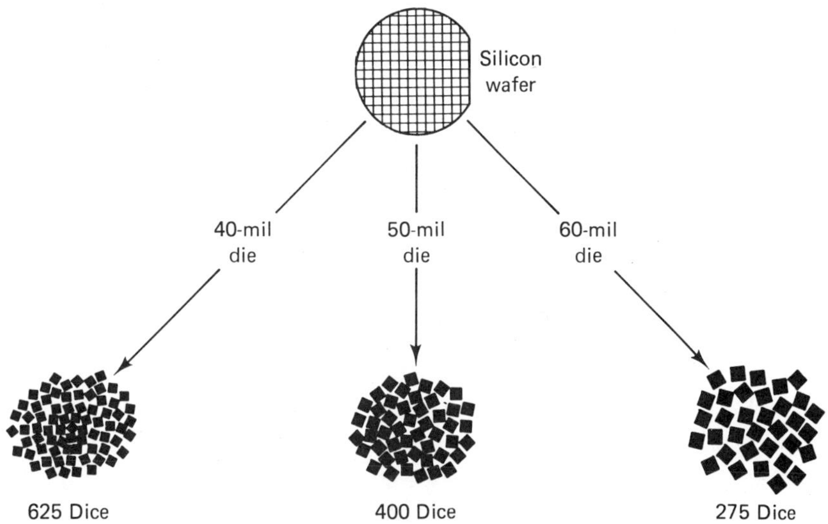

FIGURE 9-2 Cutting wafer into individual ICs.

in this case is a complete IC. It may contain 50 to 60 interconnected components. After the individual chips are tested, each is placed in a package or housing.

The number of useable IC chips developed from a production run is called the *yield.* IC yield is usually expressed as a percentage of the total possible number of chips that can be developed in a production run. The yield of a complex IC may be as low as 25%, which means that 75% of the chips in a particular run may be faulty or may not meet specifications. These chips are discarded and the materials reclaimed.

Monolithic ICs

The term *monolithic* describes the structure of an IC in which all elements of the circuit are built in a single crystal of semiconductor material. "Mono" refers to something of a single structure, and "lithic" refers to a printing process called lithography that refers to printing with a stone. Monolithic therefore refers to the silicon-diffused planar process in which all process steps are grown or formed on the surface of one silicon wafer. All contacts or connections to various components are made on the same surface. Interconnections between different components of the chip are made by depositing a metal wiring pattern on the silicon dioxide–covered surface of the wafer.

The shapes of individual components of a monolithic IC are designed so that they can all be formed simultaneously. This includes transistors, diodes, resistors, and capacitors. Specific elements of a complete circuit can therefore be diffused on a single wafer by using the same processes.

A monolithic IC is made by evaporating, depositing, and etching materials on a piece of silicon. You all have taken a hot shower and noticed a layer of water that has evaporated and then condensed on a mirror or glass door. This process is similar to the evaporating process used by IC manufacturers to deposit the vapors of such materials as chromium, nickel, or silicon dioxide on a substrate. The evaporating process develops a thin layer of N-type silicon on the surface of a P-type silicon substrate. Figure 9-3(a) shows an example of the evaporating process. This process was described as epitaxial growth in Unit 3. Epitaxial growth and evaporation achieve the same thing in IC construction.

After an N-type layer of material is formed on a P-type substrate, the wafer is placed in an oven that has a controlled atmosphere. This causes the N-type material to diffuse or be absorbed by the P-type silicon substrate. This can be compared with the absorbing of water by a dry sponge or paper towel. The towel is obviously many times larger than the silicon chip. The completed chip can easily fit into the lowercase *o* of this printed page.

The amount of heat used and the length of time that the chip is exposed to the heat determine the depth of diffusion. The area of diffusion is determined by the area of the N- or P-material deposited on the substrate. The addition of the N-material to the P-substrate material when completed is shown in Fig. 9-3(b).

Bipolar ICs

Component construction of a monolithic IC is used to determine the classification or grouping of a particular chip. Bipolar ICs utilize the same technology that is used in the construction of a bipolar transistor. The bipolar transistor is the fundamental component of this family of ICs.

N- and P-materials can be evaporated or grown on certain areas of a diffused N-material to make a bipolar transistor. Figure 9-4 shows an NPN bipolar transistor with the emitter, base, and collector leads exposed. Note the different levels of diffusion needed to form the transistor. Each deposition layer and diffusion operation requires special processing.

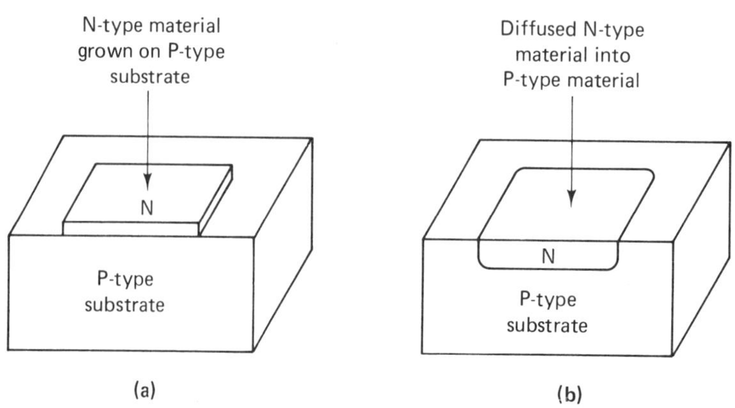

FIGURE 9-3 Epitaxial growth process: (a) N-type material grown on P substrate; (b) sectional view of diffused area.

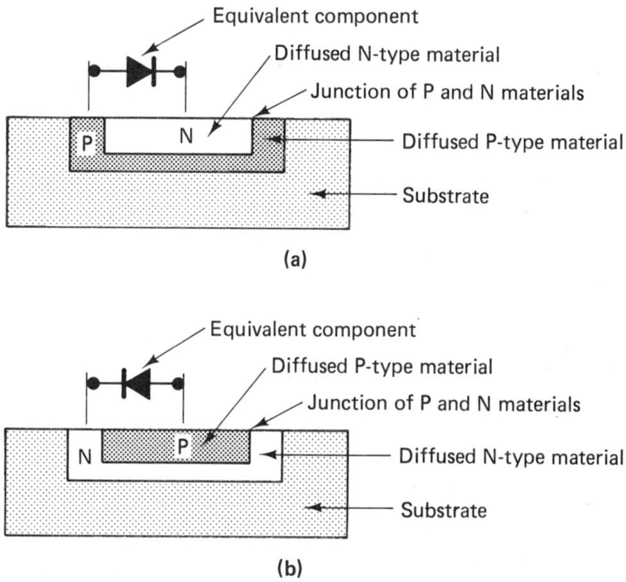

FIGURE 9-4 PN junction (diode) formed in substrate: (a) diffusing N impurities into P material; (b) diffusing P impurities into N material.

Diodes are made by diffusing N-type impurities into a portion of the diffused P-type material. By adding enough N-type impurities into the P-type material, the N-type impurities can dominate and produce an N-type semiconductor in a specific area. Figure 9-4 shows the crystal formation of a diode on a substrate. Metallized leads are attached to the respective material areas to achieve the diode structure. These may be connected to outside terminals or interconnected with other components on the chip.

Resistors are made from doped P-material having a certain internal resistance. The area of the P-material and the doping level of the material determines its resistance. If a small resistance value is required, a relatively large number of impurity atoms are diffused into the substrate. Conversely, larger resistances require fewer impurities. Figure 9-5 shows a P-material resistor constructed on a substrate. The metallic contacts attached to the P-material are deposited by a special process called *metallization.* In general, the metal must not be permitted to diffuse into the N- or P-material. It is used only to make an electrical connection or to interconnect components.

Capacitors are made by using a layer of silicon dioxide as the dielectric material. A diffused N-layer serves as one plate, and a metallized layer covering the layer serves as the other plate. Figure 9-6 shows a representation of an IC capacitor. The area of the material and the thickness of the silicon dioxide layer determine the value of the capacitance.

Leads are connected to the IC chip by a very high temperature solder. A sectional view of a typical IC is shown in Fig. 9-7. The leads are attached at points called *bonding pads.* In the illustration these connection points are labeled "evaporated jumper

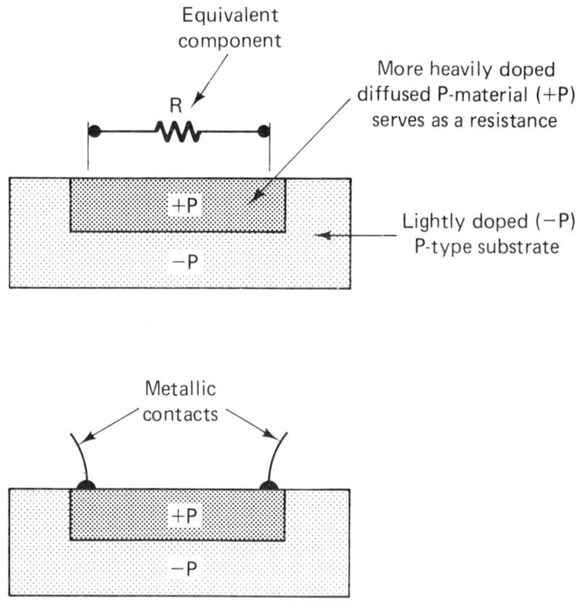

FIGURE 9-5 Resistor formation by diffusion of semiconductor material.

connections." Note the location of the capacitor, the transistor, and resistors in this sectional view. The oxide layer serves as an insulating material and a dielectric for the capacitor. This chip drawing shows only a few necessary components needed to make a complete IC.

MOS Integrated Circuits

The monolithic IC just discussed uses bipolar transistors in its construction. Bipolar ICs represent an important division of the IC market. A second major division of monolithic ICs is the metal-oxide semiconductor field-effect transistor or MOSFET. This class of ICs is being used in many applications. It is less expensive than the bipolar devices because it requires only one diffusion step to form the source and drain. A bipolar chip requires three and sometimes four diffusion steps to form a transistor. The MOSFET chip does not require isolation between the source-drain and other PN

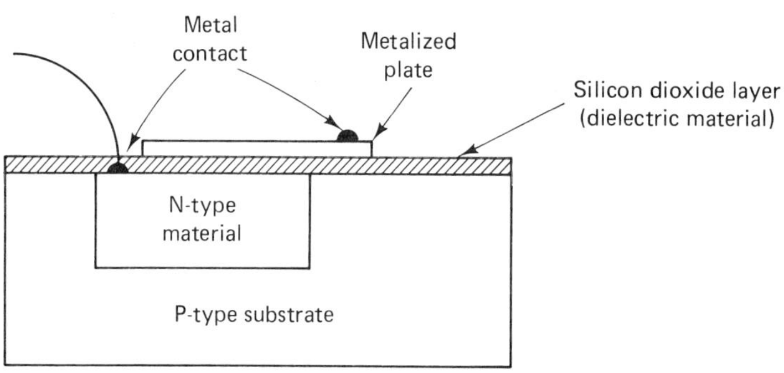

FIGURE 9-6 Representative IC capacitor.

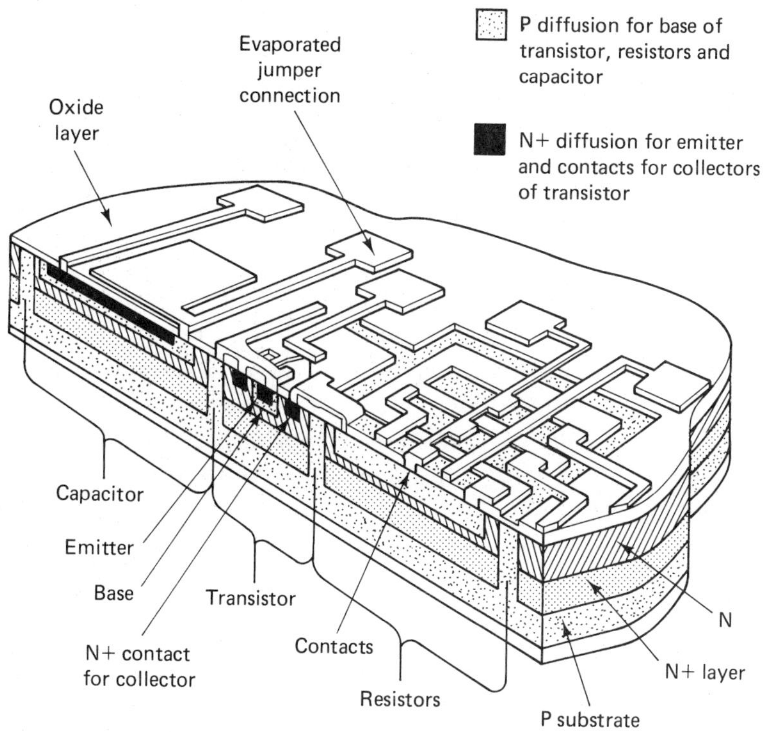

FIGURE 9-7 Sectional view of an integrated circuit.

junctions formed on the substrate. Figure 9-8 shows where two P-channel enhancement-MOSFETs are formed on the same substrate.

The package density of a MOS IC is very high compared with that of a bipolar IC. This means that more devices can be fabricated in a given area of the chip. As a rule, component density of a MOS IC is 10 or more times that of the bipolar IC. Isolation between the elements is one reason for the increased component density. MOS resistors occupy less than 1% of the space needed for a conventional diffused resistor. This high-density construction capability makes the MOS IC an extremely important device in complex ICs.

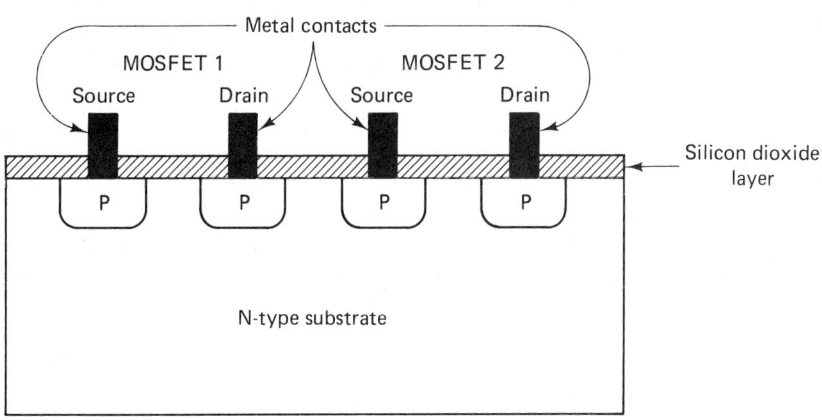

FIGURE 9-8 MOS IC formation on a substrate.

MOS ICs utilize three basic components in their construction, namely, MOSFETs, MOS capacitors, and MOS resistors. A resistor is formed by simply using an E-MOSFET with its gate connected to the drain so that the device is biased in the on state. The channel of the device then serves as a resistor whose value is controlled by its size and doping level.

The principal disadvantage of a MOSFET IC is its operating speed. This is because the MOS transistor is a high impedance device and it cannot charge the stray circuit capacitance very quickly. The operating speed problem applies primarily to only ultra high speed applications. The advantages of low cost, low power consumption and high-density packaging make the MOS IC a very attractive device for very large scale integration (VLSI) chips. VLSI chips contain over one thousand components on a chip. These chips are used in calculators, memory, and microprocessors.

Film ICs

Film ICs are an extension of the earlier discrete-component electronic circuits. These ICs are an assembly of components formed on the surface of an insulating substrate. Conductors and resistors are formed by evaporation and by screen-printed deposition areas. Transistors, diodes, and MOS capacitors are added to the circuit in discrete form. These components are attached to the circuit with a special cement. Film ICs are classified as *hybrid* devices. Hybrid refers to something that is the outgrowth of two different types or styles. In this case, hybrid is used to denote an IC that employs monolithic and film construction in its fabrication. This grouping of ICs is either of the thin-film or thick-film type.

Thin-film ICs are so named because conductors, resistors, capacitors, transistors, and diodes are formed as very thin films on a substrate. The film is in the order of a thousand angstroms thick. An angstrom is 10^{-10} meters. The thin films are deposited on substrates of glass or alumina through a vapor-deposition process. Resistors are formed by depositing nichrome, tantalum, or tin oxide in thin strips on a glass substrate. Resistance values can be effectively controlled by varying the length, width, or thickness of the deposited material. Values can be made from 10Ω to 1 MΩ. Conductors are made from a thin layer of deposited gold-nichrome. Transistors, diodes, and capacitors are generally added to the circuit in chip form. These components are secured to the substrate by a conductive epoxy. A thin-film IC structure is shown in Fig. 9-9.

Thick-film ICs are hybrid devices that are formed on films in the order of a mil or 0.001 inch by printing or silk screening conductor patterns onto the substrate. The printed material is a mixture of pulverized glass and aluminum. A wide range of resistor values and conductor path designs can be by this type of construction. Active semiconductor components are attached to the structure in discrete form. These components are attached to the chip

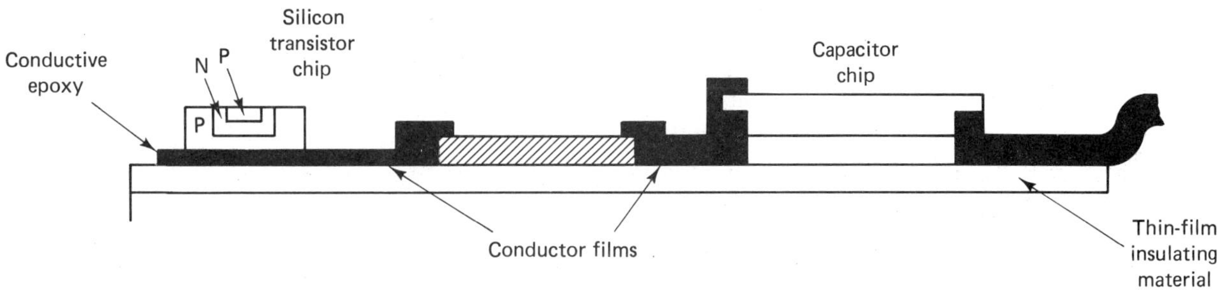

FIGURE 9-9 Thin-film IC construction.

by conductive epoxy. Thick-film ICs are similar in many respects to the thin-film type of IC.

IC Packages

The outside appearance of an IC is a major concern to the user. In this regard, there are a number of different package types and styles used to house ICs. Each package has unique advantages and in some cases disadvantages over the other types. Figure 9-10 shows some of the more popular packages for ICs. Where small size is important, a package known as the flat pack is used. This package is 0.26 inch long, 0.15 inch wide, and 0.05 inch thick. The assembled unit is hermetically sealed with metal to glass seals for each lead. The leads of this device are soldered into the circuit. A second form of housing is the TO-5 package. This package is similar to that used to house transistors. A very popular form of IC housing is the dual in-line package or DIP, which has a row of leads on each side of the package. Some of the most popular ICs are housed in DIPs and have 8, 14, 16, 24, 48, and 64 pins. The leads are on 0.1-in. centers to match mounting holes in sockets or circuit boards. A 14-pin DIP is 0.8 in. long, 0.25 in. wide, and 0.2 in. thick. The molded plastic housing is suitable for most circuit applications. Ceramic units are available for applications where

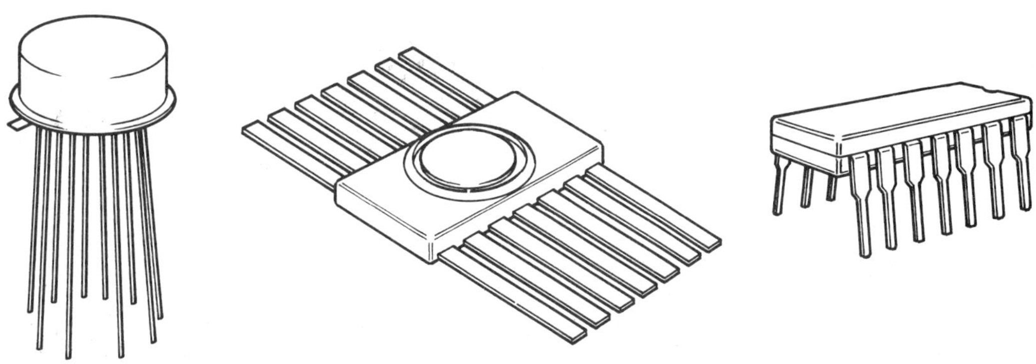

(a) (b) (c)

FIGURE 9-10 Common IC packages: (a) TO-5 housing; (b) flat pack; (c) dual in-line package.

temperature is a problem. Approximately 80% of all IC packaging is done by DIPs today.

A rather new trend in packaging has come about with the need for more pin connections and reduced size ICs. Such names as small-outline integrated circuits (SOIC), plastic leaded chip carrier (PLCC), ceramic leaded chip carrier (CLCC), leadless ceramic chip carrier (LCCC), and pin-grid array (PGA) are used to describe these packages. These new housing techniques permit ICs to be packaged in smaller units and have more connection leads. Figure 9-11 shows some of the new IC package styles. As a rule, these packages can house more pin connections because they are placed on all sides of the device. This tends to reduce the size of the package so that it more closely conforms to the actual size of the chip. DIP packages for example, are approximately 50 times larger than the chip that they house. The new packages are smaller and can accommodate more leads.

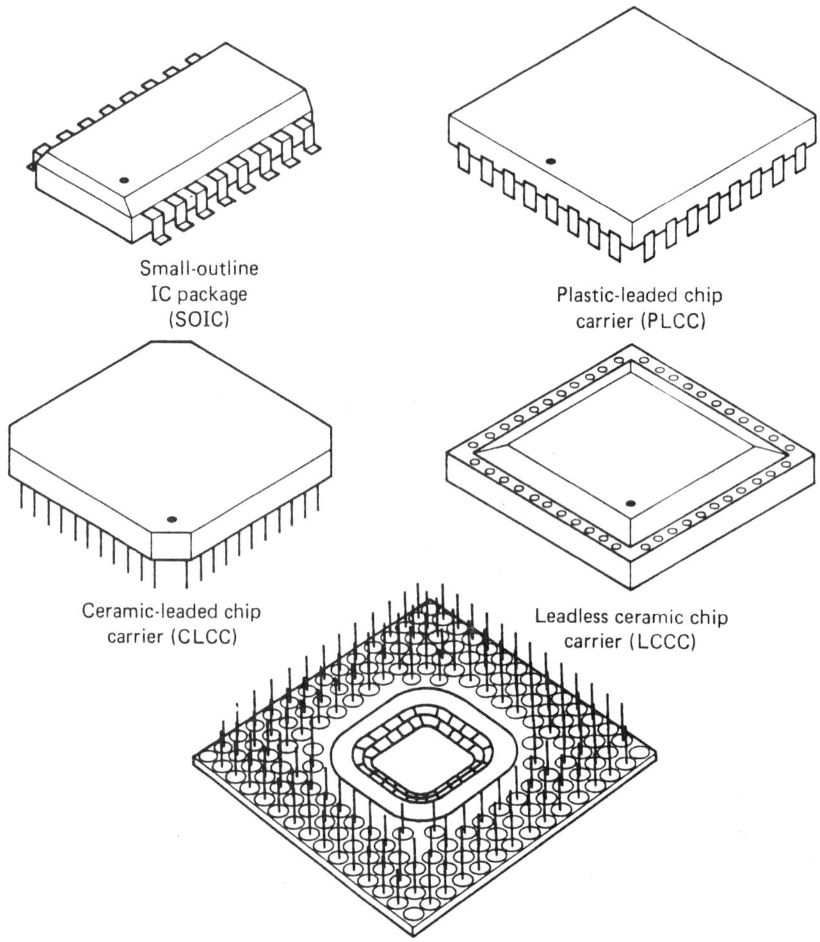

FIGURE 9-11 New IC package styles.

SELF-EXAMINATION

11. Construction of an IC starts with a circular piece of silicon called a ______.

12. The material or foundation on which an IC is constructed serves as the ____________ of the chip.
13. The small square of material that houses an IC is called a ____________.
14. The percentage of good chips developed from a production run is called ____________.
15. The construction technique that has components formed on a single piece of silicon is called ____________.
16. Through a process of ____________ a thin layer of material can be formed on a substrate.
17. ____________ is the process in which an evaporated or deposited layer of material is absorbed into another piece of material.
18. Metal contacts formed on deposited materials are achieved by a process called ____________.
19. Monolithic ICs that use bipolar transistors in their construction are called ____________ ICs.
20. Monolithic ICs that use MOSFETs as an element in their construction are called ____________ ICs.
21. Film ICs have components assembled on the surface of an ____________ substrate.
22. Today the most widely used package for housing ICs is the ____________.
23. ____________ ICs use monolithic and film construction in their fabrication.
24. The component density of an ____________ IC is much greater than that of a bipolar IC.
25. The value of a ____________ in a MOS IC is determined by length, area, and material of a channel.

ANSWERS

11. Wafer	12. Substrate
13. Chip	14. Yield
15. Monolithic	16. Evaporation
17. Diffusion	18. Metallization
19. Bipolar	20. MOS
21. Insulator	22. Dual in-line package or DIP
23. Hybrid	24. MOS
25. Resistor	

INTEGRATED CIRCUIT APPLICATIONS

There are two general classifications of ICs according to their function or application, *linear* and *digital* devices. Linear ICs develop an output signal that is proportional in some way to its input signal. Linear ICs are used to achieve amplification, regulation, and voltage comparisons.

Digital ICs contain switching circuitry whose input and output voltages are limited to two possible states of high and low. The output of a digital IC is related to the input in some logical way. Digital ICs include logic gates, flip-flops, counters, clock chips, calculators, memory, and microprocessors.

Linear and digital ICs are both widely used in modern electronic circuits. Each IC type has an important role to play in the control of an electronic circuit. Some newer ICs contain both digital and linear capabilities on a single chip. These hybrid chips will be more prevalent in the future when single-chip ICs will be used to accomplish complete system operations.

Linear ICs

Linear integrated circuits are a unique part of the IC family of electronic components. Linear amplification refers to the capability of an amplifier to operate in the linear or straight-line portion of the device's operating range. In discrete-component amplifiers, linear amplification is achieved by selecting an operating point near the center of the dynamic transfer characteristics. A bias operating point is selected near the center of the input characteristic. A signal applied to the input is increased in amplitude. The output must resemble the input with the only change being the size of the signal. Linear amplification is widely used in electronic communication circuits and home entertainment equipment.

Linear ICs are essentially self-contained amplifiers placed in an appropriate housing. These chips are usually housed in standard IC packages. Some modification of the DIP is used when a device is used for power amplification. The internal structure of the chip is generally quite small and rather complex. The physical size of the chip varies a great deal for different types of ICs. A triangle symbol is used to denote the amplification function. The point of the triangle is the output. The flat side with two leads is the input. A plus sign on the input lead denotes the noninverting input. The negative sign indicates the inverting input. All the components of the amplifier are located inside the triangle. Some chips have several complete amplifiers housed in the same enclosure.

A large number of the linear ICs in use today are called *operational amplifiers*, or *op-amps*. These devices have high ac and dc gain capabilities with stable operating characteristics. Op-amps are primarily used for small-signal amplification, waveform generation, and impedance matching. Op-amps were originally designed to perform a number of mathematical operations, such as summing, subtracting, averaging, differentiation, and integration. The amplification function of this device is an outgrowth of the original math operations.

The op-amp is essentially a high-gain amplifier that is designed to operate over a wide range of voltages. Its internal structure is small and quite complex. As a general rule, components connected outside the op-amp determine its operating capabilities. Gain can be 100,000 or more when the device is operated in an open-loop circuit. As a voltage follower, gain may be only 1. The output voltage cannot exceed the value of the supply voltage.

The fundamental operation of an op-amp is based on a differential amplifier circuit. A simplification of this circuit is shown in Fig. 9-12. Transistors Q_1 and Q_2 respond as the differential amplifier. Q_3 serves as a constant current source for the amplifier.

Normally, with no input signal applied to either Q_1 or Q_2, no output appears. If identical values of input are applied to both Q_1 and Q_2, there is still no output. In a sense, each transistor conducts the same amount. The output signal is therefore of the same value or balanced. There is no potential difference in the two outputs. This means that there is no output signal.

When different signals are applied to the input of Q_1 and Q_2, the output becomes unbalanced. The transistor receiving the greater signal will conduct more current. With current regulated by Q_3, the alternate transistor will conduct less current. A pronounced difference in the two outputs will occur. The resulting output is indicative of the conduction difference in Q_1 and Q_2.

If a signal is applied only to the inverting input of a difference amplifier, the output will be amplified and inverted. A signal applied only to the noninverting input will be amplified without in-

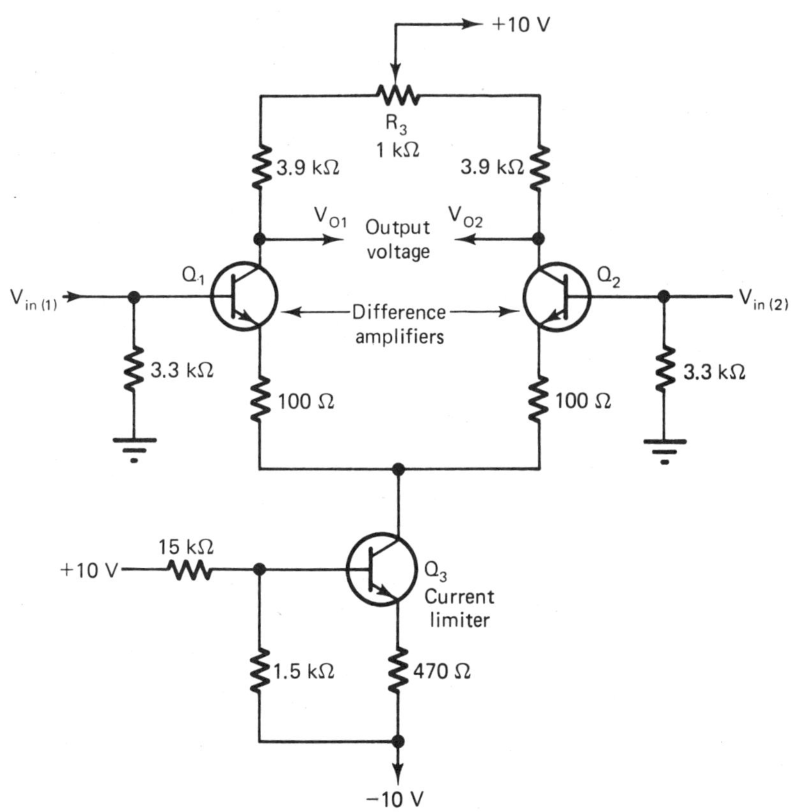

FIGURE 9-12 Transistor difference amplifier of an op-amp.

version. For the circuit to respond in this manner, the input not being used must be grounded. The output will be the difference between input and ground.

Op-amps have a number of other circuits in their construction. A high-gain amplifier, for example, follows the differential amplifier. This amplifier brings up the signal level so that very small input signals can be amplified. An emitter-follower amplifier comes after the high-gain amplifier. The emitter-follower provides the op-amp with a low-impedance output. This permits it to be connected to a variety of different output devices. The combination of all this makes the circuitry of an op-amp rather complex. As a general rule, the internal construction is not very important to a person using the device. Operation is usually controlled by the selection of a few external components.

Open-Loop Operation

Figure 9-13 shows a circuit diagram of an op-amp designed for open-loop operation. In this circuit configuration, gain is controlled by the strength of the input signal. When the input signal voltage is zero, the output voltage is zero. If an input signal is applied, the output voltage will rise accordingly. Voltage gains of 100,000 or more are typical. If the output voltage rises to the value of the source voltage, the op-amp is saturated. An input signal of only a few millivolts is needed to cause the op-amp to reach saturation.

Applications of the op-amp in an open-loop configuration are not very common. Very few circuits require a gain factor of 100,000. The input signal level for a circuit of this type must be extremely small. A simple calculation of voltage amplification demonstrates this limitation. The formula for voltage amplification is

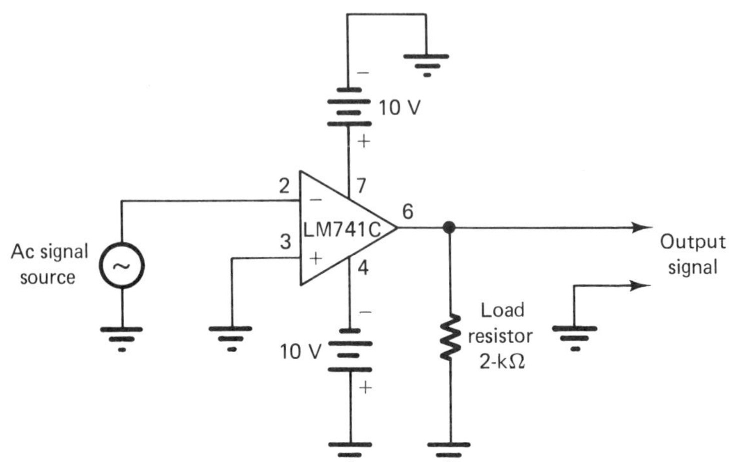

FIGURE 9-13 Op-amp connected for open-loop gain.

voltage amplification = voltage output/voltage input

or

$$A_v = \frac{V_{out}}{V_{in}}$$

For the circuit of Fig. 9-13, the A_v formula is transposed to determine the input voltage. The formula now becomes

$$V_{in} = \frac{V_{out}}{A_v}$$

We will assume, in this case, that an A_v of 100,000 is achieved by the op-amp. The output voltage (V_{out}) cannot exceed the value of the source voltage. In this circuit V_{CC} is 10 V. The input voltage value is

$$V_{in} = V_{out}/A_v = 10/100{,}000 = 0.0001 \text{ V or } 0.1 \text{ mV}$$

Saturation occurs when the input voltage reaches 0.1 mV. An ac signal of 0.1 mV p-p would also cause distortion of the output. This shows that the op-amp in the open-loop configuration is easy to distort. Closed-loop operation is generally used to alter this problem.

Closed-Loop Operation

When an op-amp is used in a closed-loop configuration the voltage gain can be adjusted to nearly any desired value. This is achieved by a feedback resistor network connected between the output and the input. Figure 9-14 shows two simplified op-amps connected to a feedback network. Part (a) shows the op-amp connected in a noninverting circuit. The output will be the same polarity as the input. Ideally, the two inputs are balanced when resistors R_1 and R_2 are equal. In an actual circuit, R_2 is a calculated value based on the parallel arrangement of R_1 and the feedback resistor R_f. Amplifier gain is based on the formula

voltage amplification
= input resistance + feedback resistance/input resistance

or

$$A_v = R_1 + \frac{R_f}{R_1}$$

For the circuit in part (a), the gain will be

$$A_v = R_1 + R_f/R_1 = 1 \text{ k}\Omega + 99 \text{ k}\Omega/1 \text{ k}\Omega = 100$$

If R_f is changed to 9.9 kΩ and R_1 to 100Ω, the gain would be reduced. How would the value differ?

Figure 9-14(b) shows the inverting circuit configuration. The input signal is applied to the negative terminal. The feedback network is composed of R_1 and R_f. R_2 returns the noninverting input

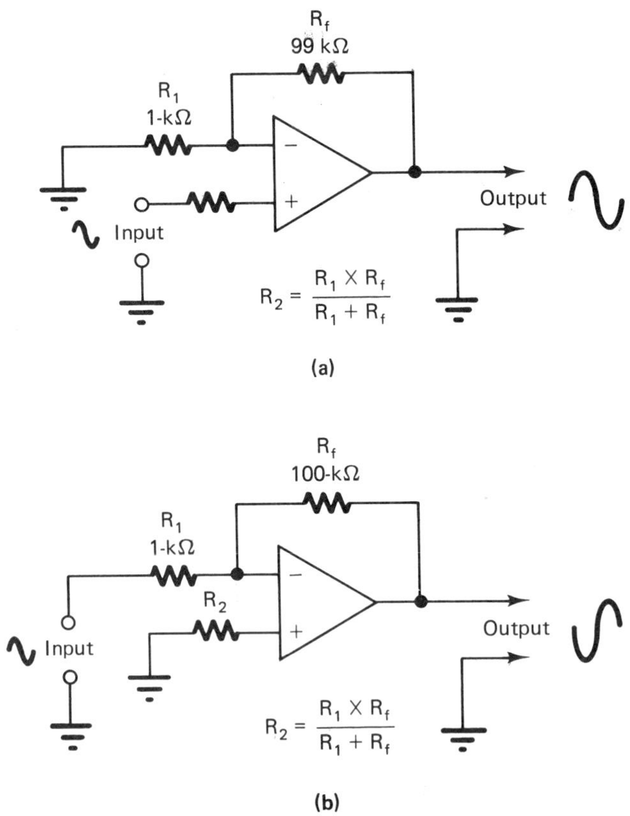

FIGURE 9-14 Closed-loop op-amp simplification: (a) noninverting amplifier; (b) inverting amplifier.

to ground. The value of R_2 is determined by the parallel value of R_1 and R_f. R_2 is generally smaller than R_1. The voltage gain of this amplifier is determined by the formula

$$\text{voltage amplification} = \frac{\text{feedback resistance}}{\text{input resistance}}$$

or

$$A_v = \frac{R_f}{R_1}$$

For the inverting op-amp of part (b) the gain will be

$$A_v = \frac{R_f}{R_1} = \frac{100 \text{ k}\Omega}{1 \text{ k}\Omega} = 100$$

Practical Op-Amp Circuits

Figure 9-15 shows a functioning op-amp circuit that is connected for noninverting amplification. The numbers near each lead are the pin numbers. The LM741 used in this circuit is housed in an eight-pin dual-in-line package. Notice that operating voltage is supplied from a split or divided power supply. Common leads on each side of the supply are connected to ground. The ground is also con-

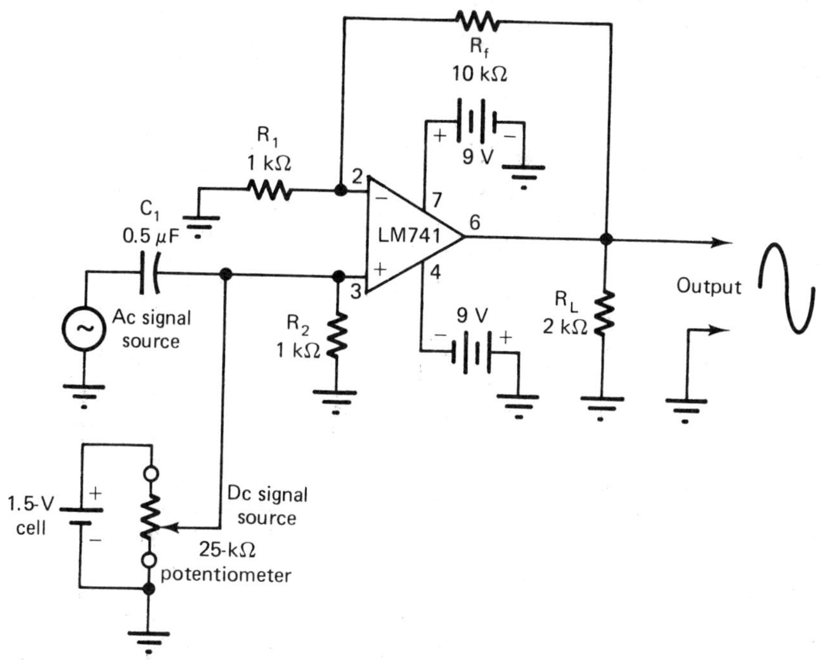

FIGURE 9-15 Noninverting op-amp circuit.

nected commonly to a number of circuit points. The output of this circuit cannot exceed 10 V dc or 20 V p-p ac. The gain achieved by this circuit closely follows the calculated values of the feedback network. When the circuit is made operational, gain and signal phase can be observed at the input and output with an oscilloscope. Input and output voltages can also be measured with a voltmeter.

A practical inverting op-amp circuit is shown in Fig. 9-16. This amplifier has the same gain capabilities as the noninverting circuit. Note the connection differences on the input circuit. The operational waveforms, signal phase, and voltage values can be observed on an oscilloscope and measured with a voltmeter. Op-amp circuits of this type are used widely today.

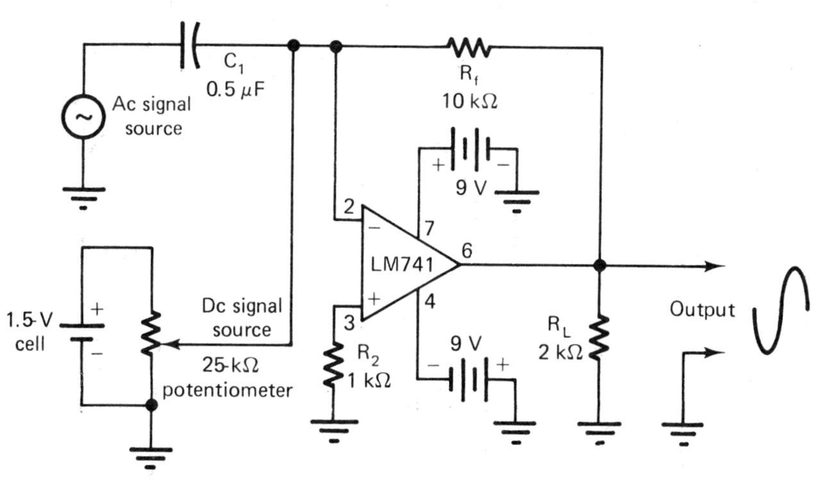

FIGURE 9-16 Inverting op-amp circuit.

Linear IC Voltage Regulators

A rather recent innovation in power supply circuitry is the linear integrated-circuit regulator. This IC accepts an unregulated dc input and develops a regulated output. Some IC regulators are designed to deliver one fixed output voltage. A typical series of regulators is available in values of 5, 8, 12, 15, 18, and 24 V. Adjustable regulated voltage outputs can also be obtained with the addition of a few extra components. Current output of 1 ampere (A) or more can be delivered to a load by a representative IC of this type.

The internal structure of an IC regulator is somewhat complex. In general, it contains a series-pass transistor, reference voltage source, feedback amplifier, and short-circuit protection. Figure 9-17 shows a block diagram of the internal structure of a representative three-terminal linear voltage regulator.

Operation of the three-terminal regulator is based on the ap-

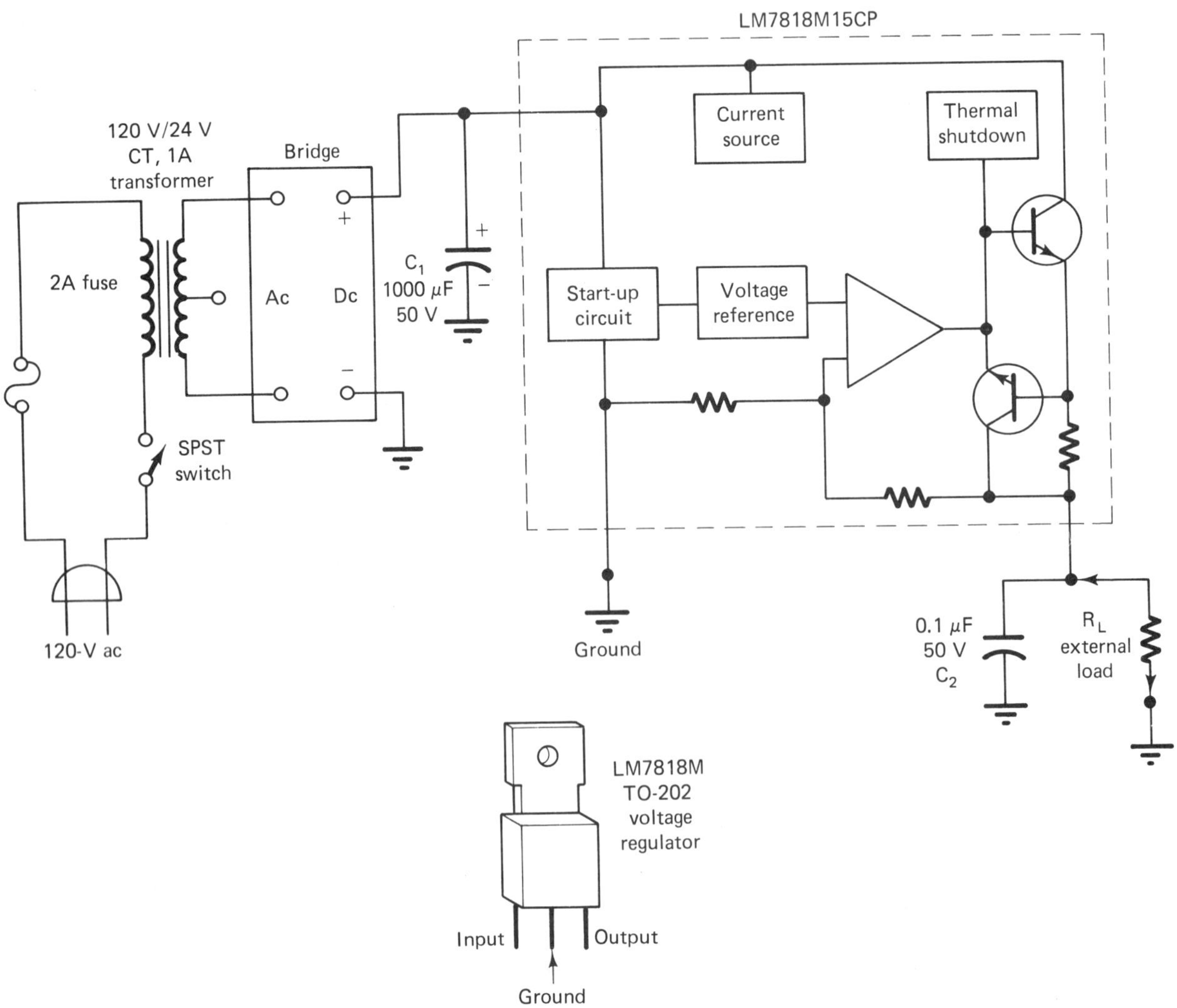

FIGURE 9-17 Three-terminal IC regulator.

plied input voltage and developed output voltage. Output voltage is compared with a reference voltage by a high-gain error amplifier. Any difference in voltage will be sensed by the error amplifier. The output of the error amplifier then controls conduction of the series-pass transistor. This transistor responds as a variable resistor. It changes inversely with the load current to maintain the output voltage at a constant value. An increase in output voltage would cause a corresponding increase in the conduction of the series-pass transistor. This in turn would cause more voltage drop across the series resistor, which would lower the output voltage. A decrease in output voltage would cause less conduction of the pass transistor. Reduced current through the series resistor will cause less voltage drop. The output voltage would therefore increase in value to compensate for the decrease in voltage.

Linear integrated-circuit regulators are usually equipped with some type of current overload protection. This function prevents damage to the series-pass transistor. The three-terminal regulator of Figure 9-17 has a thermal shut-down circuit. The temperature of the series-pass transistor is sensed. If its operation temperature exceeds 175°C, the regulator turns off. Removing an excessive load will lower the output current and permit the transistor to cool down. Operation will be restored when the temperature returns to normal. Protection of this type makes the chip virtually indestructible.

Linear IC regulator troubleshooting is rather easy to accomplish. Essentially, the IC an be open, shorted, or good. An ohmmeter test of the IC generally does not tell a great deal about its condition. If the regulator is shorted, there will be a low resistance between the input and output. An open condition cannot be tested with an ohmmeter. This means that the IC must be tested in circuit with meaningful power applied. In this regard, it must have voltage applied to the input and be capable of developing a regulated output voltage. This can be observed with a voltmeter. As a rule, this test must be performed with varying load values. An extremely heavy load may cause the regulator to shut down when its power dissipation rating is exceeded. Reducing the load should cause the IC to turn on and restore the regulation function.

SELF-EXAMINATION

26. Two general classifications of ICs according to their function or application are ________________ and ____________.
27. ____________ ICs achieve switching operations.
28. ____________ ICs develop an output that is proportional to the input signal.
29. ____________ were originally designed to achieve mathematical operations.

30. An op-amp is essentially a ____________ gain amplifier.
31. The two inputs of an op-amp are labeled ____________ and ____________.
32. The input of a typical op-amp has a ____________ amplifier.
33. A signal applied to the (−) input of an op-amp will produce a ____________ output signal.
34. A three-terminal voltage regulator is a (linear-digital) IC.

ANSWERS

26. Digital, linear
27. Digital
28. Linear
29. Op-amps
30. High
31. Inverting and noninverting
32. Differential
33. Positive (+)
34. Linear

PRECISION TIMER ICs

Special linear integrated-circuits are now available which have the capability of producing a variety of timing functions. The 555 precision timer IC was the first in this series. This IC is classified as a linear IC. It has several linear amplifiers in its construction. Figure 9-18 shows a functional block diagram of the precision timer IC.

The operation of a 555 timer is directly dependent on its internal functions. The three resistors of its block diagram, for example, serve as an internal voltage divider for the source voltage. One-third of the source appears across each resistor. Connections made at the 1/3 and 2/3 V_{CC} points serve as reference voltages for the two comparators. A comparator is an op-amp that changes states when one of its inputs exceeds the reference voltage. Comparator 2 is referenced at +1/3 V_{CC}. If a trigger voltage applied to the negative input of this comparator drops below 1/3 V_{CC}, it causes a state change. Comparator 1 is referenced at +2/3 V_{CC}. If voltage at the threshold exceeds this reference voltage, the comparator goes through a state change. Note that the output of each comparator is connected to an input of the flip-flop.

The flip-flop of the 555 timer is a bistable multivibrator. It changes states according to the voltage value of its input. If the voltage value of the threshold terminal rises in excess of +2/3 V_{CC}, it causes the flip-flop to change its state. A decrease in trigger voltage below +1/3 V_{CC} also causes comparator 2 to change its state. This means that the output of the flip-flop is controlled by the voltage values of the two comparators. A state change occurs

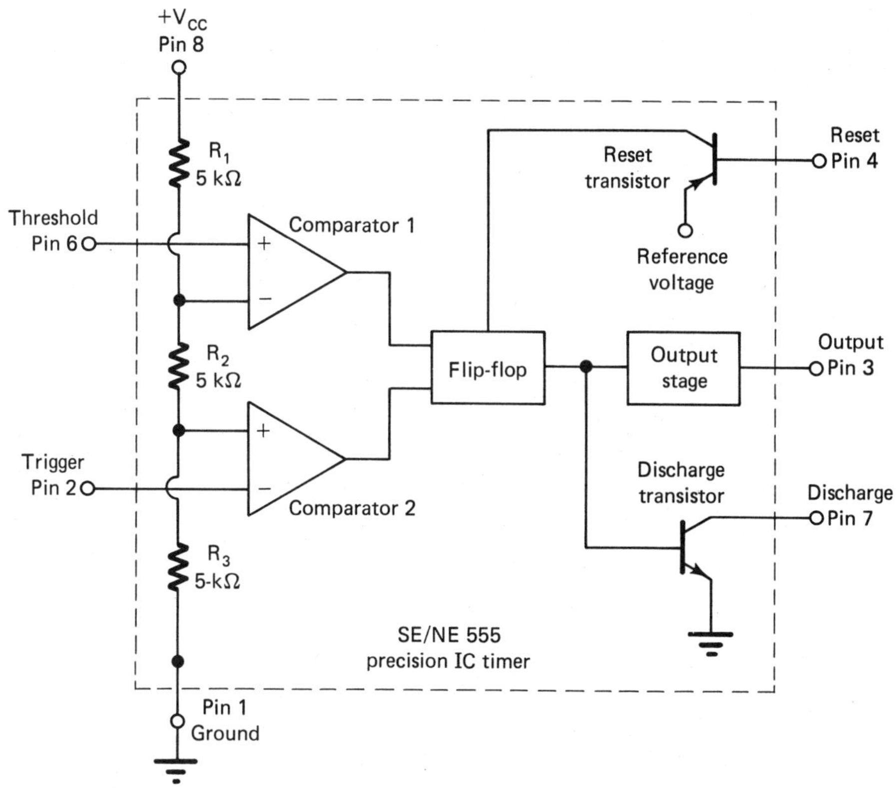

FIGURE 9-18 Functional block diagram of an SE/NE555 IC.

when the trigger input drops below +1/3 of V_{CC} or when the threshold rises above +2/3 V_{CC}.

Note that the flip-flop output of the 555 is used to drive the discharge transistor and the output stage. A high or positive flip-flop output is used to turn on both the discharge transistor and the output stage. This condition causes the discharge transistor to be conductive and causes it to respond as a low resistant short to ground. The output stage responds in the same manner. A reversal of this action takes place when the flip-flop output changes to a low or zero state. The discharge transistor then responds as an open circuit or becomes infinite. The output state swings to its high or positive V_{CC} state. In effect, the operational state of the output and discharge transistor is based on voltage applied to the trigger or threshold input terminals.

The 555 timer has two fundamental modes of operation. Practically all timing applications are in some way a variation of these operational modes. One application deals with the generation of waves. Used in this manner, the 555 responds as an astable multivibrator. The other mode of operation acts as a monostable multivibrator. Both applications use the 555 as a multivibrator. A multivibrator is defined as a generator that produces a waveform that has regenerative feedback between its two amplifiers. An astable multivibrator is not stable and changes continually between two states. Monostable has one stable state. With a trigger pulse applied, it changes to one state and automatically returns to its stable state.

An astable multivibrator is often referred to as a "clock" in system applications. It switches back and forth between two states causing a square or rectangular wave to be produced. Figure 9-19 shows a 555 connected as an astable multivibrator. In this circuit, pins 8 and 4 are connected to a dc energy source of somewhere between 5 and 15 V dc. The trigger (pin 2) and the threshold (pin 6) are connected together, which causes automatic triggering. An external capacitor (C_1) charges through resistors R_A and R_B and discharges through R_B and the internal parts of the IC. When C_1 is charging, the output is in its high state. The discharge transistor is cut off at this time. When the charge on C_1 exceeds 2/3 V_{CC} it causes the output to go to its low state. At the same point in

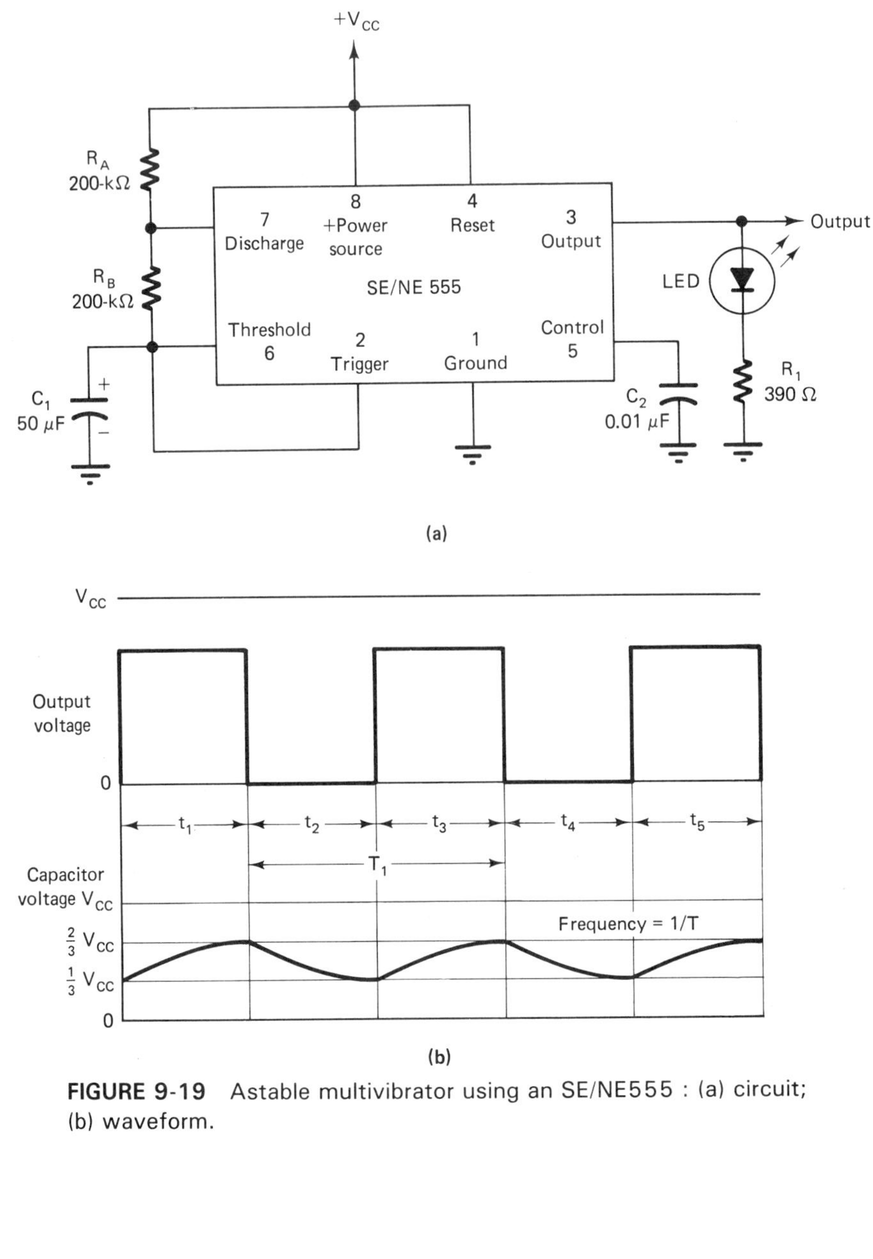

FIGURE 9-19 Astable multivibrator using an SE/NE555 : (a) circuit; (b) waveform.

time the discharge transistor becomes conductive. Note the state change of the output and the capacitor voltage (V_C) waveforms. The charge voltage of C_1 then discharges through resistor R_B. When this voltage drops below 1/3 V_{CC} it causes a change in the output and the discharge transistor. Operation is based on the charge and discharge time of the voltage across C_1. The charge time is 0.693 (R_A + R_B) C. The discharge time is 0.693 R_B C. The combined charge and discharge times make one complete operational cycle. The frequency of the wave is 1.44/(RA + 2RB)C.

Not all timing applications require a continuous wave such as that developed by the astable multivibrator. Many timing situations call for operations that last for a specified length of time. The monostable multivibrator permits the 555 timer to be used for these applications. Figure 9-20(a) shows the 555 IC connected in a monostable circuit configuration. Note that only one resistor (R_A) and one capacitor (C_2) are used in the timing part of the monostable circuit.

A monostable multivibrator is designed primarily to generate a single output pulse in response to an input trigger pulse. The duration of the output pulse depends entirely on the time constant of C_2 and R_A. When a negative going pulse is applied to the trigger input (pin 2), it causes the output (pin 3) to go high or to $+V_{CC}$. The trigger pulse causes comparator 2 to drop below its referenced value of +1/3 V_{CC}. This action causes the flip-flop to go to its low state. A negative voltage to the discharge transistor causes it to become infinite. This removes the short to ground for capacitor C_2. When the voltage of C_2 exceeds +2/3 V_{CC}, it causes comparator 1 to change states. This in turn causes the discharge transistor to again become conductive. C_2 then discharges very quickly to ground through pin 7. The output stage follows this action and drops to its low or ground state. In effect, the output stage follows the change in the trigger input level. Figure 9-20(b) compares the trigger input and output waveforms.

The duration of the trigger pulse of a multivibrator can be either longer or shorter than the generated output pulse. The resulting width of the output pulse is based on the values of R_A and C_2. The time (T) of the output is based on the formula

$$\text{Time (high)} = 1.1\, R_A \times C_2$$

This means that the time duration of the output can be altered according to the application. A short-duration input pulse can therefore be used to produce a rather long duration output pulse. Because of this, monostable multivibrators of this type are often called pulse stretchers. The monostable mode of operation is also used to produce delay and interval timing operations. Note that the starting of a monostable multivibrator is based on the application of a trigger pulse.

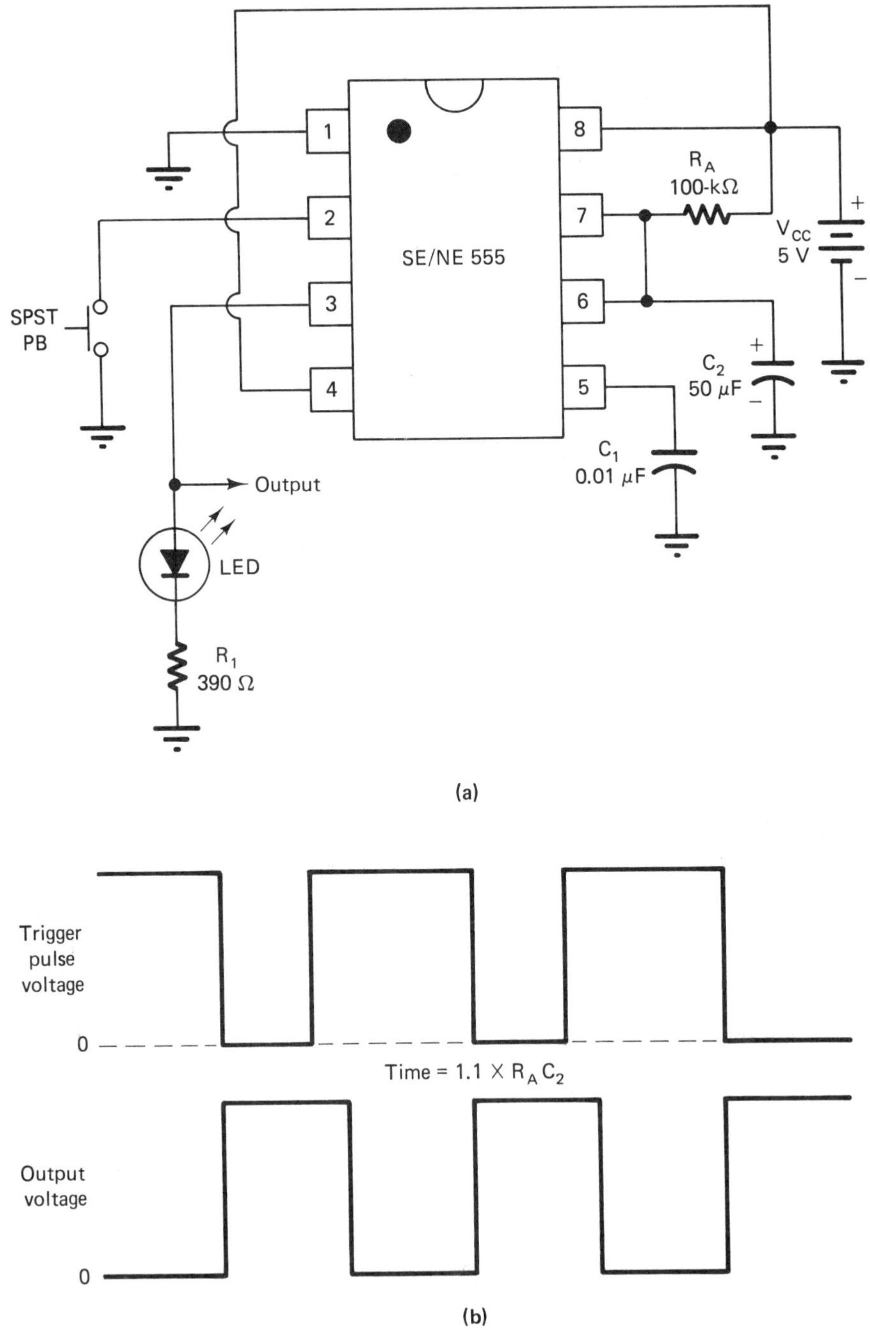

FIGURE 9-20 Monostable multivibrator operational mode: (a) circuit; (b) trigger-output waveforms.

SELF-EXAMINATION

35. The 555 IC timer block diagram shows a three ______ voltage divider connected across the power source.
36. In the IC timer block diagram, comparator 1 is referenced at ______ V_{CC}.
37. In the IC timer block diagram comparator 2 is referenced at ______ V_{CC}.

38. The output of comparators 1 and 2 are connected to a ______________ ______________.
39. The output stage and discharge transistor of a 555 are controlled by the ______________ ______________.
40. Two modes of operation of the 555 are ______________ and ______________.
41. The charge time of C_1 of Fig. 9-19 is ______________ seconds and the discharge time is ______________ seconds.
42. A monostable multivibrator generates a single output pulse in response to an input ______________ pulse.
43. The output of the monostable multivibrator of Fig. 9-20 will be high for ______________ seconds.

ANSWERS

35. Resistor
36. +2/3
37. +1/3
38. Flip-flop
39. Flip-flop
40. Astable, monostable
41. 13.86 and 6.93
42. Trigger
43. 5.5

DIGITAL ICs

Digital electronics is undoubtedly the fastest-growing area in the field of electronics today. Computers, calculators, watches, clocks, video games, instruments, and industrial control are only a few examples of digital IC applications. Most of these examples were unheard of a few years ago. Digital electronics now plays an important role in our daily lives.

Our discussion of electronic devices up to this point has centered primarily on analog applications. An analog is a method of determining one quantity in terms of another. Temperature, for example, is often determined by the position of a column of mercury in a glass tube. Voltage, current, and resistance can be determined by a coil of wire that interacts with a magnetic field. Analog devices are usually concerned with continuously changing values.

Digital electronics is considered to be a counting operation. A digital watch tells time by counting generated pulses. The resulting count is then displayed as numbers representing hours, minutes, and seconds. A computer also has an electronic clock that generates pulses. These pulses are counted, and in many cases, manipulated to perform a control function. Digital circuits can store signal data, retrieve it when needed, and make operational decisions. Signal values are generally represented by two-state data. A pulse is either present or it is not. Data are either high or low in value. There is nothing between these two values. Digital electronics is primarily dependent on the manipulation of two-state data.

Two-state data are effectively controlled by digital ICs. A large part of this control centers around the manipulation of data through logic gates. A logic gate can be formed with bipolar transistors, resistors, diodes, and MOSFETs. As a rule, a large number of logic gates are needed to perform some control functions. A computer, for example, may incorporate thousands of gates in its circuitry. The IC lends itself well to this type of circuit construction. Duplicate gates can be readily formed on an IC and housed in a standard package. The dual in-line package (DIP) is used widely to house digital electronic gates.

A number of digital IC families are available today. These families range from the earlier diode logic (DL) family to resistor-transistor logic (RTL), transistor-transistor logic (TTL), emitter-coupled logic (ECL), to complementary-symmetry metal-oxide semiconductor (CMOS) ICs. This number of IC families has caused some confusion to those not directly concerned with the changing technology of ICs. In a sense, the differences in IC families are not particularly unusual when compared to other products. Automobile manufacturers, for example, produce different types of cars which are all designed to perform the same overall function. The different IC families all achieve the same basic function but in a slightly different way. TTL and CMOS ICs tend to be the most widely used digital ICs in operation today.

Digital ICs are considered "black boxes" or functional blocks that achieve some form of control. In a manufacturer's data book they are listed according to symbol representations instead of schematic diagrams of the actual circuit components. As a rule, a person using the IC is concerned only with what the chip will accomplish and not how it actually achieves a specific operation. The internal circuitry is generally not an important consideration. The user cannot effectively alter the internal circuitry if it does not function properly. The operational characteristics of a particular chip are very important to the user. These characteristics are taken into account when the chip is selected for a specific application. Each IC family has a number of unique characteristics which distinguishes it from other families. These characteristics are considered to be above and beyond the normal functional conditions of bias, source voltage, input, and output signals:

1. *Operating temperature.* This is an indication of the normal range of operating temperature in which the IC can perform satisfactorily. Typical operating temperatures are −55°C to +125°C.
2. *Propagational delay.* This is a measure of the time difference that occurs between input and output. Typical propagational delay values are in the range of 5 to 20 ns.
3. *Total power dissipation.* This is an indication of the total

amount of power consumed by the IC when it is fully operational.

4. *Fan-out.* This is a measure of the total number of parallel circuits that can be driven by the output of a gate or functional block. Fan-out is based on the output device of the functional block.
5. *Fan-in.* This is a measure of the number of parallel input circuits that can be connected to the input of a logic gate or functional block. Each input is essentially connected in parallel with another component. Increasing the number of inputs loads down the input of the IC. The fan-in characteristic varies greatly among different IC families.
6. *Noise immunity.* In ICs, noise immunity is a built-in feature that represents the ability of a circuit to tolerate external voltages, signals, ground changes, power supply voltage variations, and transient pulses. This is based on the speed with which a chip responds and the minimum voltage change needed to initiate a state change in the output.

The transistor-transistor logic, or TTL, family of ICs is used very widely in digital circuits. Its advantages are low cost, high speed, moderate drive capabilities, low noise immunity, and availability. A broad range of these devices can perform practically any logic function. These chips are manufactured by a number of different companies and can be used interchangeably.

The circuitry of a TTL two-input positive-logic NAND gate is shown in Fig. 9-21. The term NAND refers to a logic function that is achieved by switching between two voltage levels. The output of the gate is an indication of the voltage levels of the two inputs. Instead of referring to specific voltage levels, logic gates generally refer to one level of operation as high and the other as low. Voltage is an indication of the high, or 1 state, whereas no voltage is the 0, or low, state. A table of the four possible operational states of a NAND gate is shown in Fig. 9-21(b).

A schematic diagram of the logic gate's circuitry is not generally used to identify the function of a specific chip. As a rule, the circuit may be somewhat complex. IC functions are usually expressed by logic symbols. These symbols are easy to draw and show the relationship of the input and output. Figure 9-21(c) shows the logic symbol of a NAND gate. Many ICs are capable of housing several logic gates on the same chip. A four NAND gate SN7400 package outline is shown in Fig. 9-21(d). This shows the pin-out or connections for the respective gates of the chip. Notice that this chip is a 14-pin dual in-line package. Each gate utilizes only three pins. The inputs are to pins 1 and 2 of NAND 1, and the output appears at pin 3. NAND gates 2, 3, and 4 are similar in all respects but utilize different pins. The package is numbered in a counterclockwise direction from the identification notch. Power for the chip is supplied by pins 7 and 14.

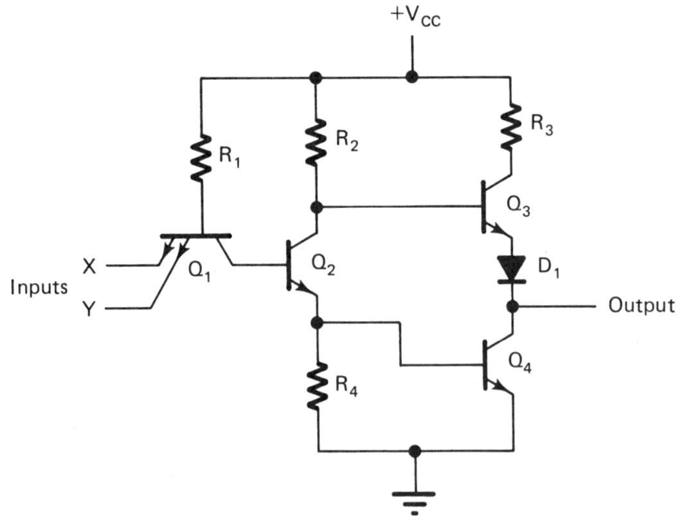

(a)

Inputs		Output
X	Y	Q_4
0	0	1
0	1	1
1	0	1
1	1	0

(b)

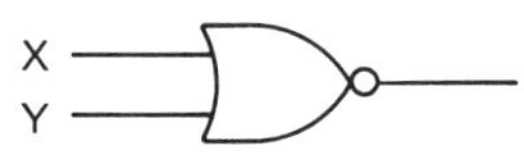

(c)

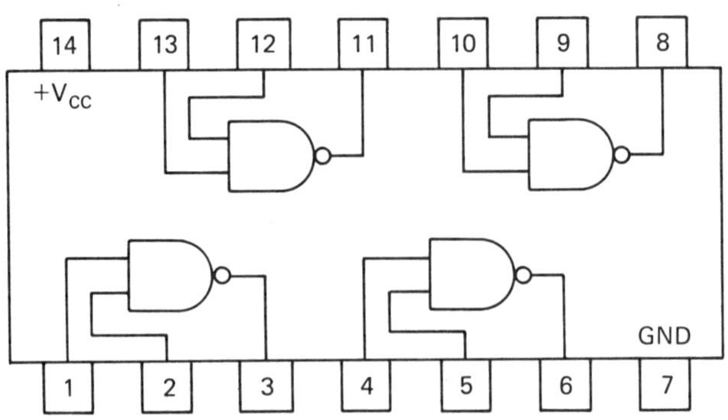

(d)

FIGURE 9-21 (a) TTL NAND gate schematic; (b) truth table; (c) logic symbol of a NAND gate; and (d) pin-out of SN7400 IC.

ICs using MOS transistors in their construction are commonly used in digital ICs today. These ICs can perform the same functions as the TTL family but with less power consumption. The most popular circuit configuration of the MOS devices is the complementary symmetry metal-oxide semiconductor or CMOS family. Complementary symmetry refers to the use of different E-MOS devices that have identical characteristics. P-channel and N-channel MOSFETs are used in the circuitry of this IC. Figure 9-22 shows a schematic diagram, pin-out connections, and an operational table for the MM74C04 hex inverter. Note that two MOSFETs are used in the formation of this gate. The upper transistor is a P-channel MOSFET. Its source is connected to the substrate and the $+V_{DD}$ source. The lower MOSFET is an N-channel device with identical characteristics. The gates of each device are connected to form the input. The output appears at the commonly connected drains. The MM74C04 has six of these inverters housed in the same package. There is one input and one output for each inverter. The operational table of an inverter shows that the output is an inversion of the input. A high or 1 input will produce a low or 0 output. Reversing the input to low or 0 will produce a high or 1 output. The inverter gate symbol has one input and one output lead. Note the pin-out connections for the six gates of the MM74C04.

CMOS IC gates have a number of distinct advantages over other families of ICs. Power dissipation is quite low for nearly all conditions of operation. The fan-in and fan-out capabilities are rather high when compared with other IC families. The noise immunity of the family is very good. Logic gates are available that can operate at power supply voltages from +1.3 to +15 V. This is a considerable advantage over other logic families that require a narrow range of supply power. CMOS ICs are not directly com-

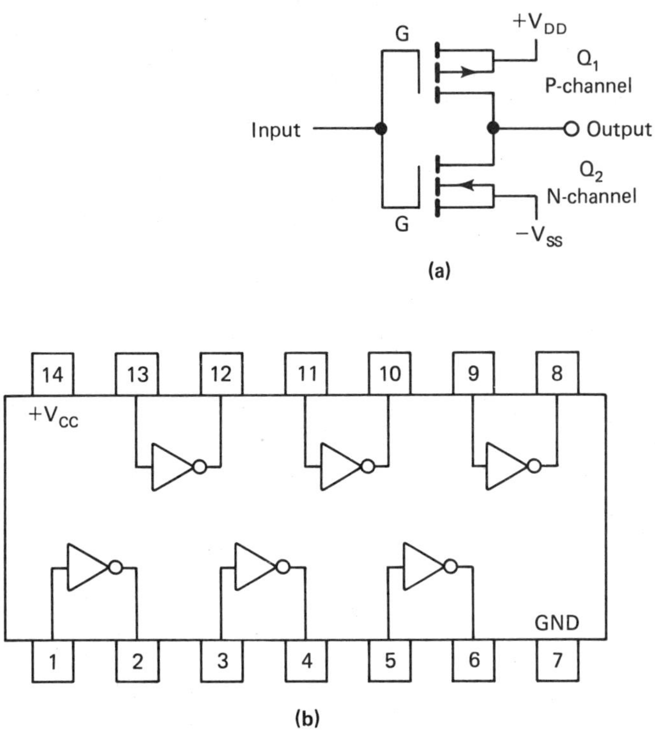

In	Out
0	1
1	0

(c)

FIGURE 9-22 CMOS hex inverter: (a) schematic; (b) pin-out of MM74C04; (c) truth table of inverter gate.

patible with other logic families and have a slower propagational time than TTL. These problems have been reduced to some extent with new MOS technology.

SELF-EXAMINATION

44. Digital electronics is considered to be a ____________ operation.
45. The data of a digital device have two operational states ____________ and ____________.
46. ____________ and ____________ IC families are the most widely used.
47. The ____________ family of ICs has two or more transistors connected between the inputs and output of a logic gate.
48. The family of ICs that employs two complementary symmetry connected MOSFETs in its construction is the ____________.
49. The pin layout of a DIP is numbered in the ____________ direction from the identification notch.
50. An IC that switches or reverses its input signal in the output is called an ____________.

ANSWERS

44. Switching
45. High (1) and low (0)
46. TTL and CMOS
47. TTL
48. CMOS
49. Counterclockwise
50. Inverter

SUMMARY

Integrated circuits are microminiature circuits built on a substrate. ICs are made of active and passive components. They have been available for a number of years. Originally ICs were developed for the space industry and military circuit applications.

The advantages of ICs are small size, cost, reliability, low power consumption, and ease of maintenance. The disadvantages are low power, low voltage, and limited component selection. The disadvantages are primarily trade-offs that make the advantages possible. The advantages outweigh the disadvantages.

The construction of ICs is not particularly new. Such things as substrates, epitaxial growth, and diffusion were used in the construction of bipolar transistors and MOSFETs. Monolithic ICs have all components built on a single crystal structure. The components of this IC are all formed on a common substrate at the same time. One group of monolithic ICs uses the bipolar transistor as its fundamental component. MOSFETs are used in the construction of another class of monolithic devices called MOS ICs. These chips are cheaper to construct and can be made smaller than the bipolar IC. Film ICs are an extension of the earlier discrete-component electronic circuits. These ICs are an assembly of components formed on an insulating material substrate. Film ICs are classified as hybrid devices. Conductors and resistors of a thin-film IC are formed by the evaporation process and located in a specific area by a screen-printed deposition process. Transistors, diodes, capacitors, and MOSFETs are added to the circuit in discrete form. Thick-film ICs are similar in many respects to the thin-film devices. The primary difference is the thickness of the film and the method of material deposition on the film.

ICs are housed in a number of unique packages. Where small size is important, the flat pack is used. Dual in-line packages are used to house over 80% of the ICs today. A number of new packages have been developed to accommodate the reduced size of ICs and to increase the number of pin connections.

The two general classifications of ICs according to their functions are linear and digital. Linear ICs develop an output which is proportional in some way to the input signal. These ICs operate in the straight-line portion of the operating range. Linear ICs are primarily used to achieve amplification. A large number of these ICs are called operational amplifiers. These amplifiers have high-gain capabilities. The input has a differential amplifier that has two inputs. One input, identified by a positive sign (+) will not invert the input signal. A negative sign (−) denotes the inverting input. The differential amplifier is followed by a high-gain amplifier. The output is typically an emitter-follower. It permits the op-amp to have high gain and low impedance output. Operation of a linear amplifier IC is usually controlled by the selection of a few external components.

A three-terminal voltage regulator is a linear IC. Operation is based on the applied input voltage and the developed output. Output voltage is compared with a reference voltage by a high-gain error amplifier. The output of the error amplifier then controls conduction of the series-pass transistor. This transistor responds as a variable resistor. It changes inversely with the load current to maintain the output voltage at a constant value.

Special ICs are now available that have the capability of producing a variety of timing functions. The 555 series of ICs was developed to accomplish these timing functions. A 555 contains a three resistor voltage divider, 2 comparators, a flip-flop, an output, a reset transistor, and a discharge transistor. Operation is based on voltage values applied to the trigger and threshold inputs. When the threshold input voltage exceeds 2/3 of V_{CC}, it causes a state change in the flip-flop. When the trigger input drops below 1/3 of V_{CC} it causes a state change in the flip-flop. The flip-flop controls the operation of the output stage, the reset transistor, and the discharge transistor. A 555 has two general modes of operation: astable and monostable. Astable operation is used in the generation of square or rectangular waveforms. Monostable operation is used to produce a wave that can be used to achieve delay or interval timing operations.

Digital ICs respond as counting devices. Two-state data are controlled by a digital IC. These data are generally manipulated through logic gates. Voltage is an indication of the high or 1 state, while no voltage is the 0 or low state.

Digital ICs are considered to be black boxes that achieve some form of control. The function of an IC is shown by logic symbols. The characteristics of a digital IC include operating temperature, propagational delay, total power dissipation, fan-out, fan-in, and noise immunity.

The TTL family of ICs is very popular today. Bipolar transistors are used in the construction of this IC. A circuit includes bipolar transistors connected directly on the input and a single transistor in the output.

ICs using MOSFETs in their construction are now commonly found in digital devices. These ICs function as TTL devices but consume less power. A very popular circuit configuration uses MOS devices connected in a complementary-symmetry fashion. These ICs have low power dissipation, high fan-in and fan-out capabilities, and good noise immunity. Logic gates are available that operate on 1.3 to 15 V. CMOS ICs are not particularly compatible with other IC families and have a slower propagational time.

EXPERIMENT 9-1
INVERTING OPERATIONAL AMPLIFIERS

Inverting operational amplifiers are used widely in electronic circuits and systems. Primarily, this type of IC is used as a high-gain amplifier. The versatility of this amplifier makes it very useful in control applications. With no external feedback circuit, an op-amp can achieve an open-loop gain that is around 100,000. With a feedback circuit, gains can be achieved from 1 to 100,000 according to the selected resistance values.

In this experiment you will construct a simple inverting operational amplifier. This circuit will amplify a dc voltage. Some calculations will be made to show voltage gain. This activity will permit you to see the operation of an inverting amplifier and predict what its output will be. Upon completion of this experiment, you should be able to design a simple operational amplifier that will achieve a specific amount of voltage gain. Measurements will permit you to see how to troubleshoot an amplifier of this type.

OBJECTIVES

1. To connect an inverting operational amplifier circuit.
2. To evaluate the operation of an inverting operational amplifier.
3. To calculate the voltage gain of an inverting operational amplifier.

EQUIPMENT

9-V transistor radio batteries or a split or divided power supply (2)
Volt-ohm-milliammeter (VOM), standard or digital
1.5-V dry cell input source
LM741C operational amplifier
1 kΩ $\frac{1}{4}$-W resistor
10 kΩ $\frac{1}{4}$-W resistor
100 kΩ $\frac{1}{4}$-W resistor
1-kΩ 1-W linear potentiometer
Circuit construction board

PROCEDURE

1. Construct the inverting dc amplifier of Fig. 9-1A. The power source for the LM 741C is derived from two 9-V batteries, a split or divided variable dc supply, or two interconnected power supplies.
2. Adjust the potentiometer of the dc input voltage source to produce 0.3-V dc when connected to the amplifier input.
3. Energize the IC by connecting the two 9-V batteries to $+V_{CC}$ and $-V_{CC}$ or turning on the split power supply. Measure and record the output voltage (V_{out}) across R_L. __________ output voltage.
4. How does the polarity of the output compare with that of the input? ____.
5. Increase the value of the input voltage to 0.6 V. Then measure and record the output voltage (V_{out}). V_{out} = __________ V.
6. The amplification ratio of an operational amplifier is determined by resistor values. R_f is the feedback resistor and R_{in} is the input resistor. The amplification factor formula is: $A_f = -R_f/R_{in}$. The negative sign of R_f denotes the inversion function.
7. Calculate the amplification factor for the circuit of Fig. 9-1A. A_f = __________.
8. The voltage gain or output voltage of an operational amplifier is determined by multiplying the amplification ratio by the input voltage. This is expressed as $V_{out} = A_f \times V_{in}$. Determine the output voltage for step 2 where

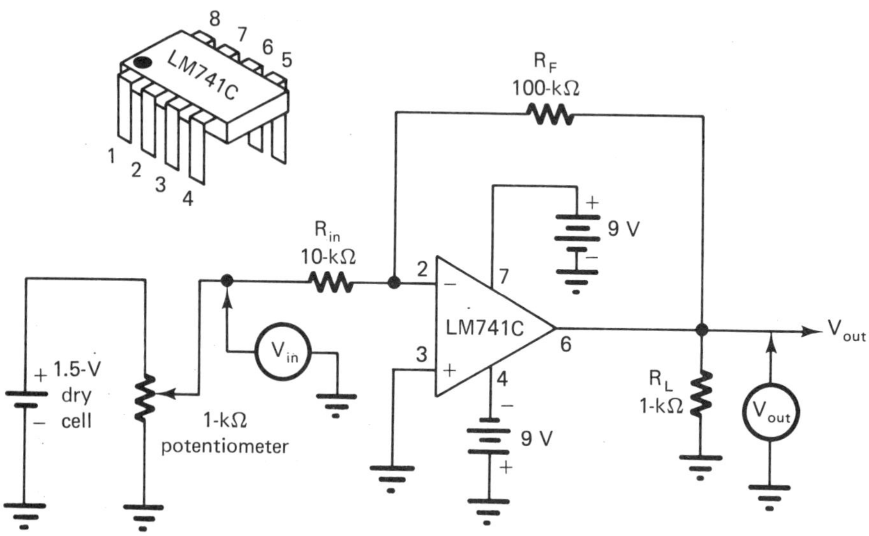

FIGURE 9-1A Inverting operational amplifier.

the input voltage is 0.3 V. V_{out} = ____________ V. How do the measured and calculated output voltages of step 3 compare? ____________________________.

9. Determine the output voltage for step 5 where the input voltage is 0.6 V. V_{out} = ____________ V. How do the measured and calculated output voltages of step 5 compare? ____________________________.
10. When troubleshooting an operational amplifier the input and output voltages are measured and the amplification factor determined. If the circuit is working properly, these values should be very close.
11. When the input voltage of an operational amplifier is large enough to exceed the amplification ratio, an amplifier becomes saturated. This is generally called *overdriving*. The output voltage cannot exceed the value of the source voltage. This means that a saturated amplifier may not produce an output voltage that corresponds with the amplification factor. The output voltage of a saturated amplifier will always be less than the desired amplification ratio.
12. Using the calculated amplification ratio of Fig. 9-1A, at what value of input voltage will saturation of the amplifier occur? ____________ V.
13. To test the point where saturation occurs, simply increase the input voltage to where it does not produce a change in output voltage. This can be measured across the load resistor. Then go back and measure the input voltage to find its value. The input voltage that produces saturation is ____________.
14. If time permits, change the value of R_f and R_{in}. Record the values of R_f and R_{in}. R_f = ____________, R_{in} = ____________.
15. Adjust the input voltage to 0.1 V. Measure and record the output voltage (V_{out}). V_{out} = ____________ V. Calculate the amplification ratio and output voltage. A_f = ____________, V_{out} = ____________.
16. Disconnect the circuit.

DISCUSSION

In this experiment, you were asked to construct an inverting operational amplifier circuit. An input voltage of 0.3 V was applied. The output voltage (V_{out}) was then measured and found to be approximately −3V. The polarity of the output voltage is opposite to that of the input voltage. The input voltage was then changed to 0.6 V and the output voltage was measured and found to be approximately 6 V. The amplification factor of the circuit was found to be 10. This shows the amplifying function of the op-amp.

When troubleshooting an operational amplifier, the input and output voltage are measured and compared with the calculated amplification value. The values should be similar. If the input voltage value exceeds the amplification ratio, the amplifier will be driven into saturation. Its actual amplification will be less than the calculated value. Using the calculated amplification ratio of 10, saturation occurs when the input voltage exceeds 0.9.

In the last part of the experiment, the resistance ratio of the amplifier circuit is changed. The new resistance ratio should be calculated, an input applied, and the output voltage measured. The amplification ratio should compare very closely with the measured voltage values.

ANALYSIS

1. What determines the amplification factor or ratio of an inverting operational amplifier?
2. What troubleshooting procedure is used to test an inverting operational amplifier?
3. What would occur if both R_f and R_{in} were of the same value?

EXPERIMENT 9-2
NONINVERTING OPERATIONAL AMPLIFIERS

Electronic circuits frequently need amplifiers that develop high voltage gain without signal inversion. The noninverting operational amplifier is designed specifically for this type of application. Both ac and dc signals may be amplified by this circuit. The input signal is applied to the noninverting input. Voltage gain is determined by the feedback resistor (R_f) and input resistor (R_{in}) connected to the inverting input. The output of this amplifier is developed across load resistor (R_L).

In this experiment you will connect a simple noninverting operational amplifier and observe its operation. The amplification factor or ratio of this amplifier is somewhat different from the one used for inverting amplifiers. The amplification ratio and voltage gain will be calculated and measured. These values will then be compared. This procedure is used to troubleshoot operational amplifiers. Saturation level will also be tested and used to evaluate the status of the output.

OBJECTIVES

1. To construct a noninverting operational amplifier circuit.
2. To test the operation of a noninverting operational amplifier.
3. To determine the voltage gain and output polarity of an operational amplifier.
4. To troubleshoot an operational amplifier.

EQUIPMENT

9-V transistor batteries or a split or divided power supply (2)
Volt-ohm-milliammeter (VOM), standard or digital
1.5-V dry cell input source
LM741C operational amplifier
1-kΩ, $\frac{1}{4}$-W resistor
10-kΩ, $\frac{1}{4}$-W resistor
100-kΩ, $\frac{1}{4}$-W resistor
1-kΩ, 1-W linear potentiometer
Circuit construction board

PROCEDURE

1. Construct the noninverting dc amplifier of Fig. 9-2A. The power source of the LM741C is derived from two 9-V batteries, a split or divided variable dc power supply, or two interconnected power supplies.
2. Adjust the potentiometer of the dc input voltage source to produce 0.2 V when connected to the amplifier input.
3. Energize the IC by connecting the two 9-V batteries to $+V_{CC}$ and $-V_{CC}$ or turning on the split-power supply. Measure and record the output voltage (V_{out}) across R_L. V_{out} = ____________ V.
4. Measure the input voltage and the output voltage with a VOM and compare the polarity of the two values. The input and output polarity are ____________.
5. The amplification ratio of a noninverting amplifier is determined by resistor values. R_f is the feedback resistor and R_{in} is the input resistor. The amplification factor formula is: $A_f = (R_f/R_{in}) + 1$.
6. Calculate the amplification factor for the circuit of Fig. 9-2A. A_f = ____________.
7. The voltage gain or output voltage of an operational amplifier is determined by multiplying the amplification ratio by the input voltage. This is expressed as $V_{out} = A_f \times V_{in}$. Determine the output voltage for step 2 where the input voltage is 0.2 V. The calculated V_{out} = ____________. How do the measured and calculated output

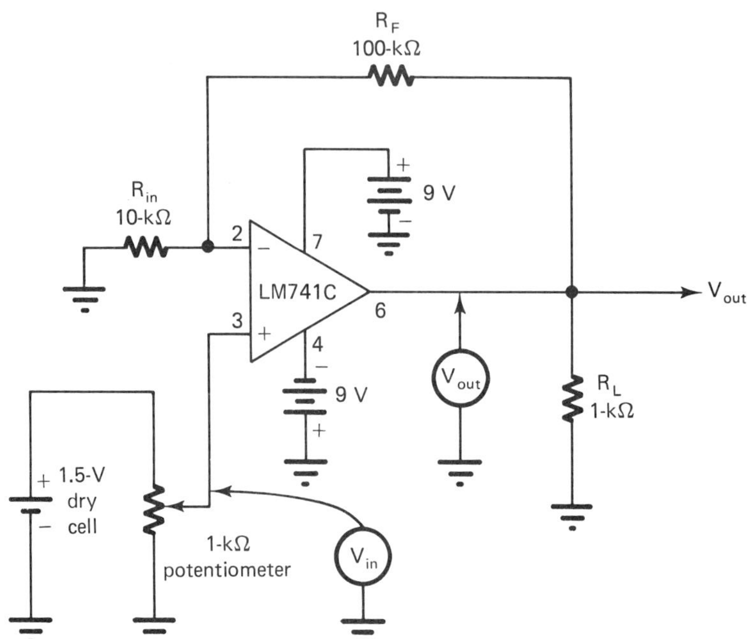

FIGURE 9-2A Noninverting operational amplifier.

values compare? ________________________________
____.

8. Adjust the dc input voltage to 0.6 V. Measure and record the output voltage. V_{out} = ____________ V. Calculate the output voltage. Calculated V_{out} = ____________. How do the measured and calculated values compare? ____________.
9. Troubleshooting a noninverting operational amplifier is accomplished by measuring the input and output voltages and comparing the voltage gain with the calculated value. The polarity of the input voltage and output should be the same. If ac is processed by the amplifier the ac input and output voltages are measured. An oscilloscope can also be used to observe the amplification status of the amplifier.
10. Test the amplifier to determine where saturation occurs. Saturation is where an increase in input voltage does not cause a corresponding change in output voltage. To determine where saturation occurs, simply measure the output voltage while increasing the value of the input voltage. The point where no change in V_{out} occurs is the point of saturation. The value of V_{in} that produces saturation is ____________ V.
11. If time permits, change the ratio of R_f and R_{in} to some other values. Record the values of R_f and R_{in}. R_f = ____________, R_{in} = ____________.
12. Adjust the input voltage to 0.7 V. Measure and record the output voltage (V_{out}). V_{out} = ____________ V. Calculate the amplification ratio and output voltage. A_f = ____________, V_{out} = ____________.
13. Disconnect the circuit.

DISCUSSION

In this experiment, you were asked to construct a noninverting operational amplifier circuit. In step 2, the dc input voltage was set to 0.2 V. The output voltage (V_{out}) measured across R_L should have been approximately 2.2 V. The polarity of V_{out} is the same as V_{in}. This shows the noninverting characteristic of the op-amp. The calculated amplification factor for this circuit is 11. This compares very closely with the measured value of V_{out}. Saturation of the noninverting amplifier occurs around 0.8 V.

When troubleshooting a noninverting operational amplifier, the input and output voltage are measured and compared with the calculated amplification value. The values should be similar. If the input voltage exceeds the amplification ratio, the amplifier goes into saturation. Its V_{out} will be decidedly less than the calculated value. Saturation for this circuit occurs when the input is greater than 0.8 V.

ANALYSIS

1. What type of troubleshooting procedure is used to determine if a noninverting operational amplifier is working properly?
2. What is amplifier saturation?
3. What distinguishes a noninverting amplifier from an inverting amplifier?

EXPERIMENT 9-3
THREE-TERMINAL IC REGULATORS

Three-terminal voltage regulators are available with several fixed output voltages making them useful in a wide range of applications. This type of IC achieves voltage regulation by sampling the output voltage and feeding back some of this voltage to control the conduction of a series transistor. The internal circuitry of the regulator is fairly complex. It is generally classified as a linear IC. The external circuitry is very simple. In some applications, it can be used without any external components other than the dc input and the load resistance connected to the output. In general, three-terminal regulators are designed for fixed voltage regulation. With some additional external components, a three-terminal regulator can be used as an adjustable voltage regulator.

OBJECTIVES

1. To examine the performance of a three-terminal IC voltage regulator.
2. To connect a three-terminal IC voltage regulator.
3. To troubleshoot a three-terminal IC voltage regulator.

EQUIPMENT

Variable dc voltage power supply 0 to 20 V
Volt-ohm-milliammeter (VOM), standard or digital
Three-terminal IC regulator, National Semiconductor LM7805CT or equivalent
100Ω, $\frac{1}{4}$-W resistor
220Ω, $\frac{1}{4}$-W resistor
1 kΩ, $\frac{1}{4}$-W resistor
1-kΩ, 1-W linear potentiometer
Circuit construction board

PROCEDURE

1. Construct the three-terminal voltage regulator circuit of Fig. 9-3A. The dc input voltage for this circuit is developed by a variable power supply.
2. Connect the VOM across the input and measure the input voltage. Adjust the input voltage to 4 V. Then mea-

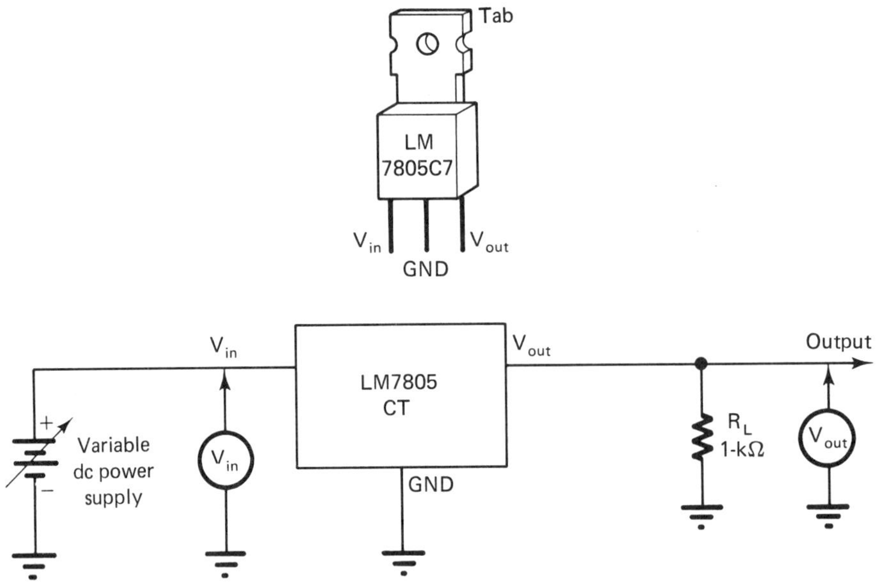

FIGURE 9-3A Three-terminal voltage regulator.

sure the output voltage across the load resistor with the VOM. Record the measured output voltage on the 1-kΩ load chart of Fig. 9-3B. Touch the regulator's tab one minute after each measurement. Show if the temperature is cool, warm, or hot by placing a check mark in the temperature column of the chart.

3. Repeat the procedure for each of the input voltage values of the chart.
4. Adjust the input voltage to 4 V, and exchange the 1 kΩ resistor for a 100Ω load resistor. Measure the output voltage across R_L and record the output voltage in the 100Ω R_L chart. Test the temperature of the regulator's tab and check its status as cool, warm, or hot.
5. Repeat the procedure for each of the input voltage values of the chart.

1-kΩ load

V_{in}	V_{out}	Temperature		
		Cool	Warm	Hot
4				
7				
10				
13				
17				
20				

100-Ω load

V_{in}	V_{out}	Temperature		
		Cool	Warm	Hot
4				
7				
10				
13				
17				
20				

FIGURE 9-3B Regulator charts.

6. Troubleshooting a regulator is generally done in the circuit while it is in operation. In this regard, a regulator will be either open or shorted. A shorted regulator will measure the same input and output voltage. This condition generally causes the regulator to be hot regardless of the input voltage. An open regulator will not produce output voltage. This condition is similar to not applying input to the regulator.
7. To test the response of a shorted regulator, connect a wire between the input and output of the regulator. Increase the input voltage to 5, 7, and 10 V while measuring the output voltage. The output voltage should be the same as the input.
8. To test the response of an open regulator, disconnect the input from the regulator. Measure the output voltage for some input voltage values. The output should be zero for all input values.
9. An ohmmeter can be used to evaluate the status of a shorted regulator when it is removed from the circuit. Disconnect the regulator from the circuit and determine the resistance between the input and output terminals. A good regulator should show infinite resistance in either polarity of the ohmmeter. A shorted or damaged regulator will show some resistance in either direction. Test the resistance of the regulator used in this activity. What is the condition of the regulator? ____________.
10. Connect the variable regulator circuit of Fig. 9-3C.
11. Turn on the power supply and adjust it to produce an input voltage of 20 V.
12. Connect the VOM across the output of the regulator. Adjust the potentiometer to its extreme counterclockwise position. Measure and record the output voltage appearing across the load resistor. Record this voltage value in the chart for the CCW position.
13. Adjust the potentiometer to the $\frac{1}{4}$ clockwise position. Measure the output voltage across R_L and record its value in the $\frac{1}{4}$ CW position of the chart.
14. Repeat the procedure for $\frac{1}{2}$ CW, $\frac{3}{4}$ CW, and CW position settings of the potentiometer.
15. Turn off the power supply and disconnect the circuit.

DISCUSSION

In this experiment, you were asked to construct a voltage regulator with a 1k ohm load resistor. The input voltage was adjusted to different values and the output voltage was measured. When the input was 4 V, the output was very low or 0. This shows the point where no regulation occurs. For input voltage values in ex-

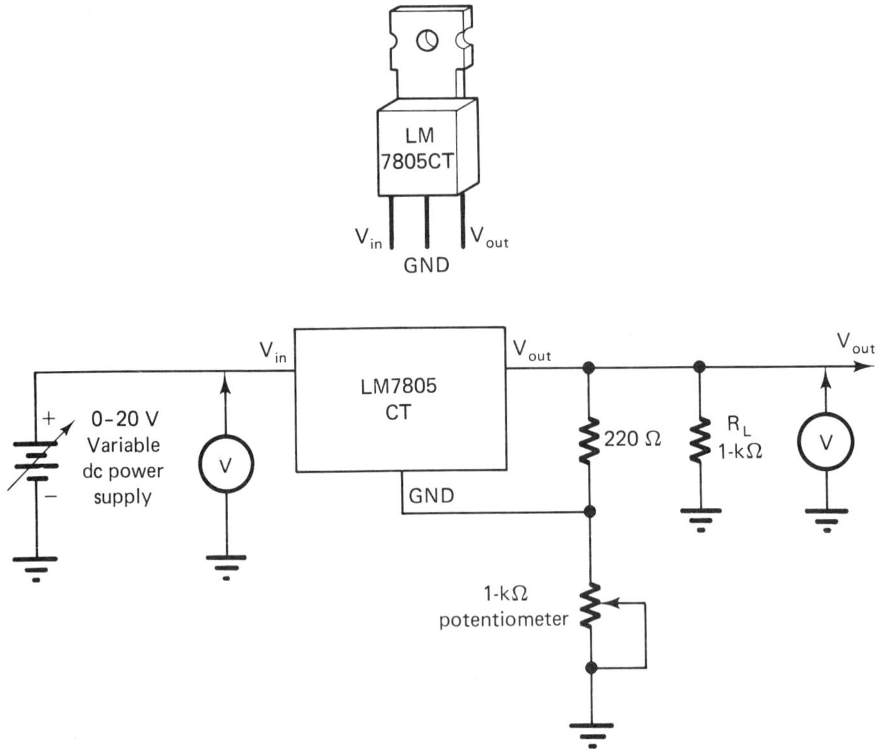

Chart

Potentiometer position	V_{out}
CCW	
$\frac{1}{4}$ CW	
$\frac{1}{2}$ CW	
$\frac{3}{4}$ CW	
CW	

FIGURE 9-3C Variable voltage regulator.

cess of 5 V, the output maintains a constant output voltage. This should be somewhere in the range of 5.3 V. For 7 V and 10 V, the regulator should be cool. With 13, 15, and 20 V of input the regulator gets warm. When the load resistor is changed to 100 ohms, the regulator is cool for 7 and 10 V. It gets warm for 13 V and hot for 17 and 20 V. As a rule, the output voltage is slightly more with a smaller load resistor or heavier load.

Troubleshooting a three-terminal regulator is generally achieved in-circuit. The input and output voltage values are measured and compared. If the output is the same as the input, the regulator is shorted. If there is no output voltage when input is applied, the regulator is open. A regulator can also be tested with an ohmmeter. This test is performed with the regulator out of the circuit. A good device will not show resistance between the input and output for either polarity of the ohmmeter. A shorted device will show some resistance between the input and output. A good device should be infinite between the input and output.

In the last part of this experiment, the regulator was connected as a variable voltage regulator. As a rule, the variable output voltage will not drop below the normal voltage regulation value. In this activity, the lowest adjust voltage is 5.3 V. The output can be adjusted to higher values by different position settings of the potentiometer. The output can be adjusted from 5.3 to 20 V.

ANALYSIS

1. What are the names associated with the three terminals of the LM7805CT regulator?
2. Why does the tab of the regulator get hot when the input voltage is increased?
3. How do you troubleshoot a three-terminal regulator?

EXPERIMENT 9-4
555 TIMER OPERATION

The operation of a 555 timer is determined by the condition status of its trigger, reset, threshold, and discharge terminals. The status of these terminals determines the condition of the output terminal (pin 3). The output will be either high (voltage) or low (no voltage). The trigger (pin 2) initiates the timing sequence of the 555. When the trigger goes low, it causes the output to go to its high state. The threshold and discharge terminals are connected to ground by a capacitor. This connection causes the status of the output to remain high after the timer has been triggered. The timer will latch and stay in this state as long as power is applied to the device. The reset terminal (pin 4) is generally connected to the high side of the power source. When pin 4 is switched low it causes the output to go to its low state. Operation of this circuit is primarily based on the conditional status of the trigger and reset inputs.

OBJECTIVES

1. To connect a 555 timer IC and to become familiar with the pin connections and terminal names.
2. To evaluate the operation of a 555 timer with respect to its trigger, reset, and output.
3. To trigger a 555 timer, reset its operational state, and see how this alters the output.

EQUIPMENT

Precision Timer IC, National LM555CN, Texas Instruments SE/NE555C, or equivalent

LED with a 2 V_F, Heath #412-40, Motorola MV5054-2, or equivalent

Variable dc power supply

Volt-ohm-milliammeter (VOM), standard or digital

100-Ω, $\frac{1}{4}$-W resistor

10-kΩ, $\frac{1}{4}$-W resistor

0.1-uF, 50 V capacitor

Circuit construction board

PROCEDURE

1. Construct the 555 timer circuit of Fig. 9-4A. Note that three pins of the 555 are attached to a wire. This wire is used to change the status of the pin from high (+5 V) to low (ground). The wire is simply connected or attached to a designated voltage value at specific times. Connect the wire according to the instructions.
2. The LED must be connected to the circuit according to the designated polarity. The longest lead usually identifies the cathode (K) with the short lead being the anode. Identify the leads of the LED with an ohmmeter.
3. Apply power to the circuit to make some general tests. The output status of the 555 is random when pins 2, 4, 6, and 7 are disconnected. If the LED lights, it indicates the "on" state of the output. When the LED is not lighting it indicates the "off" state of the output. To assure that the 555 is working properly, momentarily touch the wire of pin 2 to ground. This should cause the LED to light, which indicates that the output of the 555 is in its high state. With the VOM, measure the output voltage between pin 3 and ground. V_{out} = ____________ V.
4. If the 555 does not respond to this test, the circuit may be wired incorrectly, or the leads of the LED may be incorrectly identified or connected. Check the circuit wiring and LED polarity.
5. To test the triggering status of the 555, pin 2 (trigger) and pins 6-7 (threshold and discharge) are alternately

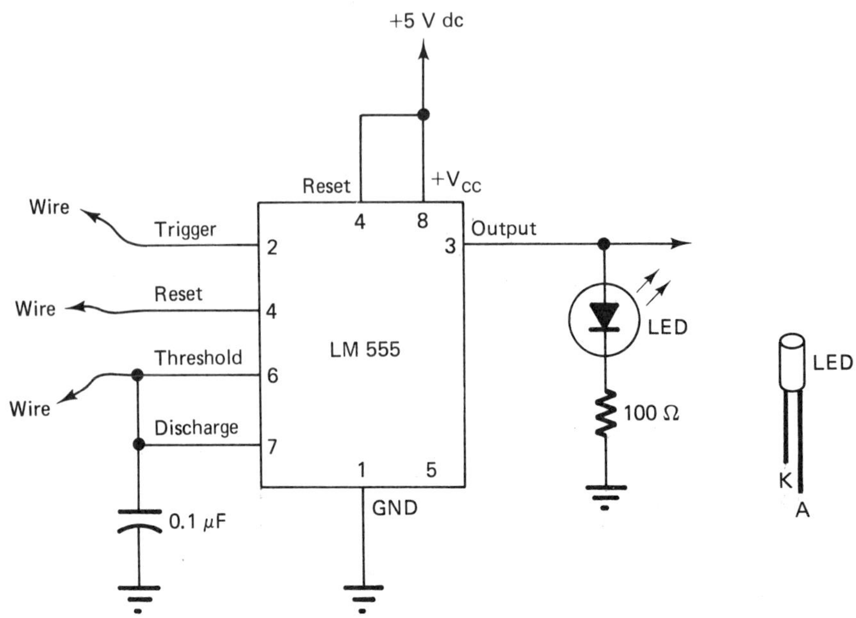

FIGURE 9-4A 555 test circuit.

connected high and low. With the LED on as in step 3, momentarily touch the free end of the wire connected to pins 6 and 7 to +5 V. This causes the threshold (pin 6) to rise above the +2/3 V_{CC} voltage reference and causes a state change of the internal flip-flop. This action causes the output to change from high to low. Measure the output voltage in the low state. The low state of the output is ____________ V.

6. To change the status of the output, momentarily connect the free end of the wire attached to pin 2 (trigger) to ground. This should cause the LED to light. This indicates that the output is again in the high state. This operation causes the trigger input to drop below the +1/3 V_{CC} reference voltage and change the status of the internal flip-flop. The flip-flop then causes the output to change from low to high.
7. Momentarily touch the wire from pins 6 and 7 to +5 V, then touch pin 2 to ground. This should cause the status of the output to switch from high to low. Repeat this procedure several times to make certain what voltage polarity is needed to cause a state change in the two inputs.
8. A rather interesting condition could occur during the switching action just described. The status of the 555, for example, can be changed by simply touching pin 2, and in some cases, pins 6 and 7 with your finger. Try to change the status of the 555 by touching these pins. The 555 can be used as a touch switch when leads are attached to metal touch pads at some remote location.
9. The reset (pin 4) of the 555 is used to return the status of the output to its starting position. Set the output of the 555 to its "on" state. Momentarily touch the wire connected to pin 4 (reset) to ground. This should cause the output to change states and go low. To return the output to its original status (high) touch the trigger wire to ground. Repeat the reset operation several times to become familiar with the procedure.
10. Be certain that you are familiar with the procedure needed to trigger the 555 into its "on" state, its "off" state, and the resetting operation. Practice these procedures several times to familiarize yourself with the process.
11. Turn off the power source and disconnect the circuit.

DISCUSSION

In this experiment, you were asked to construct a 555 timer circuit that would permit it to be triggered into conduction, change the status of its output, and reset the operation. When the circuit was initially energized, the load (pin 3) could be either high or low because of the unpredictable characteristic of the internal flip-flop. The output is turned on by connecting the trigger to ground. The "on" output voltage (V_{out}) should measure 3.8 V. With the output on, its state can be changed by connecting pins 6 and 7 to +5V. This causes the output to go to its low state. The off state output voltage is 0 V. The output can be changed, in many cases, by simply touching pin 2 and pins 6 and 7 with your fingers. This shows how the 555 is used as a touch switch.

The reset of the 555 is used to start the operational sequence in a specific position. When pin 4 is momentarily connected low or to ground, it places the output in the low state. The LED is off for this state of operation. The circuit responds essentially as a latch. It remembers that a negative-going pulse has been applied to its trigger input. The resetting function is also useful in some timing operations.

ANALYSIS

1. What voltage polarity does it take to trigger the 555 into its "on" state?
2. When the output of a 555 is on, what voltage level does it take for the threshold to change states?
3. What is the function of the reset?

EXPERIMENT 9-5
MULTIVIBRATORS

The 555 precision timer has two fundamental modes of operation. One operational mode deals with the generation of waves. In this mode, the 555 is used as an astable multivibrator. The other mode of operation is as a monostable multivibrator. The output of an astable multivibrator continually changes between two states (high and low). A monostable multivibrator has one stable state. This circuit must be triggered into conduction. It changes to one state and returns to its original state after a period of time.

Astable multivibrators are generally used as pulse generators or clocks. When a 555 is connected as an astable multivibrator, two external resistors and a capacitor are needed to complete the circuit. The output frequency can be determined by the value of these components. The reset is connected to $+V_{CC}$. The trigger and threshold are connected together and a capacitor runs to ground. The capacitor charges through resistors R_A and R_B. The capacitor discharges through R_B. Operation is based on the charge and discharge time of the capacitor. Since the threshold and trigger are connected to the capacitor, the capacitor determines the time duration of the state change. The output of the IC is connected to an LED to identify its status.

A monostable multivibrator is designed to generate a single output pulse in response to an input trigger pulse. The duration of the output pulse depends entirely on the time constant of a capacitor and a resistor. When the trigger input is made low or connected to ground, it causes the output of a 555 to go high. The trigger input causes comparator 2 to drop below its 1/3 V_{CC} reference voltage. This causes the flip-flop to change states. As a result, the output goes high and the discharge transistor becomes infinite. This causes the capacitor to begin charging to the source voltage. When the charge voltage exceeds 2/3 of V_{CC}, it causes comparator 1 to change states. This initiates a state change in the flip-flop, which in turn causes the output to go low and the discharge transistor to become conductive. The output will remain in this state until the next trigger pulse occurs.

OBJECTIVES

1. To connect 555 astable and monostable multivibrator circuits.
2. To analyze the operation of 555 astable and monostable multivibrators.

3. To calculate the output time duration of 555 multivibrators.

EQUIPMENT

Precision Timer IC, National LM555CN, Texas Instruments SE/NE555C, or equivalent
LED with a 2 V_F, Heath #412-40, Motorola MV 5054-2, or equivalent
Variable dc power supply
Volt-ohm-milliammeter (VOM), standard or digital
100-Ω, $\frac{1}{4}$-W resistor
56-kΩ, $\frac{1}{4}$-W resistor
100-kΩ, $\frac{1}{4}$-W resistor
330-kΩ, $\frac{1}{4}$-W resistor
0.1-uF, 50 V capacitor
10-uF, 50 V capacitor
Circuit construction board

PROCEDURE

Part A: Monostable Multivibrator

1. Connect the 555 monostable multivibrator circuit of Fig. 9-5A.
2. A wire is attached to the trigger input (pin 2) of the circuit. This wire is used to change the status of the trigger

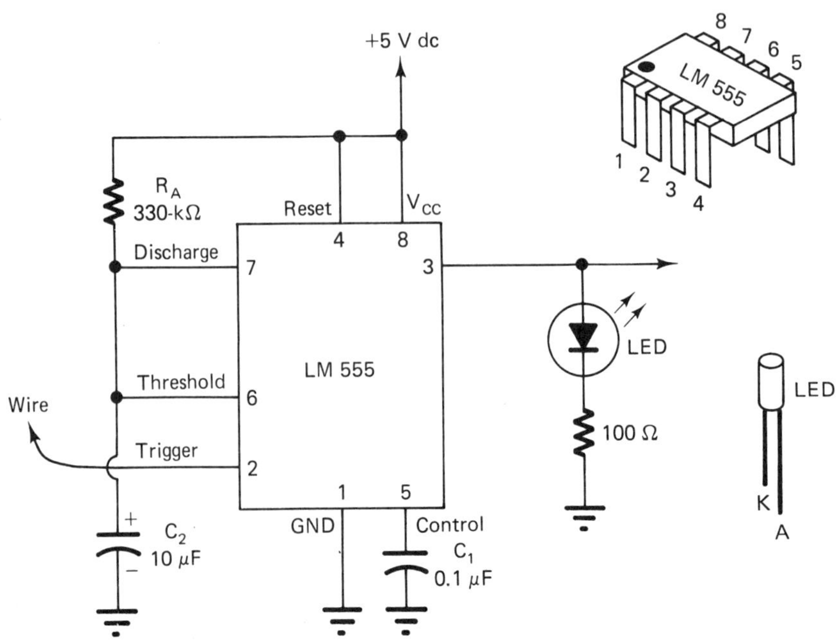

FIGURE 9-5A Monostable multivibrator.

from high to low. A monostable multivibrator must be triggered into operation for it to function. Initially, the trigger wire is not connected.

3. Apply power to the circuit. The 555 can be energized with voltages from 5 to 15 V dc. In this circuit 5V is used. Note the initial status of the output. If the LED is connected properly, it will be on when high and off when low. The initial status of the LED is ______. Measure the output voltage (pin 3) of the 555. V_{out} = ______ V.
4. Connect the wire attached to pin 2 to ground. Does this change the status of the output? ______. Measure the output voltage (pin 3) of the 555. V_{out} = ______ V.
5. How does the output respond after a few seconds of time? ______.
6. Repeat the triggering procedure by connecting pin 2 to ground. Evaluate the status of the output. The output is (high-low) initially and changes to (high-low) after connecting the trigger to ground. After a few seconds, the output changes back to its (high-low) state.
7. Connect the VOM across pin 6 and ground. The threshold voltage of pin 6 measures ______ V.
8. Trigger the 555 by momentarily connecting pin 2 to ground. Observe the change in threshold voltage. The threshold voltage is ______ to ______ V. How does this compare with the output voltage? ______.
9. The time duration that the output is high is based on the formula T = 1.1 $R_A \times C_2$. Calculate the time duration of the circuit using the values of R_A and C_2. T = ______ seconds.
10. Using a watch with a sweep hand to observe the time change, trigger the 555 and determine the length of time that it takes the output to cycle through a state change. The measured time is ______ seconds.
11. Turn off the power supply.

DISCUSSION

In the first part of the experiment you were asked to construct a monostable multivibrator and trigger it into operation. Triggering is accomplished by connecting pin 2 to ground. Initially when the circuit is energized by the power supply it causes the output to be low. The LED is off and the output voltage is 0V. Connecting pin 2 to ground causes a state change in the output. The LED is turned on and the output voltage measures 3.8 V. The output is low initially, changes to its high state, then changes back to its low state. The threshold voltage of pin 6 was then measured. This voltage changes from 0 to 3.5 V and changes back to 0 V. This is comparable to the output voltage that is low initially and changes to 3.8 V after triggering. The output voltage is two-state or either high (3.8 V) or low (0 V). The threshold voltage changes gradually from 0 to 3.5 V. The calculated time constant of the circuit is 3.3 s. The measured time constant is 3 s.

ANALYSIS

1. What is the status of the output of a multivibrator when it is energized initially?
2. When a multivibrator has been triggered, describe the status of the output sequence.

PROCEDURE

Part B: Astable Multivibrator

1. Modify the monostable multivibrator circuit so it conforms with the astable multivibrator circuit of Fig. 9-5B.
2. Turn on the power supply. If the circuit is wired correctly, the LED should be on for a few seconds and off for a few seconds on a continuous basis. The frequency is $f = 1.44/(R_A + 2R_B)C_2$. Determine the frequency of the circuit. f = ______________.
3. With a VOM, measure the output voltage of the circuit.

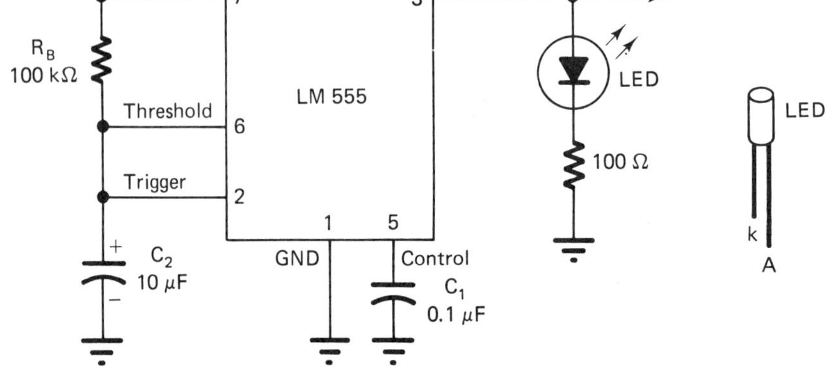

FIGURE 9-5B Astable multivibrator.

The output voltage changes from ________ to ________ V in one cycle of operation.

4. Using a watch, measure the time of one operational cycle. The measured time of one operational cycle is ________ seconds. The operational time of one cycle is $T = 1/f$. Using the calculated frequency of step 2, determine the operational time of one cycle. $T =$ ________ seconds. How do the calculated and measured time of this circuit compare? ________

5. Connect the VOM across capacitor C_2 and observe the voltage change. This voltage change appears at the trigger-threshold input. The trigger voltage (V_t) changes from ________ to ________ V. Ideally, this voltage should be 2/3 of V_{CC} on the upper level and 1/3 of V_{CC} on the lower level. Calculate the voltage range of your circuit using the value of V_{CC}. $V_t =$ ________ to ________. Compare the measured values of V_t with the ideal calculated values. The values are ________ ________.

6. Since the trigger-threshold voltage (pins 2 and 6) determines the status of the output, what is the value of voltage that it takes to cause a state change in the output? This can be determined by observing the trigger voltage that causes a state change in the LED. The state change should occur at the upper and lower extremes of the voltage across C_2. The upper voltage that causes triggering is ________ V. The lower voltage that causes triggering is ________ V.

7. When the upper voltage value across C_2 is reached, it causes the output to change (high-low). When the lower voltage across C_2 is reached, it causes the output to change (high-low).

8. The procedure described above is used to troubleshoot a 555 astable multivibrator. If the voltage change across C_2 and the output occurs very quickly, it could be observed on an oscilloscope. The waveforms observed will be a sawtooth on the trigger-threshold and a square wave on the output. If time permits, change the values of R_A, R_B, and C_2 and repeat the procedure. Smaller values of resistance and capacitance will cause an increase in frequency and a shorter operational time.

9. Turn off the power supply and disconnect the circuit.

DISCUSSION

In this part of the experiment, you were asked to construct a 555 timer connected as an astable multivibrator. This multivibrator has the reset (pin 4) connected high and the threshold-trigger (pins

6 and 2) connected together across capacitor (C_2). Capacitor C_2 charges through R_A and R_B and discharges through R_B. The operational time for one complete cycle is 0.2716 Hz. The voltage change in the output is 3.62 to 0 V. The measured time of an operational cycle is 3 to 3.5 seconds. The calculated time is 3.68 s. These times should be fairly close. The voltage change across C_2 is 3.3 to 1.7 V. The calculated voltage values are 3.32 to 1.64. The upper voltage value (3.3 V) causes the output to go low and the lower voltage value (1.7 V) causes the output to swing high. This procedure can be used to troubleshoot a 555.

ANALYSIS

1. Why does an astable multivibrator continually change states?
2. If C_2 charges through R_A and R_B and discharges only through R_B, the "on" and "off" time of the output will be __________.
3. What causes capacitor C_2 to charge and discharge in a 555 astable multivibrator?

UNIT 9 EXAMINATION
INTEGRATED CIRCUITS

Instructions: Circle the response that most correctly completes each statement.

1. An IC chip is a:
   a. Wafer on which construction is achieved
   b. Fractional part of a wafer
   c. Transistor or diode in the construction of an IC
   d. Logic gate
   e. Substrate
2. The most popular form of digital IC packaging is the:
   a. Flatpack
   b. TO-3 package
   c. Dual in-line package (DIP)
   d. Small outline integrated circuit (SOIC)
3. The foundation on which an IC is built is called a:
   a. Thick film
   b. Thin film
   c. Wafer
   d. Substrate
4. The overall production cost of an IC:
   a. Depends on the number of ICs being manufactured
   b. Is design dependent
   c. Is lower than a discrete-component circuit
   d. Depends on the size of the wafer
5. Monolithic IC construction uses:
   a. Extensive numbers of resistors and capacitors
   b. High values of resistors and capacitors
   c. Extensive numbers of transistors
   d. Discrete components on a thin film
6. Monolithic refers to:
   a. All circuit elements built on a common substrate
   b. Thin film construction technology
   c. Active components being built on discrete chips
   d. Circuit elements formed on an insulated substrate
7. MOS and CMOS ICs have an advantage over most other IC families because they:
   a. Are best suited for digital circuit construction

b. Consume less power
c. Operate with low voltage and current
d. Have low fan-in and fan-out capabilities

8. The yield of an IC is:
a. Greater with a decrease in circuit area
b. Dependent on the size of the circuit area
c. Normally very high
d. Approximately 60% of the potential number of devices on a wafer

9. Normal operation of a linear IC is set:
a. Near cutoff
b. Midway between cutoff and saturation
c. Near saturation
d. At zero 0 input level

10. In a linear op-amp symbol the negative input terminal is:
a. Connected to the power supply
b. The inverting input lead
c. The noninverting input lead
d. Grounded

11. An op-amp essentially responds as a:
a. Transistor amplifier
b. Mathematical calculator
c. High-gain amplifier
d. Logic gate

12. Fan-out of a logic gate is the:
a. Total number of package connections
b. Number of output terminals
c. Number of circuits the output can drive
d. Output voltage developed

13. The most popular and widely used digital IC families are:
a. TTL and RTL
b. MOS and RTL
c. CMOS and MOS
d. TTL and MOS
e. CMOS and TTL

14. A complementary-symmetry circuit configuration is part of:
a. The TTL IC family
b. A bipolar IC
c. MOS ICs
d. CMOS ICs

15. Digital ICs primarily respond as:
a. High-gain amplifiers

b. Mathematical devices
c. Switching devices
d. Voltage controllers

16. A 555 IC used in its astable mode of operation:
a. Produces repeated square waves
b. Responds as a gate
c. Regulates voltage
d. Is a linear amplifier

17. A monostable multivibrator with an RA of 200 kΩ and a C of 1 uF, will cause the output to change its time in the range of:
a. 0 to 1 second
b. 1.1 to 2 seconds
c. 2.1 to 3 seconds
d. 3.1 or more seconds

UNIT TEN

Optoelectronic Devices

UNIT INTRODUCTION

In the last few years the fields of optics and electronics have combined to form a new technology called *optoelectronics.* Optics is a branch of physical science that is concerned with the study of light. Optoelectronic devices deal with devices whose electronic properties are affected by light energy. The term light is used in a general way to include visible, infrared, and ultraviolet regions of the frequency spectrum.

The role of optoelectronics is expanding very rapidly in today's technology. This area was once called photoelectricity. Photocells and phototubes were used to achieve electronic control. Today the field has expanded to include a number of unique solid-state devices. These devices are capable of converting light energy into electrical energy or vice versa. A device can be used as a source of light energy or as a detector of light energy. As a rule, these devices are combined in an operating system that is used to control light energy. Light-control systems are largely responsible for the rapid expansion of this field. Light-emitting diodes (LEDs), photodiodes, phototransistors, lasers, and optocouplers are some of the devices included in the area of optoelectronics.

UNIT OBJECTIVES

Upon completion of this unit, you will be able to:

1. Describe the physical construction of several optoelectronic devices.
2. Distinguish among devices that are used to produce light or to detect changes in light energy.
3. Define some of the photometric and radiometric terms associated with optoelectronics.
4. Describe some of the basic characteristics of light energy.
5. Explain the basic operational principles of several optoelectronic devices.
6. Inspect the characteristic data of several optoelectronic devices and use these data to evaluate the performance of a specific device.
7. Identify representative optoelectronic devices by appearance and schematic symbols.

IMPORTANT TERMS

Before reading Unit 10, review the following terms. These terms provide a basic understanding of terms associated with optoelectronic devices.

Angstrom A radiometric measuring unit used to describe the wavelength of an electromagnetic wave. An angstrom is 10^{-10} meters.

Candela The unit of luminous intensity, defined as 1/60 cm of a black body at 2042 K.

Chromatic Something relating to color, color phenomena, or color response of the eye.

Detector A device that converts optical energy to electrical energy.

Foot candle Illumination level of 1 lumen per square foot. A photometric unit of measurement.

Heterochromatic Of or pertaining to something that has several different colors.

Lumen The amount of light falling on a unit surface all points of which are a unit distance from a uniform point source of 1 candela.

Luminance Photometric brightness, or the intensity of a surface in a given direction per unit of projected area of the surface as viewed from that direction.

Monochromatic Something generating or responding to only one color.

Panchromatic Something that produces or responds to all colors in the visible spectrum.

Photon A small packet or quantum of energy associated with light.

Photometric A method of describing or quantifying light associated with the visible part of the frequency spectrum.

Quantum A discrete amount of energy required to move an electron into a higher energy level.

Quantum theory A theory based on the absorption and emission of discrete amounts of energy.

Radiation A process of emitting or releasing energy in the form of waves sent through space from atoms and molecules as they undergo an internal change. Light energy is emitted or radiated from the sun.

Radiometric A method of describing or quantifying electromagnetic energy that includes both visible and invisible radiation.

Transverse wave A wave that causes particles of a medium to vibrate at right angles to the direction in which the wave is moving.

Wavelength The distance that an electromagnetic or light wave travels in one cycle.

THE NATURE OF LIGHT

Optoelectronic devices are light-sensitive devices that rely on the characteristics of light in order to function. Light is a type of radiant energy that travels through space in the form of electromagnetic waves. Radiation is the process of emitting or releasing energy in the form of waves away from the point of origin or the primary source of energy. Infrared, ultraviolet, and visible are types of radiant energy classified as light. Radio waves, microwaves, and X-rays are forms of radiant energy that are not visible. Radiant energy is usually described by a graph that shows the location of different frequencies. This graph or chart is called a *spectrum.* The electromagnetic spectrum of Fig. 10-1 shows frequencies for radio, television, infrared radiation, visible light, ultraviolet light, X-rays, and gamma rays. These differ only with respect to their frequencies or wavelengths.

The Wave Theory of Light

The wave theory of light states that light consists of waves that spread out from the primary source of light. These waves are sim-

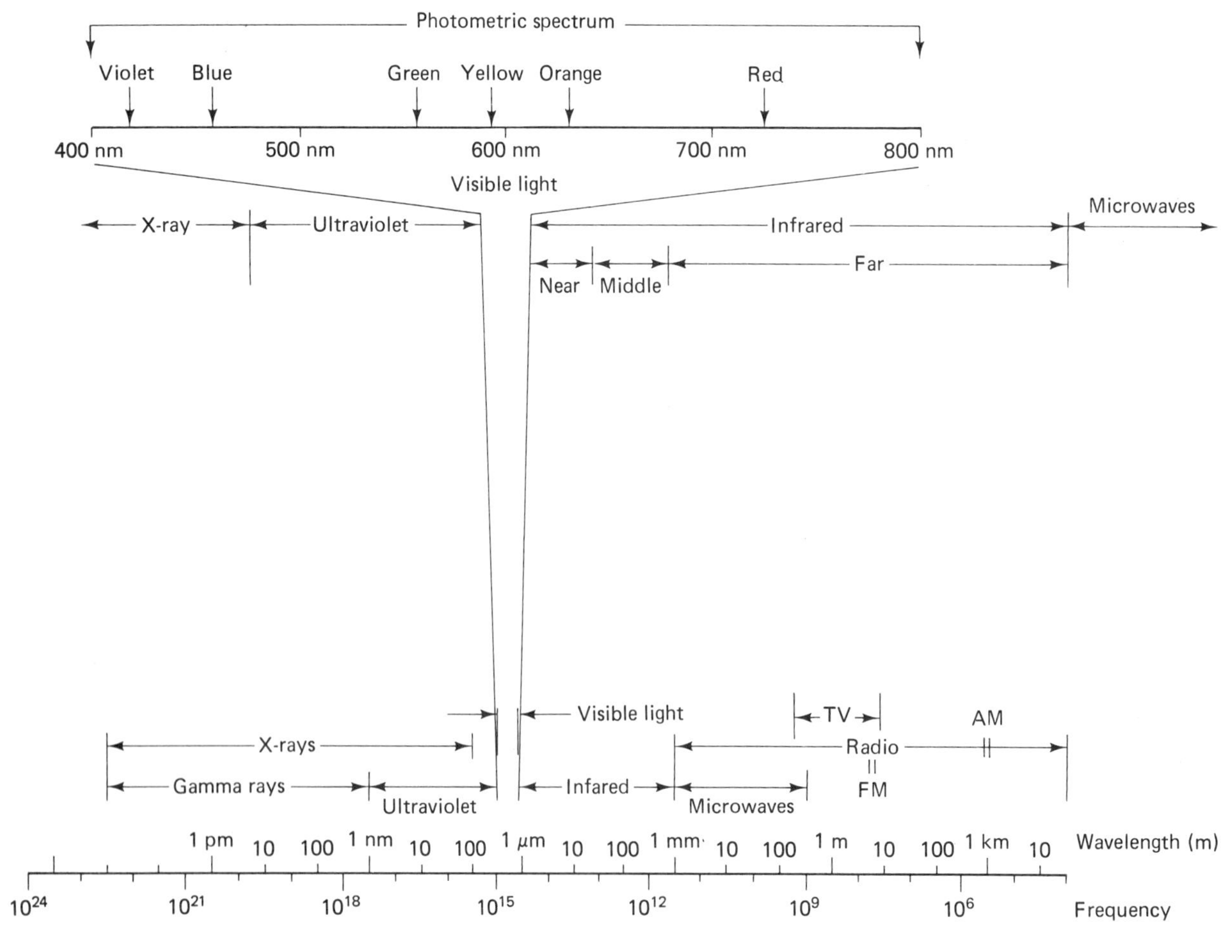

FIGURE 10-1 Electromagnetic spectrum.

ilar to those that occur in water when a pebble is dropped in a pool. Light waves are considered to be transverse. A *transverse wave* is one that causes the particles of a medium to vibrate at right angles to the direction in which the wave is moving. Each wave consists of a crest and a trough. As the wave moves forward, there is an alternate rise and fall in its amplitude.

Light is emitted from a source in the form of electromagnetic waves that are of a specific length or size. One wavelength is a measurement of the distance that a wave travels in space during one cycle of operation. Wavelength is determined by the frequency of the vibrating charge that produces it. All electromagnetic waves, regardless of wavelength, travel at 186,000 mi/s, or 300,000,000 m/s, which is the speed of light.

To understand the wave theory of light transmission, it is necessary to relate the speed of travel, angular velocity, and shape of the radiated signal. Because the speed of light is a constant, there is an inverse relationship between the number of cycles per second (frequency) of the wave and its wavelength. If the frequency is doubled, one cycle, or one wavelength, will occupy one-half as much space. In other words, the higher the frequency, the shorter the wavelength, or the lower the frequency, the longer the wavelength. The symbol most commonly used to denote wavelength is the Greek letter lambda (λ).

The relationship between wavelength and frequency is expressed mathematically as follows:

$$\lambda = \frac{300{,}000{,}000}{f}$$

where λ is the wavelength in meters, f is the frequency in hertz, and 300,000,000 is the speed of light in meters per second. To find the wavelength in meters, divide the velocity of propagation in meters per second by the frequency in hertz. Since the frequency is a very large value, it may be easier to use megahertz (MHz) or gigahertz (GHz) instead of hertz. One megahertz is 10^6 Hz, and 1 GHz is 10^9 Hz.

To demonstrate the size of different wavelengths let us compare some common frequencies. The AM radio band occupies frequencies in the range of 535 to 1620 kHz. The center of this frequency range is near 1000 kHz. The wavelength of this radio frequency is:

$$\lambda = \frac{300{,}000{,}000 \text{ m/s}}{1{,}000{,}000 \text{ Hz}}$$

$$= 300 \text{ m}$$

A meter is 39.37 in. long, so 300 m is equivalent to three football fields laid end to end.

The range of the FM frequency band is 88 to 108 MHz. The approximate center of this band is 100 MHz. The wavelength of a 100-MHz FM signal is:

$$\lambda = \frac{300{,}000{,}000 \text{ m/s}}{100{,}000{,}000 \text{ Hz}}$$
$$= 3 \text{ m}$$

Three meters is the approximate height of a basketball goal, or 10 ft.

Visible light occupies the frequency range between 390,000 GHz (0.39×10^{15} Hz) for red and 760,000 GHz (0.76×10^{15} Hz) for violet. The wavelength of red light is:

$$\lambda = \frac{300{,}000{,}000 \text{ m/s}}{390{,}000{,}000{,}000{,}000 \text{ Hz}}$$
$$= 0.77\ \mu\text{m, or } 7700\ \text{Å}$$

The wavelength of violet light is:

$$\lambda = \frac{300{,}000{,}000 \text{ m/s}}{760{,}000{,}000{,}000{,}000 \text{ Hz}}$$
$$= 0.4 \text{ micrometers } (\mu\text{m}) \text{ or } 4000\ \text{Å}$$

Thus, the wavelength of light is extremely short compared with radio frequencies. Wavelengths are more commonly used than frequencies to describe light. Thus the wavelengths of visible light range from 4×10^{-7} m (400 nm) for violet to 7.7×10^{-7} m (770 nm) for red.

The human eye responds to electromagnetic waves in the visible light band of frequencies, between 400 and 770 nm. Each color of light is associated with a different frequency or wavelength. In the order of increasing frequency or decreasing wavelength, colors range from red, orange, yellow, green, blue, to violet. The response of the human eye to visible light is frequency sensitive, as shown in Fig. 10-2. The greatest sensitivity is near 500 nm, and the poorest sensitivity is around 400 nm at the lower wavelengths and 770 nm at the higher wavelengths. Our eyes perceive various degrees

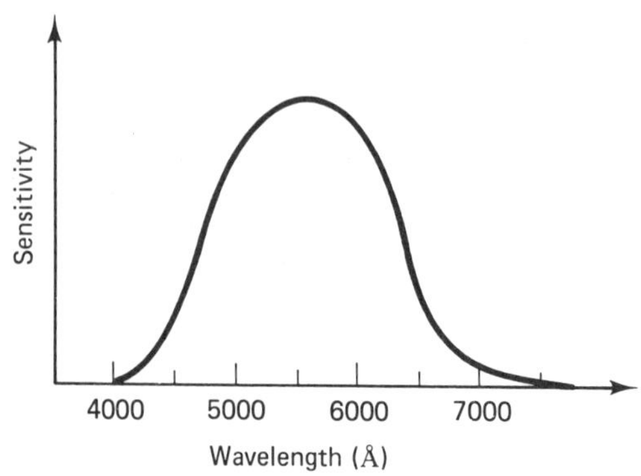

FIGURE 10-2 Response of human eye.

of brightness due to their response to different intensities of energy at different wavelengths of light.

The Quantum Theory

The wave theory of light transmission is used to analyze the response of the human eye to light energy. It is also used to explain how light bends when it passes through glass or water and many other phenomena. Today, we know that light also behaves as though it consists of bundles of energy when it interacts with matter. This model, which is known as the *quantum theory,* is used to explain how light energy is transferred to an optoelectronic device.

The quantum theory states that light is emitted by the atoms of a luminous body in separate packets of energy called quanta or photons. The energy of each bundle depends on the wavelength of the light. Each atom has a distinctive spectrum of wavelengths that it can radiate. These spectrums depend on the atoms' distribution of electrons. Photons travel at the same speed, namely, the speed of light, but they may have different amounts of energy. The energy content of a photon determines the color that will be perceived by the brain when light strikes the eye.

Earlier, we learned that discrete amounts of energy applied to an atom can cause it to release electrons. A loss of electron energy can cause the electron to fall into a lower energy level. Today, we believe that photons are released from atoms when electrons fall to a lower energy level. The photon has energy that is directly related to the frequency of the light wave it produces. This energy is expressed by the equation $e = hf$, where e is the quantum of energy (joules), h is Planck's universal constant (6.62×10^{-34} joules/s) and f is the frequency (hertz).

The quantum theory shows that light transmission has some distinct characteristics that are important in optoelectronics. It is made of discrete bundles of energy that travel from the source in a wave pattern. The wave pattern has a distinct frequency that determines its color. Light should therefore be considered a system that has a source of energy, a path, a control, and a load.

Terms and Units of Measurement

The optoelectronics field is unusual in that it uses two distinct systems of measurement to evaluate the operation of an optoelectronic device or system. The *photometric* system deals with visible light energy and applies specifically to the response of the human eye. The *radiometric* system deals with the entire optical spectrum, including ultraviolet, visible, and infrared light. Wavelengths in this part of the frequency spectrum range from 0.005 to 4000 μ. Visible light is only a small part of the total radiometric spectrum. The two systems are similar in many respects but cannot be used interchangeably. It is often difficult to convert units of measure from one system to the other.

Radiometric Systems

A beam of radiant energy has several measurable characteristics that are important in the optoelectronic field. Optical energy can be radiated through space in a wide range of patterns, can be emitted from a material, or can be reflected from a surface. The radiation process must be specified in order for measurements to be meaningful. Terms such as flux, illumination, intensity, and luminance are used. These terms apply to both radiometric and photometric measurements. Radiometric terms are distinguished from photometric by being preceded by the word *radiant.* In the radiometric system, energy travels from an energy source in the form of electromagnetic waves. Radiant flux or radiant power is the flow rate of radiant energy per unit of time and is represented by the Greek letter phi (ϕ). It is measured in joules per second, or watts.

The term *radiant intensity* is used to indicate the amount of electromagnetic energy produced by a specific source. Radiant intensity is a measure of the radiant power per solid angle unit, or watts per steradian. The solid angle of a sphere is shown in Fig. 10-3. A sphere contains 4π steradians. A steradian (sr) is the unit of measure of a solid angle originating at the center of a sphere with a radius of 1 m subtended by 1 m^2 on the surface of the sphere. A steradian is generally considered to be a dimensionless unit like the radian. It is best expressed as a solid angular measurement. Keep in mind that this measurement has a light source at the center of the sphere and it indicates the amount of intensity produced on 1 sr of outer surface.

Radiant incidence, or *irradiance,* is a measure of the radiant energy that strikes the surface of a specific area. It is measured in watts per unit area. The unit area most comonly used is the square meter or centimeter. Incidence is primarily used to describe the amount of radiation applied to the active surface of a detector. If the sensitivity of the detector is known, incidence will indicate if a particular signal will be detected.

Radiant exitance is a measure of the radiant power emitted or released by a specific surface. Radiant exitance is also called *radiant emittance* or simply *emittance.* It is measured in watts per

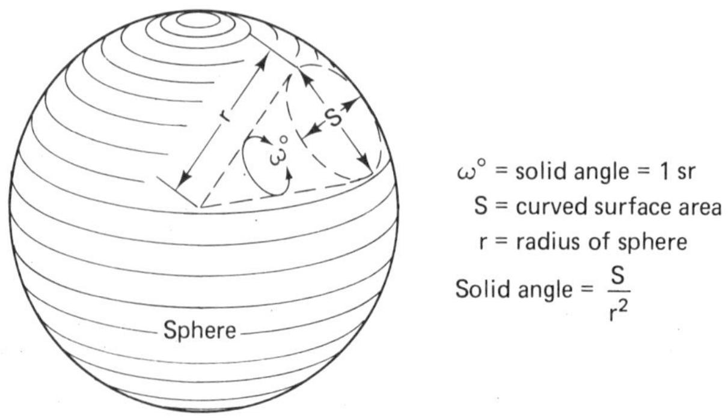

FIGURE 10-3 Solid angle of a sphere.

square meter or square centimeter. Emittance is an indication of the optical power reflected from a surface. The reflectance of a surface varies with wavelength. For example, a radiant exitance of 3.2 mW/cm^2 will result when a surface with 50% reflectance is irradiated with 6.4 mW/cm^2.

Radiance is a measure of the radiant intensity that is either leaving, passing through, or arriving at a specific surface area. It is measured in watts per steradian meter squared and is determined by dividing the radiant energy intensity from a source by the projected area.

Photometric Systems

As we mentioned previously, photometry deals with the measurement of electromagnetic energy that falls in the visible part of the frequency spectrum. Photometric terms are distinguished from radiometric terms by the prefix luminous. Luminous intensity, luminous flux, luminous exitance, and luminous power are some of the common photometric units of measurement.

One characteristic of light that is important in many applications is intensity. *Intensity* is a measure of the amount of energy contained in an electromagnetic wave. In the photometric system the amount of light produced by a source is called its *luminous intensity.* The unit of luminous intensity is a standard light source called a *candela* (cd). One candela is the amount of light that is emitted from a $\frac{1}{60}$-cm^2 hole in one side of a platinum-lined ceramic box after it has been heated to 1772°C, or 3222°F, which causes the platinum to melt. The glow of the melting platinum is emitted from the hole. The candela is a more exacting replacement for the older candle or foot-candle measurement of intensity.

Luminous flux is another important unit of the photometric system. Luminous flux is sometimes called *luminous power.* It specifically refers to the rate at which light energy flows per unit time. The unit of luminous flux is the *lumen* (lm) and is denoted by the Greek letter phi (ϕ). The lumen is the basic unit of measurement for the photometric system. The amount of light falling on a unit surface all points of which are a unit distance from a uniform point source of 1 cd is 1 lm. The lumen is used only to describe visible light energy. It takes 680 lm to equal 1 W of radiant flux, based on the wavelength to which the human eye is most sensitive.

Illuminance, also called *incidence,* is the density of luminous power that it takes to illuminate a surface. The unit of illuminance is the *lux.* One lux is equal to 1 lm/m^2. Illuminance is frequently used to express the amount of light received by a solid-state device. In the United States the foot-candle is used to denote illuminance. One foot-candle is the equivalent of 1 lm illuminating 1 ft^2 of a surface. A foot-candle equals 10.76 lux.

Luminous exitance is a measure of the amount of luminous flux given off or reflected by a surface. It is measured in lumens per square meter and is used to denote the reflecting capability of a specific surface. The reflectance of the surface varies with the wavelength of the illuminance. The unit of luminous exitance was once called luminous emittance or simply emittance.

SELF-EXAMINATION

1. Optoelectronics combines the technologies of ________ and ____________.
2. ____________ is the process of emitting or releasing energy in the form of electromagnetic waves.
3. A graph or chart that shows the location of different electromagnetic wave frequencies and wavelengths is called a ____________.
4. A ____________ wave causes particles to move at right angles to the direction of motion.
5. Electromagnetic waves travel at ____________ miles per second or ____________ meters per second.
6. One ____________ is a measure of the distance that an electromagnetic wave travels in space during one cycle of operation.
7. Higher frequencies cause the wavelength to be ____________.
8. Visible light ranges from a wavelength of ____________ (Å) for violet to ____________ (Å) for red.
9. The color ____________ has the lowest frequency, whereas the color ____________ has the highest frequency.
10. The color ____________ has the longest wavelength, whereas the color ____________ has the shortest wavelength.
11. The quantum theory is based on the emission of energy packets called ____________.
12. ____________ systems deal with the entire electromagnetic spectrum.
13. ____________ systems deal with visible light.
14. Radiant ____________ is measured in joules per second, or watts.
15. Radiant ____________ is used to measure the amount of electromagnetic energy produced by a source.
16. Radiant ____________ is a measure of the density of radiant power applied to a surface.
17. Radiant ____________ is a measure of the radiant power released from a surface.
18. Luminous ____________ is a measure of the amount of visible light produced by a source.

19. The unit of luminous intensity is the ____________.
20. The unit of luminous flux is the ____________.
21. The unit of illuminance is the ____________.

ANSWERS

1. Optics, electronics	2. Radiation
3. Spectrum	4. Transverse
5. 186,000, 300,000,000	6. Wavelength
7. Shorter	8. 4000, 7700
9. Red, violet	10. Red, violet
11. Photons	12. Radiometric
13. Photometric	14. Flux
15. Intensity	16. Incidence
17. Exitance	18. Intensity
19. Candela	20. Lumens
21. Lux	

RADIATION SOURCES

A variety of radiation sources are used in optoelectronic systems. Light-emitting diodes (LEDs), tungsten lamps, neon lamps, fluorescent lamps, and xenon tubes are some of the more common sources of electromagnetic radiation. As a rule, these sources radiate energy in the visible part of the electromagnetic spectrum. Some LEDs, however, are purposely designed to radiate only infrared energy. Emitters that generate infrared wavelengths or energy outside the visible part of the spectrum are said to be radiometric. Since all optoelectronic sources can be classified as radiometric, optoelectronic specifications may be stated in either radiometric or photometric terms.

Photometric sources are frequently classified by the amount or range of visible light emitted. In this regard, sources are described as being *panchromatic, heterochromatic,* or *monochromatic.* The term *chromatic* in these expressions refers to a specific color or various colors of the spectrum. An incandescent lamp is an example of a panchromatic source. Its radiation extends over a very large portion of the optical spectrum. A wide range of visible colors can be produced by an incandescent source.

Heterochromatic sources produce a very limited number of different colors. Mercury arc lamps, for example, produce colors that are predominantly red or orange. Specifically, this particular source generates light in the region of 6500 Å.

A monochromatic source radiates energy of only one specific wavelength or in a very narrow part of the spectrum. A sodium-vapor lamp, for example, radiates energy at 5000 Å. Figure 10-4 shows the spectral response of several different light sources. Notice the chromatic differences in these sources.

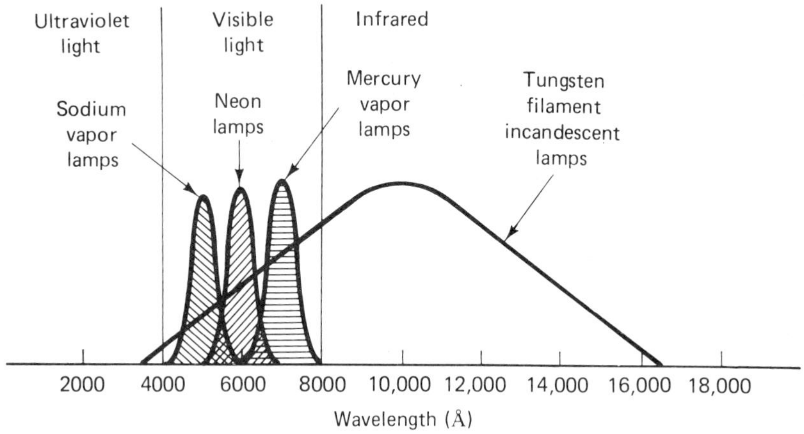

FIGURE 10-4 Spectral response of radiation sources.

Incandescent Lamps

An incandescent lamp is frequently used as the radiation source of an optoelectronic system. This type of source is readily available, is reliable, and has a consistent level of operation. Tungsten filament lamps are used widely for this application. The radiation output of this source peaks at approximately 10,000 Å, as shown in Fig. 10-4. It also radiates electromagnetic energy in the visible light region.

When incandescent lamps are used as a light source, they are generally operated at the lowest voltage value that is consistent with good performance. Low-voltage operation extends the life span of the lamp because the operating life of a lamp is equal to the twelfth power of the ratio of the rated voltage to the operating voltage. Lamp life in an optoelectronic system ranges from 5000 hours to something in excess of 50,000 hours, depending on the lamp and its operating voltage.

Operation of an incandescent lamp is based on the passage of current through an electrical conductor. When electrons pass through a conductor, they encounter a certain amount of resistance. This opposition to electron flow causes heat to be produced in the conductor. When enough heat is produced, the conductor changes its appearance. At low values of heat there may be no change in the outward appearance of the material. As the temperature rises, the color of the material changes. It is possible to see a dull orange glow, bright orange, yellow, and a very intense white glow called incandescence. Tungsten lamps operating at 3500°C produce a very high level light intensity. These lamps have a very short life expectancy. Operating the same lamp at a lower temperature will increase its life expectancy. The wavelengths of light produced by a tungsten filament lamp are shown in the spectral response of Fig. 10-4.

Light-Emitting Diodes

A semiconductor optoelectronic device that is used rather extensively as a light source is the LED, or solid-state lamp (SSL). The LED contains a P-N junction that emits light when forward biased. A schematic symbol, crystal structure, and typical package of an LED are shown in Fig. 10-5. The housing or lens of an LED is transparent and focuses on the P-N junction.

When an LED is forward biased, electrons cross the junction and combine with holes, which causes them to fall out of conduction and return to the valence band. The energy possessed by each free electron is then released. Some of this appears as heat energy, and the rest of it is given off as light energy. Special materials such as gallium arsenide (GaAs), gallium phosphide (GaP), and gallium arsenide phosphide (GaAsP) cause this reaction when used in a P-N junction. These materials are purposely used because they permit various forms of radiant energy to be produced.

When an LED is forward biased, the applied electrical energy causes the radiation of electromagnetic energy. The spectral response of an LED varies according to the type of semiconductor material used in its construction. Combinations of certain materials and varying doping levels are used to produce LEDs that emit many different wavelengths of energy and consequently different colors. Some of these combinations and the resulting wavelengths are listed:

GaAs	8800 Å (infrared)
GaP	5500 to 7000 Å (green to red light)
GaAsP	5800 to 6600 Å (amber to red light)

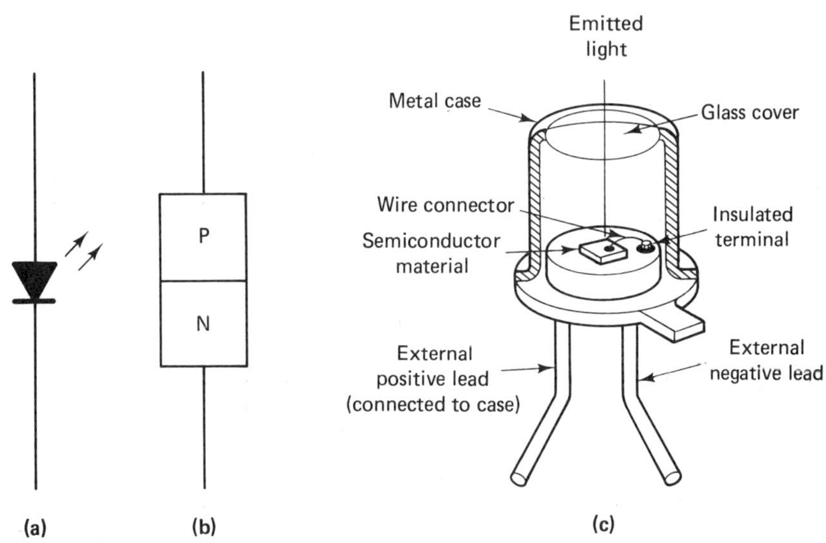

FIGURE 10-5 Light-emitting diode: (a) LED symbol; (b) crystal structure; (c) package.

When GaAs is used in the construction of an LED, it emits infrared energy, which is not visible to the human eye. Devices of this type are called *infrared emitting diodes*, or *IREDs*. IREDs are used in systems that must not be influenced by ambient light or normal room light.

Spectral response curves for different LEDs are shown in Fig. 10-6. Compare these curves and their peak response with the color response curve shown in Fig. 10-4. This comparison shows that the wavelengths of radiated energy range from the visible to the infrared region of the frequency spectrum. Specific combinations of different phosphors are added to produce the desired color effects. LEDs are very sensitive to temperature changes, which cause the spectral response of the LED to shift to longer wavelengths. An increase in temperature also reduces the amount of radiant output power.

A unique feature of the LED as a light source is its rapid switching time. Compared with an incandescent lamp, an LED can be switched on and off very rapidly. Switching rates of several thousand hertz can be achieved today, with no significant delay time. High-speed switching of an LED can be achieved with a square-wave generator or a trigger pulse. No heat is required by an LED to produce light. Radiation is produced only when the P-N junction is forward biased, and holes and electrons recombine at the surface of the junction.

LED Characteristics. The operational characteristics of an LED are very similar to those of a typical diode. The current increases very rapidly, with the voltage remaining nearly constant after forward conduction occurs. Figure 10-7 shows an *I-V* characteristic for a GaAs LED. There is a limit to the forward current and reverse voltage that an LED can withstand without being damaged. Most LEDs can withstand a forward current of 50 to 100 mA; higher current values require a heat sink to prevent thermal damage to the junction. The peak reverse voltage or the PRV rating of an LED is usually quite low compared with a conventional silicon diode. Only a few volts can be tolerated without causing permanent damage to the junction. Typical PRV ratings are less than 50 V for an LED.

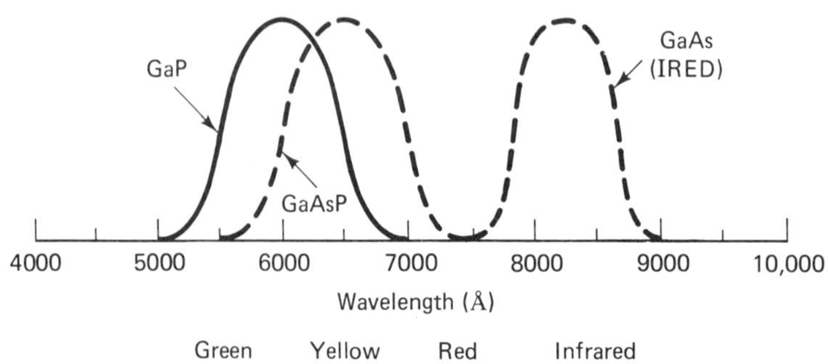

FIGURE 10-6 LED spectral response curves.

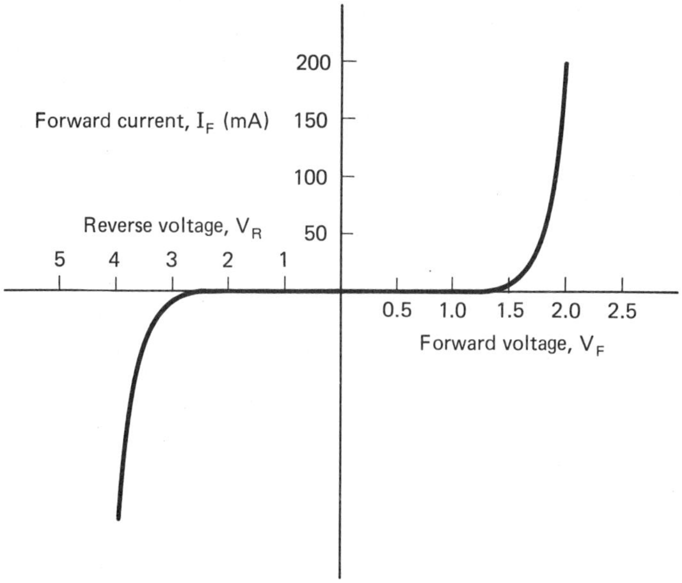

FIGURE 10-7 *I-V* graph of a GaAs LED.

The light-output characteristic of an LED is an important operational consideration. Figure 10-8 shows a graph of the output of a red GaAsP LED. This graph was made by pointing an LED into a calibrated photometer and applying current to the LED. The current was then monitored and recorded for each level of light intensity. Notice that the light output is linear until approximately 65 mA is reached. Beyond 80 mA the light output falls well below the peak value due to the development of excess heat across the junction. This particular LED is intended for low-current applications. Its forward current rating is 50 mA. Operation is permitted beyond the upper current rating, with reduced efficiency.

LED Applications. LEDs are used primarily as indicating devices for electronic circuits. When the LED is forward biased, it produces light. Reverse biasing does not produce light. In an actual circuit, very little current flows or light is produced until the forward bias voltage is greater than the forward voltage (V_F) of the LED. Typical V_F values are in the range of 2 V. In the reverse-bias direction, very little current flows, and no light is produced.

Most applications of the LED require that it be protected from excessive forward current flow. This is achieved by connecting the LED in series with a current-limiting resistor. Figure 10-9 shows an LED connected to a voltage source and a series resistor. The value of the series resistor can be calculated from Ohm's law. The forward voltage or V_F of the LED is fairly constant when it goes into conduction. The voltage across the series resistor (V_{RS}) is the difference between the source voltage (V_{CC}) and the forward voltage (V_F) of the diode, or $V_{RS} = V_{CC} - V_F$. A safe forward current (I_F) is chosen that will produce a reasonable

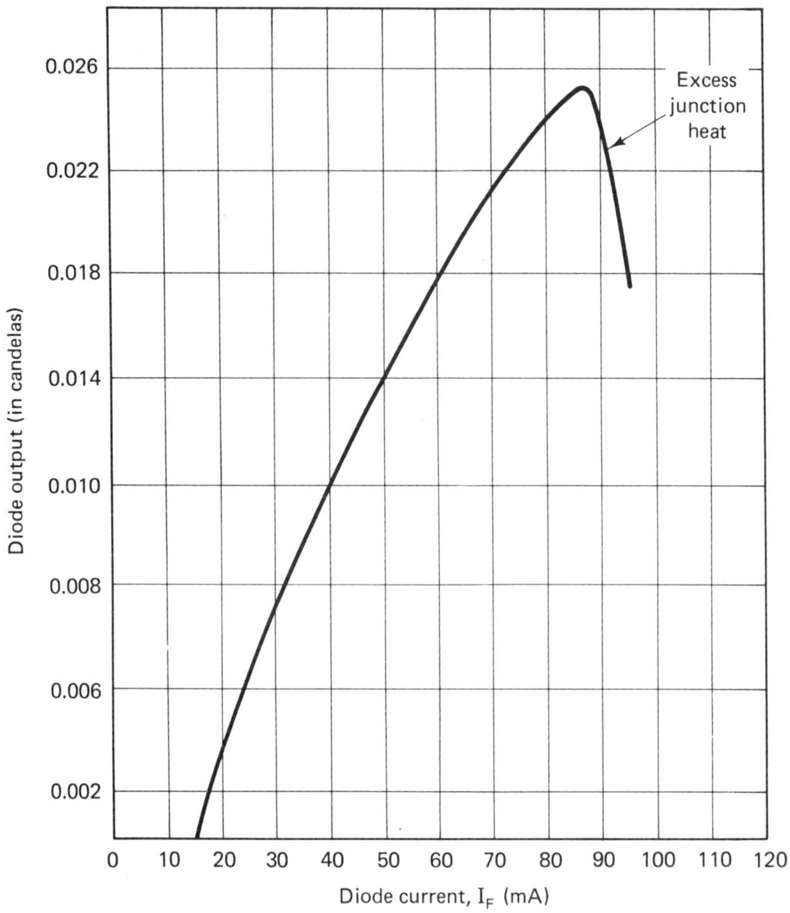

FIGURE 10-8 Light output versus forward current characteristic of a GaAsP (LED).

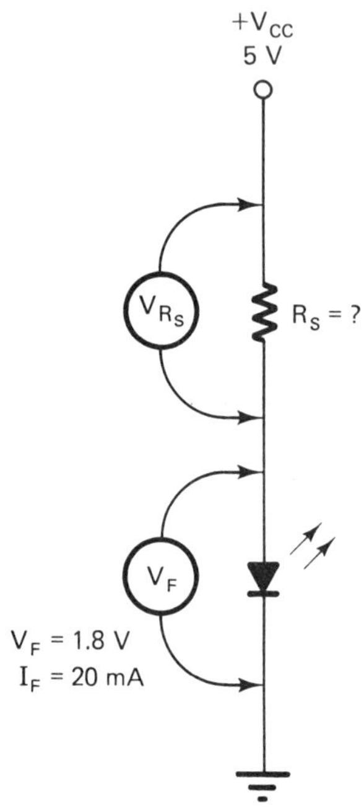

$$R_S = \frac{V_{CC} - V_F}{I_F} = \frac{V_{R_S}}{I_F}$$

$$= \frac{5\text{ V} - 1.8\text{ V}}{20\text{ mA}} = \frac{3.2\text{ V}}{20 \times 10^{-3}}$$

$$= 160\ \Omega$$

FIGURE 10-9 Current-limiting resistor calulation for an LED.

level of light. This current, which also flows through R_s, is then divided into V_{RS} to find the value of R_s, as expressed by the formula $R_s = V_{Rs}/I_F$.

LEDs are commonly used in seven-segment and 5 × 7 dot-matrix displays. The LEDs of these devices produce visible light when forward biased and no light when reverse biased. As a result of this two-state condition, discrete segments or dots can be illuminated when diodes are energized. Typically, the positive side of

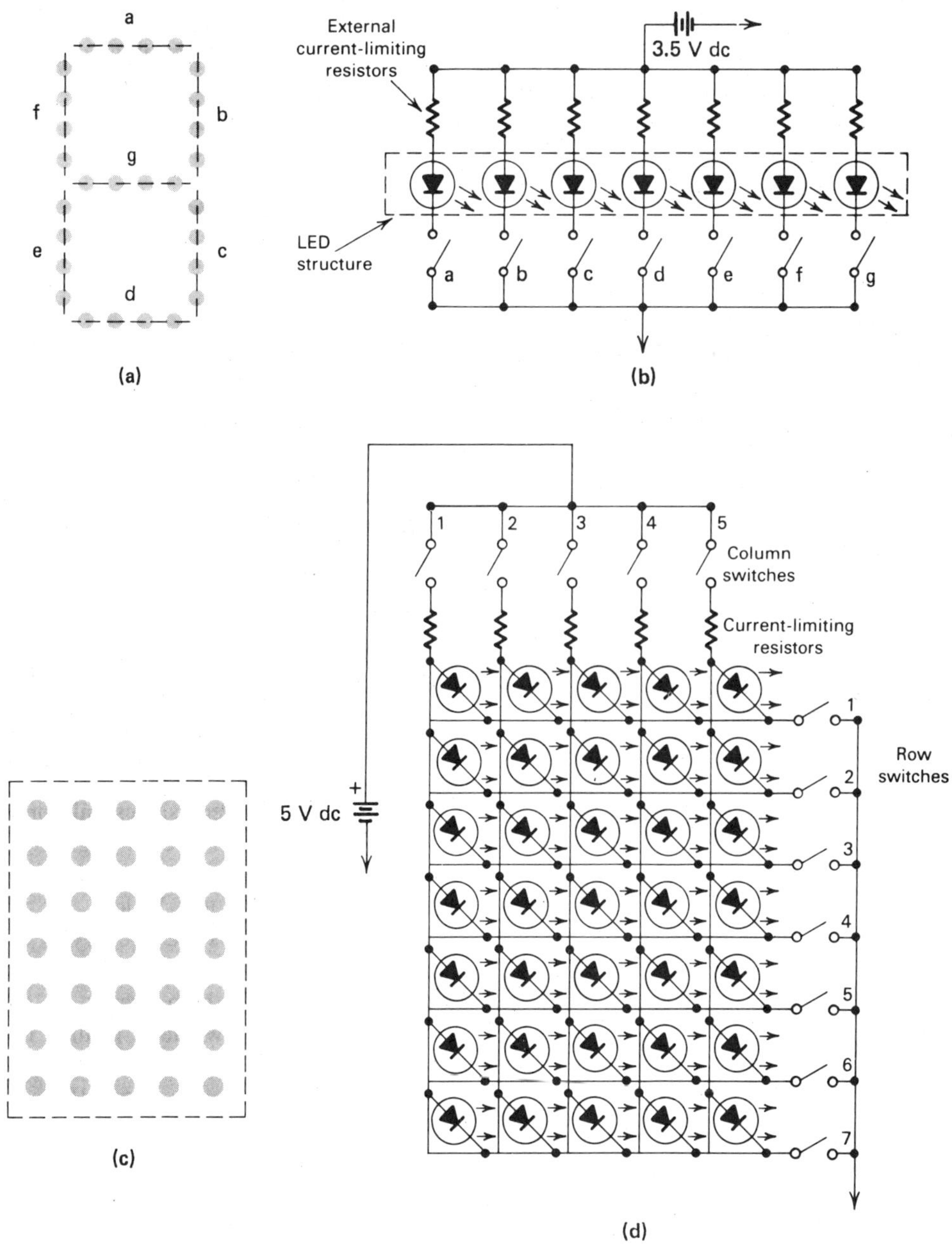

FIGURE 10-10 LED display devices: (a) seven-segment LED display with four diodes per segment; (b) diagram showing connections of segments; (c) 5 × 7 LED dot-matrix display; (d) 5 × 7 LED dot-matrix diagram.

the energy source is applied to the anode of each diode through a current-limiting resistor. The cathode of the respective diode is then grounded by switching. When the circuit is complete, the diode is energized, thus producing light.

Seven-segment LED display devices often contain four or more discrete diodes connected in parallel to form a segment. This type of construction usually requires only one current-limiting resistor for each segment. The amount of voltage needed to produce illumination is typically 3.5 to 5 V dc.

Figure 10-10 shows the circuitry of seven-segment and 5 × 7 dot-matrix LED display devices. The LEDs in both circuits are similar in all respects. The switching method needed to energize specific diodes is somewhat different. In the seven-segment device of Fig. 10-10(b) each segment is controlled by a single switch. The dot-matrix circuit of Fig. 10-10(d), by comparison, is controlled by two or more switches. A discrete diode can be energized by two or more switches, such as row 4, column 5. A complete vertical row would require one column switch and all seven row switches. A complete horizontal row would be energized by one row switch and all five column switches. As a general rule, dot-matrix display devices are used to produce letter displays more so than numbers. LED display devices are used more frequently in circuit applications today than all other devices combined.

Another application combines an LED with other light-sensitive devices such as photodiodes and phototransistors to form an optocoupler. *Optoelectronic couplers,* which are sometimes re-

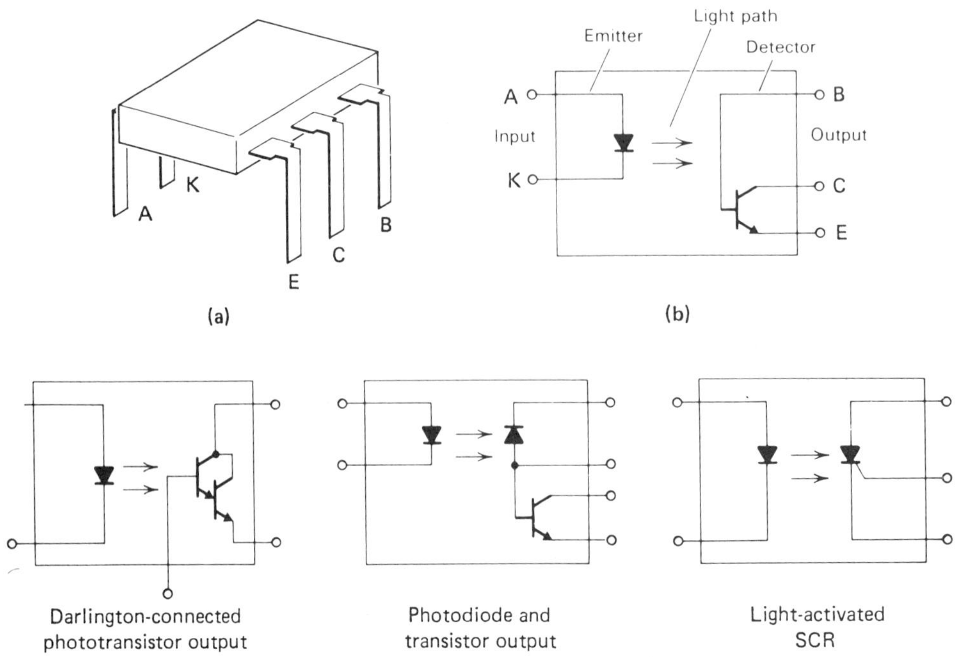

FIGURE 10-11 Optocouplers: (a) package; (b) equivalent circuit; (c) other configurations.

ferred to as *optoisolators*, provide a one-way transfer path for electrical signals. Optoelectronic couplers are commonly manufactured in dual-in-line packages as shown in Fig. 10-11(a). Note the equivalent circuit of the optoelectronic coupler in Fig. 10-11(b). An LED or an IRED is combined with a transistor amplifier in the same package. The input of this device is the LED; the output is the transistor circuit. The collector current of the transistor varies in direct proportion to the current through the LED; therefore, when an input signal is applied to the LED, it emits light in proportion to the applied signal voltage. The light causes the transistor to conduct, thus increasing the collector current. This is achieved with no electrical connection between the devices that comprise the coupler. Isolation of the two circuits is therefore achieved.

SELF-EXAMINATION

22. The source of an optoelectronic system is responsible for generating some form of ____________ energy.
23. Radiant energy produced by the source of an optoelectronic system must be picked up by a ____________.
24. An incandescent lamp is an example of a ____________ source.
25. ____________ lamps produce light when current passes through the conductive material of the filament.
26. An LED produces light when ____________ biased.
27. When GaAs, GaP, or GaAsP are used in the construction of an LED, the LED will produce different ________ of radiant energy.
28. IREDs produce ____________ energy.
29. A unique operational feature of an ____________ source of radiant energy is that it can be switched on and off very rapidly.
30. Most LEDs can withstand a forward current of ____________ mA.
31. When an excessive amount of forward current passes through an LED, its output intensity is reduced because of ____________.
32. When an LED is connected to an energy source, it must be protected from excessive current with a ____________ resistor.
33. A ____________ display has several LEDs formed into bars or segments.
34. A ____________ display has individual LEDs formed in a 5 × 7 matrix.

35. In an optocoupler an LED serves as a ________ of light energy.
36. The primary function of an optocoupler is to ________ the input from the output.

ANSWERS

22. Radiant	23. Detector or receiver
24. Panchromatic	25. Incandescent
26. Forward	27. Wavelengths
28. Infrared	29. LED
30. 50 to 100	31. Heat
32. Current-limiting	33. Seven-segment
34. Dot-matrix	35. Source
36. Isolate	

OPTOELECTRONIC DETECTORS

The detector of an optoelectronic system is classified as a *transducer.* Detectors are designed to change radiant energy into electrical energy, that is, their function is opposite that of the source. A number of detection devices are available today, and they are divided into three general categories: (1) photoemissive, (2) photoconductive, and (3) photovoltaic. *Photoemissive* devices emit electrons in the presence of light. Vacuum and gaseous phototubes are typical photoemissive devices. *Photoconductive* devices are designed so that their resistance decreases when light becomes more intense and increases when light intensity decreases. Photodiodes, phototransistors, and photoresistive devices are common examples of photoconductive devices. *Photovoltaic* devices convert light energy into electrical energy. When a photovoltaic device is illuminated, an electrical potential is created by the device. Most optoelectronic detectors fit into one of these categories. However, there is such a diversity of semiconductor optoelectronic devices that it is more desirable to discuss each device individually.

Photoemission

The phototube shown in Fig. 10-12 is similar in appearance to a vacuum tube or gaseous tube. When light strikes the cathode of a phototube, photons of light are absorbed by the surface of the cathode. The absorption of this energy causes electrons on the surface of the cathode to gain enough energy to leave the cathode. This phenomenon is called *photoelectric emission.* The energy of the electrons depends on the wavelength of the light striking it. A spectral response curve is plotted by the manufacturer for each

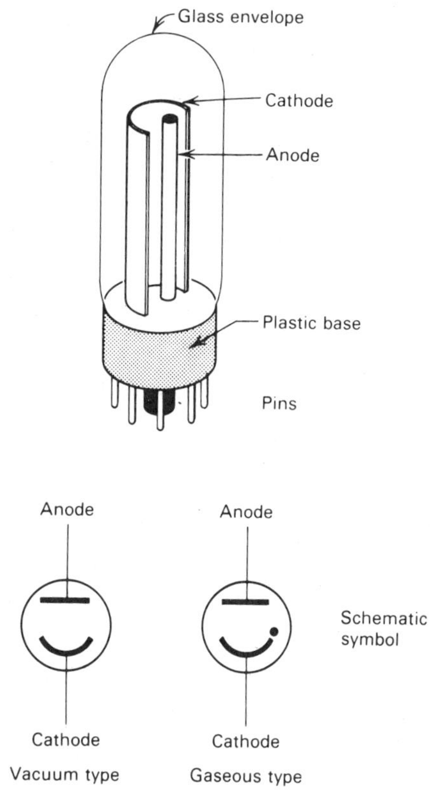

FIGURE 10-12 Photoemissive cell.

type of device produced. A representative spectral response curve is shown in Fig. 10-13.

A phototube may be connected in a circuit as shown in Fig. 10-14. When light of the proper wavelength is focused on the cathode, electrons are emitted and travel to the positively charged anode or plate, where they cause a plate current (I_p) to flow, which produces a voltage drop across the load (R_L). The plate current, caused by various combinations of light and plate voltage, may be determined by using a phototube characteristic curve supplied by the manufacturer. Such a curve is shown in Fig. 10-15. Pho-

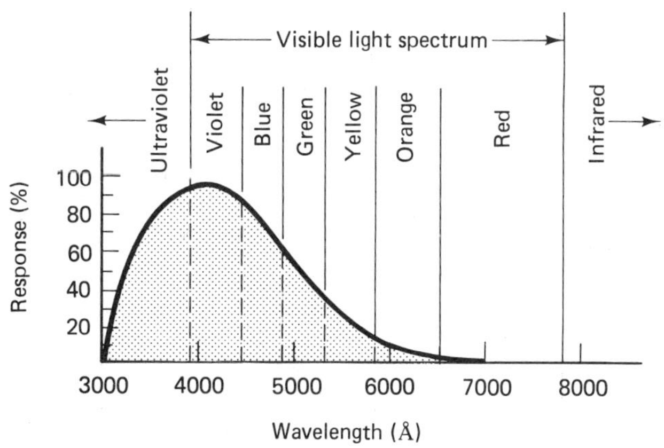

FIGURE 10-13 Spectral response curve for a phototube.

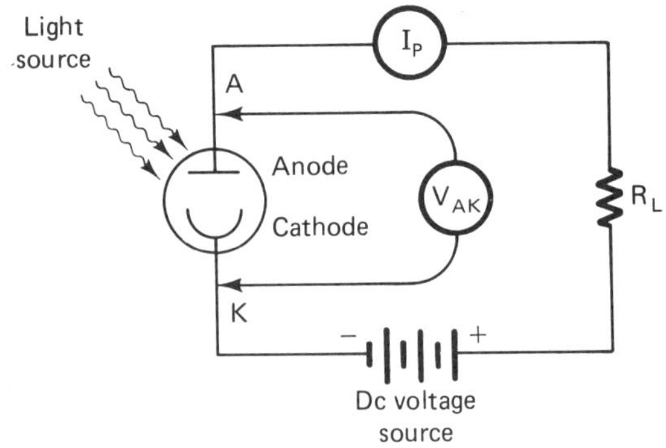

FIGURE 10-14 Phototube circuit action.

toemissive tubes may be of the vacuum type or filled with a gas. A gas phototube is usually more sensitive and thus requires less light to produce a given amount of anode current. A phototube circuit may be calibrated in order to measure specific values of light. This permits the tube to be used in instruments such as light-exposure meters.

At one time nearly all photoelectric control was achieved by photoemissive tubes. The unusually high voltage needed to energize a phototube is rather uncharacteristic of most electronic devices today. In general, this type of device has been replaced by solid-state photodiodes. As a rule, circuit control can be achieved more efficiently with lower voltage values. The photoemissive principle is important, however, because it is the basis of photomultiplier tube operation.

Photomultiplier Tubes. Most phototubes have a very small output when the light intensity is low. The photomultiplier tube, shown in Fig. 10-16, overcomes this disadvantage. Note the internal construction of this tube. Light that strikes the cathode causes electrons to be emitted and focused toward positive plates called *dynodes*. Each dynode possesses a successively higher positive charge to attract these electrons. The anode has the highest positive potential. Through the principle of secondary emission, a large number of electrons are produced by each successive dynode.

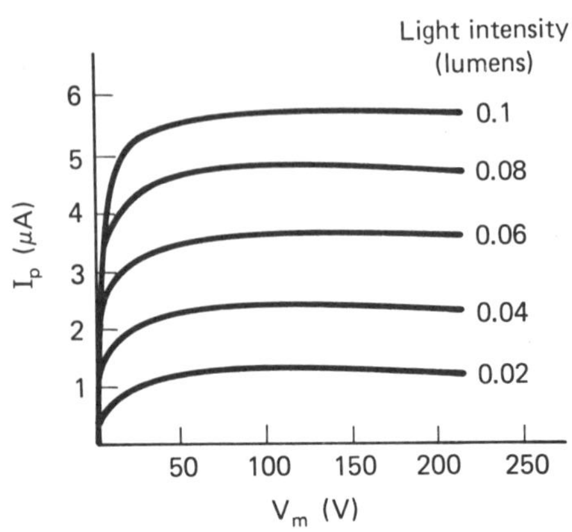

FIGURE 10-15 Phototube characteristic curves.

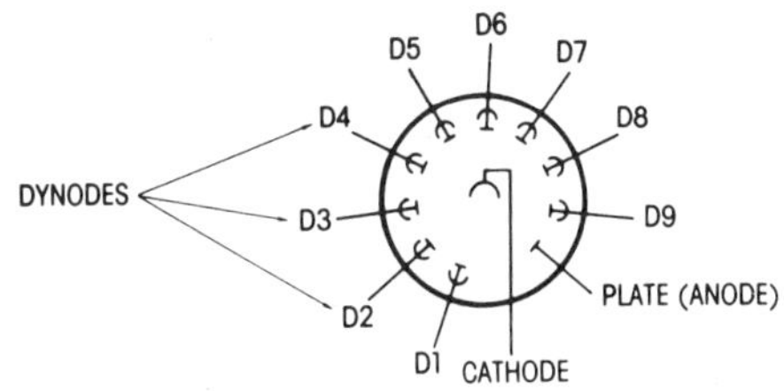

(A) Schematic symbol.

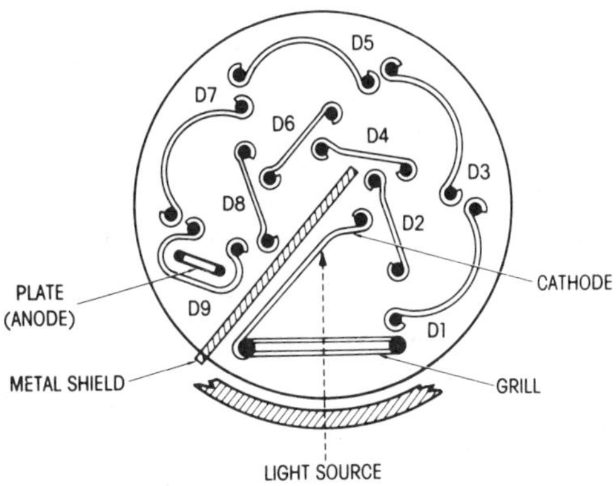

(B) Construction.

(C) Photograph.

FIGURE 10-16 Multiplier phototube.

Most multiplier phototubes use from 9 to 14 dynodes to produce a high output at low light levels.

Thus a small amount of light can cause enough electrons to be emitted to produce a significant output. Photomultiplier tubes have a fairly high level of signal multiplication, which is extremely important in detecting low levels of light.

Photoconductive Devices

An optoelectronic device that changes light intensity to electrical conductivity is classified as a *photoconductive* device. Conductance or conductivity is the reciprocal of resistance. Conductivity is a measure of the ease with which current carriers pass through a material. Because of the relationship between conductance and resistance, these devices are usually described as *photoresistive cells,* or light-dependent resistors. Photoresistive cells are essentially semiconductor devices. Light-sensitive materials such as cadmium sulfide (CdS), cadmium selenide, (CdSe), and cadmium telluride (CdTe) are used in the construction of these devices. The material used in device construction determines how it will respond to different levels of light and different wavelengths of radiant energy. CdS cells have a response very similar to that of the human eye. They respond best to yellow-green light, which has a wavelength of 5500 Å. CdS cells are more sensitive to red, or 7000 Å, whereas CdTe is best suited for infrared light, or 8000 Å.

Cadmium sulfide cells are currently used in more applications than the other materials. This popularity is primarily related to its high sensitivity to light. The resistance of a CdS crystal in the dark may be from 10,000 to 100,000 times greater than its resistance when exposed to an intense light. The sensitivity of the material is improved by increasing its surface area. Most CdS cells are constructed with a geometric pattern of the material on a glass substrate. Figure 10-17 shows a pattern designed to have maximum surface area coverage. This wafer is then hermetically sealed in a glass or metal housing. A glass or plastic window covers the CdS pattern area. Light energy passing through the window causes the material to change resistance.

The operation of a photoconductive cell is based on the response of its material to light energy. When photons of light strike the light-sensitive material, they cause valence electrons to break their atomic bonding. These electrons are then free to take part in the conduction process. For each free electron produced, a corre-

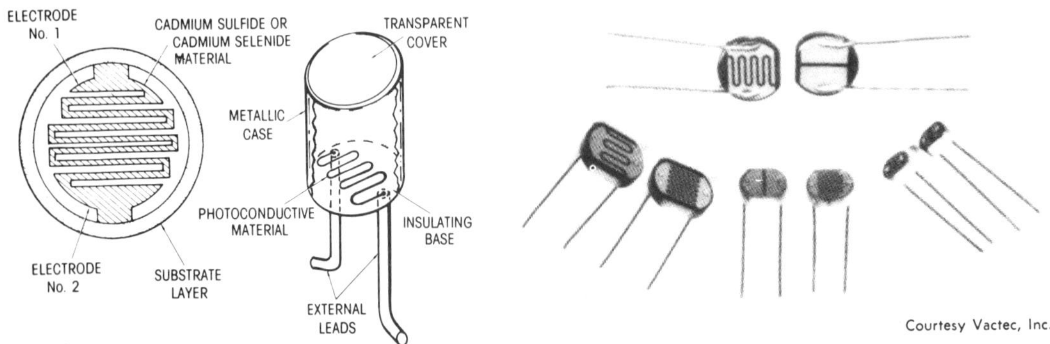

(A) Top view. (B) Cutaway view. (C) Photograph.

FIGURE 10-17 Cadmium sulfide photoconductive cell: (a) top view; (b) cutaway view; (c) photograph.

sponding hole is established in the covalent bonded structure. When conduction takes place, electrons move in the reverse direction. Conduction is based on the number of current carriers moving in the material. High levels of light intensity cause the material to be low resistant. A typical photoconductive cell will change resistance to 100 Ω when the light intensity is 100 lux, which is the equivalent of 9 foot-candles. Reduced light intensity causes an increase in material resistance. In total darkness the resistance of the cell may be 100 kΩ. When the light intensity is reduced, electrons fall out of conduction and recombine with holes. The intensity of the light and its wavelength control the conductivity of the material. The total change in resistance due to a small change in light intensity is the unique advantage of this device.

The symbol for a CdS cell or light-dependent resistor (LDR) is shown in Fig. 10-18. This transducer is represented by two different symbols. The symbol with arrows shows light energy being directed to the resistive material. The alternate symbol shows a resistor with the Greek letter lambda (λ), which is a common designation for wavelength. This symbol shows that the resistor responds to light wavelengths. When these devices are connected in a circuit, as indicated, an increase in light intensity causes a reduction in resistance. This causes a corresponding increase in device current and a reduction in voltage across the LDR. In a sense, this type of device responds as a variable resistor that is controlled by light intensity.

An application of the photoconductive cell is shown in Fig. 10-19. In this circuit the SCR conducts when light is focused on the light-dependent resistor. When light strikes the photocell, its resistance decreases. The potential at point A becomes more positive and causes gate current (I_G) to flow. A sufficient amount of gate current triggers the SCR into conduction. When the SCR conducts, the load device is activated. The SCR then conducts until its anode circuit is opened. The variable resistor (R_1) is used as a sensitivity adjustment to control the level of light required to cause the SCR to conduct.

Photodiodes. Photodiodes are also classified as photoconductive devices. This type of device has light-sensitive P- and N-materials in its construction. A P-N junction is formed by the two materials. Normally, a photodiode is connected in the reverse-

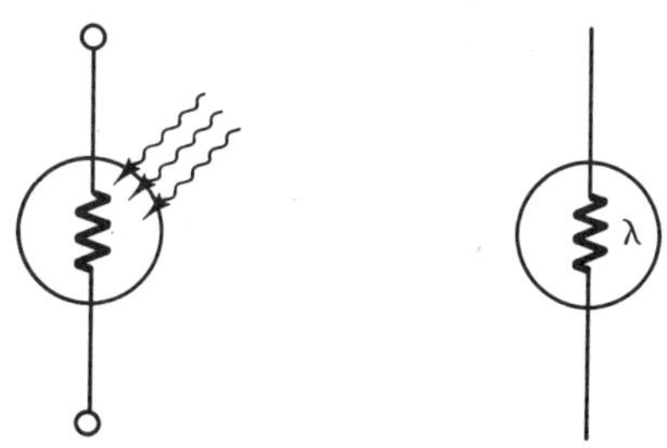

FIGURE 10-18 Schematic symbols of a CdS cell.

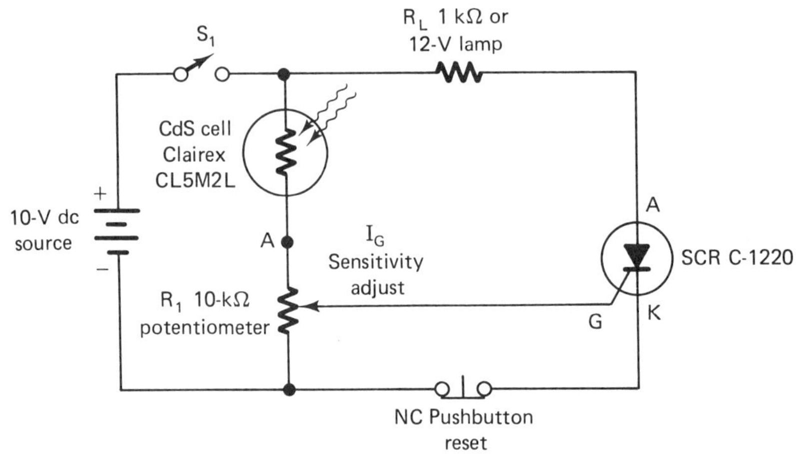

FIGURE 10-19 Photoconductive cell application.

biased mode of operation. Without light, there is an extremely high reverse resistance. Any conduction current that flows is called *dark current.* As a rule, dark current is due to the thermal generation of current carriers. Only a few nanoamperes (nA) of dark current occur in a reverse-biased photodiode at low light levels.

When a photodiode is illuminated, radiant energy causes the valence electrons of the P- and N-materials to go into conduction. In effect, this causes electrons and holes to be returned to the current carrier depletion region near the reverse-biased P-N junction. The return of current carriers increases conductivity and lowers junction resistance. A photodiode will change its conductivity with light intensity only when the junction is reverse biased. Forward biasing the same P-N junction causes conductivity that does not change with light intensity. Photodiodes are energized in a circuit with reverse-biased dc.

PIN Photodiodes. A rather recent innovation in photodiode construction is the PIN device. As shown in Fig. 10-20, a regular photodiode has a P-N junction in its construction. Light applied to the junction causes current carriers to return to the depletion region. This type of diode has a rather significant junction capacitance (C_j), which tends to slow down the response of the diode to changes in light intensity.

A PIN diode, as shown in Fig. 10-20(b), has a layer of undoped semiconductor material between the P- and N-materials of the junction. The letter I denotes this intrinsic or undoped semiconductor material. The added I layer physically increases the width of the depletion region of the P-N junction, which lowers the junction capacitance of the device. A PIN diode therefore has lower C_j and a faster response to changes in light intensity. As a rule, this type of construction also reduces the dark current of the device to a much lower value.

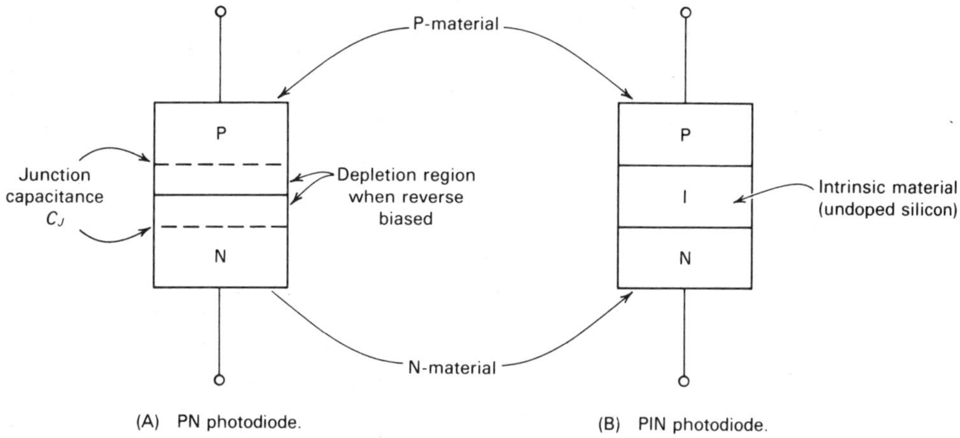

FIGURE 10-20 Comparison of PN and PIN photodiodes: (a) PN photodiode; (b) PIN photodiode.

Phototransistors. The phototransistor is a conductivity device that has two P-N junctions in its construction, similar to that of a conventional bipolar transistor. The collector-base junction therefore responds as a photodiode. Light energy applied to this part of the device generates current carriers in the base region. This photocurrent controls the emitter-collector current. Essentially, a small change in base current is used to control a larger collector current. A phototransistor is therefore capable of amplification. This makes the phototransistor more sensitive to changes in light intensity than the photodiode.

A cross-sectional view of a phototransistor is shown in Fig. 10-21. In this structure, light energy must be directed toward the base area. The entire assembly is housed in an enclosure that has a lens or window centered over the base region. Many devices of this type have the base lead omitted. A floating-base device has only emitter and collector lead connections. Base current is generated by light energy and controls current flow between the emitter and collector. With an external base lead, the phototransistor has additional control over the collector current. A three-lead phototransistor can also be used as a photodiode by disconnecting the emitter and using only the base-collector leads. Figure 10-22 shows

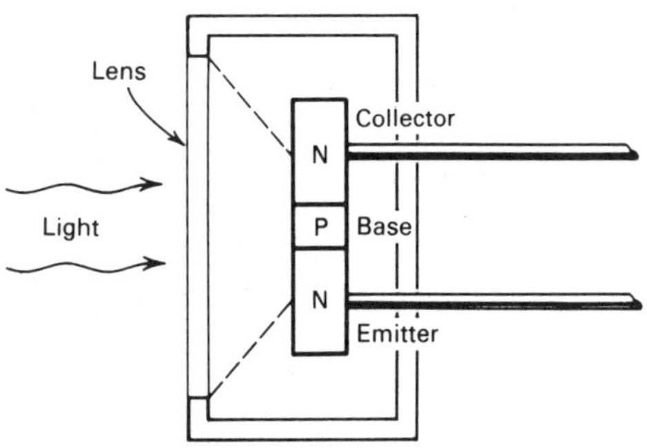

FIGURE 10-21 Cross-sectional view of a phototransistor.

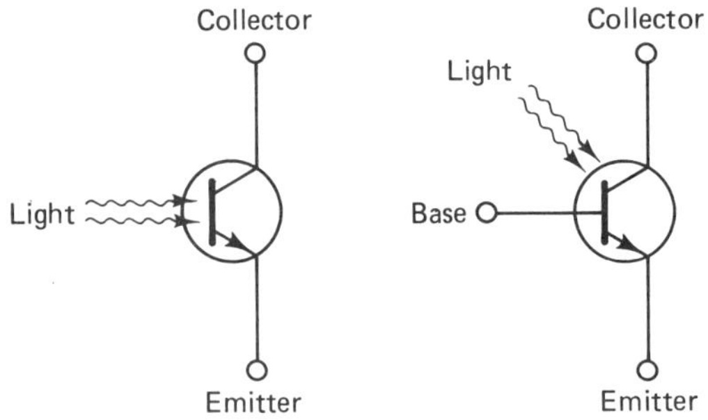

(a) (b)

FIGURE 10-22 Schematic symbols of a phototransistor: (a) floating-base; (b) three-lead.

schematic symbols for the floating-base and three-lead phototransistors.

A phototransistor equivalent circuit and characteristic curve for a three-lead device are shown in Fig. 10-23. The collector-base junction is reverse biased. When light is focused on the collector-base junction, base current flows. This base current is amplified by the transistor. It is possible to bias the base to control the sensitivity of the device.

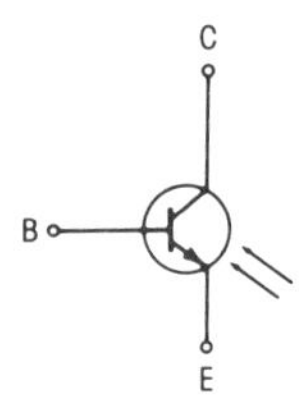

(A) Symbol.

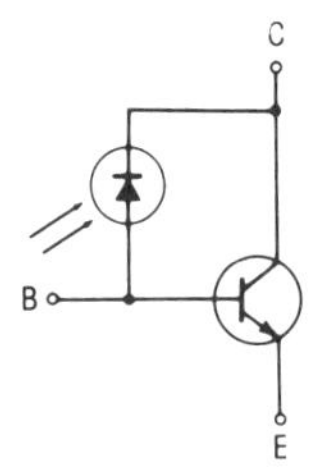

(B) Equivalent circuit.

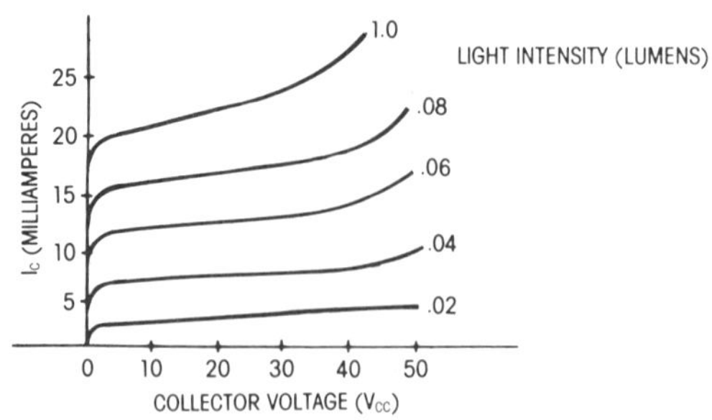

(C) Characteristic curves.

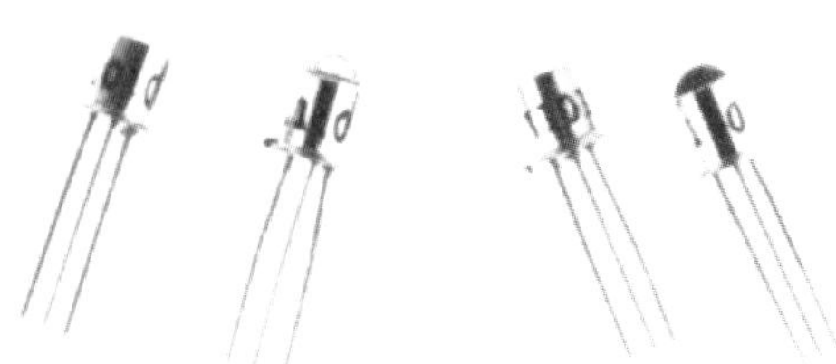

(D) Photograph.

FIGURE 10-23 Phototransistor: (a) symbol; (b) equivalent circuit; (c) characteristic curves; (d) photograph. [Courtesy of EG&G Vactec.]

Photo-Darlington transistors are now available through a number of manufacturers. This type of device is composed of two directly coupled phototransistors on a single structure. It is characterized as having high input impedance, lower output impedance, and high sensitivity. The response time of a photo-Darlington connected transistor is slower than the phototransistor and photodiode.

Phototransistors in general are not very widely used as discrete-component control devices. As a rule, other optoelectronic devices can achieve equal, and in many cases much better, control than a single phototransistor. Phototransistors are, however, very commonly used in optocoupling devices today. This application permits the interfacing of a computer output to a high-current or high-voltage source with excellent isolation. Light energy is coupled from an LED to a phototransistor without a direct electrical connection between the two devices. This permits electrical signal transfer between equipment without the hazards of a direct electrical connection.

Photovoltaic Devices

The photovoltaic cell, commonly called a *solar cell*, is used to convert light energy into electrical energy. Since this process represents a direct energy conversion, a great deal of research has been conducted in an attempt to convert large amounts of light energy into electrical energy. A common application of the solar cell is in photographic light-exposure meters. The electrical output of the solar cell is proportional to the amount of light falling onto its surface. The output of the solar cell is used to energize a light-intensity meter.

The construction of a photovoltaic cell is shown in Fig. 10-24. This selenium cell has a layer of selenium deposited on a metal base, then a layer of cadmium. In the fabrication, one layer of cadmium selenide (CdSe) and another layer of cadmium oxide (CdO)

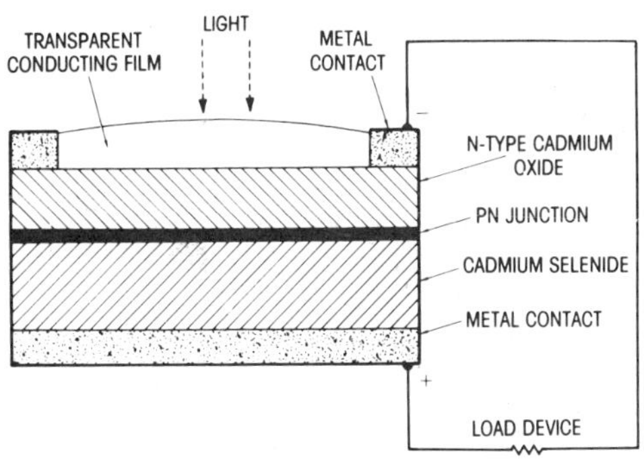

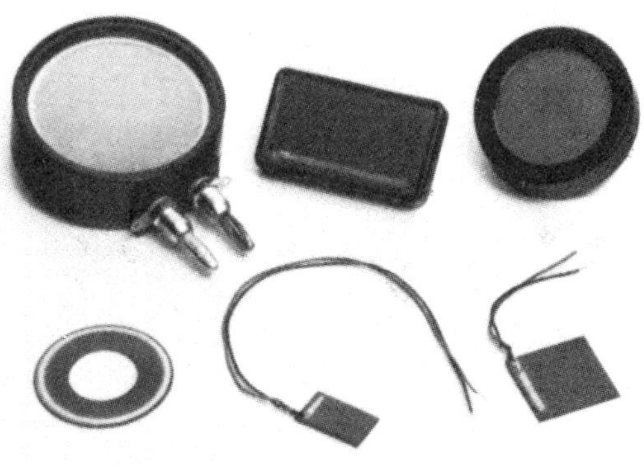

FIGURE 10-24 Selenium photovoltaic cells: (a) structure; (b) photograph. [Courtesy of EG&G Vactec.]

are produced. A transparent conductive film is placed over the CdO, and a section of conductive alloy is then placed on the film. External leads are connected to the conductive material around the CdO layer and the metal base. When light strikes the CdO layer, electrons are emitted and move toward the external load device. A deficiency of electrons is then created in this region, which is filled by electrons from the CdSe. Electrons from the metal contact then flow into the CdSe, causing the metal contact plate to become positive. Thus light energy causes a difference in potential between the two external terminals.

Selenium cells have a very low efficiency for converting light energy into electrical voltage. Typical efficiency ratings are less than 1%. A great deal of light and an extremely large active area are needed to produce electrical energy of any significant value. Selenium photovoltaic cells, however, have a spectral response that is very similar to that of the human eye. In many applications the spectral response is much more significant than a high level of output efficiency.

Silicon photovoltaic cells are much more commonly used today than the selenium units. Silicon cells have efficiencies as high as 15%. The more common silicon photovoltaic cell is shown in Fig. 10-25. When no light is focused onto the silicon cell, it operates similarly to a conventional P-N junction diode. When light strikes the cell, a voltage is developed across the external leads. The more intense the light, the greater the potential difference across the cell.

Photovoltaic cells are used in a variety of applications. Although their electrical output is low, they may be used with amplifying devices to develop an output that will drive a relay or some other load device. One such application is shown in Fig. 10-26. In this circuit the output of the photovoltaic cell is amplified by transistor Q_1, so that when light strikes the photovoltaic cell, the base current of the transistor increases. The increase in

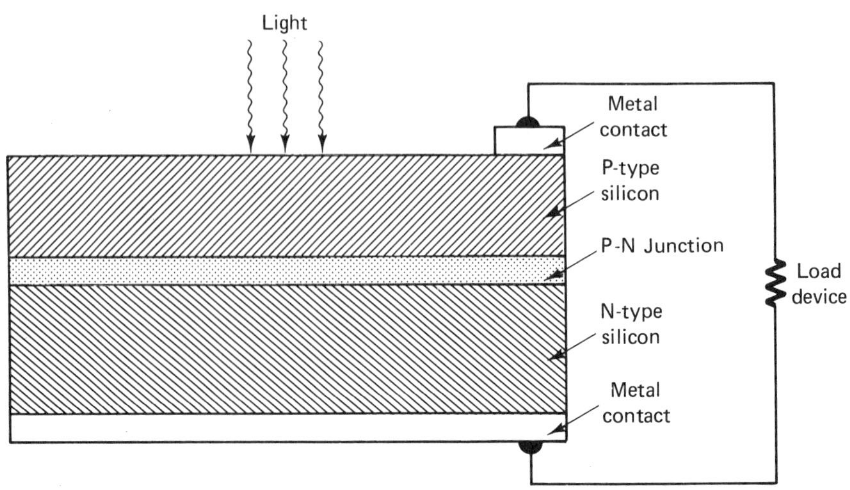

FIGURE 10-25 Silicon photovoltaic cell.

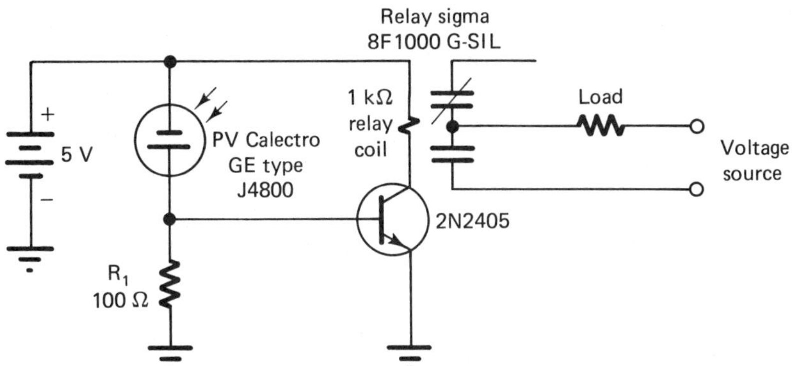

FIGURE 10-26 Photovoltaic cell relay control circuit.

I_B is amplified by Q_1, causing a collector current that is sufficient to activate the relay. The load connected across the relay terminals could be turned either off or on by the presence of light on the photovoltaic cell.

SELF-EXAMINATION

37. An optoelectronic __________ is designed to change light energy into electrical energy.
38. Optoelectronic detectors are divided into three general groups of devices: __________, __________, and __________.
39. When photons of light energy strike the __________ of a photoemissive tube, it emits electrons.
40. Operation of a photomultiplier tube is based on the principle of __________ emission from dynode plates.
41. A __________ cell changes light intensity into electrical conductivity.
42. __________ is the ease with which current carriers pass through a material.
43. Cadmium sulfide (CdS) cells are __________ devices.
44. A high level of light intensity applied to a CdS cell will cause it to be __________ resistant.
45. A low level of light intensity applied to a CdS cell will cause it to be __________ resistant.
46. A light-dependent resistor (LDR) is classified as a __________ device.
47. Photodiodes are classified as __________ devices.
48. In order for a photodiode to be responsive to changes in light intensity, it must be __________ biased.
49. Light energy applied to a phototransistor generates current carriers in the __________ region.
50. A solar cell is an example of a __________ device.

51. The efficiency of a silicon photovoltaic cell is ________ than that of a selenium photovoltaic cell.
52. The output voltage of a photovoltaic cell circuit configuration increases when they are connected in ________.
53. The current capability of a photovoltaic cell circuit configuration increases when they are connected in ________.

ANSWERS

37. Detector	38. Photoemissive, photoconductive, and photovoltaic
39. Cathode	40. Secondary
41. Photoconductive	42. Conductance
43. Photoconductive	44. Low
45. High	46. Photoconductive
47. Photoconductive	48. Reverse
49. Base	50. Photovoltaic
51. Higher	52. Series
53. Parallel	

SUMMARY

Optoelectronics is a new technology that combines the fields of optics and electronics. Optoelectronic devices have electrical properties that are affected by light. The term light includes visible, infrared, and ultraviolet regions of the frequency spectrum. Light is a form of radiant energy.

One theory of radiant energy transmission considers light to travel as transverse waves that cause particles of a medium to vibrate at right angles to the direction of motion of the waves. Each wave has an alternate rise and fall in its amplitude. These waves travel at 186,000 miles per second, or 300,000,000 meters per second, which is considered to be the speed of light. There is an inverse relationship between the length of a wave and its frequency. Wavelength is a measure of how far the wave travels during one cycle of operation.

The quantum theory of light considers light to be emitted in discrete packets of energy called photons. The energy of a photon depends on the wavelength of the light. Each atom radiates a distinctive spectrum of wavelengths that is related to the atomic structure of the atom. Photons travel at the same speed, namely, the speed of light, but they may have different amounts of energy. Today it is believed that light is made up of photons that travel in a wave pattern.

Radiant energy is made up of the entire optical spectrum, which includes ultraviolet, visible, and infrared light. This energy can be radiated through space in a wide range of patterns, emitted from a material, or reflected from a surface. Radiant flux is the flow rate of radiant energy per unit of time. Radiant intensity indicates the amount of electromagnetic energy produced by a specific source. Radiant incidence refers to the density of radiant power that strikes a surface. Radiant exitance is a measure of the emittance from a specific surface. Radiance is a measure of the radiant intensity that is either passing, leaving, or arriving at a specific surface area.

Photometry deals with electromagnetic energy that falls in the visible part of the electromagnetic spectrum. Photometric terms are preceded by the word luminous. The unit of luminous intensity is the candela. Luminous flux is the rate at which light energy flows per unit of time and is measured in lumens. Illuminance is the density of luminous power that illuminates a surface and is measured in lux. Luminous exitance is a measure of the amount of luminous flux given off or reflected by a surface.

An optoelectronic system is distinguished from other systems by its source, which must generate some form of radiant energy, and its detector, which responds to or receives radiant energy. Most optoelectronic systems respond to radiant energy that falls in the visible part of the spectrum.

The source of an optoelectronic system is usually classified according to the amount of visible light emitted. Sources can be panchromatic, heterochromatic, or monochromatic. An incandescent lamp is an example of a panchromatic source, which generates light over a large part of the visible spectrum. Mercury arc lamps are heterochromatic. This lamp radiates energy in a narrow part of the spectrum that is red to orange and peaks at 6500 Å. Sodium vapor lamps are monochromatic. They generate light that is predominantly of one wavelength. Sodium vapor sources peak at 5000 Å.

Light-emitting diodes (LEDs) are solid-state sources that have a P-N junction that emits light when forward biased. When holes and electrons combine in the depletion region of the P-N junction, they cause energy to be released in the form of heat and photons. Material used to make the P-N junction determines the wavelength of energy released. LEDs are used singly, formed into seven-segment displays, and used in 5×7 dot-matrix displays. The LED must be connected in series with a current-limiting resistor to prevent excessive current from damaging the junction.

The detector of an optoelectronic system is designed to change radiant energy into electrical energy. A detector is functionally the reverse of a source. Detectors are photoemissive, photoconductive, or photovoltaic.

Photoemissive detectors give off or emit electrons when photons of light energy are absorbed by the light-sensitive material of the cathode. Photomultiplier tubes are photoemissive devices that respond to the secondary emission of several dynode plates that increase the output of the device.

Photoconductive detectors change light intensity into electrical conductivity. Conductance is the reciprocal of resistance. Cadmium sulfide (CdS) cells change resistance when exposed to light. With a high level of light intensity, the resistance of a CdS cell will drop from 100,000 Ω to 100 Ω.

Photodiodes are photoconductive devices that have a light-sensitive P-N junction. Normally, a photodiode is connected in the reverse-bias direction. When light is applied to the junction, current carriers return to the depletion region and cause conduction. A PIN photodiode has a layer of intrinsic semiconductor material between the P- and N-materials. This construction lowers the junction capacitance, which causes it to have a faster response time. Phototransistors are photoconductive devices that have two P-N junctions. The collector-base junction responds as a photodiode. Light applied to this junction causes current carriers to be injected into the base region. This photocurrent controls the emitter-collector current. A small change in base current is used to control a large value of collector-emitter current. A phototransistor has an amplification capability, which makes it more sensitive to changes in light intensity.

Photovoltaic devices are designed to change light energy directly into electrical energy. These devices have two different layers of silicon or selenium formed on a substrate. When light is applied, it causes a difference in potential to appear across the junction. This cell develops voltage directly from light energy. Selenium photovoltaic cells have an efficiency of 1%, whereas silicon devices have an efficiency of 15%. The voltage of a photovoltaic cell can be amplified if there is a need for more output.

EXPERIMENT 10-1
LIGHT-EMITTING DIODE TESTING

An LED is a diode and responds in the same way as other diodes. When it is forward biased, current carriers recombine at the PN junction and produce current and light. The value of forward-bias voltage determines the amount of conduction produced by the LED. Normally this current can be very large. As a rule, the LED must be protected from excessive current. A series resistor is used to limit the forward current to a reasonable value.

An ohmmeter can be used to test an LED and define its status. When the source of an ohmmeter forward biases an LED, it will indicate low resistance. If the ohmmeter source is large enough in value, it may cause the LED to respond by producing light. Reverse biasing an LED will produce an infinite or extremely high resistance reading on the ohmmeter. The ohmmeter test of an LED is essentially a good-bad test. A good LED will show low resistance when forward biased. The forward-bias resistance is generally much higher than that of a standard silicon diode. When reverse biased, the LED must show an infinite or extremely high resistance value. If an LED does not respond in this way, it is considered to be faulty.

The characteristics of an LED are similar to those of a standard silicon diode. When an LED is forward biased, voltage across it will increase in value until conduction is established. Typically, the forward voltage of an LED is in the range of 2 V. This value is maintained after it has been forward biased. The forward current rises in value with an increase in source voltage.

In the reverse-bias direction an increase in source voltage will cause the diode voltage to increase in value. Normally there is very little or no reverse current.

OBJECTIVES

1. To construct an LED circuit and test its operation.
2. To test an LED with an ohmmeter and evaluate its status.
3. To determine the value of a series current-limiting resistor for an LED circuit.
4. To determine the operating characteristics of an LED.

EQUIPMENT

LED with a 2 V_F, Heath #412-40, Motorola MV5054-2, or equivalent
1-kΩ, $\frac{1}{4}$-W resistor
1-kΩ, 1-W potentiometer (linear taper)
Variable dc source
Volt-ohm-milliammeter (VOM), standard or digital
Circuit construction board

PROCEDURE

1. Place the LED in the circuit construction board and prepare the VOM to measure resistance. Connect the ohmmeter leads to the LED as indicated in Fig. 10-1A. Note the location of the anode, cathode, and identifying mark (flat side) of the LED. If the LED has leads of different lengths, the longest lead is the anode. The flat side of the housing is oriented nearest the cathode.
2. Measure and record the forward resistance and reverse resistance of the LED: forward resistance = ________; reverse resistance = ____________.
3. If other LEDs are available, perform this test procedure on another device. As a rule, there are rather significant differences among the resistance values of LEDs.
4. Connect the LED characteristic test circuit of Fig. 10-1B. Adjust the dc source to approximately 10 V as measured across the potentiometer. Use the VOM to measure V_s, V_{Rs}, and V_F. Calculate the forward current (I_F) by using the voltage across the series resistor (V_{Rs}) and the value of the series resistor (1 kΩ).
5. Adjust the potentiometer to produce a V_s of 1 V. Measure and record the values of V_{Rs}, and V_F for this V_s value. Adjust V_s to 2 V and repeat the procedure. Repeat the procedure for each of the V_s values in the chart.
6. Using the value of V_{Rs} and R_s, calculate the forward cur-

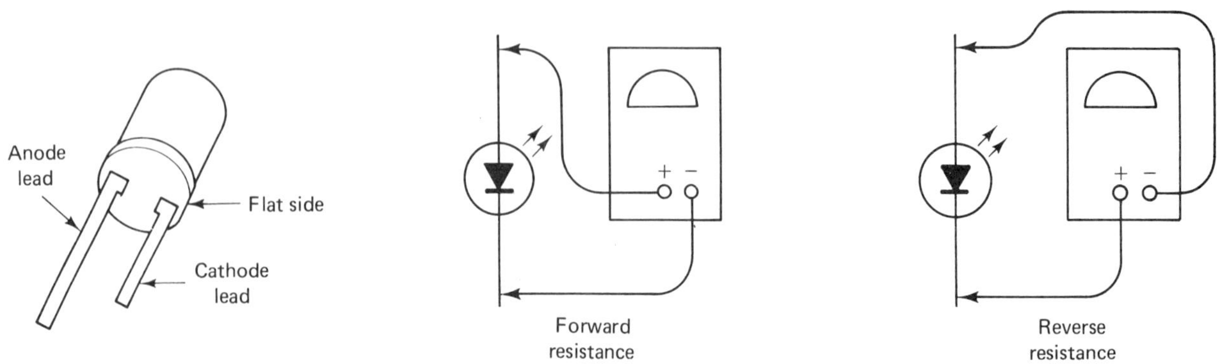

FIGURE 10-1A LED ohmmeter test procedure.

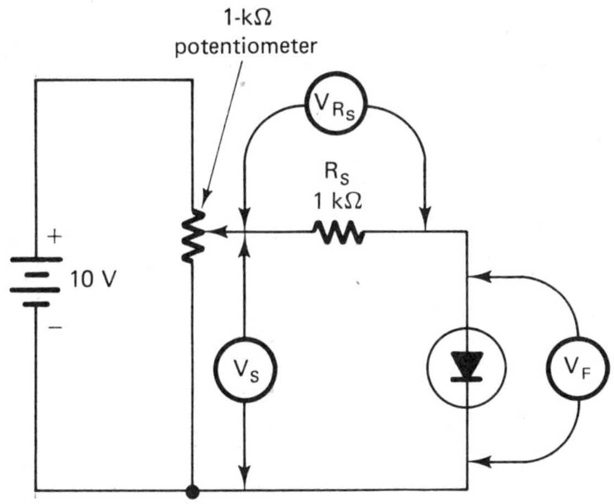

Forward characteristic chart

V_S	V_{R_S}	V_F	I_F
1 V			
2 V			
3 V			
4 V			
5 V			
6 V			
7 V			
8 V			
9 V			
10 V			

FIGURE 10-1B LED characteristic test circuit and chart.

rent for each V_s of the chart. Record the calculated values in the chart.

7. On the graph of Fig. 10-1C plot the forward current-voltage characteristics of the LED. The horizontal part of the graph represents V_F and the vertical I_F.
8. Return the V_s voltage to 1 V and reverse the diode lead connections of the circuit so that the LED is connected in the reverse-bias direction.
9. With a V_s of 1 V measure and record the V_{Rs} and V_F of the diode. Record these measured values in the reverse characteristic chart of Fig. 10-1D.

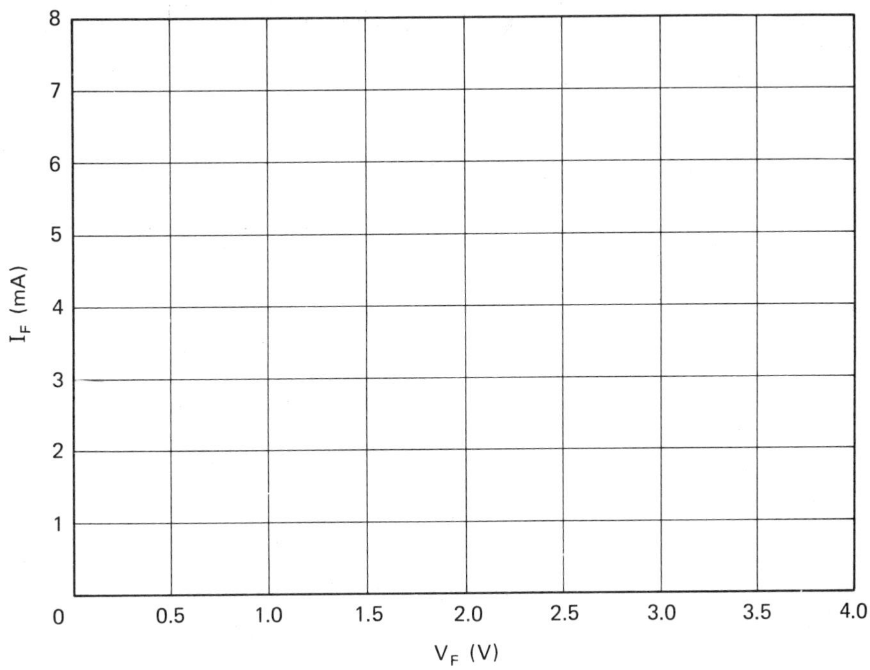

FIGURE 10-1C I_F-V_F plotting graph.

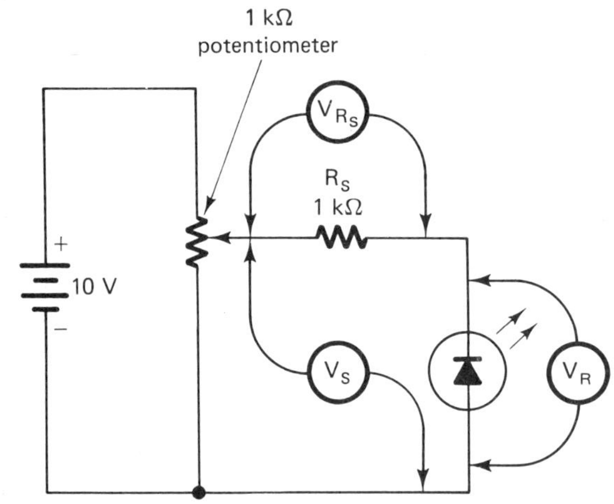

V_S	V_{R_S}	V_R	I_R
1 V			
2 V			
3 V			
4 V			
5 V			
6 V			
7 V			
8 V			
9 V			
10 V			

FIGURE 10-1D LED reverse characteristic test circuit and chart.

10. Calculate the reverse current for each value of V_s. If no noticeable voltage is measured across resistor R_s, assume that there is no significant reverse current.
11. Turn off the power source and disconnect the circuit.

DISCUSSION

In this experiment you constructed a test circuit that evaluated the forward conduction characteristics of an LED. In the first part of the activity you were asked to measure the resistance of the LED with an ohmmeter. This should have shown the LED to have resistance in the forward direction and to have infinite resistance in the reverse direction. The forward resistance is generally quite high compared with that of a conventional diode. A typical forward resistance value is 100 kΩ.

In the second part of the experiment you were asked to construct an LED test circuit that showed the forward current and voltage characteristics. The forward current values were calculated by using the measured voltage across the series resistor R_s. The forward current increased rather slowly for V_F values below 1 V. After 1 V the I_F increased very steadily. The range of I_F was 0 to 10 mA. An I_F versus V_F graph was then plotted showing the forward characteristics of the LED.

ANALYSIS

1. How does V_F of an LED respond when forward biased?
2. How does the V_R of an LED respond with changes in V_s?

EXPERIMENT 10-2
LED OPERATION

When an LED is forward biased, current carriers return to the depletion region and combine to cause current flow. When an electron fills a hole, it causes current flow and generates photons of light energy. The light energy is released from the junction and causes the device to respond as a light source. When the junction is reverse biased, the depletion region becomes wider. This causes reduced current flow and stops the emission of photons.

When an LED is forward biased, it must be connected in series with a resistor. This resistor limits the forward current flow to a safe operating level. Selection of the current-limiting resistor is determined by Ohm's law. In this regard, $R_s = V_s - V_F/I_F$. V_s is the source voltage, V_F is the forward voltage, and I_F is the forward current. The series resistor is of no concern when the LED is reverse biased.

OBJECTIVES

1. To construct an LED circuit that will turn on when forward biased and turn off when reverse biased.
2. To show the procedure for calculating the value of a current-limiting resistor.

EQUIPMENT

LED, Heath #412-40, Motorola MV 5054-2, or equivalent
220-Ω, $\frac{1}{4}$-W resistor
470-Ω, $\frac{1}{4}$-W resistor
1-kΩ, $\frac{1}{4}$-W resistor
Variable dc power supply
Volt-ohm-milliammeter (VOM), standard or digital
Circuit construction board

PROCEDURE

1. Refer to the LED circuit of Fig. 10-2A. With the given values of V_s of 5 V, V_F of 2 V, and an I_F of 10 mA, calculate the value of R_s. The formula is $R_s = V_s - V_F/I_F$.
2. Select a resistor that closely approximates the calculated value from the three values listed for this activity.

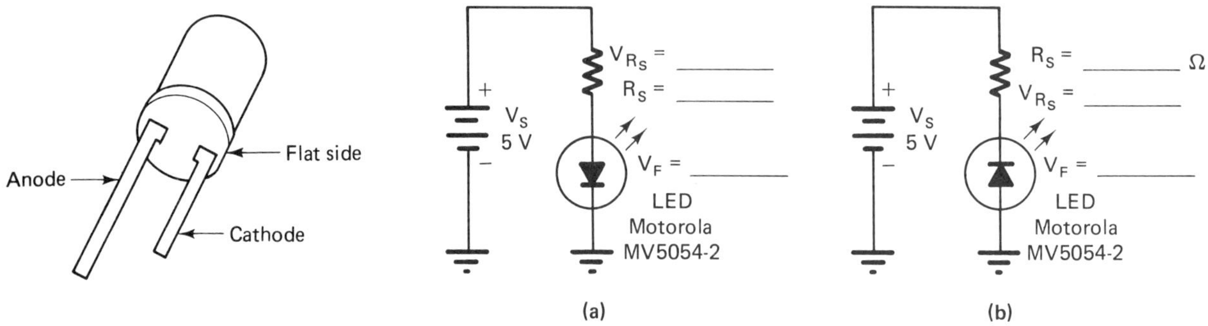

FIGURE 10-2A LED operation: (a) forward biased; (b) reverse biased.

Connect this resistor in series with the LED and connect it to the dc source as indicated in the schematic of Fig. 10-2A(a).

3. Turn on the dc source and observe the operation of the LED. The LED is (on, off).
4. Measure and record the actual value of V_{Rs} and V_F. Calculate the value of I_F using the value of R_s and V_{Rs}. The calculated value of I_F is ________ mA.
5. Turn off the power source and switch the leads of the LED to conform with the circuit of Fig. 10-2A(b). Turn on the power source and describe the response of the LED. The LED is (on, off).
6. Measure and record in the provided space of Fig. 10-2A(b) V_{Rs} and V_R. Calculate the value of I_R using V_{Rs} and R_s. I_R = ________ mA.
7. Turn off the dc voltage source.
8. Switch the LED leads so that it is connected as in Fig. 10-2A(a).
9. Assume now that the source voltage is to be increased to 12 V and that 10 mA of I_F is needed to make the LED respond. To compensate for this change, a new current-limiting resistor is needed. Calculate the value of the new current-limiting resistor: R_s = ________ Ω.
10. Select a resistor that closely approximates the calculated value. The selected resistor value is ________ Ω.
11. Connect the circuit of Fig. 10-2A(a) using the new value of the current-limiting resistor and source voltage. Turn on the dc source voltage and observe the operation of the LED. The LED is (on, off).
12. Measure the voltage across the LED and the voltage across the current-limiting resistor: V_F = ________; V_{Rs} = ________. Calculate the forward current flow of the circuit: I_F = ________ mA. How closely does the calculated I_F compare with the given I_F value? ________
13. Turn off the dc power source and disconnect the circuit.

DISCUSSION

In this experiment you were asked to construct a typical LED circuit connected to a dc source. A current-limiting resistor was selected to produce a safe operating current value. The circuit was then energized and the LED forward voltage and R_s voltage were measured. The values of R_s and V_{Rs} were used to calculate the value of forward current (I_F). The value of I_F should have closely approximated the given value of 10 mA used to determine the series resistor. The selected series resistor should have been 220 Ω. The LED should have produced light when energized.

When the LED leads were switched as in Fig. 10-2A(b), it caused reverse biasing. The V_F voltage increased in value to approximate V_s, and V_{Rs} dropped to zero. This indicated that no current was passing through the current-limiting resistor and the LED was off.

The last part of the experiment followed the same procedure with a new current-limiting resistor to accommodate a different source voltage value. The value of this resistor should have been approximately 1 kΩ.

ANALYSIS

1. How does the value of the series resistor limit the current flow of an LED circuit?
2. If the source voltage supplied to an LED is increased in value, what will occur?

EXPERIMENT 10-3
PHOTODIODE OPERATION

The photodiode is a unique optoelectronic device that is used as the control element of an electronic system. A photodiode is normally connected in the reverse-bias direction. When light is applied to the junction, it causes current carriers to return to the depletion region. When these current carriers enter the depletion region, they recombine to produce current flow. This current flow is an increase over the normal reverse current. The intensity of light and its wavelength determine the resulting conductivity of a photodiode.

In this experiment one P-N junction of a phototransistor is used as a photodiode. The phototransistor must have three leads in order for this function to be achieved. The photodiode junction will respond as a conventional two-lead phototransistor.

OBJECTIVES

1. To measure the forward and reverse resistance of a photodiode with an ohmmeter.
2. To determine the electrical response of a photodiode to light when it is forward and reverse biased.
3. To build a photodiode circuit and determine how it will respond as a control element.
4. To change a phototransistor into a photodiode by selecting the proper combination of leads.

EQUIPMENT

Phototransistor, General Electric L14F1, Heath 417-851, or equivalent

1-kΩ, $\frac{1}{4}$-W resistor

1-kΩ, 1-W potentiometer (linear taper)

12-V lamp, ALCO #2182, GTE Sylvania 12ES, Heath 412-601, or equivalent

Volt-ohm-milliammeter (VOM), standard or digital

Variable dc power supply

Circuit construction board

PROCEDURE

1. In the first part of this experiment a VOM is used to evaluate the lead assignment of a phototransistor. When the lead resistance has been tested, the phototransistor will be used as a photodiode by selecting the base-collector leads. One of the transistor's two junctions is used as a photodiode.
2. Refer to the phototransistor outline of Fig. 10-3A. Connect the positive lead of the ohmmeter to the base and alternately switch the negative lead between the emitter and the collector. The forward emitter-base resistance is ______________ Ω, whereas the forward base-collector resistance is ______________ Ω. Connect the negative lead of the ohmmeter to the base and alternately switch the positive lead between the collector and the emitter. The reverse emitter-base resistance is ______________ Ω, while the base-collector resistance is ______________ Ω.
3. If the polarity of the ohmmeter's source is known, the forward and reverse resistance procedure just performed will identify the transistor.
4. The phototransistor of this activity is an NPN device and should have had lower forward than reverse resistance. Normally, the forward resistance is several thousand ohms, whereas the reverse resistance is infinite. A good transistor will respond in this manner, whereas an open device will show infinite resistance in both directions. This is generally a good indication of the response of a usable transistor.
5. Connect the transistor circuit of Fig. 10-3B. In this case the transistor is connected as a diode by using only the base and collector leads. The intensity of the lamp is changed by adjusting the potentiometer. Position the lamp so that its light is directed to the light-sensitive area of the transistor. Adjust the lamp to its brightest level of intensity.
6. Observe the current flow indicated by the VOM. Adjust the position of the lamp to produce the highest current reading. Record the highest current flow obtained: current flow = ______________ μA.

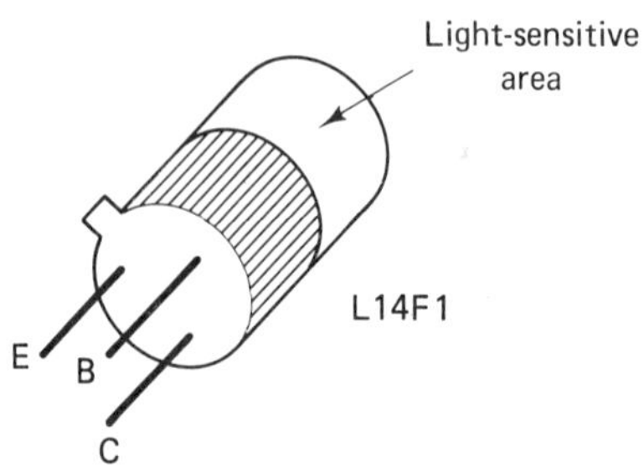

FIGURE 10-3A Phototransistor lead assignments.

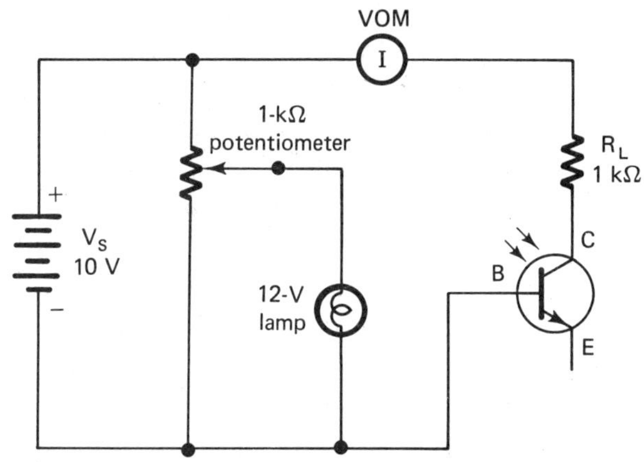

FIGURE 10-3B Photodiode test circuit.

7. Vary the intensity of the lamp by altering the potentiometer. Note how the intensity of the lamp alters the photocurrent of the circuit. Describe the response of lamp intensity and photocurrent. ______________________________.

8. Remove the VOM from the circuit and complete the circuit path between R_s and the source. Prepare the VOM to measure voltage. Measure the voltage across R_s and across the base-collector junction with the light source at its most intense level: V_{Rs} = __________; V_{BC} = __________.

9. Reduce the intensity of the lamp so that there is no light. Measure and record V_{Rs} and V_{BC}: V_{Rs} = __________; V_{BC} = __________.
10. Prepare the VOM to measure current, and connect it in series with R_s as in the original circuit of Fig. 10-3B.
11. Reverse the collector-base leads of the transistor so that the junction is forward biased. This will produce a much greater current indication on the VOM. Vary the intensity of the light source and alter the position of the transistor so that it gets maximum light exposure on the sensitive area. There should be a very noticeable effect on the response of the diode connected in the forward-biased direction. How does the photodiode respond to changes in light intensity? __________________.
12. Disconnect the circuit.

DISCUSSION

In this experiment you were asked to examine the junction status of a phototransistor with an ohmmeter. The ohmmeter has a source of dc energy in its circuitry when connected to the phototransistor's junctions. A phototransistor is like a conventional transistor in that a forward-biased P-N junction will show low resistance, and a reverse-biased junction will show high resistance. The base-emitter and base-collector junctions showed approximately 1.5 to 10 kΩ of resistance when connected to the ohmmeter. When the polarity of the ohmmeter was reversed, the same two junctions showed infinite resistance.

In the second part of this activity the base-collector junction of the phototransistor was used as a photodiode. When connected in the reverse-biased direction, there was very little current flow through the junction. When light was applied to the junction's light-sensitive area, there was a significant increase in current. Typically, this current should have been in the 100-μA range. The voltage measured across R_s was 0 V without light and 0.2 V with light. The measured voltage across the base-collector junction was approximately 10 V without light and 9.7 V with light applied.

Reversing the polarity of the base-collector junction so that it was forward biased caused a substantial increase in current flow. A measured current value was 9 mA. This current was significantly greater, but did not change value when exposed to changes in light intensity. This showed that a forward-biased photodiode is not sensitive to changes in light intensity.

ANALYSIS

1. Discuss the relationship of photocurrent and light intensity in the operation of a photodiode.
2. Why is a photodiode responsive to changes in light intensity when reverse biased and not when forward biased?

EXPERIMENT 10-4
PHOTOTRANSISTOR OPERATION

A phototransistor has a photodiode as the base-collector junction and a silicon diode as the base-emitter junction. A three-terminal phototransistor can be used with the base floating (unconnected base) or with the base energized by the source. The floating-base type of circuit is widely used because of its simplicity. The sensitivity of a connected-base circuit can be adjusted to different levels of light intensity.

A phototransistor can achieve the same type of control as a photodiode. A phototransistor has an amplification capability that a photodiode does not. A small change in light intensity applied to the base is capable of controlling a rather large value of collector-emitter current. This capability makes it possible for the phototransistor to achieve some degree of amplification. In a strict sense, there is no measurable base current when the transistor is connected with a floating base. The gain of a phototransistor is quite high.

OBJECTIVES

1. To construct a phototransistor circuit.
2. To evaluate the operation of a phototransistor circuit.
3. To measure photocurrent and circuit voltage to show how a phototransistor will respond to different levels of light intensity.

EQUIPMENT

Phototransistor, General Electric L14F1, Heath 417-851, or equivalent

1-kΩ, $\frac{1}{4}$-W resistor

1-kΩ, 1-W potentiometer (linear taper)

12-V lamp, ALCO #2182, GTE Sylvania 12-ES, Heath 412-601, or equivalent

Volt-ohm-milliammeter (VOM), standard or digital

Circuit construction board

Variable dc power supply

PROCEDURE

1. Connect the light source (lamp) of the circuit in Fig. 10-4A. Apply energy to the lamp and vary the potentiometer. This should cause the intensity of the lamp to vary from full intensity to off. Position the sensitive area of the phototransistor near the light source. Prepare the VOM to measure resistance and connect it across the emitter-collector of the phototransistor. The positive side of the ohmmeter should be connected to the collector and the negative to the emitter.
2. With the light off measure the resistance of the phototransistor: emitter-collector resistance = ________ Ω.
3. Turn on the light and adjust it to the highest level of intensity. Measure the resistance of the phototransistor: emitter-collector resistance with a high level of light intensity = ____________ Ω.
4. Reverse the ohmmeter leads so that the collector is negative and the emitter is positive. Measure and record the resistance of the phototransistor with light and without light. With light the emitter-collector resistance is ____________ Ω. Without light the emitter-collector resistance is ____________ Ω.
5. Disconnect the ohmmeter from the phototransistor and prepare the VOM to measure current. Connect the VOM in series with the load resistor to measure photocurrent as indicated in Fig. 10-4B.
6. With the highest intensity of light produced by the light source, measure and record the photocurrent. Transistor photocurrent with a high level of light intensity is ____________ mA.
7. Turn off the light source. Transistor photocurrent with no light applied is ____________ μA.
8. Block the light-sensitive area of the phototransistor with a piece of black plastic tape or some opaque material.

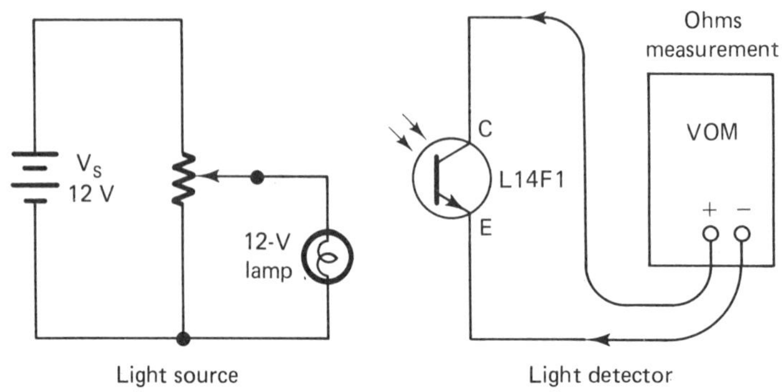

FIGURE 10-4A Resistance test circuit of a phototransistor.

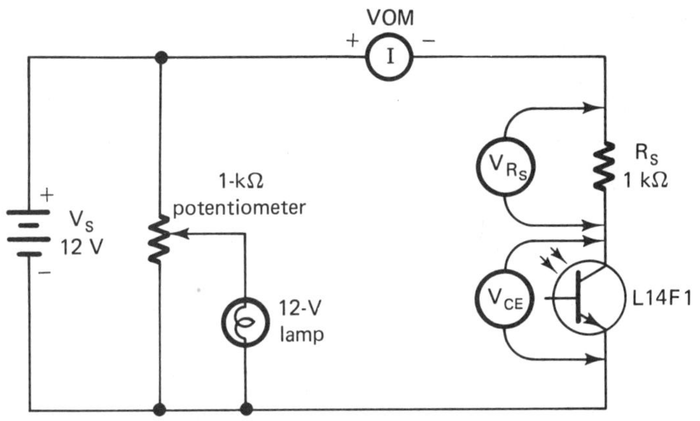

FIGURE 10-4B Phototransistor test circuit.

The dark current of a phototransistor is ________ μA.

9. Remove the tape from the phototransistor. Disconnect the VOM ammeter from the circuit and complete the path where the meter was located.
10. Prepare the VOM to measure voltage. Adjust the light source to its highest level of intensity. Measure and record the voltage across the series resistor (V_{Rs}) and across the phototransistor (V_{CE}): V_{Rs} = ________ V; V_{CE} = ________ V.
11. Turn off the light source. Measure and record the voltage across the series resistor and the phototransistor: V_{Rs} = ________ V; V_{CE} = ________ V.
12. Turn on the light source and adjust it to a high level of intensity. Momentarily block light from the sensitive area of the phototransistor and observe its response to voltage changes. How would you describe the response time of the phototransistor? ________.
13. Disconnect the circuit.

DISCUSSION

In the first part of this experiment you were asked to measure the emitter-collector resistance of a phototransistor with an ohmmeter. Typical resistance values should have been in the range of 80 kΩ to 100 kΩ with light and infinite without light. This showed how the resistance of the transistor changed for different levels of light.

In the second part of the experiment you were asked to measure photocurrent for different levels of light. Typical current values were 0 μA for no light, 80 μA for ambient light, and 9.2 mA for the most intense level of light. The collector-emitter voltage measured across the transistor should have varied from 0.8 V for the most intense light to 10 V with no light. The voltage across the resistor was the difference between the source voltage and V_{CE}.

This experiment should have given you an idea of how the voltage and current of a phototransistor responds to different levels of light intensity. The photocurrent controlled by the transistor is significantly greater than that controlled by the photodiode of the previous activity. A phototransistor is capable of amplification, while a photodiode is a conductivity control device.

ANALYSIS

1. Why is the base of the phototransistor not used in this activity?
2. How does a change in light intensity cause a change in V_{CE}?
3. Why does the V_{CE} of a phototransistor decrease in value when the intensity of light increases?

UNIT 10 EXAMINATION
OPTOELECTRONIC DEVICES

Instructions: Circle the response that most correctly completes each statement.

1. Discrete bundles of energy produced by a source of light are called:
   a. Ions
   b. Electrons
   c. Neutrons
   d. Photons
2. Light is best described as a form of:
   a. Work
   b. Energy
   c. Waves
   d. Force
   e. Heat
3. Radiation is described as:
   a. Energy transfer
   b. Energy conversion
   c. A form of work
   d. A frequency
   e. A series of waves
4. A photometric system deals with:
   a. The response of the human eye
   b. All electromagnetic radiation
   c. Only infrared energy
   d. Only ultraviolet energy
   e. A combination of infrared and ultraviolet energy
5. Radiometric systems deal with:
   a. Only visible light
   b. All electromagnetic energy
   c. Only AM, FM, and TV electromagnetic energy
   d. Only gamma rays, X-rays, and microwaves
   e. Only the invisible part of the electromagnetic spectrum
6. The human eye responds to wavelengths of electromagnetic energy that are in the range of:
   a. 300 to 3000 Å
   b. 3000 Å to 4000 Å

c. 4000 Å to 7700 Å
d. 7700 Å to 10000 Å
e. 10000 Å and above

7. The sensitivity of the human eye peaks at the:
   a. Red part of the visual region of the spectrum
   b. Violet region
   c. Green area of the spectrum
   d. White part of the spectrum
   e. Black or absence of light
8. An LED source produces light when:
   a. Holes and electrons combine in the depletion region
   b. The PN junction is reverse biased
   c. The depletion region becomes wider
   d. Electrons are emitted from the junction surface
   e. The PN junction becomes hot
9. A photomultiplier tube responds to:
   a. The conductivity of current through a specific material
   b. The emission of electrons from a cathode
   c. The secondary emission of electrons from dynodes
   d. Thermionic emission
   e. Electrostatic excitation
10. An LED with a V_F of 2 V is to be operated from a 10-V source at 10 mA. In which range of resistance values will the current-limiting resistor be found?
    a. 1 to 100 Ω
    b. 101 to 300 Ω
    c. 301 to 500 Ω
    d. 501 to 700 Ω
    e. 701 to 900 Ω
    f. 901 ohms or higher
11. Under normal operating conditions a GaAs LED will have a forward voltage between:
    a. 0 and 1 V
    b. 1.1 and 3 V
    c. 3.1 and 5 V
    d. 5.1 and 7 V
    e. 7.1 and 9 V
12. Which of the following is not classified as a photoconductive device?
    a. A PN photodiode
    b. A PIN photodiode
    c. A phototransistor

d. A light-dependent resistor
e. A photovoltaic cell

13. Phototransistors respond much like a regular transistor except that light energy:
a. Acts as a switch to turn on the transistor
b. Produces base current
c. Alters the leakage current
d. Changes the base voltage
e. Alters the emitter current

14. An ohmmeter connected across an LED shows 100 kΩ of resistance in the forward direction and infinite resistance in the reverse direction. This indicates:
a. That the device is shorted
b. An open condition
c. A faulty device
d. A good device
e. That the condition of the device is questionable

15. Light striking a properly connected photodiode will cause an increase in:
a. Forward current
b. Resistance
c. Reverse current
d. Reverse voltage
e. Forward voltage across the device

APPENDIX A

The Elements

PERIODIC TABLE OF THE ELEMENTS

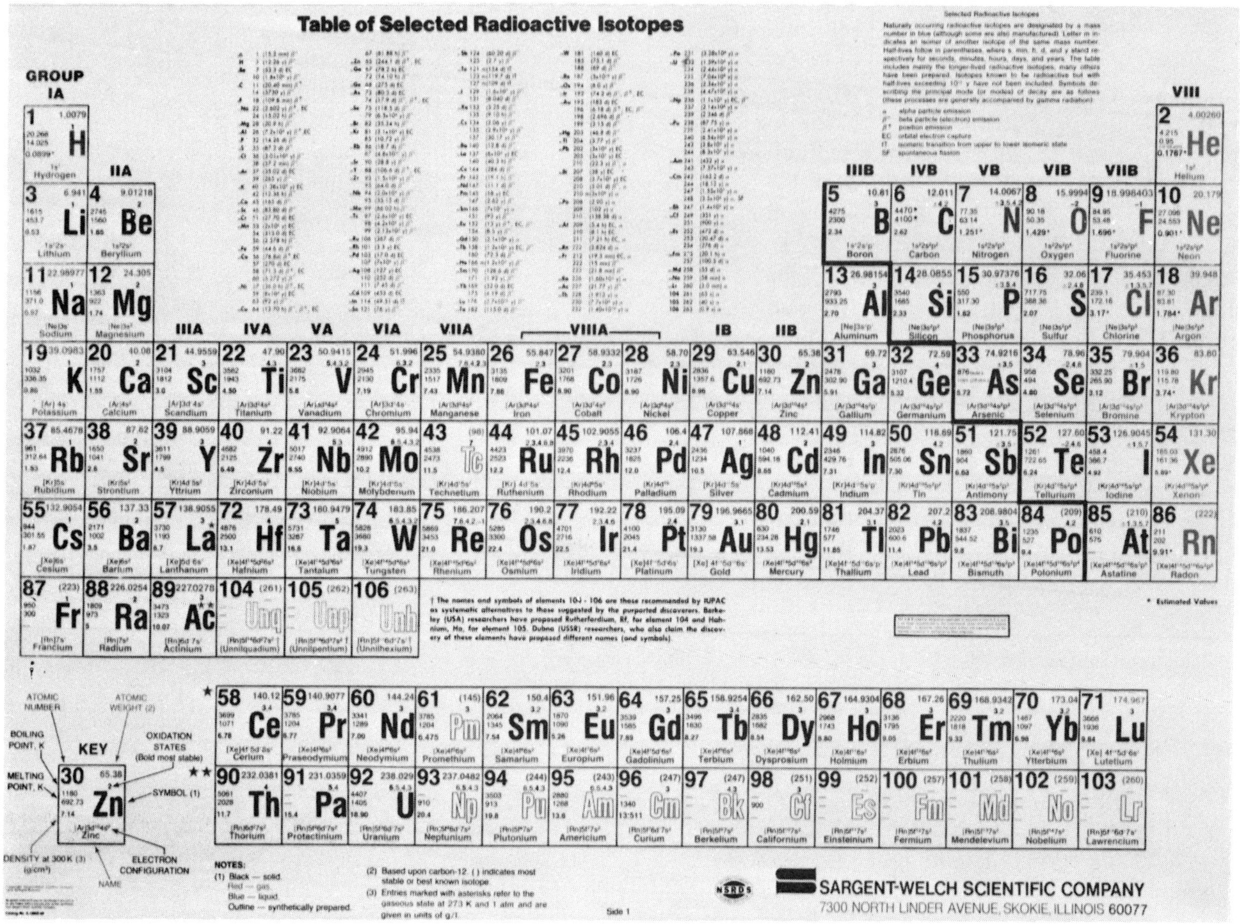

An Alphabetical List of the Elements

Element	*Symbol*	*Atomic no.*	*Atomic weight*
Actinium	Ac	89	227*
Aluminum	Al	13	26.97
Americium	Am	95	243*
Antimony	Sb	51	121.76
Argon	Ar	18	39.944
Arsenic	As	33	74.91
Astatine	At	85	210*
Barium	Ba	56	137.36
Berkelium	Bk	97	247*
Beryllium	Be	4	9.013
Bismuth	Bi	83	209.00
Boron	B	5	10.82
Bromine	Br	35	79.916
Cadmium	Cd	48	112.41
Calcium	Ca	20	40.08
Californium	Cf	98	251*
Carbon	C	6	12.01
Cerium	Ce	58	140.13
Cesium	Cs	55	132.91
Chlorine	Cl	17	35.457
Chromium	Cr	24	52.01
Cobalt	Co	27	58.94
Copper	Cu	29	63.54
Curium	Cm	96	247
Dysprosium	Dy	66	162.46
Einsteinium	E	99	254*
Erbium	Er	68	167.2
Europium	Eu	63	152.0
Fermium	Fm	100	255*
Fluorine	F	9	19.00
Francium	Fr	87	233*
Gadolinium	Gd	64	156.9
Gallium	Ga	31	69.72
Germanium	Ge	32	72.60
Gold	Au	79	197.0

Element	*Symbol*	*Atomic no.*	*Atomic weight*
Hafnium	Hf	72	178.6
Hahmium	Ha	105	262*
Helium	He	2	4.003
Holmium	Ho	67	164.94
Hydrogen	H	1	1.0080
Indium	In	49	114.76
Iodine	I	53	126.91
Iridium	Ir	77	192.2
Iron	Fe	26	55.85
Krypton	Kr	36	83.8
Lanthanum	La	57	138.92
Lawrencium	Lw	103	257*
Lead	Pb	82	207.21
Lithium	Li	3	6.940
Lutetium	Lu	71	174.99
Magnesium	Mg	12	24.32
Manganese	Mn	25	54.94
Mendelevium	Mv	101	256*
Mercury	Hg	80	200.61
Molybdenum	Mo	42	95.95
Neodymium	Nd	60	144.27
Neon	Ne	10	20.183
Neptunium	Np	93	237*
Nickel	Ni	28	58.69
Niobium	Nb	41	92.91
Nitrogen	N	7	14.008
Nobelium	No	102	253
Osmium	Os	76	190.2
Oxygen	O	8	16.000
Palladium	Pd	46	106.7
Phosphorus	P	15	30.975
Platinum	Pt	78	195.23
Plutonium	Pu	94	244
Polonium	Po	84	210
Potassium	K	19	39.100

Element	*Symbol*	*Atomic no.*	*Atomic weight*
Praseodymium	Pr	59	140.92
Promethium	Pm	61	145*
Protactinium	Pa	91	231*
Redium	Ra	88	226.05
Rodon	Rn	86	222
Rhenium	Re	75	186.31
Rhodium	Rh	45	102.91
Rubidium	Rb	37	85.48
Rutherium	Ru	44	101.1
Rutherfordium	Rf	104	260*
or	or		
Kurchatonium	Ku		
Samarium	Sm	62	150.43
Scandium	Sc	21	44.96
Selenium	Se	34	78.96
Silicon	Si	14	28.09
Silver	Ag	47	107.880
Sodium	Na	11	22.997
Strontium	Sr	38	87.63
Sulfur	S	16	32.066
Tantalum	Ta	73	180.95
Technetium	Tc	43	97*
Tellurium	Te	52	127.61
Terbium	Tb	65	158.93
Thallium	Tl	81	204.39
Thorium	Th	90	232.12
Thulium	Tm	69	168.94
Tin	Sn	50	118.70
Titanium	Ti	22	47.90
Tungsten	W	74	183.92
Uranium	U	92	238.07
Vanadium	V	23	50.95
Xenon	Xe	54	131.3
Ytterbium	Yb	70	173.04
Yttrium	Y	39	88.92
Zinc	Zn	30	65.38
Zirconium	Zr	40	91.22

*Mass number of the longest-lived of the known available forms of the element.

APPENDIX B

Electronic Symbols

SYMBOLS FOR ELECTRICITY

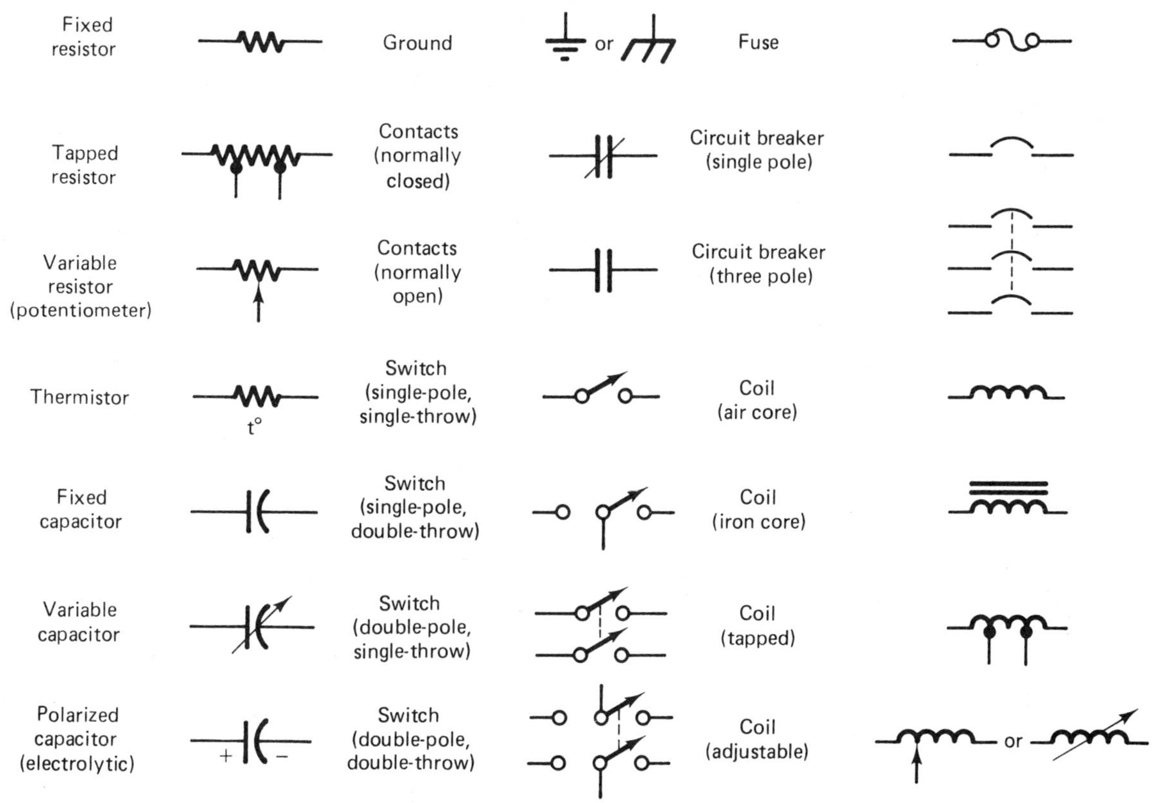

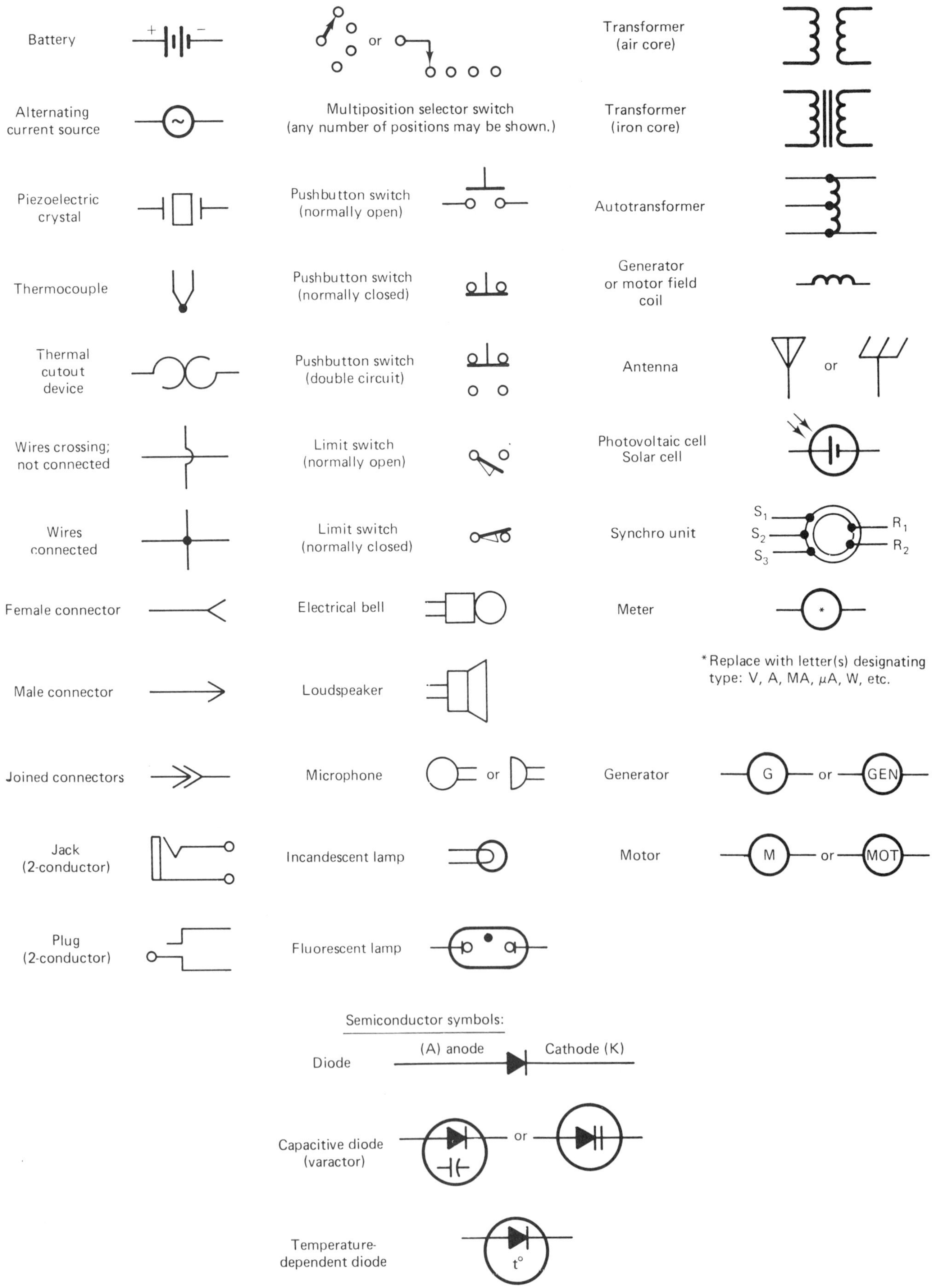
Battery
+
−
or
Multiposition selector switch
(any number of positions may be shown.)
Transformer
(air core)
Alternating current source
Transformer
(iron core)
Piezoelectric crystal
Pushbutton switch
(normally open)
Autotransformer
Thermocouple
Pushbutton switch
(normally closed)
Generator or motor field coil
Thermal cutout device
Pushbutton switch
(double circuit)
Antenna
or
Wires crossing; not connected
Limit switch
(normally open)
Photovoltaic cell
Solar cell
Wires connected
Limit switch
(normally closed)
Synchro unit
S1
S2
S3
R1
R2
Female connector
Electrical bell
Meter
*
*Replace with letter(s) designating type: V, A, MA, μA, W, etc.
Male connector
Loudspeaker
Joined connectors
Microphone
or
Generator
G
or
GEN
Jack
(2-conductor)
Incandescent lamp
Motor
M
or
MOT
Plug
(2-conductor)
Fluorescent lamp
Semiconductor symbols:
Diode
(A) anode
Cathode (K)
Capacitive diode
(varactor)
or
Temperature-dependent diode
t°

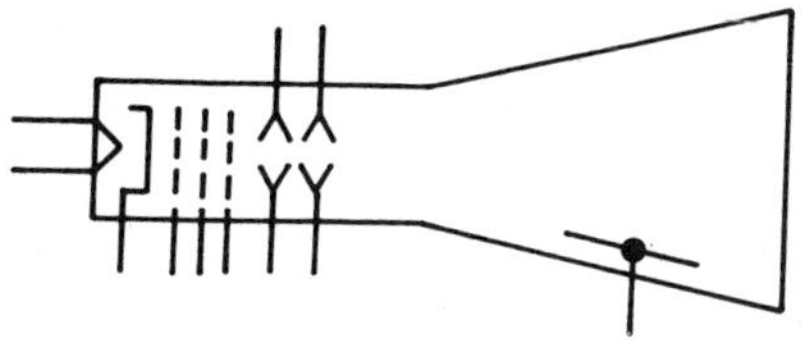

Cathode-ray tube
(electrostatic deflection)

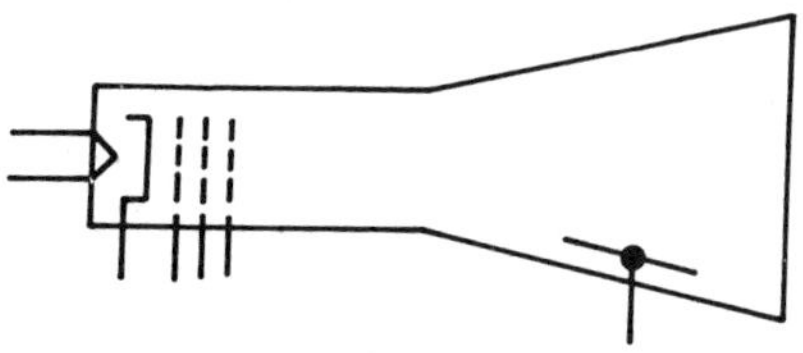

Cathode-ray tube
(electromagnetic deflection)

Thyristor,
bidirectional-
diode type

Thyristor,
bidirectional-
triode type (triac)

Bipolar
transistor

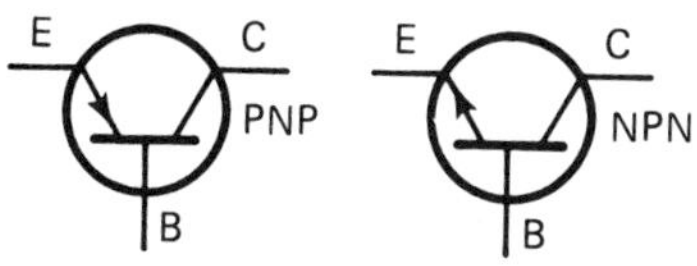

Photodiode

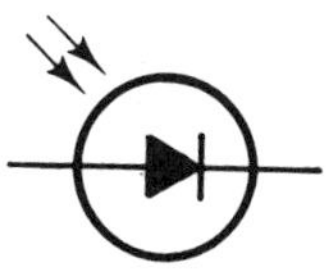

Light emitting
diode (LED)

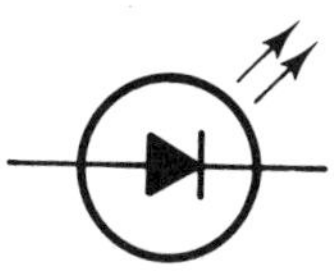

Zener
diode

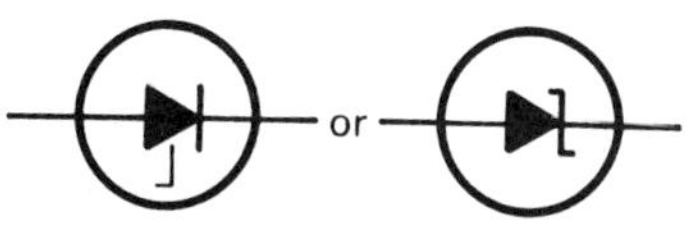

Thyrector
diode

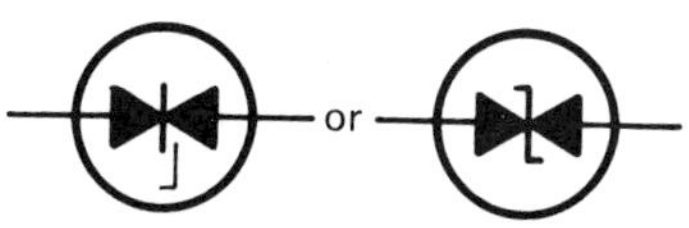

Tunnel
diode

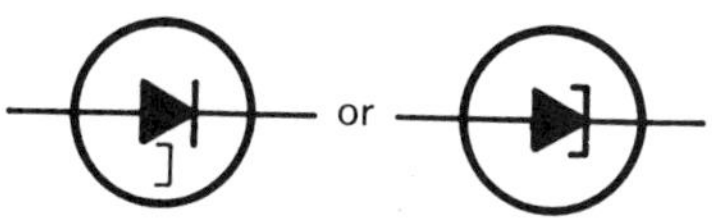

Trigger diac.
unidirectional

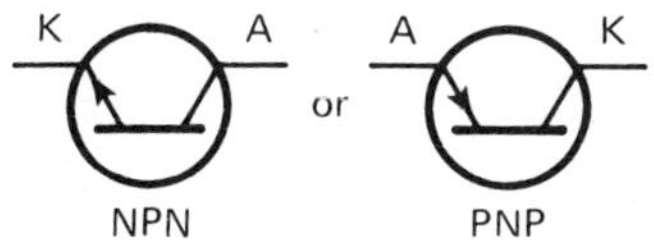

Thyristor,
reverse-
blocking-
diode type

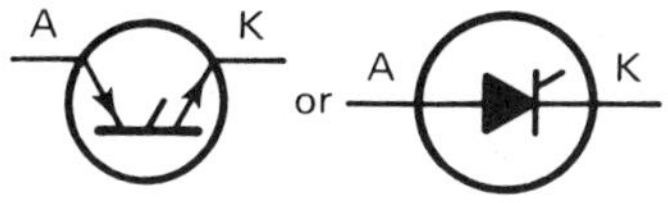

Thyristor,
reverse-blocking-
triode type
(solid-state thyratron, or SCR)

P-channel MOSFET,
enhancement type

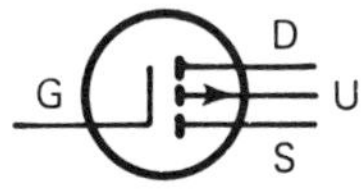

Vacuum-type
diode

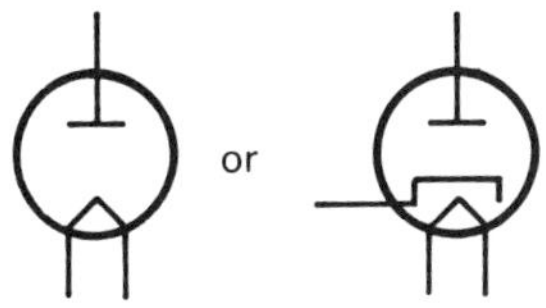

Gas-filled
diode

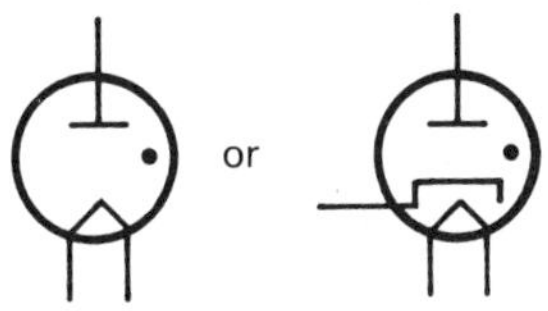

Phototransistor

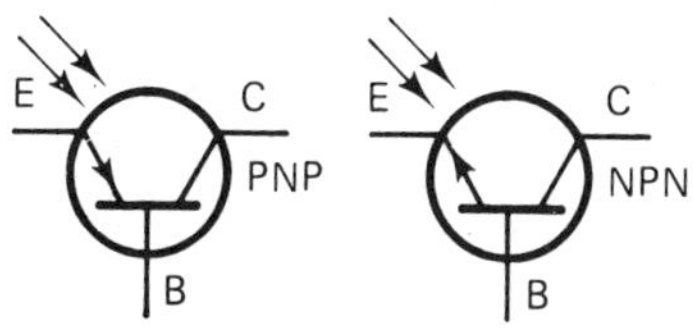

Darlington
transistor

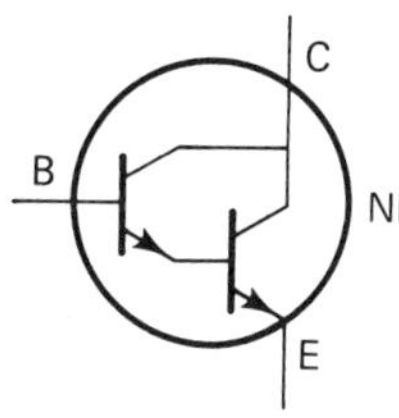

Unijunction
transistor

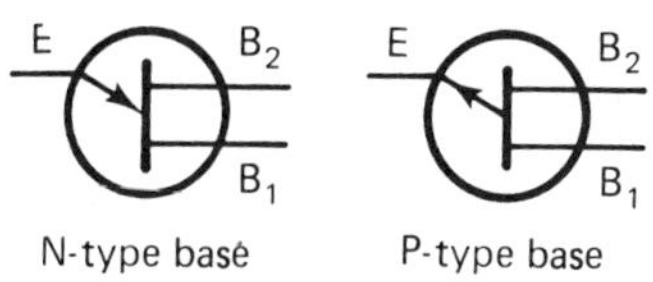

N-channel
JFET

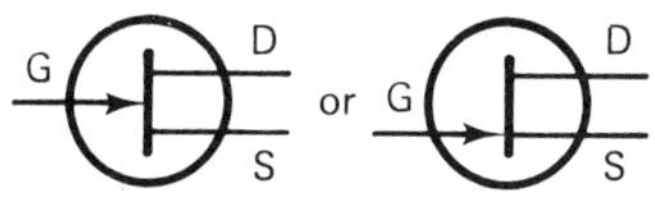

P-channel
JFET

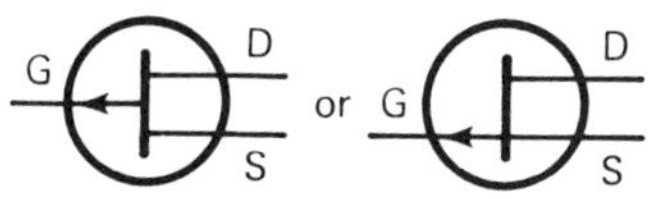

N-channel MOSFET, depletion type

N-channel MOSFET, enhancement type

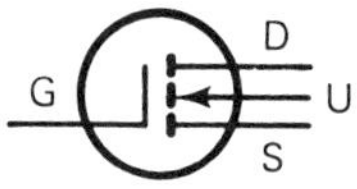

P-channel MOSFET, depletion type

AND function

OR function

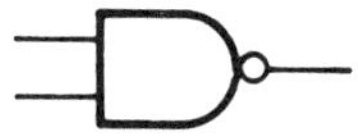

NAND function

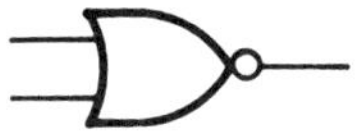

NOR function

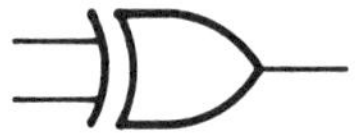

Exclusive OR function

Inverter (NOT) function

Operational amplifier

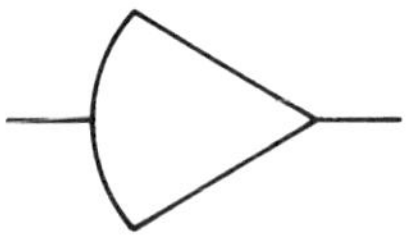

or

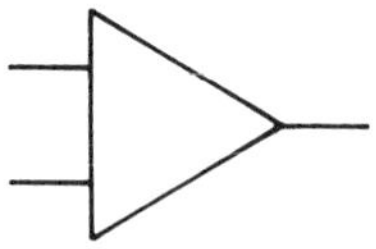

Flip-flop (general)

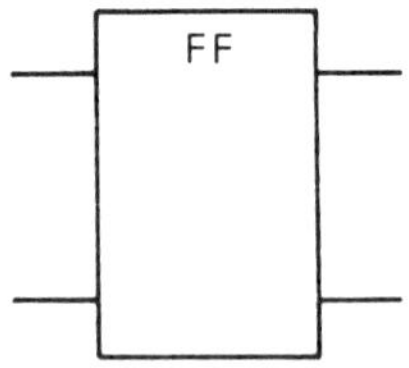

APPENDIX C

Electronic Equipment and Parts Sales

Listed below are several of the names and addresses of several companies that sell electronic equipment and parts. You can write to these companies and obtain catalogs and price lists for purchasing the equipment you need.

All Electronics Corp.
905 S. Vermont Ave.
P.O. Box 20406
Los Angeles, CA 90006
800-826-5432

Allied Electronics
1355 N. McLean Blvd.
Elgin, IL 60120
800-433-5700

Brodhead-Garrett
4560 E. 71st St.
Cleveland, OH 44105
216-341-0248

Cal West Supply Inc.
31320 Via Colinas, Suite 105
Westlake Village, CA 91362
800-892-8000

Circuit Specialist Co.
P.O. Box 3047
Scottsdale, AZ 85257

Digi-Key Corp.
P.O. Box 677
Thief River Falls, MN 56701

Edlie Electronics
2700 Hempstead Twp
Levittown, L.I., NY 11756-1443
800-645-4722

ETCO Electronics
North Country Shopping Ctr.
Plattsburgh, NY 12901

F.W. Bell
6120G Hanging Moss Rd.
Orlando, FL 32807

Heath Co.
Benton Harbor, MI 49022
800-253-0570

Hewlett-Packard
1501G Page Mill Rd.
Palo Alto, CA 94305

Hickok Teaching Systems
2 Wheeling Ave.
Woburn, Mass. 01801
617-935-5850

Hughes-Peters
4865 Duck Creek Rd.
P.O. Box 27119
Cincinnati, OH 45227
800-543-4483

Jameco Electronics
1355 Shoreway Rd.
Belmont, CA 94002

Kelvin Electronics, Inc.
P.O. Box #8
1900 New Hwy.
Farmingdale, NY 11735
800-645-9212

Lab Volt
Buck Engineering Co.
Farmingdale, N.J. 07727

MCM Electronics
858 E. Congress Park Dr.
Centerville, OH 45459

Merlin P. Jones & Assoc.
P.O. Box 12685
Lake Park, FL 33403-0685
305-848-8236

Mouser Electronics
2401 Hwy. 287 N
Mansfield, TX 76063

Omnitron Electronics
770 Amsterdam Ave.
New York, NY 10025
800-223-0826

Priority One Electronics
21622 Plumer St.
Chatsworth, CA 91311
800-423-5922

RNJ Electronics
805 Albany Ave.
Lindenhurst, NY 11757
800-645-5833

Satco
924 S. 19th Ave.
Minneapolis, MN 55404
800-328-4644

Techni-Tool
P.O. Box 368
Plymouth Meeting, PA 1946
215-825-4990

Tektronix, Inc.
P.O. Box 1700
Beaverton, OR 97075

Triplett Corp.
1G Triplett Dr.
Bluffon, OH 45817

URI Electronics
P.I. Burks Co.
842 S. 7th St.
Louisville, KY 40203

APPENDIX D

Soldering Techniques

Soldering is an important skill for electrical technicians. Good soldering is important for the proper operation of equipment.

Solder is an alloy of tin and lead. The solder that is most used is 60/40 solder. This means that it is made from 60% tin and 40% lead. Solder melts at a temperature of about 400°F.

For solder to adhere to a joint, the parts must be hot enough to melt the solder. The parts must be kept clean to allow the solder to flow evenly. Rosin flux is contained inside the solder. It is called rosin-core solder.

A good mechanical joint must be made when soldering. Heat is then applied until the materials are hot. When they are hot, solder is applied to the joint. The heat of the metal parts (not the soldering tool) is used to melt the solder. Only a small amount of heat should be used. Solder should be used sparingly. The joint should appear smooth and shiny. If it does not, it could be a "cold" solder joint. Be careful not to move the parts when the joint is cooling. This could cause a "cold" joint.

When parts that could be damaged by heat are soldered, be very careful not to overheat them. Semiconductor components, such as diodes and transistors, are very heat sensitive. One way to prevent heat damage is to use a *heat sink,* such as a pair of pliers. A heat sink is clamped to a wire between the joint and the device being soldered. A heat sink absorbs heat and protects delicate devices. Printed circuit boards, such as the one shown in Fig.

D-1, are also very sensitive to heat. Care should be taken not to damage PC boards when soldering parts onto them.

Several types of soldering irons and soldering guns are shown in Figs. D-2 to D-7. Small, low-wattage irons should be used with PC boards and semiconductor devices.

Below are some rules for good soldering:

1. Be sure that the tip of the soldering iron is clean and tinned.
2. Be sure that all the parts to be soldered are heated. Place the tip of the soldering iron so that the wires and the soldering terminal are heated evenly.
3. Be sure not to overheat the parts.
4. Do not melt the solder onto the joint. Let the solder flow onto the joint.
5. Use the right kind and size of solder and soldering tools.
6. Use the right amount of solder to do the job but not enough to leave a "blob."
7. Be sure not to allow parts to move before the solder joint cools.

Figure D-1 Printed circuit (PC) board. [Courtesy of TRW/UTC Transformers.]

Figure D-2 Low-wattage soldering iron kit with several soldering tips. [Courtesy of The Cooper Group.]

Figure D-4 Cordless soldering iron kit. [Courtesy of The Cooper Group.]

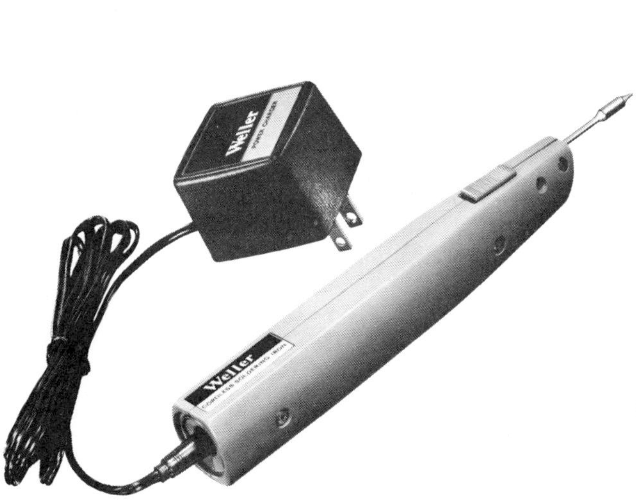

Figure D-3 Cordless soldering iron with charger. [Courtesy of The Cooper Group.]

Figure D-5 High-wattage soldering gun kit with 60/40 solder and spare tips. [Courtesy of The Cooper Group.]

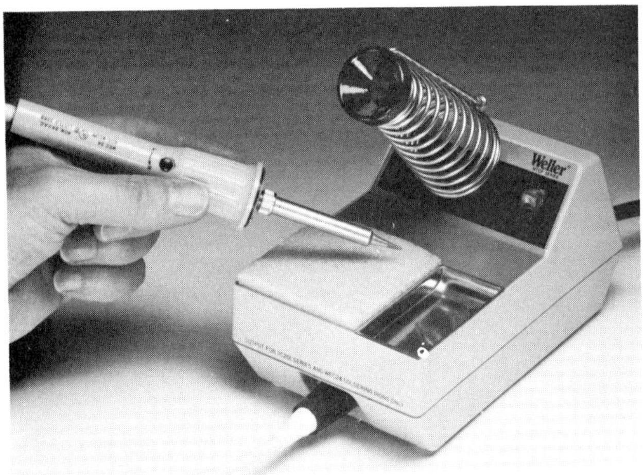

Figure D-6 Soldering iron with temperature control station and cleaning pad. [Courtesy of The Cooper Group.]

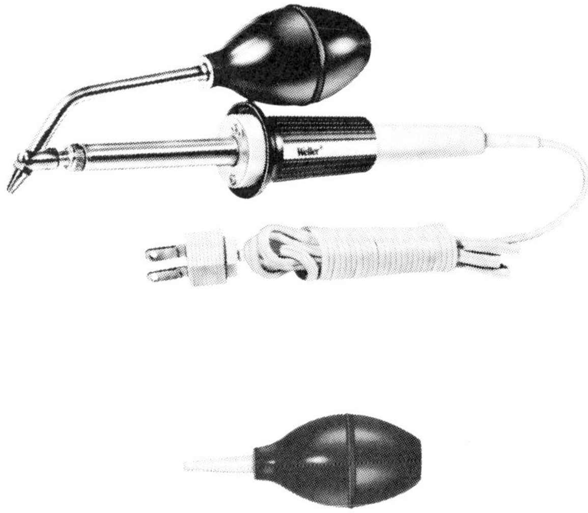

(b)

Figure D-7 (a) Desoldering tool; (b) desoldering bulb. [Courtesy of The Cooper Group.]

Index